Handbook of
Package Design
Research

Handbook of Package Design Research

Edited by WALTER STERN

A WILEY-INTERSCIENCE PUBLICATION

JOHN WILEY & SONS   New York • Chichester • Brisbane • Toronto

Library of Congress Cataloging in Publication Data

Main entry under title:

Handbook of package design research.

    "A Wiley-Interscience publication."
    Includes index.
    1. Packaging—Addresses, essays, lectures.
2.   Design, Industrial—Addresses, essays, lectures.
3.   Marketing research—Addresses, essays, lectures.
I.   Stern, Walter
HF5770.H27   658.5′64′072   80-39935
ISBN 0-471-05901-3

Printed in the United States of America

10 9 8 7 6 5 4 3 2 1

Preface

Package design research—the use of scientific methods to evaluate the degree to which a product's personality, claims, and benefits are communicated to the consumer through the structure and graphics of its package—has been practiced in some form or other since the early 1950s. However, in spite of the fact that its use and the budgets devoted to it have recently grown at a phenomenal rate, no formal reports or literature have yet been published that would aid the marketer in the assessment of the various methods available to him in certain situations. The profession of package design research thus shows some striking similarities in the stages of its development to that of package engineering—the science of developing and testing packaging structures that will successfully protect the product throughout its entire distribution cycle.

Package engineering started in a similarly informal manner as an attempt to reduce damage ratios through packaging improvement, primarily through the actions of certain specially trained individuals in the organizations of the common carriers concerned with damage claims, and of the mail order companies to whom safe transit of merchandise could spell the difference between profit and loss. Learning by trial and error and establishing a nucleus of expertise mostly through an apprenticeship approach to teaching, the profession gradually established a core of knowledge that was given its first compendium by the publication, in the early forties, of a Package Engineering Handbook.

The demands of world-wide distribution of all kinds of goods under the most extraordinarily difficult and critical conditions presented by the logistics of World War II served as a powerful stimulus to the further development of this science and resulted eventually in the structure of university courses leading to B.S. and M.S. degrees in the package engineering sciences. Today, numerous handbooks and reference works are available, and more than 30 universities, learning centers, and government bureaus teach courses in all aspects of this craft/science. Formal examinations are held by a number of states to license packaging engineers, and a world-wide interrelated network of package engineering societies has been created.

Package design research has arrived at the same point in its development that package engineering reached in the early forties. Except for some well attended but isolated seminars, no regular courses have yet been established to teach this highly specialized discipline of consumer market research. Except for a few articles, no literature is available to those who would want to learn more about this field. Moreover, no group or professional society has yet emerged to combine the widely scattered expertise and knowledge into a cohesive organization facilitating interchange of insights.

It is particularly important to establish a basic fund of experience and knowledge in this field because its tools are used to an ever increasing degree to justify, or at least evaluate, the staggering investments demanded by today's advertising and sales promotion budgets for retail products, and because of the steady rise in product/package development costs. Because package design research provides advance insights into the degree to which your marketing objectives and your product's performance claims and benefits are communicated by the package, and because market test failures can be analyzed by package design research in such a manner that defects or flaws can be pinpointed for improvement action, a handbook in which all aspects and facets of this field are discussed is timely and essential for those who deal in products and their packaging.

The book's organization is simple. Part I explores the complex ways in which consumers interact with packaging, the formulation of marketing objectives and how they are aided by package design, and the role of package design research in product development. It provides a basic but thorough introduction to design research and serves as a general orientation, particularly to those readers whose contact with the field has been limited.

Part II outlines in considerable detail the various methods of investigation that are used today to assess packaging effectiveness. It covers the most frequently used approaches.

Part III covers the place of design assessment in product development, deals with the interaction of package and product in marketing, and cites case histories demonstrating the manner in which design research can be used even at the product concept stage.

Part IV is concerned with the manner in which package design research is integrated into the corporate marketing structure, and shows how its use can significantly influence the entire corporate marketing strategy.

Part V explores the use of package evaluation techniques in segmented and highly specialized markets and product categories where routine application of the tools discussed in Part II may render misleading or skewed results.

Part VI investigates the use of package design research in other countries whose marketing problems differ from those in the United States, and explores the problems that arise when a package design or design format is used in several countries with differing marketing, ethnic, economical, or cultural environments. It also reviews the situations and problems that may arise from multinational marketing plans and from the need to arrive at meaningfully coordinated results when studies are conducted in a number of countries.

Part VII outlines new and often highly experimental test designs that are in some cases used only on a laboratory basis and may well serve to predict what will occur in design research within the next decade.

The authors of this handbook are active in a considerable variety of occupations. It was felt by the editor that package design research techniques and experiences should be described not only by consultant consumer research organizations specializing in this area but also by those in the corporate structure who use these consultant services, by the advertising agencies and corporate product management who plan the product and packaging strategies that are to be tested, by package design groups, and by consultancies in the areas of behavioral studies from which many of the package design research tools derived. Moreover, because design research is used on a world-wide basis wherever major budgets are assigned to market planning, authors from other countries were invited to contribute to this overview.

It will be apparent to the reader almost at the very outset that package design research is neither an art nor a science but a highly creative combination of both. Thus a fairly simple investigative approach such as the focused group interview may be used in many widely varying ways by different research groups and in different situations. Because of this necessity to explore not only the techniques but also the ways in which they may be designed, applied, and analyzed, I felt that in a number of areas not one but several authors should deal with identical subjects. This may in some cases result in overlapping information. However, at the same time it develops in the reader the necessary insight that, unlike physical package performance testing (which may, for instance, use an American Society for Testing and Materials' standard method to test susceptibility of polyethylene bottles to soot accumulation, regardless of

the country and conditions in which the test application occurs), the application of package design research tests depends in its use and execution entirely on the experiences and convictions of the individual or group who execute them. Thus while physical package performance testing groups may vary in the thoroughness, accuracy, and "finish" with which tests are executed and reports are rendered, the difference between any two individuals, groups, or organizations who conduct package design research can result in different end results unless very careful planning, monitoring, and auditing are used.

There are basically two ways to write, coordinate, and edit a handbook that is the result of contributions from a considerable number of authors (and possibly in a number of different languages). One approach is an almost complete rewrite in order to establish a uniform tone of delivery, organization, treatment, and syntax. This will result in a text of considerable homogeneity and integrity that will, in effect, provide the impression of having been written by one author. The other approach would be to edit for a certain consistency of format and style but to leave untouched the author's individual vocabulary, expressions, and even idiosyncrasies. Because in the area of package design research individual attitudes and values regarding a certain test pattern are highly important factors and may vary considerably, the second method is utilized here to give the reader a clear and direct impression of the author's voice and personality in interaction with standard methodology. Just as a package design organization's attitude toward problem solving is often an expression of the personality and philosophy of its director, so the research approach to investigation and solving of problems of evaluation is a direct reflection of the person who directs the program. Only the program director's individual way of expressing his or her beliefs and insights is important in the treatment a certain body of knowledge receives.

This handbook is thus essentially the product of many minds and provides the reader with direct access to an impressive array of perceptions, experiences, and judgments of individuals whose credentials in this field are outstanding. Credit for this compendium of seasoned knowledge and expertise is shared by all members of the handbook's Board of Contributing Authors. All members are not only theorists but also practitioners of product and package design research.

To give our reader an opportunity to get to know the Board's members, we have provided detailed biographical backgrounds for each of the contributors whose chapters follow.

Because all segments of the marketing and research communities have shared equally in the creation of this volume, it is my hope that this handbook will become a standard reference work for marketing decisions both here and abroad and that its usefulness can be maintained and widened by future revisions that keep pace with the ever changing marketing environments.

WALTER STERN

Wilmette, Illinois
March 1981

Contents

Introduction

Package Design Research: The State of the Art

Walter Stern

Today it is obvious to most experienced marketers that package design can detract from a marketing effort or contribute immeasurably to its success. For this reason, corporate packaging decisions are increasingly based on research rather than hunch or opinion. When millions of dollars are spent on new product introductions or line extensions, the marketing field needs facts to guide them, and so the use of package testing has in recent years increased at a vast rate. Yet not a single reference work is available to brief the users of this highly specialized area of consumer research.

When a leading manufacturer of chewing gums in the United States decided to introduce a new brand of sugarless gum, their brand went up against well entrenched competitors with truly impressive market shares; it also was positioned to compete against the manufacturer's own brands. In most new product and package development projects, the decision to go to market follows years of expensive preparation and testing. In the case of the sugarless gum these tests took place not only in the laboratory but also in the European market—a route that is not unusual in the candy field. What is important in this context is that that research triggered the starting gun for staggering expenditures in advertising and promotional support that ran to over 10 million dollars in the first year alone.

When less than two years ago this country's most dominant tobacco products company decided to introduce the first of what eventually became a whole family of low-"tar" cigarettes positioned for richness of taste, the budget amounted to almost 40 million dollars, an investment based largely on the results of painstaking research into the product, its advertising, and its package.

A certain portion of that monumental budget was spent on package design research to evaluate, assess, and validate design decisions made during the conceptual stage, before test market, and prior to national market rollout. In a critical new product introduction such as this, in which impressive budgets are bet on product and package success, the designer's experienced judgment or the packaging committee's educated determinations are just not considered sufficiently reliable by corporate market management to justify financial commitments of such magnitude. When you roll your dice under the watchful eyes of a board of directors, the computer printout is preferred over the gambler's intuition.

However, while advertising research is by now a well documented science with an impressive array of published literature and reference material, and while the development of new products and product rejuvenation have for quite a few years been guided by well established research disciplines, the subject of package design research is simply too young for that kind of backup. Let me cite the following example.

A *Handbook on Market Research,* considered a veritable bible without which no brand manager would start his day at the office, was published several years ago, is revised every year, and runs to 1440 pages of 6 point type; 91 research consultants contributed chapters to that comprehensive reference volume. However, not a single chapter is devoted to package design research, which has, in this entire prolific coverage

of the state of the art of market research, been given the space of one single paragraph. The paragraph simply states that the field of package research is one that in recent years has grown in vast proportions; vast proportions notwithstanding, that's it on package design research.

"Package research," of course, has been practiced for over 40 years. But, in our terminology, package research is used for the process of testing a package's physical performance characteristics, especially how well it performs its functional requirements during its entire distribution and use life. "Package design research," however, operates in an entirely different area governed largely not by recording test instruments but by perceptions, emotions, and by the entire vague and largely uncharted area of the consumer's psychological involvement with products and their packaging. Its testing methods and its analytical methodology are based not on physics, but on psychology and the behavioral sciences.

Package design research will not predict how well a package will do in the market. It will, however, examine precisely to what extent a certain design has succeeded in communicating the marketing objectives on which its marketing platform was based. It is an analytical instrument, not a crystal ball. Yet in spite of its trappings of impressive instrumentation and professional lingo, it is still at this stage largely an art rather than a science.

Two concepts serve as the basis for all package design research:

1   Consumers generally do not distinguish clearly between a product and its package, and many products are packages (and many packages are products).

2   Consumers relate emotionally not to the facts (the realities) of the products/packages they are involved with,

but rather to their "perceived reality."

Let me attempt to illuminate these two concepts with some examples.

There are a great number of items offered for retail sale that are quite obviously resisting categorization as either product or package. Consumers do not differentiate between razor blades and their plastic dispenser, between a solid room freshener and its housing, between a hair spray formulation, its propellant, and its aerosol dispenser. Portion control packaged marmelade is considered a different product from the same marmelade if packed in a conventional jar. An antiperspirant applied by a roll-on dispenser is considered a different product from the same formulation when applied as a spray.

These are some of the more obvious examples. But this tendency of the consumer to consider the product and its package an integral entity goes a great deal further.

A West Coast university's school of marketing recently conducted a number of taste panel experiments which tried to rank five national brands of beer by such quality ingredients as blandness or heartiness; light or heavy feel; sweet, tart, or bitter taste; body; color; and aftertaste. Respondents on the taste panel poured their beers out of bottles merely marked by a letter of the alphabet for distinction. A definite ranking sequence was established in which brand B was rated tops and brand A was at the very bottom of desirability.

The test was repeated four weeks later with results that duplicated those of the first test. It was thus a totally reliable taste analysis except for one puzzling factor. When the test was repeated a third time, but with bottles labeled with typed brand names such as Schlitz, Budweiser, and Pabst, the taste rating was rearranged to the point that now A was tops and C at the bottom of the group of five.

The point illustrating the consumer's tendency to perceive package and product as one, however, emerged in the final test when the panel, for the first time, was confronted by the products in their conventional retail packaging. In test after test, product C was rated tops in quality; D was at the low end of the scale. Why?

The experiment demonstrated impressively the intimate relationship between the package and the manner in which its product is perceived or experienced. Not, of course, just the graphics, the structure, the material, or the functions of the package, but all experienced communications elements that surround it: its advertising umbrella, its promotional aura, where it has been seen, how the consumer was introduced to it, how his or her peers relate to it—all these elements determine the consumer's product perception.

In a frequently quoted example, a major toiletry marketer had come to a final decision on the design of a new roll-on deodorant label; the only thing to be settled was the color scheme of that label, and three final contenders were to be evaluated. The three were applied to containers distributed for in-home trial. The rationale: we are sending you for tryout three slightly different formulations of a new deodorant; please evaluate them for effectiveness, fragrance, and ease of use.

An overwhelming percentage of the users voted for the product whose label was executed in color scheme B. It dried almost immediately after application; it had a pleasant but unobtrusive fragrance; it effectively protected the wearer from underarm odor and wetness for up to 12 hours. Color scheme C of the identical product did not fare as well. There was much criticism of the strong aroma of the product, and its effective antiperspirant action lasted for only a few hours.

Color scheme A? Well, color scheme A almost involved the company in a series of lawsuits because a number of users had developed an irritating underarm rash—three had actually visited a dermatologist for a professional prognosis.

However, regardless of label color schemes, there was only one product, one formulation, one scent, one strength involved in these three tests. Here we are entering the mystifying realm of what I would like to call "perceived reality."

A truck driver pulls his 18-wheeler up to a roadside bar and grill, kills the ignition, climbs stiffly down from his cab, and enters the alcohol dispensing premises. He walks up to the bar, is welcomed cheerfully by his professional brotherhood, and orders a beer. He is soon an indistinguishable member of that solid fraternity and feels the desire to light a cigarette. He pulls out a pack, lights up, passes the pack around. At least he tries to, but everybody has withdrawn from his vicinity and he suddenly feels an icy exclusion from the friendly crowd. His brand is Eve.

Eve is a slim cigarette made from a blend of tobaccos strikingly similar to that of many other brands on the market, with similar average moisture content and similar tobacco aggregate residues. Yet the package contributes a personality to that product that is so distinctly unique in its positioning that its implications on the poor trucker's personality are devastating. The reality: just another cigarette. The perceived reality: a totally feminine product.

Now let's reverse the example; let's consider a brand with a wholesomely masculine personality—one any truck driver would be proud to pull out of his shirt pocket, for instance, Marlboro. Supposing in our reversed example the mothers' bridge club gathers around the coffee table for their final pretournament session and one of them pulls a pack of Marlboros from her purse. What happens? Why, nothing at all, why should anything happen?

Package design research truth number 1: People react to a product's perceived reality rather than to its actual, factual attributes. Truth number 2: Don't ever jump to conclusions in package design.

Let us look at one more example of perceived reality and the way it is supported (or even created) by its packaging: We are looking at two cans of cat food. Both contain 6 ounces of product (tuna). The guaranteed analysis of protein, crude fat, crude fiber, moisture, and ash content is identical on both. The two most prevalent ingredients in both are tuna and water. Both contain the same group of vitamin supplements, and thus both seem to offer complete and balanced nutrition for cats. Cats, as a matter of fact, don't distinguish between the two, because cats go by realities.

But consumers don't. One can is marketed by a leading pet food producer. It sports a six-color, process illustrated label, and its brand name is endorsed by a corporate logo that stands for widely acclaimed achievements in animal nutrition experimentation and product development. The other is a generic brand of a large supermarket chain; its label is printed in two-color line art, and it sells for over 50% less than its more ambitious companion product. What are the perceived realities brought out by package design research?

Briefly, the imagery played back by tests that we will describe in detail in the following indicates that inflationary pressures and the possibility of saving over $100 per year by buying the generic brand are powerful persuaders toward buying the cheaper brand, especially because Puss does not know the difference. But the consumer who does so does not rest easy; sooner or later she will, with a sigh of relieved conscience, revert to the premium product. The perceived reality is that while the generic brand is good for her pocketbook, she gambles with her pet's health in a highly selfish and irresponsible manner.

Present and pending packaging legislation demand that any package illustration faithfully portray the package's contents. This is a totally factual approach until you begin to consider the consumer's involvement with perceived realities. When she buys a muffin mix, does she buy the small heap of whitish powder that the bag in the box contains? Is that what the box illustration should show? Or does she buy the deep-rooted satisfaction of being able to dish up that plate of crisp, aromatic, crunchy but soft, steaming hot muffins when the kids storm in, back from school and hungry as lions? Isn't this the perceived reality she purchases, the gratification for which she puts down her hard earned money?

Should that foil envelope of chili seasoning mix portray in faithful color photography that same small heap of powder, this time red instead of white? Or should it project the perceived reality of serving up, in a rustic handpainted casserole, those steaming hot portions of tantalizing, delectable chili to dad and the kids when they come in from the cold, half frozen from building that snowman? Which is she buying, the ingredient or the gratification of achievement? Both, of course. But the reality she perceives in the product/package is the latter not the first.

Ask the housewife. She knew exactly what she was buying: a dehydrated "all-in-one" seasoning mix. Yet she also bought the deeply gratifying feeling of being an able, knowledgeable, and skilled provider. That, ultimately, is what she paid her 31 cents for. In terms of the perceived reality, the sizzle is often more important than the steak.

Thus in the realm of perceived realities the package's design, structure, materials, function, copy, advertising, and the product itself are often confused and intermingled by an intelligent, generally well informed consumer population that does not always arrive at a buying decision through rational, factual value and performance evaluation but through deep-seated emotional appeals. In a recent 1000-person random sample of "product defects" that consumers remembered from the past year of buying health and beauty aids, for instance, 70% of the "bad product" complaints were, on further analysis, found to be caused by faulty dispensers.

This confusion in the consumer's mind is another demonstration of the perception of package and product as entity. When a shopper complains, for example, about being deceived by a wall cleaner package, she may mean just that (e.g., "it looks twice as big as the competitor's but contains only 30% more product")—a package size deception. However, she may also mean "I have to scrub much harder than is implied by the television commercial duplicated in the package's illustration" (a promotional deception, not a package deception) or even "the trigger sprayer spatters" (no deception whatever but a mechanical package—or is it product?—fault).

These, then, are the murky areas in which package design research operates to probe the perceived realities projected by the package. Specifically, it will answer the following questions.

WHAT IS THE PRODUCT'S CATEGORY?

One of the most important marketing objectives to be achieved by any packaging is to communicate clearly and unequivocally the product's category. In doing so successfully it will help to focus the product's appeal on the consumer target group for which it has been positioned. However, if it miscues, no amount of plain English package copy will rectify the situation, and a confused consumer may turn into a nonbuyer. For example, take a 12 ounce (hypothetical) aerosol container. The brand name, in elegant script is "CareFree." The illustration is the head of a ravishing young blond, shining hair freely waving in the spring breeze. The product, however, isn't a hair spray. The package design badly miscued: the product is a room air freshener.

Or suppose your marketing strategy for your flavored fruit drink mix attempts to get away from competitive pressures by positioning it not with the other canned mixes but in the dairy case, packaged in a typical Purepak or Excello milk carton. Badly miscued by the double shift of container style and sales environment, your consumer badly misreads her cues with resulting rejection rather than purchase.

So we see that not only the graphics, but also container style, color, and even material telegraph subconscious information about the product's category. Departing from the norm may be innovative and it will certainly help achieve deep differentiation from competition, but sometimes at the hazard of badly misread product category cues. If the milk container is printed in red and black, the milk contained is perceived as "homogenized grade A," and if you want to sell chocolate milk in the traditional "skim milk" color schemes of blue and white it'll be uphill all the way. Homogenized grade A milk in a color scheme containing green will be misread as buttermilk; homogenized grade A milk in an amber widemouth jar won't communicate anything but utter confusion even if you proclaim the product's category in 14 point type.

A small 2 × 2½ inch white paper packet contains an institutional serving of sugar. If you put pepper in it, you're in trouble. If it's pink, it will most likely contain an artificial sweetener. These are instantaneous and deeply ingrained size and color cues that can be explored or confirmed by package design research. You can certainly attempt to sell cookies in a milk carton, jelly in a yogurt cup, or clarinet reeds in a fliptop box. Your packaging will be highly unique, but it will take a sizable communications budget to overcome the intentional confusion of the puzzled consumer.

Yet clear and simple product category and position projection is one of the most basic requirements of good package design. Let us say you are selling a dentifrice. What's its positioning? Is it to be perceived as just another toothpaste, only cheaper? Is it a stain remover? A tooth whitener? A breath freshener? A tooth cleaner for especially sensitive teeth? Does it prevent cavities? Does it make you more attractive, preferably to the opposite sex? Or is it an especially strong stain remover that is also a denture cleaner? Is it based on a dental hygienist's cleaning compound? Or is it none of these, but really a denture cleaner? Or denture adhesive? Or denture adhesive for worn dentures? Or does it counteract gum erosion?

These are subtle variations in product positioning that must be communicated by the package if their competitive stance is to be successfully supported at the retail store. Package design research will determine if this communication has been achieved effectively, or if not, why not.

Positioning is especially important in product segmentation: unless the product is perceived correctly in its attributes, claims, benefits and personality, it will not be identified by the consumer segment it is targeted for. To mention another example, your candy product should be positioned through packaging, store placement, pricing, and possibly advertising in such a way that it is clearly and unequivocally perceived in its right segment. Is it a super deluxe top department store candy, such as Lady Godiva? Is it positioned as an exotic import with all the superb quality "you can get only in Europe these days"? Is it a diet food? Is it typical party candy (adult) or typical party candy (teens) or typical party candy (kids)? Or is it "trick or treat" candy for kids, where quantity is more important than quality?

Does it have adult masculine overtones so that the vice president can sneak a bite of it out of his desk drawer? Or is it really not candy at all but rather a filled and enrobed cookie? If so, is it sold with the cookies? Or with the candies? Or at the check-out counter?

Product perception as cued by its packaging is an important field of inquiry for package design research. Even an expert would have difficulty distinguishing between Johnny Walker Black Label whisky and Johnny Walker Red Label whisky, and, as we indicated before, it has been impossible for even sophisticated and knowledgeable beer drinkers to distinguish one brand from another unaided. It's often the package that introduces that distinction, and it's the advertising that stands behind the package that shapes the product's unique identity. The Coors drinker and the Marlboro smoker make statements about themselves. The package design aids in making that statement and is thus responsible for a significant part of the consumer's emotional involvement and ultimate satisfaction with the product.

We have seen how the package aids in positioning and defining the product's category, character, and competitive claims by its shape, construction, color scheme, and graphics. Let's examine these graphics a little further.

One of the most important statements package design can make is that of the brand name. Not only is the brand name often the one single most important element that provides a direct tie-in between advertising and the displayed package on the retail shelf, but by its execution it assumes a character of its own that reinforces that of the product. These functions, the way it conveys the product's character both quickly and convincingly, and its all-important capability to stand out in the welter of competitive store environment, can be accurately measured by package design research. Especially the perceived product personality conveyed by the brand name's lettering style, color, execution, and placement can be assessed in considerable detail. Is it feminine? Masculine? For kids? Strong? Subtle? Elegant? Sophisticated? Medicinal? Therapeutic? A good value? A bargain? Expensive but worth it? Fun?

Or, to get back to the candy vernacular for a minute: Does it compete with Lady Godiva? Or a Bloomingdale's imported specialty? And so forth. Brand name execution, price bracket, and product character often go hand in hand. How well does the brand name within the balance of the package's perceived realities succeed in projecting the product's personality and value level to the consumer segment at which it is focused?

Package design research is also an excellent tool to measure how fast and how well the product's competitive performance claims and benefits are communicated by the design's material, structure, style, texture, size, graphics, copy, and illustrations. In perfect claim communication, the consumer will be aware of why and how the product is different from and better than any competitive product by merely looking at it. Successful claim communication can be achieved either by literally conveying all of the product's perceived and real advantages or by establishing an instantaneous, extensive recall of previously communicated advertising claims. A design of classic elegance that simply mentions the brand name Chanel Nº5, for instance, would fall into the latter category.

Before we begin to examine the many tools of package design research that are today at the disposal of the packaging and marketing communities, let us look for a minute at what package testing can and cannot do. Simply stated, your corporation would hope that it would indicate how well the product will sell, but that is exactly the kind of answer that is not provided by package assessment. Package assessment will, rather, establish how well the package will communicate the objectives specified by the design platform. It will thus indicate how well the package will perform as your marketer and salesman; how well it will succeed, however, depends obviously on a whole array of other factors as well, such as distribution, promotion, advertising, pricing, competitive environment, and dealing.

Package design research is contraindicated if it will not influence any decision making. It is also contraindicated if the gains that can be achieved by its insights are outweighed by the cost of the research program. Finally, it is to be distrusted when the marketing objectives it is to check are too ambiguous or vague, when its testing techniques are too synthetic and too far removed from the typical shopping environment, and, last not least, when intramural rivalry and what I like to call "special interest skew" result in sets of figures or reports that are really little more than pseudo-scientific justifications for decisions already made. In that case package design research becomes merely a seal of approval for preconceptions that, for all their specious derivation, become indisputable fact simply because they are rendered in a "research report."

The present state of the art of package design research is such that it will perform best if it serves to compare the degree to which several design candidates will, in all probability, fulfill the design objectives outlined in the original brief. It will furnish valuable insights into the efficacy of the various design components (always with the understanding that "design" may concern the material, structural functions, shape, color, texture, and graphics of the package). It will thus give *relative* assurance of the design's eventual marketing success. It will also perform excellently as an analytical tool to probe why a package is not performing as it should. Moreover, it will probe into the various equities established by the design of presently marketed packages in order to determine how much the design can be changed without losing loyal customers. Unfortunately, as pointed out earlier, because of understandable corporate pressures for that advance assurance, package design is sometimes programmed, applied, and interpreted with somewhat less scientific discipline, insight, and, above all, detachment than would be desirable.

Today, the tools available to the package design researcher are many—some of them basic, some highly sophisticated, some well established, some quite innovative and yet untried and unproven. One of the most important decisions to be made at the start of the research phase is therefore the selection of research methods that are pertinent to the problem. Also, because package testing relies heavily on the planning, application techniques, and final analysis of the research group that conducts it and is therefore just as much an art as a science, the selection of the right research group is of paramount importance.

While the most logical grouping of research techniques in this area would be by whether they probe what the consumer sees or what the consumer perceives, a simpler and possibly less confusing grouping would probably use the criterion of whether tests are optical or verbal. In these two categories some of the more frequently used techniques are the following.

OCULAR TESTS

Eye Movement Tracking

For many years researchers have been trying to determine exactly what a person's eyes see, how long they dwell on each element of what they see, and to which new element they move then. Whether we are considering a package front panel, a typical shelf display in a store, a floor stand, or a highway sign such as that of a gasoline station or a motel, it is important to understand the visual sequence in which its elements are seen, how thoroughly they are observed, and especially which important elements are apparently skipped entirely.

During World War II, a group of industrial designers charged with improving the instrument clusters of antiaircraft batteries on British destroyers designed a radically new method of observing exactly what the gunnery personnel, under great emotional stress, saw; how much attention they paid to certain important information bits; and what they tended to overlook because of the physical arrangement of dials, indicators, digits, and symbols.

The instrumentation consisted of a helmet to be worn by the respondents to be tested while they were placed in front of the indicator dials, with stressful situations such as enemy aircraft attacks being simulated by a fairly realistic display of sights and sounds. The helmet contained toward the front a motion picture camera that operated at eye level in

such a manner that its lens saw the identical field of vision observed by the man who was fitted with the helmet. Simultaneously an infrared light beam was first directed onto one of the respondent's pupils and then reflected through a prism and mirror arrangement back into the film camera in such a manner that its location was superimposed on the film image portraying what the respondent saw. In this manner, the film showed not only the total field of vision being observed but at the same time the focal point of the respondent's gaze and thus how long it rested on each element seen and where it moved next.

In the late 1950s this rather cumbersome but revealing approach was adapted to a study of what exactly was seen by a typical supermarket shopper, if such a person exists, by equipping a young housewife with the helmet camera, connected to a battery pack in her shopping cart, and sending her through supermarket aisles with a shopping list. Here, for the first time, the camera recorded the visual experience of a shopper looking for a brand, finding a new product, examining it and comparing it to its competition, rejecting or accepting it. It also recorded what on the observed package was seen first, what next, what required repeated return of the viewer's gaze for reexamination (or verification)— all bits of information that were heretofore simply not available.

Today, at the newly established Consumer Research Center of the University of Chicago and at various research centers around the country, eye movement tracking is largely automated and continuously recorded by computer with resulting videotape for CR display or computer print-outs that can be analyzed for similarities between respondent's reactions on a statistical basis.

Another approach has entirely dispensed with the respondent's role by storing in a memory bank the reactions to certain pictorial and type elements and to various graphics contrast levels obtained in thousands of tests so that package panels or shelf display situations can be examined and eye movement tracks established without the use of human reaction (and at sizable reductions in research budgets).

In a typical examination for findability (visibility), legibility, and eye movement, such a test on say a frankfurter package would show that the respondent's eye would move smoothly over the package's front panel, dwelling first on the brand mark (or brand name), then on to a special panel or spot qualifying the franks as "All Beef," then on to the product category statement (which could be subordinated in the package design because the structure of the package and its marketing environment would make the product category self-evident). Furthermore, the test would show that attention was "adequately held" at each dwell point.

The general rating for such a design, which apparently succeeded very well in communicating all important points in the right sequence and with the appropriate emphasis, would be "excellent." Note that this test reflects such psychological elements as emotional involvement, perception, and comprehension only indirectly and that there is no probing that would allow the respondent to verbalize her reactions. As a matter of fact none of the ocular tests per se allow for verbiation; any emotional response is reflected only by inference through quantitative measurements.

Pupillometry

Researchers in the behavioral sciences determined long ago that there is a very real relationship between basic emotional responses and the size of your pupils. All measurements of emotional involvement, which measure the degree of commitment reflected by a respondent's verbal responses, involve the interaction between emotions and the autonomic nervous system and can thus also be used to determine veracity, but only one of these methods (for others see the following pages) involves ocular measurement.

Unfortunately the recording of pupil dilation on track with verbal responses or comments involves heavy instrumentation and thus tends to create a highly atypical interview situation that could conceivably set up emotional stress in itself. It is also not always easy to eliminate all extraneous light level variations (and the resultant variations in pupil diameter). However, those with extensive experience in pupillometry believe it furnishes a fairly reliable means to gauge how deep-seated the respondent's statements are (that is, "does she really mean it or is she just trying to be nice to the interviewer?"). Pupillometry is most often used as an adjunct to other tests (such as eye movement tracking) to give them an added dimension of reliability.

Tachistoscopy

One of the oldest optical measurement devices in this very young field is the tachistoscope, which is used to measure legibility, conspicuity, recognition, and recall. It serves best to measure the clarity of certain visual communications elements or design elements by exposing them to the respondent on a regular projection screen in increasingly longer, split second intervals. An electronic flash timer connected with a projector's aperture is often used in connection with 35mm slides of the design to be tested (whether a design element, a total package panel, or package; whether a display or a whole gondola filled with packages). Viewed at 3 foot viewing distance the image may appear roughly life-size (or actual package size).

Exposures typically start at $^1/_{100}$ second and are gradually increased until the product category is identified. Then the product is identified, the brand is identified, and the other elements to be tested have registered. The test thus compares degrees of legibility, contrast, recognition of graphic elements, correct interpretation of illustrations, and so on. The test is a very helpful measurement of both predesign and postdesign effectiveness, which assumes that some important degree of communication must be accomplished even if a shopper glances only fleetingly at a package.

Because its results are quantitative, because it can be easily administered, and because its cost is a direct product of the number of respondents involved, it is frequently used to screen out the weakest contenders before the more expensive and subtle research investigation starts. Its danger is, of course, implied in its simplicity. It will tell you nothing whatever about those package design elements that communicate product character, stimulate involvement, and often trigger the buying decision. However, it will tell you whether you have been seen, observed, and understood. And that's a great deal.

Angle Meter

Used as an ancillary test based on the assumption that the shopper walking down a supermarket aisle will not see your package face on but rather at first at a very acute and then gradually at a less acute angle, the instrument simply exposes the package at varying angles, recording the degrees at which various communications are recognized.

Blur Meter

Based on the assumption that many shoppers have inadequately corrected vision because they do not wear glasses or contact lenses when they should, this ancillary test uses an optical projection device that displays the package to be tested in various degrees of out-of-focus adjustment to respondents with 20/20 vision. It is an instrument that favors simplicity of layout, large lettering and logotype sizes, and maximum contrasts in a manner similar to that of the tachistoscope. It can also be utilized to test recognition because it has been proven on innumerable occasions that not only can small fractions of well-known trademarks furnish recognition of the whole, but also an out-of-focus condition can still result in accurate identification provided the mark or logotype is well known. Familiarity is therefore an important adjunct in this test.

VERBAL TESTS

Because verbal tests can in most cases dispense with instrumentation, they can be brought not only to the consumer but also to the shopping site or the home. For the same reason, they are much more flexible in their location. Eye movement tracking, for instance, requires major installations that are extremely cumbersome to move. Testing therefore takes place in major marketing centers such as New York or Chicago. Verbal tests, however, can be conducted at hotels and meeting places, in shopping malls, at advertising agencies or manufacturer's offices, in clubs, churches, homes, and, in some cases, even on the sidewalk. Certain interview techniques are predominantly conducted by telephone over WATS lines and can in this manner establish statistically significant results at relatively low cost. They are thus not only used to provide important insights on a qualitative level but often will furnish quantitative answers that, to a digit oriented business society, are vastly preferred over the complicated, subtle, and sometimes elusive language in which some qualitative research reports must be worded.

Focus Group (or Focused Group Discussion)

In any serious discussion of package design research the focus group technique is probably mentioned most often. Unfortunately, it is also often misapplied and misunderstood.

Originally developed to test preliminary new product concepts, the technique is conducted exactly as its name implies. It usually employs a group of carefully selected consumers (mostly more than 6; rarely more than 10) who have a demonstrated interest in the product area to be discussed and have some valid opinions formed in actual shopping and use experience. This immediately indicates one of the factors that determines the cost of conducting focus group interviews: the easier it is to assemble such groups, the less expensive is the procedure. Housewives who serve potato salad to their families are easier to sample than surgeons with pacemaker implant experience.

The group gathers in a relaxing setting—the approach is unstructured to some extent although the moderator who will lead the ensuing discussion usually has worked out, prior to the meeting, an outline of what is to be covered. Often cigarettes or refreshments are present. Because the group's verbal reactions will be carefully analyzed later on, they are usually recorded, sometimes on video tape. The group will of course be advised of this recording arrangement, but it has been the general experience of research groups that this in no way interferes with or inhibits free and frank discussions.

The moderator usually introduces him(her)self, offers refreshments, reviews the participants' screening sheets

for names, and points out the various visible equipment items, but usually omits to mention that the room may be equipped with a one-way mirror that allows clients of the research group to follow the proceedings without being seen. The moderator will stress that the group's interest is mostly in individual opinions, thoughts, impressions, or reactions and that honesty and frankness are appreciated.

A brief warm-up discussion of the product area in general usually allows the participants to get to know each other better. Gradually the moderator will then delve into specific product problem or opportunity areas to get reactions that may or may not result in a consensus. Focus groups are an ideal tool for the development of consumer feelings and comments that cannot be developed through any other corporate discussion process, and they are invaluable in highlighting potential strengths and weaknesses in products as represented by their packaging. They should never be used to develop quantitative information. Thus the statement that in 3 groups of 10 respondents each a total of 15 (or 50%) felt the product claims lacked believability is irrelevant and misleading because it is made without having the basis of a statistically significant sample.

It is generally felt that the direct solicitation of comments about a package's design does not often produce valid results because it positions the focus group as design critic rather than as potential or present product user. Observations about packaging are therefore generally obtained by discussing the product itself—in an indirect manner.

Focus group analyses are usually furnished in the form of a summary of deductions, based on a detailed and repeated replay of the recorded or video tapes. Clients are frequently invited to watch the interviews and discussions because of the clues that verbal inter-action can give to the hidden viewer.

Programmed Interviews

Programmed interviews consist of a se-ries of carefully phrased questions that probe the potential shopper's anticipa-tions of product quality, character, and performance, based on package design. Because not only the shopper's percep-tions but also emotional and rational reactions are reflected by the answers, the tests often render qualitative as well as quantitative results. Should certain functional traits of a package be evalu-ated (for instance, a special opening or dispensing feature), the shopper would be invited to handle the package or a fairly detailed, functional prototype of it in a simulated use situation.

Depth Interviews

Although respondents for the interview-ing techniques described above are se-lected on the basis of demographics alone (demographics: profile of the sam-pled section of the total consumer target group universe in terms of vital statis-tics, such as age, sex, marital status, and income), candidates for depth interviews are in addition selected by screening their psychographics (psychographics: a life style profile of attitudes and factors contributing to consumer habits and be-havior, such as educational background, occupation, travel and entertainment habits, readership, and beverage con-sumption). Although depth interviews may explore roughly the same areas as the focus group explores, they would modify and evaluate all statements on the basis of fairly detailed personality profiles of the respondents. Where a focus group rarely takes much more time than an hour, depth interviews may last several hours and are usually conducted

on a one-to-one basis rather than in groups.

They tend to be administered by trained psychologists, and their interpre-tation of how well a package design will meet certain marketing objectives is con-sequently highly judgmental. It is there-fore a research method that gets, more than any other we have discussed, fairly deeply into the opinion-as-basis-for-de-cisions area that package design research in general tries to guard against. In other words, at this stage the tools of research come to resemble to some extent the experience-based judgment of the pack-age designer on which they originally intended to improve by replacing indi-vidual assessment with unbiased, scien-tific evaluation.

This is especially true if tests attempt to compare new, unknown brands with brands now being marketed, in which case not only the packages themselves but also (with known brands) the prod-uct's entire aura of advertising and its imagery is set by the test to compete, if you will, with a totally new image that has not as yet been incorporated in the respondent's learning curve. But let us deal with that topic later on.

Semantic Differential

A respondent's reaction to certain stim-uli is usually placed only vaguely within a whole spectrum of possible reactions. In order to force the respondent to be more specific about the product attri-butes projected by the packaging to be tested, semantic differential tests are de-signed to rate accurately certain product characteristics by viewing or handling the package, but not the product itself. The rating is accomplished by inviting the respondent to place a checkmark on a scale of adjectives extending into both the positive and negative ranges. For example, the respondent may be given, for viewing only, a bread package. She will then be asked to rate the kind of bread she would expect to buy in this package on the degree to which its contents are perceived in the areas of flavor, softness, freshness, nutritional elements and dietary attributes, texture, kind of crust, color, keeping quality, and so on.

Typically, a texture scale in the above example would ask the respondent to rate the bread in the package she views as to where she expects its texture to fall on a scale from harsh and abrasive through grainy and crunchy to soft, to spongy, to nondescript. Because she will not be able to taste or even handle the product itself, her reactions will be based entirely on the visual clues transmitted by the package design.

Because all responses can be expressed in binary language, they can be profiled through a computer program and the various profiles can be compared and finally combined into a multidimensional product description based solely on its packaging. In addition, responses are factor-analyzed to determine major dimensions of the stimuli and responses. It is then possible to determine the primary meaning clusters which appear to surround each of the stimuli and thus gain a direct input to design effectiveness evaluation.

Such quantitative analysis is possible in all forced choice tests, of which the semantic differential is one, and this is one reason that forced associative selection is such a popular test methodology.

Attitude Study Interviews

These are usually conducted at sites where potential shoppers of the researched product or service category can be found, such as supermarkets, bowling alleys, shopping malls, taverns, dentist's waiting rooms, subway stations—depending on the research subject. The respondents are usually consumers who purchased or used the product or service or who intend to do so. Interviewers would typically attempt to gain insights in these areas:

- Brands currently bought, and brands bought in the past.
- Evidence of brand switching and reasons.
- Evidence of brands discontinued and reasons.
- Imagery conveyed by new designs.
- Demographic or even psychographic information.

The outstanding advantage of this interview technique is its closeness to the "scene of action" and the fact that the sterile and dissociated laboratory setting is avoided. At the same time, of course, it cannot be expected to supply the rigid controls available through a laboratory setting but may be skewed in its results by a whole number of factors, such as time of year, time of day, weather, or noise (whether ambient or artificially introduced).

Unless the consumer's reactions in all these tests are caused by the autonomic nervous system (such as in pupillometry) and are thus involuntary, they are of course subject to questions regarding their validity. Does she really feel this way? Or is she acting out a role she believes is expected of her? Or is she reacting to subtle peer pressure? Or does she rationally mean what she says, but subconsciously reject the notion? How deep, in other words, is her commitment to her statements? Is she really emotionally involved or is she merely supplying intelligent responses?

During the last few years a number of validators have been developed that tend to probe into these areas of emotional involvement in a manner similar to that of the pupillometer. All are based on involuntary responses and are presumed to be totally beyond the conscious control of the respondent.

An emotional reaction to an external stimulus is produced by the hypothalamus area in the human brain, usually by stimulating a response within the autonomic nerve system. This exemplifies itself by such actions as pupil dilation, increased respiration or perspiration, or changes in the tension of the larynx caused by the glottis. The validators used by package design research have been used to confirm or negate verbal statements by respondents by measuring changes in these areas through instrumentation that can serve to separate "lip service" from commitment from outright lie.

Validators determine two factors: the number of questioned respondents whose answers are caused by true emotional involvement and the depth of such involvement. In addition to pupillometry well-known techniques such as psychogalvanometry are used to measure electric skin discharges; in a more experimental recent development brain waves have been measured, and voice pitch has been computer analyzed.

We mentioned previously, if briefly, some of the difficulties encountered in comparing known with unknown brands. In these cases the overwhelming bias in favor of the "known" versus the "unknown" is sometimes overcome by testing so-called "naive" subjects, occasionally by conducting the tests in a country or area in which the competing brands are not sold or advertised, so that the "known" brand is just as unknown as the newly tested brand.

This brings us to some of the stickier problems faced by package design research, and some of the very real limitations that appear to me to exist.

In quantitative package design research the first important limitation seems to lie in the field of sampling and hinges on the relationship between a representative sample and its universe (universe = total population being investigated in a certain study; sample = a representative number of respondents selected from a universe from whom or about whom information is to be obtained during a study). The universe from which, for instance, a nationwide representative sample of adults is drawn is the total adult population of the nation. Correspondingly, the universe from which a representative sample of adult housewives living in Chicago is drawn is the total population of adult housewives living in Chicago. As long as the universe from which we are to draw our sample is well articulated by the marketing plan's description of the product's target consumer population, probability statistics provide us with the mathematical means to draw a highly accurate sample from that universe. At the same time, these statistics can be employed to indicate the expected margin of error that must be allowed between sample responses and universe responses were it possible to question the entire universe.

Thus in theory we have at our disposal a highly reliable method to determine answers to questions from a sample that may be only a small fraction in dimension when compared to the total universe. Unfortunately this is not the case when dealing with the everyday pragmatisms of package design research. Rather, the situation may well develop something like this.

The client says to the researcher, "Here are three package designs. I'm already over my budget, but if you can keep the cost down, I'd like to test them on about 200 Chicago housewives." So the researcher submits his cost estimate, and the number of respondents is promptly cut in half.

I am sure most of us have had similar experiences sometime in our professional careers, but now note what exactly has happened. We have, to start out with, not had any articulate and detailed description of the target consumer group (the specific universe) from which we were to draw a representative sample. Instead we were given the assignment to "test 100 housewives." Now keep in mind that small number and also keep in mind the client's budgetary restraints, and you cannot possibly conceive of a research group that could follow the precepts of probability statistics and select a representative sample from the nation's housewives under these conditions.

But you also trust your client's acumen sufficiently to know that he could not possibly be interested in the opinions and reactions of only the specific 100 Chicago housewives that were mentioned in the budget. They won't ever buy enough of his product to make any difference—and just as likely may not buy any of it. Like many intelligent marketers not too well acquainted with design research he may vaguely assume that somehow the responses of those 100 housewives will tell him how hundreds of thousands of other women would respond if they were questioned. In other words, he is not interested, of course, in the answers from the sample—he is interested in what answers he would get from the universe.

All right, so what happens now? The researcher does not want to lose his client so he selects 100 housewives, not by the probability method, which has to be based on a universe, but by applying his experience and insight. He takes leave of his statistical science and substitutes judgment and common sense in order to escape his dilemma. Of course he is soon faced with this question: What universe (what larger section of the total population) do his 100 housewives represent? For if they are not representative, their answers cannot be deemed representative either. His answer must honestly be "I do not know."

He could possibly prevaricate by saying "I am fairly sure that this sample could be considered representative of a fairly large sector of Chicago housewives." But he has no scientifically accurate way of identifying the universe represented by his small sample. Moreover, it appears to be essential that the evolving research results are very carefully weighed in view of the rather slipshod manner in which we arrived at them. But they almost never are. Numbers are numbers, and a pretty set of figures is too seductive to business management to spoil it with petty qualifications.

There are other limits to package design research that cannot be overcome by an adequate infusion of budget. One of those raises its ugly head frequently when you start analyzing your test results, and that is the definition of the simple word "preference." Respondents preferred design solution B over design solution A by a 3 to 1 margin. Good, now we can make decisions. Except that we really do not know what "preference" means.

Are we talking about visibility? Legibility? Command of attention in a typical store display environment? Attractiveness on the kitchen counter? Easy fit into cupboard shelves? Enticing reuse possibilities? Ease of handling such as opening, closing, dispensing? As we mentioned before, how do you accommodate the traditions, habits, nostalgias, even rituals associated with an old package when you want to evaluate the merits of a proposed redesign?

What we are in effect discovering here is the vexing phenomenon that the ways in which people may relate to a product and its package are very likely too numerous to allow comprehensive testing with repeatable results and conclusions because, as we have pointed out on several occasions, even the most reliable research methodology still involves judgment, the "art" part of the science-art of package design research.

A package may do superbly well in the ten criteria that it was tested on and perform miserably on the eleventh that wasn't tested. The dilemma always reminds me of a respondent who said "I know! I remember it. It's that highly attractive white package that always ruined my fingernails when I tried to open it! I stopped buying it."

If flawed statistics and vague semantics are ever present snares in the woods of design research, one of the most frustrating problems is one we have referred to repeatedly, that of familiarity and how to equate for it. A number of years ago in conducting a seminar on design research we had a discussion of just that area, and the comments went somewhat like the following:

First of all, familiarity is something we speak of with great ease, and we have a number of ways of defining it. Unfortunately even our colloquialisms seem to have opposite meanings, so that on one hand we might be saying that "familiarity breeds contempt" but on the other that "repetition equals reputation." Similarly we might assume that novelty is usually a delightful experience and one of the stronger reasons that the public accepts our new packaging. Then again when research does not turn out as hoped, we might shrug it off by indicating that the design "is too new for the public."

What we are really concerned with here is the entire body of psychological history known as the psychology of learning. In spite of the fact that package design research is to a large extent based on psychology and the behavioral sciences and in spite of the fact that certain principles of learning have been for many years widely applied in the fields of teaching and training, this all-important aspect of how products are perceived has been largely ignored by design research.

The principles of learning as they are known today are few and are easily spelled out. One is that repetition (or practice) is an effective way of learning new skills. Another is that spaced practice is usually more effective than continuous practice. Eight half hour tennis lessons are better than one 4 hour lesson.

It has also been well known for many years that comprehensive learning of a whole unit is better than learning by bits and pieces. It's easier to learn a backhand by practicing the whole motion than by concentrating one day on the stance, another day on the wrist, another on addressing the ball, and a fourth on follow-through. A final very important principle is that forgetting can be greatly reduced by overlearning, that is by continuing your lessons over and over even after you are sure you have mastered the subject.

The application of these principles to package design is important not because of their practice but because of their omissions. The general conduct of design research is practiced somewhat on the basis of instantaneous affections. Whether we use instrumentation in a laboratory setting, focus groups, or mall intercept interviews, we allow in most cases only a preference judgment or comment or reaction after one single, sudden exposure. First and last chance, period. In comparison with everyday life situations that appears to be, let's admit it, a highly unrealistic procedure.

When a package enters the stores, the typical consumer often has a considerable number of occasions to relate to it. She might see it at first on television advertising, for instance, which establishes the products benefits, character, claims. She may then see it in different stores and different outlet types, displayed under vastly varying conditions, and she may meet it again and again for weeks or even months. She may not even discover it until her fifth exposure; after she has seen it seven times more it may occur to her that she has seen it before on a few occasions; on the fourteenth exposure, something suddenly triggers her interest; on the fifteenth, she decides she is interested in it. However, she may not actually buy it until the nineteenth exposure, and after she's had it in her home 247 times she may decide it is dull and stodgy and a "nothing product," or she may be disillusioned in its performance simply because a new product came along.

I am sure you can see what we're driving at. You get to like things by living with them; you feel comfortable with them because they are familiar—as comfortable, warm, and friendly as the proverbial old shoe. There is, to put it succinctly, a learning process involved in establishing product preferences, and its essence is repetition.

How do we account for the learning process in design research? Well, I hope somebody in our readership contradicts me, because it's a worrisome detail, but from my experience I do not think we equate for the learning factor at all. We try to eliminate it from our procedures by not testing the old versus the new, but we cannot eliminate the mental and emotional comparison inadvertently drawn by the consumer, and so all we have left in our defense is the old saw that any controlled tests can only approximate reality.

Approximating reality brings up one more limitation we have to deal with when conducting package design research. I call it degree of verisimilitude; it deals with the degree of reality simulation that is affordable in testing new or modified packaging. It has been my experience that only too often the tested prototypes used in design research are synthetic and artificial in appearance, especially when new structural ap-

proaches or materials are involved in addition to new graphics. Their feel, function, and copy treatment thus offer only very limited verisimilitude.

Why is that? Almost invariably because all economics of package production are based on mass production. Let us look at a typical example by assuming a research program utilizing a 200 respondent sample in three different locations for a new packaging approach with novel visual and functional communications characteristics to be tested. Let's also assume the not infrequent provision that the package is to contain a food product, such as a pie filling, rendered on the front panel in breathtaking color photography, and that its very special advantage over competitive products is some kind of push-button feature, whatever that may be.

Now 200 respondents times 3 locations means 600 prototypes because once a package has been opened by push button it cannot be used again on another respondent, and it can't be put back on the shelf for display. All right, so you order 600 prototypes.

Except for the fact that you may have to spend $12,000 to $15,000 for color separations and plates to print those 600 packages, and that nobody, but nobody will print you just 600 packages. 60,000, yes. But even 6,000 is highly doubtful and you may spend more than a dollar for every one of them.

So you order the packages to be dummied up by hand, including that magic push button; you use one-up dies and inexpensive silk screening to reproduce the line artwork and hope that the colors will match the original design at least approximately, and you glue C-prints of your photography transparency illustration to the front panel of the packages. Of course you will have to leave those packages empty because the product development people are far behind and anyhow the filling is produced in Austria. So the empty packages feel like what they are, empty dummies.

In addition, packages that are cut, scored, perforated, and glued by hand don't feel or behave in the effortless "push-button" manner of packages that are produced on commercial high-speed box making equipment. Also a C-print looks like a C-print and not at all like the printed and varnished six-color illustration shown by competitive packs. You shrug your shoulders, and admit that nothing is perfect in this world. But neither are your test results.

What I have been describing here in a somewhat facetious way is a dilemma faced by all research based on mock-ups, renderings, concept sketches, dummies, prototypes, breadboard models: the need to establish an exact relation between the degree of achieved verisimilitude and the reliability of the emerging research results.

Let us say that at one end of the verisimilitude scale is the device of merely describing the package verbally. Let us say that at the other end is an extensive market test with commercially produced packages. Now if we feel that mere verbal description is inadequate, then where on that scale do we begin to obtain valid and meaningful results on which we can base management decisions? Or, in plain language, how close do we really have to get to the real thing to get real answers?

No careful introduction to package design research can afford to omit exposing some of its problems along with the many valid tools available today to assess packaging effectiveness in the marketplace. For the first time in the history of package development a designer can make design recommendations not based on opinion or experience alone but on factual research results; for the first time management can made investment decisions and launch new products on the basis of research that uses efficacious methods to evaluate alternatives. Package design research varies in quality just as much as the designs it researches; only experience in its use will establish a perfect consultancy relationship in this area. However, today corporations using these tools are gradually reducing the catastrophic rate of new product failure and increasing their batting average in product rejuvenation and line extensions. One reason for these profitable improvements is package design research.

PART ONE

Planning Package
Design Research

CHAPTER ONE

Determining Communication Objectives for Package Design

Herbert M. Meyers

Not many years ago a package was simply a container surrounding and protecting the product, to be used for storage at the point of origin, transportation to the point of sale, and distribution of the product to the ultimate consumer. However, packaging is no longer the sole province of the technician. Today, communication is the name of the game.

Product managers and other executives responsible for the marketing of products have become increasingly aware of the importance of packaging as an integral part in the process of communicating their marketing objectives for a specific product to the consumer.

In the past 20 or 30 years the package has increasingly become one of the most important and crucial elements in the marketing mix, equaling and occasionally surpassing in importance the product's advertising and sales promotion.

In fact, while advertising and sales promotion can stimulate the consumer to look for the product, it is only when the consumer faces the package at the point of purchase that the sale can be finalized. If the product is purchased, the package has succeeded in communicating the marketing objectives developed by product management. If the product is not purchased, the reason is, more often than not, the failure of the marketer to communicate effectively. It is possible—even likely—that this failure to communicate began at an early stage of market planning, including the failure of the marketer to communicate his packaging objectives to the designer. This chapter will examine methods for determining communication objectives that the package designer can translate into appropriate and successful packaging design.

THE ROLE OF PACKAGING IN PRODUCT MARKETING

Since 1960, packaging has undergone substantial changes in relation to equally substantial influences on the marketing of products. Some of these changes are:

- Today's consumer is better educated, more discriminating, and belongs to a smaller, younger household.
- Many households now have two wage earners—husband and wife—creating more discretionary income and changing many family traditions and lifestyles.
- At the same time, the ever-increasing number of products available and the constant addition of competitive brands, give the consumer an increasing choice of buying one product or the other, and let *the retailer* determine which product to display and which to neglect. In fact, in many instances the marketer is more concerned with the retailer than the consumer, on the theory that once the product is in the store the consumer will be a captive audience.

On the other hand, to convince the retailer to stock the product is one thing, but to convince the consumer to purchase and repurchase the product is an-

other. It is no longer sufficient merely to put the product on the shelf or hang it on a pegboard display. The simplistic preoccupation with pure shelf impact, which surged during the 1950s and 1960s, has given way to the need for establishing a sophisticated marketing strategy for the product. This strategy includes a thorough understanding of the consumer's needs, the perceived level of quality of the product, its price, the retail environment in which it is sold, and the advertising and promotional support that it receives. All have to work synergetically to create consumer demand and thereby sell the product. In this complicated process the package often plays a pivotal role in communicating the total marketing strategy.

To operate in this communications mix, an effective package must be able to do more than merely attract attention. It must be informative, it must clearly identify the product by its brand name, and it must communicate its real benefits—whether a new flavor, a more convenient method of usage, an economical advantage, or an emotional benefit—which may distinguish your product from the competitive one.

Also, to an ever-increasing degree, packages are required to furnish substantial amounts of detailed product information demanded by the consumer, such as content, weight, direction for preparation or use, ingredients, nutritional information, chemical content, origin of manufacture, and many legal requirements.

Finally, the package must *perform* effectively when it is handled and used. Its size, shape, and materials have a direct bearing on whether it's easy to hold, carry, open, store, ship, and display—and whether it will generate repeat purchases.

To achieve ultimate effectiveness of combining these and other important marketing criteria in a single package is often a difficult and challenging task, and a successful conclusion very often depends on the ability of the marketer to determine appropriate communications objectives for the design of the package.

REASONS FOR PACKAGE DESIGN

Most package design projects fall into one of three categories:

1 Redesign of an existing package or a line of packages.
2 Package design for line extensions.
3 Package design for new products.

The communications objectives for each of these categories can be substantially different from each other because their objectives are basically different.

Redesign of an Existing Package or Line of Packages

The "when, why, and how" of redesigning packages is probably one of the most difficult decisions that a responsible marketer has to face. Designing a package for a line extension or a new product is a relatively simple decision process and often a foregone conclusion.

Whether and when to *redesign* the package of an existing product, however, requires experience and a sense of timing and understanding of the potential benefits and limitations of package redesign. The redesign of a package occurs for a great variety of reasons. Among the most frequent reasons are the following:

- Product improvements.
- Competitive pressure.
- Changes in product content (net weight or quantity).
- Economics (reaction to price changes).

- Restructuring of a product line.
- Changes in brand strategy.
- Changes in corporate strategy.
- Updating an old package or package line.
- New uses for the products.
- New packaging materials and technology.

Each of these requires a different method of communication; therefore, each will be interpreted differently by the package designer. For example, to communicate improvements of an existing well-known product, the package design changes may be limited to copy emphasis, resulting in virtually unnoticeable design modifications in order to retain the franchise of the well-known product.

On the other hand, updating an old package may be related to repositioning the brand (such as skewing the product to younger, mobile consumers), which may require substantial surgery on the package in order to relate to the new marketing strategy. For this reason, in redesigning an existing package, it is particularly important to determine precise communications objectives. In the event that there are multiple communication objectives, some of which may contradict each other, it is important to determine communication priorities (i.e., which communication element is the most important, which is next important, and which is least important), which will be reflected in the design of the package. Consider the following example.

In researching the packaging of a well-known pancake mix, it was determined that the packages clearly identified the well-known *brand* on each of several packages, but that the consumer was confused as to the differences between the similar tasting, similar looking products and their different methods of preparation. It was decided to redesign the packages, with emphasis on altering the priority of visual communications to ensure the following:

1 Primary emphasis on clear, verbal product differentiation, identifying the benefits of each product.
2 Secondary emphasis on improved photography, to further communicate the differences between the products (and, incidentally, achieve improved appetite appeal).
3 Substantially different background colors to reinforce product differentiation.
4 Brand identification.

In this example, the marketer succeeded in improving a well-known product line by reversing the package design priorities from emphasis on *brand identification* to primary emphasis on *product identification*. (See Fig. 1.)

In redesigning packages for well-known products or product lines, carefully developed communications objectives and a clear understanding of communication priorities will lead to successful package design solutions.

Package Design for Line Extensions

Compared to the redesign of a line of existing packages, developing communication objectives for line extensions generally is a simpler procedure.

This is the case particularly if line extensions were anticipated at the time when the packages for the line were originally designed. At that time the designer is able to incorporate design elements that can later be modified to accommodate additional products without major alterations of the package appearance. For example, if several products in the initial product line were differentiated by several background colors, the new product additions would be phased in easily by specifying additional background colors.

On the other hand, it happens occasionally that line extensions run into complications. Such complications may occur for a variety of reasons, for example:

1 Product extensions were not anticipated at the time the product line was originally launched, requiring major

Figure 1 The recently improved packages for Pillsbury Hungry Jack Pancakes emphasize product differences by clear copy statements (''Just add water,'' ''Extra Lights,'' etc.), as well as strong background colors and distinctly different illustrations on each of the packages. (Packages designed by Gerstman + Meyers Inc)

package design surgery on the original packages.

2 Technical requirements may complicate the ability of adding additional packages to a line of products.

3 Verbal identification of the new products may resemble those of the existing products, requiring a copy reevaluation of all packages in the line.

For these and other reasons it is often necessary to reconsider communications objectives for a product line even if a single additional product is added. It may be necessary to redesign portions of the existing packages in order to make the launch of the additional products more meaningful. For example:

- In 1977, a well-known manufacturer of pet foods saw the need for adding an additional flavor to a well-known line of cat foods with six flavors. The products were packaged in horizontally displayed folding cartons. In considering the additional flavor, the manufacturer was concerned that the retailer would balk at providing additional shelf space for the already extensive product line. The rethinking of communications objectives eventually led to redesigning the packages in a *vertical* display format.

The new package dimensions made it possible to display seven flavors in the same linear shelf space previously required for six horizontal packages. (See Figs. 2 and 3.)

In some cases, it is necessary to rethink communications objectives for a line of packages to emphasize a restructure of a product line, or the availability of a new product. For example, four products in a leading line of five products were "slow movers." The manufacturer decided to phase out the four slow movers, but worried about hurting the sale of the remaining product by losing valuable shelf facings in the supermarket. Reviewing communications objectives for the line, it was decided to retain the

Figure 3 The redesigned Purina Tender Vittles packages utilize the same display space for seven flavors as did the previous six flavors, in addition to making the new designs visually more effective. (Packages designed by Gerstman + Meyers Inc)

Figure 2 Previous Purina Tender Vittles packages displayed six flavors horizontally.

product that was selling well, develop two entirely new products, and thus replace the "sick" ones. The packages for the two new products were to relate visually to the original product, yet be noticeably different in order to create excitement among retailers and consumers. As a result of careful development of communications objectives for the design of the two new packages, this scheme worked perfectly. The new products are healthy and growing alongside the original product, while the weak products have been quietly phased out. (See Figs. 4 and 5.)

Clear communications objectives developed for line extensions makes it possible to phase products in and out of lines with a minimum of risk and frequently with substantial marketing benefits.

Package Design for New Products

Developing communications objectives for new products is the most challenging and the most dangerous task. Both the manufacturer and the designer are entering untried fields. Yet this situation is becoming increasingly important as more and more new products are introduced into the market.

The development of packages for new products is generally divided into three categories:

1 New products in existing product categories.
2 New products in nonexisting product categories.
3 New products in line extensions.

It is important to understand that each of these categories require different communications objectives. A package for a new product in an existing product category may be based, at least to some extent, on the experience derived from

Figure 4 In 1974, the newly designed Purina Cat Chow packages emphasized the five flavors in the line through large cat portraits on brightly colored backgrounds. (Packages designed by Gerstman + Meyers Inc)

competitive products. The development of packages in a nonexisting product category enters virgin territory and is fraught with obvious risks. Such packages will usually benefit from substantial marketing research on which communications objectives can be based. The more that is known about the potential

Figure 5 In 1979, the Purina Cat Chow line retains the "Original Blend" flavor and replaces marginal ones with two new flavors. The packages, instead of merely differentiating the flavor variations, now strongly "romance" the specific character of each flavor. (Packages designed by Gerstman + Meyers Inc)

reaction by the retailer and consumer as to the benefits, as well as the disadvantages, of a new product, the easier it will be to determine precise communication objectives for the design of the package. This is especially important if the product requires new packaging materials or packaging techniques. For example, the manufacturer of an extensive line of plumbing and heating repair products well known among professional plumbers desired to introduce a number of existing products in the rapidly expanding amateur do-it-yourself market. Communications research pointed out potential objections by the plumbing and heating professionals to the marketing of products that could effectively compete with their expertise, especially by a company that had derived most of its profits from sales to their trade. Carefully structured communications objectives eventually led to the development of a line of products that were marketed under a new brand name, and to packages whose structures and visual appearance substantially differed from the professional line. In this way the marketer was able to disassociate, both verbally and visually, the amateur products from the professional line, eliminating the poten-

tial risks of loosing their strong franchise with the professional customers. (See Figs. 6 and 7.)

INITIATING COMMUNICATIONS OBJECTIVES FOR PACKAGE DESIGN

The initiation of communications objectives for packaging design can originate at several sources, including:

- Top management.
- Marketing/sales management.
- Design director/packaging coordinator.
- Purchasing.
- Outside sources (ad agency and/or marketing consultant).
- Package design consultant.

It is important to recognize the differences between these sources of communications objectives. It is most likely that the communications objectives initiated by each of the preceding will reflect particular business interests of the source.

The initiation of communications ob-

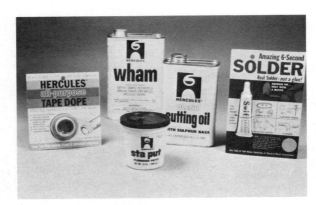

Figure 6 Prior to introducing the new U-Can line, Hercules Chemical Company tried to enter the do-it-yourself plumbing repair market with packages geared to the trade. The confusing packages contributed to the consumer's apprehension of handling potentially difficult plumbing repairs.

jectives for package design by top management, including the president of the company, was an accepted premise around 1960. But this has substantially disappeared. Today's management has shifted most of the responsibility for package design decisions to marketing and sales management, who are directly responsible for the profitability of their products.

From there, the initiation of communications objectives for package design will vary substantially from company to company, and in some cases these objectives are a combination of the concerns of several departments.

- If marketing/sales management initiates the objectives, these will usually emphasize an understanding of the sales strategy and objectives for the products but will, on occasion, neglect the practical, that is, manufacturing aspects of the package development.

- On the other hand, if the purchasing department initiates the communications objectives, emphasis will be on manufacturing requirements, especially cost parameters. Frequently this will anticipate a very conservative package development approach, utilizing existing and easily available packaging systems and materials and a minimum of visual exploration.

Neither of the above approaches is totally recommendable. In today's complicated marketing mix it is important that communications objectives for packaging be developed in consultation with *all* who are responsible for the

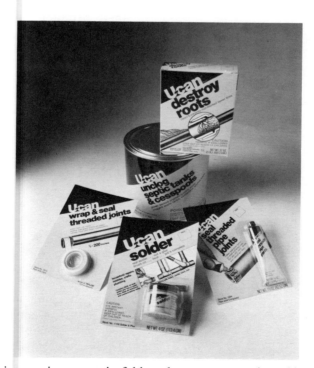

Figure 7 By utilizing a unique, meaningful brand name, copy and graphics, the new packages for the U-Can line communicate to the consumer that do-it-yourself plumbing repairs are easy (Packages designed by Gerstman + Meyers Inc).

marketing of a product or a product line, including marketing, sales, purchasing, as well as external services, such as marketing and design consultants.

Some marketers have packaging co-ordinators or design directors on their staff. Ideally, these individuals help to coordinate the various interests of these departments and will be instrumental in developing clear and precise communications objectives for packaging, combining all interests and requirements including those of marketing and manufacturing.

In any event, it is important to the designer that clear and precise marketing communications objectives are developed and communicated to him. The clearer these are, the more likely it is that his efforts result in the best possible package design solutions. Clear definition of all marketing criteria helps the design team to meet the client's objectives without waste of time and expensive revision, which inflate design and production costs. Revisions resulting from marketers' initial indecision and vagueness of communications objectives can easily add 10–50% or more in design service costs to the initially projected budget.

This can best be prevented, and the project can be launched quickly and efficiently, when the initiator (product manager, design coordinator or whoever) develops tightly specified communication and package design objectives, sometimes called a "Package Design Brief."

DEVELOPING COMMUNICATIONS OBJECTIVES FOR PACKAGE DESIGN

The package design brief should address itself specifically to three important elements that will guide the design program:

1   Recognition requirements.
2   Image communication requirements.
3   Technical requirements.

Recognition Requirements

Recognition requirements are those package design elements by which the consumer recognizes the brand, product, or product manufacturer. This may be a product logo, a corporate logo, a company color, a certain style of packaging used by the marketer or by an entire product category, or a variety of other elements. For example:

- General Electric's logo immediately identifies any product from this well-known manufacturer and thereby gives all GE consumer and industrial products, regardless of package design, an instant stamp of recognition.
- All Kellogg's cereals effectively use their product logo to identify all products and packages. The unique script styling of the logo, and its consistent application over many years, is immediately recognized by all consumers who use Kellogg cereals, despite constantly changing packaging graphics. Even if only the initial is utilized, such as in the case of Special K, the brand recognition survives.
- Company colors, though less frequently utilized for recognition purposes, can effectively identify products under certain conditions. Examples of this method of communicating product recognition are the packages of the Stouffer's frozen foods line (utilizing a dominant orange panel on all packages) and the red/white cans of Campbell's soups. These colors trigger instant brand recognition in the consumer's mind.
- Another effective method of com-

municating brand and product recognition is through the use of unique packaging structures. This has been utilized most effectively in glass packages. There are many examples of easily identifiable bottle contours, such as those of Coca-Cola, Tanqueray gin, and Chanel Nº 5 perfume, to name just a few.

Image Communications Requirements

Image communications requirements are those package design elements that are designed to influence the consumer's image perception of the product. This image could range from perception of quality to emphasis on lightness, strength, female orientation, or primarily male appeal, and so on. For example:

* A snack product packaged in a folding carton will appeal to a consumer totally different from the one that would be attracted if the same product were packaged in a transparent bag.
* A garden tool packaged on a two-color blister pack will communicate a different image than if it were packaged in a container with a full-color photograph of the product.
* A soft drink in a two-liter plastic bottle is an accepted method of packaging for the economy-minded consumer, but could you visualize the communications effect of Perrier mineral water in a plastic bottle?
* National brands of canned grocery products create good product communications by using mouth-watering photographs for the majority of products. The recent phenomenon of "no brand" groceries emphasize black and white, nonpictorial labels to communicate clearly an economy image.

Many product categories utilize certain types of packages that have created an "image mold" from which it is difficult to break out. For example:

* The Wish Bone brand of salad dressing introduced several years ago utilizes a bottle shape that today is imitated, or almost imitated, by virtually every salad dressing manufacturer.
  As a result of such incestuous image communications, considerable courage and a strong marketing concept is required to introduce a salad dressing in a uniquely different package.
* The dairy industry markets their products in gabled containers by which all dairy products can be recognized instantaneously. As a result it would be difficult today to introduce dairy products in new and unique containers.
* The most famous example of "image molding" is that of the "inscrutable" catsup bottle which has undergone numerous attempts of redesign over the years only to be defeated by reluctant consumers who apparently refuse to abandon their image of how the catsup bottle should look and function. (See Fig. 8.)

In other industries, however, the diametrically opposite situation exists. The cosmetics and toiletry industry, for example, utilizes packaging in a most flexible manner to create a myriad of images for their many brands and products by utilizing unusual bottle shapes, elaborate closure devices, and a vast variety of labeling and secondary packaging. In so doing they have been eminently successful in suggesting a wide variety of psychological implications, most of them geared to subtle—or sometimes not so subtle—sexual connotations.

Figure 8 Packages for salad dressing, milk, and tomato catsup are among those that have created industry related "image molds" which it is difficult to break.

It is particularly important that image communications objectives are transmitted to the designer with the greatest amount of detail. The more the designer can learn about the image objectives that the marketer wishes to communicate, the better will he be equipped to develop appropriate package design solutions that will fulfill the desired image communication requirements.

Technical Requirements

Technical requirements are those that relate to the package manufacturing aspects. These include:

- Information as to in-plant equipment on which the packaging will be formed, filled, closed, moved, as well as materials requirements, printing requirements, and so on.
- In the event that complicated materials requirements are involved, such as in the packaging of many chemicals, the compatibility of some materials versus other materials (if known) should be communicated to the designer.

- If structural package design is required, important details such as product viscosity, odors, product protection, required shelf life, transportability, storability, and other product related information are essential to the designer.
- Printing information, the type of printing, the number of available colors, special requirements such as types of coatings, are also important for the designer to know before he starts developing the packaging graphics.
- If a line of products is packaged in a variety of different types of packages, it is important to identify the technical parameters for each of the package types. Neglecting to communicate details will invariably result in lost time, unnecessary expenses, and the frustration of everyone involved.

Developing precise and carefully weighted communication requirements as described above—recognition requirements, image communication requirements, and technical requirements—is perhaps the most important link in the chain of communication objectives for package design. The packaging requirements established in advance of the design development program, and communicated to the designer will be a yardstick throughout the program and keep it on its desired path. They will serve as a guide without the need for scattering design explorations in too many directions, and thus are an important catalyst for saving time and money.

Putting Communication Requirements in Writing

Last, but not least, is the recognition of the fact that in the fast-moving world of marketing development changes in marketing personnel frequently occur when a package design development for a given product or product line is well on its way. If a list of communication requirements and objectives exists in writing, it will be easy for the new marketing personnel to orient themselves as to the work already completed, and the intended strategy developed by their predecessors. Thus it is possible to develop an intelligent strategy on continuing, modifying, or abandoning the existing package development strategy.

COMMUNICATING THE PACKAGE DESIGN OBJECTIVES TO THE DESIGNER

Once the desired communication objectives have been developed, it is important to communicate them to the designer in the most detailed manner possible. Again, it is essential to encapsulate communications objectives *in writing*. In addition to the list of design objectives — recognition requirements, image communications requirements and technical requirements, discussed in the preceding pages — a *written* discussion of marketing and package design objectives should be included with the package design brief, which should include the following:

1 *A History of the Product*   If possible, provide old packages (or pictures of these), annual reports and documents containing historical data, or any other source of information as to what led to the current package.

2 *Description of Product Benefits*   The benefits of your product, as well as the disadvantages of the product, should be thoroughly and openly discussed. It is important that the designer be familiar with all product benefits so that he can communicate some of these on the packages; it is also important to know the disadvantages so that he is warned against emphasizing potentially negative elements. So long as both advantages and disadvantages have been clearly communicated to the designer, he will operate within these parameters to create the package design most beneficial for your product.

3 *Degrees of Copy Emphasis*   It is important for the designer to know the exact degrees of copy emphasis ranging from product identification, brand identification, product designation, identification of flavors or other product varieties, secondary copy, promotional copy, and mandatory (legally required) copy.

4 *Point of Sale*   To be effective at the point of sale — whether in supermarkets, mass merchandising stores, department stores, specialized stores, dispensers, or any other type of sales environment — a thorough understanding of the conditions the product will encounter in the retail environment will guide the designer to develop the packaging that will operate best under given conditions. This may include specifications as to how the product is displayed (standing, lying down, hanging, in dump displays, etc.).

5 *Retail Environment*   Unless the retail environment is an obvious one, such as a supermarket, a detailed description of the conditions, the retailer's attitude toward the product, distribution problems, stocking and restocking conditions, price application methods, and so on, should be thoroughly discussed with the designer. The more he knows about how the product operates in

its retail environment, the better he can assist the marketer to develop meaningful design solutions.

6 *Marketing Plan* Current and future marketing plans for the product should be discussed freely and openly with the designer. It is important that the designer be considered a trusted and intimate part of the marketing team, not an outsider, so that he will be able to develop packaging that will satisfy not only current marketing needs but also provide for anticipated future marketing situations.

7 *Advertising Plans* The designer should be familiar with your plans for advertising and sales promotion for your product. It is always helpful to provide television commercials, print ads, and promotion material relating to the product. If such material is not available or current, provide planned advertising and promotion activities in the form of TV storyboards, animatics, and/or preliminary ad layouts that will help in communicating such information.

8 *Corporate or Brand Policies* If any corporate or brand policies exist relating to the product at hand or to all products marketed by the company, these should be communicated to the designer. Otherwise the designer will work in a vacuum, which will, sooner or later, impede the package design development. Recommendations that will have to be altered to conform to the corporate and brand policies waste time and money.

9 *Competitive Activity* Among the most important information for the designer is the knowledge of competitive activities. All major competitors should be discussed in full detail. Competitive marketing activ-

ities and packaging should be reviewed and evaluated. This is especially important if competitive packages are not easily available (e.g., packages for medical products, drugs, chemicals, or institutional services) or if the products are available only in local or regional areas, which may necessitate substantial travel. The more you can communicate competitive activities to the designer—especially the strengths and weaknesses of competitive products—the better informed the designer will be.

10 *List of Competitive Products* A list of competitive products, in their order of importance, is an important information ingredient for the designer. If possible, sales/volume information about competitive products in relation to your own products will help the designer to develop your packages in their proper context.

The importance of a thorough orientation of the package designer, prior to proceeding with his assignment, cannot be overemphasized. All professional designers will agree that the more information they have, the better they are able to accomplish meaningful design solutions in the most economical manner.

SKEWING THE PACKAGE TO THE RIGHT TARGET AUDIENCE

Another important aspect is telling the designer to which audience the package must communicate. If the package tries to communicate to too broad an audience, the design of the package is likely to end up communicating to none.

Too often this is a neglected area of the communications objectives. Too frequently the audiences are defined too broadly, for example, "men, women, and children from 10–49." In terms of communications objectives such all-inclusive definitions are meaningless and, more often than not, indicate a lack of in-depth information by the marketing management on where the really important market is located.

Target audiences should be defined precisely. Market management must discipline themselves to draw precise parameters of potential purchasers to which their products should be addressed, such as young/old, child/family, male/female, young/mature, everyday use/occasional use, economy/quality, inexpensive/high priced, and urban/suburban.

Knowledge of the precise target audience will help to arrive at the most appropriate package design solution.

USING RESEARCH TO DETERMINE COMMUNICATION OBJECTIVES FOR RETAIL PACKAGES

Although this handbook will discuss package design research in considerable detail, we would like to emphasize the potential helpfulness of predesign research in determining communication objectives for retail packages. Among the many possible research methods, three are considered by us the most helpful, if not indispensable, aids to understanding package design definitions and opportunities.

Focus Group Interviews

Perhaps the most productive method of determining communications objectives for packaging are focus group interviews. Such interviews, with carefully selected respondents, can give the package designer insight into how the poten-

tial purchasers of the products are likely to react to various visual and verbal stimuli. It is often helpful to invite the designer to be an observer during the interviews, so that he can gain first-hand knowledge and understanding of the consumers' reaction to various product and packaging concepts that the focus groups may discuss. In addition, many designers will review portions of the taped interviews with their staff so that they, too, will have the benefit of the group sessions.

Store Audits

An indispensible component of developing communication objectives for retail package design is the audit of stores where your products and competitive products are sold. This is the "real world," where your package, if it is to be successful, must gain an advantage over competitive packages in terms of shelf visibility, image communications, and emotional appeal.

The store audit should be conducted not only in a convenient location, for example near the designer's or the marketer's office, but at as many appropriate points of sale, and in as many locations, as time and budget will allow. Even a limited store audit is better than none. To design a package without a store audit is like crossing the ocean without a compass.

Packaging Evaluation

Another important ingredient in determining communication objectives for retail packaging is an evaluation of all pertinent packages. If the assignment calls for the redesign of a currently marketed package or packaging line, such packages should be thoroughly analyzed as to their current strengths and weaknesses, including their size, shape, function, brand identification, product identification, color, pictorial elements, and copy. Competitive packages should be evaluated in a similar manner, and their strengths and weaknesses should be compared to those of your own products.

DEVELOPING PRELIMINARY DESIGN CONCEPTS TO DETERMINE COMMUNICATION OBJECTIVES FOR PACKAGE DESIGN

A very helpful method of determining difficult communications objectives (with which many marketing executives are not familiar) is to develop preliminary design explorations for several concept alternatives. This is similar to the concept research conducted by advertising agencies to evaluate the potential of various copy themes.

For example, in redesigning a package or a line of well-known packages the question of how much change should occur on the package is often a difficult one to answer without looking at actual alternatives. A successful technique to focus on precise communications objectives is to develop a broad range of preliminary alternatives ranging from designs closely resembling the existing packages to more uniquely different design concepts. This makes it much easier to visualize the effectiveness of various design alternatives in relation to the desired communication objectives.

SETTING UP TIME SCHEDULES

One of the often neglected areas of developing communications objectives for packaging is a carefully structured time schedule. While time schedules do not specifically determine the appearance of a package, they will substantially influ-

ence the care with which the package has been developed.

In an article by J. Gordon Lippincott and Walter P. Margulies, published some time ago in *Harvard Business Review,* they state:

Package planning should be undertaken simultaneously with advertising, sales, and distribution planning—often with the development of the product itself. . . .

Unfortunately, in actual practice, designers are usually called in after all other decisions have been made and the overall plan is already crystallized. The package is treated as an appendage to merchandising rather than one of its integral parts. . . .

It is never too soon to involve a design consultant in your planning. The closer the designer can work with the client at the inception of the project, the better he will be able to participate in the development of the entire marketing process and to keep the product positioning in sharp focus. This knowledge invariably translates into better design solutions.

It is also important to set up a tight time schedule that will list each step in the package development process, from start-up date to delivery to the plant. A typical time schedule may look as follows:

Timetable	*Week of*
Request for proposal	May 21
Assignment approval	June 5
Orientation meeting / begin project	June 11
First design alternatives	July 2
Design refinements	July 30
Market research results	Aug. 20
Mechanicals/artwork completed	Sept. 10
Mechanicals to purchasing	Oct. 8
Packaging at plant	Dec. 17
Plant start-up	Dec. 26

IN SUMMARY

In his book, *Secrets of Marketing Success,* Louis Cheskin sums up the process of determining communications objectives as follows: "The only way to achieve effective communications is to provide the creative person—copywriter, artist, or designer—with objective information about the problem. . . ." While there is no fail-safe method of handling the process of developing communications objectives, it may help to keep the following guidelines in mind:

1   *Be sure of your marketing objectives before you begin a package design development program.* Carefully prepared preliminary analyses of your objectives are the best insurance for achieving quicker and more economical design solutions.

2   *Define package design objectives clearly by developing a written list of specific criteria.* It will be helpful to divide your objectives into three well-defined areas:
   - Recognition requirements.
   - Image communication requirements.
   - Technical requirements.

3   *Prepare a written statement of objectives to communicate your package design objectives to the designer.* Clearly define your own objectives and communication priorities, inform and reach agreement with other participants in your marketing program, and communicate clearly your objectives to the designer.

4   *Be sure that your objectives are realistic.* Don't expect the package to compensate for the shortcomings of your product. Establish the strengths of your product, communicate them to the designer, and let the designer translate them into package design solutions.

5   *Be sure your objectives are meaningful.* Criteria based on generalities will contribute little to achieve meaningful design solutions. Objectives that cover too broad a base and are laced with platitudes such as "we want to sell more products" communicate nothing meaningful to the designer.

6   *Provide all available marketing information to the designer in writing.* This information includes product history, product benefits, marketing plans, advertising and promotion plans, available research, and information about competitive products and any other pertinent information. Remember, the seeds of your carefully planned objectives will grow more meaningful package design solutions.

7   *Allow for research in developing communication objectives.* If possible, include some preliminary focus group interviews, but never begin a package design program without a careful audit of the sales environment and an evaluation of your and competitive packages.

8   *Include the designer at an early stage of your communication planning.* The inception of your project is an ideal starting point.

9   *Set up a realistic time schedule for your package design program.* It is unrealistic to expect the designer to come up with instant design solutions, to make up for time lost by delays or indecisions within your own company. Too long a time span, however, is not recommended

either. It allows for too many temptations to make frequent design alterations and may thus eventually dilute the final package design effectiveness. Give the designer a reasonable time parameter in which he can help you effectively to achieve your ultimate objectives.

10 *Don't settle for the obvious answers.* Before you decide on a final package design direction, explore all alternatives based on carefully structured communications objectives. It may take a little longer and may cost a little more, but is is your best insurance against overlooking the best solution.

BIBLIOGRAPHY

Louis Cheskin, *Secrets of Marketing Success,* Trident Press, New York, 1967.

J. Gordon Lippincott and Walter P. Margulies, "Packaging in Top-Level Planning," *Harvard Business Review,* 1956.

Aaron I. Brody, "Base Package Design Decisions on Scientific Facts, not on Myth," *Package Development & Systems,* May/June 1974.

Herbert M. Meyers, "To Redesign or not to Redesign, That is the Question," *Package Development & Systems,* January/February 1972.

Richard Gerstman and Herbert M. Meyers, "Where Do the Design Dollars Go?," *Packaging Digest,* February 1977.

Tallmadge Starr, "The Value of Packaging Research," October 1978. Unpublished Thesis, Pratt Institute.

CHAPTER TWO

Planning Package Design Research

Arline M. Lowenthal
and Cheryl N. Berkey

Packaging research is a vital, but sometimes neglected, aspect of the total marketing of a product. It is the link between the efforts of the advertising, promotion, and public relations function that produces to a large extent the environment in which the product is sold, and the product development function that creates the product itself as used by the consumer.

The relevance of packaging varies slightly with the nature of the product. For example, the packaging of a direct mail product is not seen at the point of sale; a product used over a long period of time may be separated from its wrappings soon after purchase. The following points, however, are those that differ-

entiate packaging from other aspects of promotion and advertising:

1 The product package is frequently seen at the point of sale. It must therefore be immediately recognizable as the item that has been advertised and promoted.

2 At the point of sale the package can usually be compared to other brands in the same category and must, therefore, be able to withstand the competition of several packages.

3 After purchase, the package is exposed to the consumer in a manner isolated from its competitors. By making a sufficient impact in this

39

situation, it can promote brand loyalty and repurchase.

4   The package is commonly observed close to or simultaneously with the use of the product. It should therefore be sufficiently compatible with the product's performance so that no sense of disappointment or frustration is transferred to the consumer's perception of its image.

5   Because several types of advertising and promotion may be presented during the life of a specific package design, the package must be capable of relating to a variety of promotional approaches without appearing to be inconsistent with the related advertising.

In planning our research, we must address the following questions:

1   What general concerns must we consider as we plan the role that packaging research will play in the overall development and launching of our packaged product?

2   What is our product? In what way or variety of ways can we present it so that the public will use it the most?

3   Is our package an effective marketing tool? Will it help to sell the product and create the identity and image that is most beneficial to our product?

4   What is our market, and does our packaged product appeal to these purchasers and users?

To obtain the greatest value from the research we conduct to answer these questions, we must maintain two important attitudes: open-mindedness and objectivity.

Whatever type of research we are conducting, we are wasting our resources if we do not allow feedback from the marketplace and if we do not carefully consider this feedback, whether or not it supports our preconceived notions. We must be prepared to accept, if our carefully conducted research tells us it is so, that the various alternatives we are considering, whether they be graphics, materials, logos, closures, or any other aspect of our package, are not viable in marketing our product. In some circumstances we may be forced to employ some feature that is less than ideal. What we must never do is assume that because one of a number of options is "better," it is necessarily "good."

In this chapter the types of research described will be categorized by the major headings of qualitative and quantitative research. Our definition is by question number and then by degree of "open-endedness" of the questionnaire.

In the case of qualitative research as it relates to package design research we will be tapping the creative resources and input of a few carefully selected respondents with regard to the topic being investigated at each juncture. Ideas of a serendipitous nature often surface in this setting, leading to a finer set of research findings and thus a better package design in the end.

Quantitative research should do exactly what it states: quantify the feelings and opinions of qualified respondents selected to answer the list of both open- and closed-ended questions. The number of respondents, stores, or whatever that compose your sample should be decided by the fact that the final statistics must be projectable. This projectability, usually referred to as a "confidence level," states that if you want to know what 95% of the population of your universe thinks of a certain idea, you achieve this projectability by interviewing a certain number of respondents. A ± feature is also built into the tables used to determine sample size. If for example, you want to know that 90–100% of the respondents' feelings could be projected to the total universe, you would strive for a level of confidence of 95 ± 5%.

Qualitative research, then, would be defined as reporting the reactions of fewer respondents but in greater depth and with greater creativity, whereas, with quantitative research, although parts of the questionnaire would be constructed in an open-ended fashion, the number of respondents interviewed would directly correlate to a 95% level of confidence ± 5% and the quantity, or number, would be the important key.

Not all research involves asking questions. Other effective ways of obtaining the information we need include discussions such as those in a focus group situation, product searches, and consumer observations. In planning research, however, we must start by asking ourselves what it is we need to know.

GENERAL CONSIDERATIONS

The first concern that must be addressed when planning research in general, and package design research in particular, is the basic parameters that will govern our research program. These are principally time, cost, prototype usage, and coordination of the myriad of departments and people involved in the project.

Time

Often the latter three elements involve creating a "hurry up and wait" situation on timing. In a case where contents will be packaged in the designed container, the actual decisions on package design must be finalized long before the product will hit the consumer market. Many times product schedules are made three or more months in advance, and the purchase of containers on which the

design will be printed must be made several months in advance of the actual sale time of the product. These factors, of course, dictate that to be effective, package design research must be conducted in the early stages of product development.

Too often manufacturers will allow realistic time for the proper marketing schedule of a product with the one exception of the necessary research, and particularly the package design research. For this reason much package design research is done "after the fact," and the corrections can be so costly as to negate the optimum marketing of that product or line for that production run.

One example of this occurred many years back when a manufacturer of a men's grooming product made a corporate decision to retool its production line completely by switching to aerosol containers and proceeded to hire an artist to design a label for the new package. No research was conducted although a multimillion dollar decision had been made. After the newly packaged product had been on the market a few months, the projected sales figures were not met. A research study was ordered; the outcome was that the package design was not functionally correct for the product, nor, as the research showed, were the artwork and graphics compatible with the product image.

Another aspect of timing when planning package design research is that of scheduling the various phases you have selected so as to assist the decision makers in deciding on the ultimate package to be used. It is not unheard of for a package design project to start 2 years prior to product introduction, and it is nice to think of working in a structure such as this. However, in the real world only a few of the largest companies have the market planning insight or the resources to plan this far in advance.

Moreover, even in this situation the important details of time and prototype availability may be inadvertently overlooked or forgotten.

Although most research groups are accommodating when a time pressure presents itself, it must be realized that a certain amount of time is necessary to provide useful data. Let us caution that even if a company is smaller than those referred to in the preceding, good research, conscientiously conducted, should be written into any overall marketing plan. Good research *can* be done even though the timing and budget are limited.

Cost

When considering package design research for a nationally distributed product, the question of budget is paramount. If one can't afford a heavy research project, extensive in cost and area, it is best to do a limited study *but* to do it correctly. For example, if you want a national study covering eight markets and you need at least 400 interviews in each market to achieve a 95% level of confidence ± 5%, but your budget doesn't cover this requirement, decide whether you really need eight markets and are willing to go with less interviews per market, which will provide a lower level of confidence (and thus a reduced projectability). The alternative would be to use fewer markets, for instance, three, and retain the 95% level of confidence ± 5%. If the results of these three markets vary to a great degree then exploration of the differences would be necessary.

We believe that the objective of 95% level of confidence ± 5% supercedes the inclusion of a larger number of markets into the study. One can always conduct additional qualitative research in the other markets to account for variations or regional differences.

Prototype Availability

When dealing with "one-of-a-kind" prototypes of package designs, a time, cost, and scheduling crunch may arise. The expense involved in producing certain package prototypes can be infinitely greater than the cost of producing the final product. Because only a few prototypes are produced, the length of the in-field time must be expanded dramatically. The cost of transporting the prototype from market to market will also be increased due to the irreplaceable nature of the prototype being researched.

It is not uncommon, for example, for the prototype to be hand carried from market to market aboard an airplane by a representative of either the client or of the research group or field service, rather than utilizing a quality package forwarding company.

Since in many instances this prototype is the very vehicle being tested and researched, it is the pivot and must be given special attention as the "jewel" of the operation.

Coordination of Various Departments and Personnel

This is a crucial area of concern since you will be masterminding scheduling involving people of many disciplines and temperaments. Groups from inside the company as well as outside specialists such as research consultants will be included in this coordination. Such incidentals as travel schedules, data collection coordination, and maintaining all of the project due dates and deadlines depend on carefully prepared communications and coordination of the project.

The following is a list of possible departments/personnel/consultants with whom you may be concerned while coordinating a given project. Some of these may be overlapping and, at times, represent outside services or consultants.

1 Corporate representatives.
2 Product/brand manager.
3 Industrial engineers (package designers).
4 Creative group (writers).
5 Graphics group.
6 Marketing group.
7 Outside research consultants and suppliers.

There are probably many satellite interests involved also, and after giving some thought to this area one can see that coordination is a most important task.

PRODUCT SEARCH AND EVALUATION

The next question we must ask ourselves is "What is the product?" We do not, of course, mean just the generic type, such as shampoo, garden hose, beer, or detergent, but rather the substance of the product we are packaging.

Is it dry, liquid, paste? Is it powder, granules? Is it large, is it small? Is it one object, like a cake, or lots of objects, like cookies? Does it smell? Will heat, moisture affect it? Is it heavy, so that we must build handles into our package so it can be transported? Or is it brittle, so we must protect it? Do people eat it, wear it, put it on their walls, or put it in toilet bowls?

The answers to these questions are not necessarily to be found through opinion research, although it may be that creative qualitative research methodologies will provide some interesting insights if we are able to keep a sufficiently open mind to notice and use them. More valuable material may be found in our objective consideration of the product

and in the ways in which other companies package their goods. It is not just our immediate competitors with whom we should be concerned. That would simply limit our horizons and possibly open us to the risk of plagiarism. What we should be looking at are the possible ways in which the *type* of product we have can be packaged: a liquid is a liquid and a powder is a powder, regardless of its purpose or the way in which it has traditionally been packaged.

Research Now

Secondary research, that is, product search, is called for at this point. At this stage, products with similar qualities will be objectively considered with a wide open, positive, brainstorming approach. For example, if our product is flour, we should examine other products with similar qualities, that is, powders susceptible to moisture and insects, used and sold in various quantities. Among the products we may consider are: dried milk, talcum powder, and antiseptic powder, which have a similar consistency; sugar, wallpaper paste, and potting soil, which are used in various quantities; cookies, rice, and pasta, which are susceptible to moisture and insects. Obviously, not all of these products are equally relevant to our product. They may, however, give us ideas that we can use in similar or amended form.

Initially, some research of a simple nature should be conducted on the product's competitors. Are competitive products similar? Are there practical reasons for this similarity? If so, is your product limited in the same way? If you select a radically different package shape, will you meet resistance to a new design? Can advertising and promotion overcome that resistance? "Novelty" style packaging has a major advantage in creating brand recognition, but it has corresponding limitations. A novelty package may give your product strong identity and provide recognition and be a major asset to advertising and promotion efforts, if it is very carefully selected. On the other hand, it may tie your product to that type of image, limiting your ability to change this image without total loss of identity.

The ability of your package to be produced in larger or smaller sizes must be considered. If various sizes of the same product appear radically different, a costly process of developing an individual market recognition for each size of package may be necessary.

At this point, attention should also be given to innovations in color and graphics. Particular points to consider are the suitability of a specific color to the product, and the materials and shapes that lend themselves to the product being packaged.

This type of research can be very simple. An expedition to the local supermarket, drug store, or hardware store, may be rewarding; in other cases, a careful examination of catalogs will be helpful. What is important is that we approach this study with an open mind receptive to ideas that may be less than obvious.

A more formal research process would involve qualitative research such as focus group discussions, analyzing the results with an eye to "different" ideas, and a positive attitude to creative thought patterns. Quantitative research is not called for at this point in our program, since the objective here is to gather creative ideas, not to measure reactions to a limited number of concepts.

At this stage we have a collection of ideas about possible packaging features, such as shapes, colors, materials, sizes, lids, spouts, fastenings, and handles. These ideas then must be passed to our package designers and engineers to be translated into feasible design models that we may use in later stages of our research program.

The precise manner in which these ideas are presented may vary, but it can be counterproductive to present a series of features in a connected format, since this may bias the design engineers and limit creativity.

EVALUATING THE PACKAGE AS AN EFFECTIVE MARKETING TOOL

The perfect product would be one that is alone in its market, is top quality and an excellent value for the money, and yet is highly profitable to the manufacturer and distributor. In an ideal situation we would only need to manufacture and distribute our product and the customers would flock to its place in the store. In the marketplace we normally face, of course, the tastes and values of our customers are always variable and changing. Quality and value often conflict with profitability, and however hard we work to convince the market that we have the perfect product, our competitors are working just as hard to defeat us. The principal weapon we have in our armory is the image we create for our product through advertising and promotion, which includes the packaging of our product.

The two key questions at this stage are:

What is the image we want to give our product?
What is the image that will create the best consumer attitude toward our product?

Although these questions appear similar and are closely connected, the answers to them may be quite different. For the purpose of planning an effective research program, it is valuable to consider them as two separate entities.

The Package Identity

The package identity is made up of those aspects of packaging that in a group of similar and competitive products identify our particular product, for example, the product name, the logo or trademark, and the product description. Color, too, can be an important area of identification, although in some cases colors will vary to denote the different varieties of one product. It is these aspects of identity that will remain with the product over the years.

Selecting the name of the product and its logo or trademark is not primarily a part of package design, although the name and trademark will be an integral part of the product, the package, and the image they present. Moreover, the logo may well be used over a long period of time, possibly after the rest of the package has been redesigned; therefore, an effective logo that can give continuity to the product is a long term benefit. If the logo has not already been carefully selected, it requires research in its own right.

Similarly, the color scheme of the package may have a long life, even if redesign causes the actual use of the colors to be varied.

Shape is another basic feature to which considerable attention should be given. A number of factors are relevant to this decision, many of them concerned with functional aspects of the package, such as efficiency of distribution, product protection, and ease of use. These are discussed in more detail below. At this stage, however, we are principally concerned with the impact the shape of the package will have on the image of the product.

Package shape and design, color, graphics, promotion and advertising, and image may all change over the years, but the identity of the product will remain basically the same throughout the life of the product.

In some cases, the various features of product identity may be predetermined. For example, if the package creates a new look for an existing product, probably we will not wish to alter all the packaging features in order to retain continuity of our brand identity, but instead will concentrate on a new image for the product. If the product is new but a line extension of an existing brand, we may wish to use certain line identification features, such as logo or color scheme.

Nevertheless it is important to ensure, through careful research, that logo and color are appropriate to the product. However powerful the arguments appear to be for staying with certain aspects of an already proven packaging format, it is essential to keep an open mind until research has definitely established that these features will be successful with this particular product.

Research Now

The research role at this stage is to answer the following two questions:

Are the features of the product identity, separately or together, suitable to the product we are packaging?
Is the product identity acceptable or attractive to the consumer?

These questions do not necessarily have to be answered at this stage in the design project. If time is at a premium, or if the budget is limited, we may wait and test the various ideas generated by the marketing, graphics, and advertising departments at the same time that we test some other aspects of the packaging. If we delay, however, we may face three very real problems.

1   If the entire identity concept is wrong for our product, we will have wasted valuable time that could have been used to redesign our identity.

2   If we test many variables at one time, we will require a larger and more complex study in order to be sure we have accurate feedback on individual items.

3   We cannot always be sure that respondents are reacting objectively to the features we are testing. For example, respondents' reactions to identical products can be affected by the color of the container, even though they believe they are honestly evaluating only the contents of the container.

Assuming that we decide to research our product identity, our methodology will be largely quantitative, however, if time and budget allow, qualitative input will also be valuable. We need to interview a sufficient number of respondents to ensure that our results will be projectable, but we also may benefit from picking the brains of our respondents.

Image

The image is that which creates in the mind of the consumer a sense of benefit or value in excess of, or different from, that which they would experience without a created image. For example, a piece of jewelry from Cartier carries an image of wealth and elegance; buying or wearing such an item implies elegance and wealth in the owner. A comparable item of similar value purchased from a department store may not carry these ego-boosting implications and would probably only be purchased for a lower price. In this case our packaging, in a box with Cartier's insignia, is implying the image that advertising and public relations have created. Another example

is a cake mix that implies by its image that it will produce a better cake, thus enhancing the consumer's abilities as a homemaker. Items that are promoted as low priced and good value may give the consumer a sense of having achieved a purchase of especially good value, enhancing a sense of good money management. In that case very simple packaging may produce a sense of not having wasted money on expensive trimmings.

This sense of image can be created in a variety of ways. Sometimes a very direct manner is used, as in the case of a brand of English preserves promoted as "the most expensive jam you can buy."

In other cases, a more subtle approach is used, as when a household product is advertised by a young and attractive person, implying that the user will in some mysterious way become younger and more attractive, even though the product is quite unrelated to personal appearance.

Research Now

Ideally, at this point in our research program, we should have prepared and tested the following items:

1   The general form of the package.
2   The identifying features of our product package.
3   The advertising approach being used.
4   The product itself.

This would enable us to test the image and attitude in a simple manner, with reduced danger of confusing various attributes of our entire marketing approach. In practice, however, we may wish to test, retest, or finalize any of the above items at this stage. This is not incompatible with attitude and image evaluation but, as mentioned previously, we must take care not to confuse reactions to the various features.

The following research steps are now called for:

1 To test and evaluate the image created in the mind of the potential purchasers by the package design, that is, the reaction of the consumer up to and at the point of purchase.

2 To test and evaluate the attitude of the consumer when the product is in the use situation, such as the home or office, that is, following the point of purchase up to and beyond the use of the product.

At this point in our research we must give consideration to two opposing points of view.

The first school of thought maintains that image is everything, that the performance of the product need not measure up to the image if the image is sufficiently attractive. For example, a T-shirt endorsed with a designers insignia will sell for vastly more than an identical garment without that insignia. In the same way, a carefully placed line extension of a highly successful product may benefit from the positive image of the original product. The psychological factors operating here would appear to be:

1 That the consumer believes that a product must be superior if sold under a status-oriented label even if this superiority is not overtly apparent to the purchaser.

2 That if the product itself is not actually superior, the status of owning such an item is valuable in itself.

This argument has validity only if at least one of the two conditions are met:

1 The brand image is extremely strong.

2 The product is such that the qualities of the image are critical (or potentially critical) to the performance of the product.

For example, attractive appearance is a major element in the function of clothing; effectiveness is a major element of a pharmaceutical product. Taste on the other hand, is an important element in food, but nutrition, texture, and an appetizing appearance are of equal importance. Therefore, if our image of attractiveness or taste is sufficiently strong, image will sell the product, even if the product would not be considered to have these qualities without the help of its image.

The opposing school of thought maintains that, because the consumer is discriminating and value conscious, the performance of the product is as important as the image, and the product expectations created by the image should be in keeping with the attitudes created by the product performance. If these two factors are not similar, the sales of the product will suffer. If the expectation level is higher than the performance level, disappointment will reduce repurchase. If the image is lower than the performance, fewer first purchases will be made, and fewer repurchases also, since a low image may influence opinions of performance.

If one adheres to the first school of thought, it follows that it is not necessary to test package and product together, since if the image is sufficiently high and the product performance adequate or better, an optimum situation has been reached.

If, however, one attaches importance to the relationship between image and performance, it becomes necessary to test the following:

1 Image.

2 Product.

3 Image and product together.

Regardless of whether or not it is desired to research the combination of product and image, it is extremely important to research the image in isolation in a manner that precludes the possibility that product performance, the container, or any other factor will influence the consumer reaction to image. Product identification, of course, cannot be excluded, since it contributes to image.

Our research at this stage should produce data in three areas:

1   Among the potential images we are considering, which do the respondents find most conducive to trial and purchase?

2   Why do they find this image most conducive? This may produce valuable advertising input.

3   How do our possible images compare with the existing images of our competitors?

In order to ensure the accuracy of our findings, the following basic principles of research must be carefully observed.

1   Absolute confidentiality and anonymity—respondents may easily be influenced by the knowledge of the company for whom a study is being conducted.

2   Careful screening of respondents. It is important that the sample selected include both consumers loyal to rival brands and those who do not have a brand loyalty for this product. This factor must, of course, be considered in the analysis of the results.

IDENTIFYING OUR MARKET

Identifying the market for a product, although central to the overall marketing approach, is not strictly the concern of package design research. It is important, however, that this information be available to the design/research team. Unless we know to whom our package needs to appeal we cannot effectively measure or evaluate this appeal. Specifically we need this information to answer three questions when planning our research.

What Do the Users of Our Product Have in Common: Depending on the nature of the product and the extent to which there is presently a market for our product, our parameters may be broad or specific. For example, a broad-based food product, such as flour or sugar, would offer us a market to include any person who normally does the food shopping or cooking for a household, regardless of age, sex, income, or family size. An expensive skin preparation for mature skin, on the other hand, would tend to limit us to older women with high disposable income.

Depending on our product, and the aspects of it we wish to investigate, we may wish to use only a subgroup of users for our study. For example, a disposable diaper study will concern itself with those who care for babies and very young children. This group will not only include mothers, but pediatricians, nursery nurses, day care workers, fathers, and so on. Because we are considering only the package at this time, we can start by excluding pediatricians from our sample since they will be concerned with the effect of the product rather than its packaging. We may also choose to exclude fathers, since only a relatively small group of them will be concerned with the brand selection and purchase of diapers. Nursery nurses, day care workers, and mothers, however, are equally likely to be concerned with how the diapers are packaged. At this stage we must decide which of these groups we want to study.

Using nursery nurses and day care

workers may involve us in a more costly and time-consuming study. These people may be harder to locate in the first place, and it may be necessary to convince hospitals to use our experimental diaper in their entire nursery. If they already use our diaper, this may be no problem; if not, they may be reluctant. In either case it may take us some time to set up the study.

There could be benefits, however. If we were able to state in our advertising that "83% of hospital nursery nurses found our Brand A diapers easier to use," this could be valuable. In a nursery situation, more babies would have their diapers changed over a given period than in the average home environment. Convenience would therefore be more critical in this situation, and more packages would be tested over a shorter period of time.

Using mothers with babies at home, we would have a study that would be more simple to set up and that might take less time to complete. We would also be working in a more sensitive market, since mothers can change their brand every time they purchase diapers, whereas diapers used in hospitals are usually purchased on contract, and the choice of brand is not easily or speedily changed by the nurses' preferences.

As can be seen by this example, these decisions are important to the design of the study and should be made by the marketing and research department or agency together.

What Differentiates the Users of Our Product: What variations can we expect to find in our market that will affect our product package?

Continuing from our previous example, we may have decided to limit our study to mothers with babies at home. This group will vary in a variety of ways, and we must now determine which of these variations we wish to take into

account. The age of the mother and her income, for example, are not likely to have much effect on how she selects a package of diapers. The age of her child, on the other hand, may make quite a difference; a newborn baby can be relied on to stay fairly still while she opens up a new package of diapers, but a ten month old baby may have crawled off into another room if it takes her too long to get the diaper out. How often does she use disposable diapers? If she uses them all the time, a stand-up box on the changing table may be very convenient. If she only uses them for travel, a box may be awkward to transport and she may prefer a soft plastic bag that is easier to pack. A mother diapering her third baby may have much more decisive, and informed, views on diaper packaging than the mother dealing with her first-born.

It is important that we decide which variations of usage are really relevant to our study, but care should be taken not to complicate our study with unnecessary factors. The sex of the baby, for example, can be relevant to the effectiveness of the diaper itself, but quite irrelevant to the package in which it comes.

What is Our Market Incidence: How many people in the entire population are potential users of our product? Can we expect to find local or regional variations in the incidence? We need the answer to these questions for two reasons:

- To establish how large a sample of our universe (or total population) we must interview in order to make our study projectable.
- To determine how difficult in time and cost it will be to locate the number of respondents we need.

Variations in incidence will be dependent on the product type. The occurrence of small children for our diaper study will be relatively consistent throughout the country, although it may be slightly higher in some regions, such as the West Coast. Local variations must be taken into consideration, however. It will not be helpful for an East Coast agency to arbitrarily select block clusters in San Jose if these turn out to contain only retirement communities or "adults only" apartment houses.

An ethnic food such as corn tortillas would create different problems. On the West Coast there would be a higher incidence than in the Mid West, and the ethnic structure of communities would increase or decrease incidence dramatically.

With low incidence products, special methods such as personal referrals may have to be used. Product type will again be relevant—people tend to know if their friends use contact lenses and will not hesitate to pass on that knowledge. The use of hemorrhoid preparations is unlikely to be known outside the immediate family, however, and other approaches would have to be used.

FUNCTIONALITY

Many textbooks refer to the "cracker barrel" presentation of goods as if this was a "prepackaging" concept. This is not strictly true. If we examine a little more closely this method and the way it has evolved into modern day packaging, we may get a clearer idea of what the functional requirements of packaging are today.

To start with, the manufacturer had to package his products in some way in order to ship them from the area of manufacture and to protect them as they traveled to the point of sale. He did this in bulk, because it was easier and cheaper, and he used the method most suited to the product, a barrel for fragile crack-

ers, a sack for flour. The storekeeper then "repackaged" goods in the quantities desired by the customer; the customer then "repackaged" the goods into a convenient container for use. This is the history of the canister set.

This process has many disadvantages. The goods were not always well protected (crackers can get broken in a barrel, flour in a sack is not protected against moisture or insects); it is time-consuming for the storekeeper to weigh out individual quantities while other customers wait; it was not always easy to convey goods home safely in the grocer's container (e.g., a paper bag of eggs is fraught with potential disaster); and it was inconvenient for the user to have to transfer the goods to another container before using them.

By degrees, some of the steps in this chain were eliminated as the manufacturer and the storekeeper evolved methods of packaging some of their products in ways that provided more protection to the goods and more convenience to the customer. Along the way, it was discovered that the actual package could be used as a means of promoting products.

Almost all saleable products are now packaged in some form before arriving at the point of use or consumption. The point at which they acquire this packaging, however, varies widely, and many products are actually still purchased "from the cracker barrel."

At one extreme, we have the product that arrives at the point of consumption in exactly the same form as it leaves the manufacturer. For example, margarine frequently appears on a consumer's table in the same tub or squeeze jar in which it left the manufacturer. We have eliminated the storekeeper's work in cutting off a piece from a large block, wrapping it in grease-proof paper, then putting it in a paper bag; the messy possibilities of the margarine melting on the way home have been guarded against by a tightly sealed container, the homemaker no longer has to unwrap her purchase and place it on a butter dish before bringing it to the table.

Produce, on the other hand, is shipped from supplier to the marketplace in bulk and offered to the public in an unpackaged form, just like the cracker barrel. The customer or storekeeper packages the desired quantity at the point of purchase. Other products, such as candy, are individually packaged for protection and hygiene and are then packaged into a container that will keep the items together as a convenience for customers or storekeepers. Multiple packages are also a progression of this concept, where a product is available in an individual form such as a can of soda, but also packaged into six-packs and then into cases for the greater convenience of shopper and storekeeper.

Research Now

Once the product has been completed, the details of the physical package design are the province of the packaging engineer, who should have at his disposal all the information that has been gathered, such as the intended use of the product and the demands imposed by the nature of the product. Once design has been completed, it is the responsibility of the research team to test functionality in the following areas:

1 Does the package protect the product under all normal conditions, such as during shipping, storage prior to sale, normal handling in the store, storage in the home?
2 Is the product easy and convenient to use?
3 What features should be changed to improve the protection, convenience, and ease of use?

Several very real problems often face the researcher at this point. Two of the primary ones are the cost of prototypes and confidentiality.

The construction of a realistic prototype package can cost several thousand dollars. Few companies are willing to risk retooling for an untested design or to spend the money on having a sufficient number of prototypes constructed for in-field testing. Similarly, when big money is riding on the development of an innovative concept, there is a natural hesitation about exposing it to a situation where someone connected with a rival company might become aware of it. It is a standard practice to screen respondents for family members or neighbors in associated industries, advertising, research, or the media, but it is impossible to be certain that some chance visitor to a test respondent's home will not expose a confidential idea to a rival concern. Two alternatives, which may be used singly or in combination, present themselves.

1 To send the prototype, possibly hand carried, to each of the test areas, where research can be conducted at a central location. Depending on the nature of the package, only observation or limited handling of the item may be possible.
2 To test different components of a package design separately. For example, suppose a manufacturer is planning to market a food product to be heated in a plastic bag designed to make it easy to remove from the hot water. For the reasons described previously, he is unwilling to make up a quantity of the specially shaped bags. He does, however, want to test both the bag shape and the plastic material of which the bags are to be made. He has a few prototypes of the shaped bags, and at low cost he can make

conventional bags of the new material. Therefore, he might decide on a two-stage study:

- At a central location, respondents would be asked to remove the test bag from pans of hot or cold water to test shape.
- In an in-home test, respondents would use the product packaged in conventional bags of the new material to test whether it remained sealed and did not become soft, stretch, or melt.

Through these two research phases, the manufacturer would gain valuable information about the qualities that concern him, without excessive cost and with a reduced risk of confidentiality loss. Confidentiality would be preserved further by insuring that the two phases were conducted in separate cities to reduce the risk of a rival hearing of both tests and making a connection between them.

Another problem arises if a product is not yet in production or in the store and is therefore difficult to test. To some extent it is possible to simulate handling, shipping, storage, exposure to different climates, and so on with prototypes, although it may be difficult to reproduce the real situation.

If the package is already on the market or if a similarly packaged product is available, a "shopping test" may be conducted. This is a method that can be used to monitor the performance of rival products also.

In this type of test, items are purchased from local stores in the selected cities from specified or random locations. They are then packed and shipped in a variety of ways back to the client company or research agency. Performance of the attributes in question can then be measured. If necessary, items could also be repurchased from consumers, naturally at a somewhat inflated cost to cover inconvenience.

Other problems may arise when the product is not available for testing with the package. In some cases, the product does not have a direct relation to the package, and the important feature can be tested with either a dummy product or without the product. For example, the pull tab on a beer can has no relation to the contents of the can, and tests could be made with cans filled with water, soda, or even air (although in this case the research should give due allowance to the effect of the weightless can, spillage potential, comfort of drinking, or ease of pouring). In other situations, the contents of the package may be very relevant to the package, as in the case of packaging a thick cream hair conditioner. If a dummy product is used, care should be taken to ensure that the consistency of the products is identical, and that the product reacts in a similar way under varied circumstances such as in the warm environment of a shower.

The features of functionality that must be assessed will vary with the nature of the product. The following are some of the questions it may be necessary to answer:

Is the package easy to open, if necessary with one hand? Can it be tightly resealed easily? Does it protect the contents under all reasonable conditions? Is it easy to store? Is it stable on its base? Can the package label be seen easily? Will the printing run if it gets wet? Can the package be gripped easily? With soapy hands? Do the contents come out easily, in the right quantity? Does it spill easily? Does it need special features, like a child-proof cap or a drip lip? Is it too large? Too small? Too heavy? Too awkward? Can different flavors or varieties be identified easily?

These are the types of detailed questions we must ask when we are up close to our package, but it would be a mistake not to stand a little farther away also and ask some more general questions:

Are these features important? Which of these features are sufficiently important to have a major effect on sales, on useability? Are there any broader problems that affect our packaging concept? Resistance to aerosols because of possible Freon content, for example, or a feeling that fancy packaging must be raising the price?

Are we actually trying to solve a problem that does not need solving? Does the market need a resealable flour bag, or do people put flour in a canister anyway? Perhaps we should sell our flour in a reusable canister? Or just sell canisters?

In designing our data collection tool, we must therefore incorporate these three features:

1 A detailed evaluation of the individual features of our package.
2 A comparative ranking, scaling, or rating of features.
3 An opportunity for the respondent to reject all or part of our total package concept because of factors outside the package itself.

The following anecdote, while it concerns product more than package, does serve to illustrate the need for research and the ease with which the obvious can be overlooked.

Some years ago HRH the Duke of Edinburgh opened a Design Exhibition in England and afterwards visited some of the stands. One of the exhibits was promoting a new metal teapot. The Duke, fond of his "cuppa," was interested. He is also noted for being a practical man.

After inspecting the teapot he asked for hot water. The request was unexpected and hot water not easy to obtain in the exhibition hall, but the Duke waited patiently until it was brought and poured into the teapot. The metal handle promptly became far too hot to use. His suspicions confirmed, the Duke moved on, leaving the exhibitors highly embarrassed at their failure to conduct research under realistic conditions.

CHAPTER THREE

Planning Design Strategy through Package Design Research

Francis P. Tobolski

THE PRODUCTION, DISTRIBUTION, AND MARKETING ROLES OF PACKAGING

Packaging development review offers a unique vantage of the entire business process in firms manufacturing and/or distributing packaged goods. For these companies, the package is integrally and intimately involved in the production and marketing of products.

Even if a company were primarily categorized as "production oriented," that is as one that does not totally relate to market needs,[1] its packaging would be employed in important product protection and containment roles. Beyond that application, however, is the pattern of procedure of the "marketing oriented" firm: the designing of packaging primarily for *consumer* acceptance and convenience, as a selling tool, with protection as needed.

While both orientations may not have the same emphasis and positioning for packaging, the package in both actually functions as part of the communications or promotion mix and the physical distribution mix and as a product related item. Many times it is not considered in these terms, and planning—in either orientation—may not integrate the package into related strategies.

Originally, the package was designed

primarily to contain and protect. Developments on this theme have resulted in many techniques relating to the forming and filling of the container, protecting the product in shipment, and delivering the product in an undamaged state. This technical, containment and protection view of the product/package relationship has long been familiar. It is undergoing constant development and investigation.[2]

It is when the marketing mix components—and all companies have marketing mixes, attended to or not—are viewed in their interaction that the best understanding of their contribution and packaging's related contribution can be achieved. If the package is viewed as a related element of the product mix, the promotional blend, distribution and the price component, the contributions of packaging can be researched, developed, and nurtured.

This holistic approach to packaging development, however, is not easily assimilated. McCarthy[3] correctly states that modern packaging tries to do both a product protection "packing" job as well as a promotional job. Engel et al.[4] position packaging in a category called "Supplemental Communications," listing under that title (1) public relations and (2) such sales promotion activities as (a) packaging, (b) trade fairs and exhibitions, (c) sampling, (d) premiums and trading stamps, and (e) price incentives. These authors further state, however, that "apart from the obvious function of provision of physical protection for the product, packages also serve to identify the product and to convey meaning about it."

Another source, cited by Engel et al.[5] for a definition of sales promotion, is the American Marketing Association's *Glossary of Marketing Terms,* developed in 1960.[6] This glossary does not even include packaging!

Practitioners in marketing and communications have recognized the role of packaging, however. Leo Burnett[7] placed it as a prime advertising medium: "Your package is your number one display piece. . . . In many cases it is your No. 1 advertising medium viewed by far more people than the readers of *Reader's Digest, Life,* and *Look* combined." Albert Kner,[8] a pioneer package designer, stated that "A good package has to be market, sales, consumer *and* product oriented. All those things have to be in the package design. . . . I think the very first requirement. . . . is that the customer, the consumer, has the right to know exactly what is in the package. So you have to make a clear cut statement on the package."

In these latter definitions, the package is primarily seen as a communications medium; it is a tool to inform the consumer. "Only about half the 8,000 items on the shelves of the typical American supermarket are promoted by media advertising. The others have to catch the customer's eye and, simply by (the packaging) standing there, persuade him or her to try them."[9]

As a result, packaging has been explained as having a marketing mix position in contemporary marketing strategy.[10] A package may be related to objectives positioning the product on a given *price/value* continuum. It can be measured on its added value or *expensiveness* projections, for example. It can be evaluated for its connection to different *distribution* roles: the distribution graphics communication function of model or part number or its modified function in a hypermarket environment. It can be viewed in its *promotional mix* connections—its central use in advertising, its use in lieu of advertising, its sales or consumer deal promotion role—which are easily seen in the marketplace. These all interact with the heavy product partnership and definition role. All are heavily communications oriented.

Viewed in this marketing mix context the package design strategy development should be related in statements regarding packaging as a configuration within a coordinated communications program. It should be developed to communicate the precise information and imagery regarding the product to the proper segment or segments of the consumer/user audience.

In developing packaging design strategies, we wish to inform, persuade, and remind. In order to inform, we need the proper message. In order to persuade, we must know the target markets and their expectancies and needs. In order to remind, we must know the equities of our product in the minds of our consumers. We need to connect with the consumers to help in all the dynamics in purchase and repurchase of the packaged product.

Obviously these objectives can only be accomplished within an established distribution framework with the cooperation of the intermediate channel members. The functions of the case/shipper in the warehouse, the role of the case in inventory control, color coding for mixed pallets, and so on are all involved in the full potential of the package as a distribution facilitator.

Research from a number of sources can help the marketing manager to plan for these multifaceted roles of packaging. Direction and assistance for the marketing manager evolves from a number of research sources for a number of research purposes.

THE FUNCTION OF RESEARCH IN STRATEGY PLANNING

All the tasks of marketing management require a managerial approach that consists of "analysis, planning, implementation, organization and control."[11] It is precisely to help in these tasks that marketing research is conducted. "Marketing research is undertaken to guide managers in their analysis, planning, implementation and control of programs to satisfy customer and organizational goals."[12]

Zaltman and Burger also list the areas of marketing research as those:

1 Related to products and services.
2 Related to markets.
3 Related to policy.
4 Related to sales methods.
5 Related to analysis of mass media effort.[13]

While packaging is specifically listed by these authors in the first products and services category, portions of the remaining areas are also applicable to packaging strategy planning. For example, packaging could correctly be viewed as a promotion medium (category 5), a sales/distribution element (category 4) or a consumer demand facilitator (category 2).

Green and Frank, in their managerial guide treatment of marketing research,[14] state that the "functions of marketing research naturally derive from the *problems* of marketing. . . . The purpose of the activity is to provide information useful for the identification and solution of marketing problems." If the development of packaging within the marketing mix is a reference, the solution of marketing problems such as those listed by these authors (e.g., What to sell? To whom to sell? When to sell it? How to sell it?) will also be inputs into the development of communications strategy and planning attending to the communications portions of marketing problems.

The questions that are basic to market communications planning for package design strategy—especially when the *what to sell* has been decided—are

To whom are we talking?
What are we trying to say?
How are we saying it?[15]

Marketing research is fundamentally useful in defining all of these areas. However, the application of the findings is especially productive if there is an interdisciplinary connection and cooperation among the partners in the strategy development and execution, that is, between the marketing researcher, the designer, and the marketing manager. The researcher will develop the proper data and information. He* will augment the basic research with specific packaging studies and these will add to the package design strategy, affording the package designer needed direction. The integration of the findings by the manager in his total strategy will relate to his program coordination responsibility.

Product/Package Failure Without Research

The lack of marketing, design, and research interaction is seen too frequently in inefficient programs that have not had proper strategic planning and execution, insufficient—if any—levels of marketing intelligence input, and poor cooperation of the three participants in the packaging strategy and design development.

Overly long and inefficient development programs are the result of minimal or no directional research. In many instances, the programs also lack decision augmenting data. In many more instances, the semblance of objective, exhaustive, and systematic research is included in a single last phase, a "red

*"He," "his," and "him" are used only for simplicity.

flag" phase. These often confuse the participants further, creating a frustrating back-to-the-drawing-board situation. The researcher in these instances is relegated to a message carrier role, the bearer of bad news, the nonacceptance of the package/product. Even civilized contemporary cultures have ways of dealing with such bearers of ill tidings.

Blum and Appel[16] documented one early study in which designers, advertising and marketing executives, and eventually consumers, male and female, were employed. Utilizing Stephenson's Q sort technique,[17] the consumers, designers and management arranged 18 package designs into seven categories for selected evaluative criteria. Appropriateness for gift giving and expensiveness were two of the evaluation areas. There was little agreement between the participants. "The consumers on one hand and the management and designers on the other . . . were apparently using conflicting criteria in evaluating the designs." The authors concluded that "Had the packaging decision been made on the recommendation of the design firm and . . . the client's marketing management, the net effect would have been to select designs which would have had the least appeal so far as the consumers sampled were concerned."

There are many instances, to be sure, where an eventual research cross check is not made, where the salience of criteria is not known, but where designs are developed and adopted nevertheless. In some instances, other marketing efforts can offset an inefficient or poorly positioned package. In many instances, the product will not attain its potential. In others, the new product will fail or the older product will phase through its decline prematurely. In fact, it may be that " . . . package design failure alone accounts for as many as half of all new product failures."[18]

In a recent case involving a package design format change,[19] no detailed objectives were given to the design personnel of the client company. Illustrations were enlarged while product definition graphic elements were reduced. Only *after* the design changes were adopted were diagnostic laboratory tests conducted. Information processing tests (including product identification legibility) and product projections and expectancies measurements (including attitudinal profiles) indicated weaknesses in the new design compared to the former design and competition. These weaknesses translated to a 30 to 60% drop in volume in subsequent controlled store tests conducted with matched panels of stores.

There are, conversely, success stories *with* research based strategies where the additional information on the target market and the message content and execution helped in risk reduction. One such case involved a candy manufacturer who experienced a 40% increase in sales.[20] The president of the firm was quoted: "Credit for the sales increase has to go to our new packaging; there is no other explanation. Nothing else changed except the price, which in effect went up since we are giving less candy for the same money. All the credit has to go to the improved communications of the packages and the better display techniques which they incorporate."

The interdisciplinary program that resulted in the new packages for the candy line included:

1   Market research of the acceptance and utilization of the (then) present packages and product displays.
2   Redesign strategies development.
3   Package structure and related machinery changes.
4   Creation of three graphic prototypes for new packaging and product displays.

5   Postdesign research indicating the most acceptable design execution.
6   The development of a new corporate identity program that expanded the graphics changes from packaging into a complete corporate identity program including collateral materials and the often overlooked corrugated shipping containers.

THE MARKETING MANAGER'S DECISION SEQUENCE ACTIVITIES

The marketing manager—whatever his title (brand manager, product manager, marketing director, etc.)—must consistently use research throughout his main responsibilities of planning, execution, and control. Research will help him know the size and composition of his market and help him locate his market geographically, psychographically, and competitively.

In the marketing manager's planning and related strategy development process, research will offer inputs at all stages.

STAGE 1 In the first stage of the decision sequence as listed by Ray[21] and outlined by Engel et al.,[22] a Situation Analysis is conducted. In this analysis, internal company factors, demand factors, competition, and legal considerations are reviewed and studied. Taking a promotional planning orientation, and realizing that the product, price, and distribution components must interrelate, the marketing manager must plan for integration of his promotional elements. Packaging will—or should be—part of this integration. As Ray further states:

The situation analysis is explained in communication terms. Although product, price and channels are not part of the communication mix, they and the other parts of the situation are considered because of their communications implications. For example, if products are developed through perceptual mapping procedures, some researchers suggest that the nature of the communication message is predetermined; i.e., all messages should only communicate those product characteristics which are salient to consumers.[23]

STAGE 2   Objectives are evolved from the Stage 1 situation analysis. Demand factors, for example, will indicate the need for new product development or changes in the current demand situation for an existing product. In concrete terms, this Establishment of Objectives Stage states (a) target market objectives and (b) communications message objectives. In a thorough objectives statement, the packaging component will be included. These are objectives that will be communicated both in general promotional terms and in specific packaging terms to the package designer.

The definition of target markets—which can be accomplished in a variety of dimensions—will require critical input from marketing research. Marketing research will be the instrument for defining the market and the target segments that seem to have the most potential. In existing products, research will define the users of the manager's brand and the users of other brands. For new product development, analysis of the consumer markets will help to indicate new product potentials and, in fact, may be a key factor in new product development.

The objectives of the communications message will also be clarified when we have analyzed research that indicates "whether it is necessary to increase awareness, stimulate trial, change attitudes, and so on."[22] Differentially, the package design will be directed toward helping to meet these objectives.

STAGE 3   The third section—Determination of Budget—is somewhat outside the purview of this chapter. However

complex the promotional budgetary factors are—there are some improved computer planning models to help in the process—packaging planning should be part of the stage.

Experience has indicated that much of the production needs of the company—as reviewed in the situation analysis—will direct the packaging formats, media, and configurations and therefore the packaging budget. Increasingly, separation and allocation of production costs and promotional costs is being accomplished. Dependent on the individual company's control process, the proper allocation of the promotional costs of packaging can be easily accomplished. While modifications of the budget are probably inevitable, the accountability and control aspects of the planning process can include packaging as a variable.

STAGE 4   In the Management of Program Elements, the specific inputs of packaging research can be appreciated for their effects on package design planning and strategy. Earlier steps—although generally useful and important—have not focused on specific elements. Now, all of the elements of the promotional mix are considered, and integration and blend decisions are necessary. Historically, personal selling, advertising, and, to some extent, trade factors, have had major emphases, probably due to their dominance in their budget allocations. Packaging—which has historically been considered a production expense and supplemental communications at best—can be and has been considered positively in the mix, employing the same procedures for strategy formulation.

The review of the creative platform for other promotional blend elements is obviously helpful in packaging planning. If advertising has a role of any dimension—and not all products receive significant media budgets—the manager must co-ordinate the blend. While packaging design has been considered from vantages other than the content and design of advertising messages—primarily due to the package's immediacy in the purchase behavior and its close melding with the product and its usage—the audiences are the same. The execution of the package design—and the strategy leading to that execution—must of necessity be integrated with the other program elements. However, packaging's unique marketing mix connections require its own specialized platform. Packaging research is an especially productive input to this platform/strategy.

The Need for Specialized Package Design Strategy

Specially organized, specially designed, and specifically designated *packaging objectives* research must augment the marketing manager's available marketing information and data.

The role of the package in the distribution process and in the wholesaler and retailer environments must be considered. Trade predilections on package size and configuration cannot be ignored.

The unique positioning features of the product should be delineated in relation to the package's product positioning role, its communications function at the retail level, and its involvement in product usage, for example, storage and dispensing.

Most importantly, the various functions and roles of the package must be analyzed and assessed within a consumer expectancy model. This approach does not assume the salience of the consumers' evaluative criteria, but instead measures this salience and employs its proper emphasis throughout the creative development and assessment process. The package's effects on product positioning and brand imagery must be determined so that the designer is

properly directed and aware of where the package design enhances or detracts from the salient imagery. In competitive positioning, the relative brand/product/package strengths and weaknesses should be probed and understood.

This is especially relevant when we look beyond the "packing," containment, protection functions of packaging to the information imparting role of the package at the purchase instant, its communications role in a shelf-competitive environment, the product imagery enhancement function at the point of purchase and the point of use. In these, packaging "may be 'better' than advertising." The packages "may actually be seen by many more potential customers than the company's advertising"[24] especially when there is no significant advertising.[9] Even when there is a substantial or significant media advertising budget for television, more and more consumers—probably the early product triers and innovators—are by-passing the commercials, fast-forwarding their video cassette recordings or totally by-passing the commercial telecasts while they play their video games.[10]

The Package Designer's Contribution to Design Strategy and Execution

There is the obvious additional contributor to design strategy and, of course, execution: the creative designer. He is eminently interested in various research results as a contributor/user. He acts as an interface with the marketing manager in this dual role. In his interpretation of design principles and in his use of design elements, he is directed by the target market definition and message content delineation.

Together with the marketing manager/product strategist, he must understand the audience for his designs. He must understand how the package must be positioned. He must be fully conversant with the findings of predesign research and, more importantly, their implications for his creative direction. They form the creative corridor for his creative development. Further, the diagnostic information obtained from during-design research and postdesign research helps in the fine tuning of the designs and in the final decision process of which he is a part. Eventually, his knowledge of printing and production parameters and the constraints of package engineering will allow the translation of his concepts into production realities.

Packaging design research techniques—part of the arsenal of the marketing, perceptual, attitudinal, and behavioral research techniques employed by the professional market researcher—serve to augment the analytical and creative process approaches of the designer. The designer knows the means to communicate the message, the mood, and the atmosphere suggested by the strategy. He knows how to work with the constancies of perception. Laws of perception, evaluation procedures, experimental research designs, and test scores may not fall within the confines of the designers' discipline; but trained color sense, size and figure relationships, form facility and composition, and layout and knowledge of typography do.

The creative designer, the researcher and the marketing manager may not understand all aspects of the other professions, their procedures and their responsibilities, but they can appreciate them and their ultimate interaction in marketing in general and in packaging design strategy and execution in particular.

PHASING AND STEPS FOR PACKAGING STRATEGY DEVELOPMENT

What then are the minimal packaging research inputs into the decision sequence for the creation of strategies for package design development? What are some of the critical path sequence activities in developing package design directions for new products and for revitalizing brands and products during their life cycles?[25]

Product Life Cycle Considerations

The brand or product's life cycle positioning has a major role in determining its overall promotional strategy for a brand and, relatedly, the design strategy for the brand's packaging. At different stages of a product's life cycle, trial generation, brand preference strengthening, maintenance of consumer loyalty, and so on are indicated.[26]

Packaging planning and strategy research should begin as soon as the new product screening process is begun. Packaging audits should then continue throughout the cycle so that the package—as well as other elements—can be employed to "recycle" the brand. A. C. Nielsen Co., in a five year study of the "vehicles" for recycles of a large number of brands, found a combination of product innovation and advertising changes was used in 41% of the cases. "Among product innovations, during the five years studied, new packaging was the most popular, accounting for about a third of the product innovations."[27]

Two separate outlines will be presented below: one indicating the research necessary for *new product* packaging development and the other outlining various phases for tracking *existing product* packaging. While both sections follow an overall model, they obviously should be considered as flexible guides.

The introduction of a new flanker product—for example, a new flavor or variety—may not necessarily require all of the developmental steps of the original product. However, even an added flavor or variety package may require much preparatory planning. The entire process of introducing flanker Buttermilk and

Buckwheat frozen waffles to support Quaker Oats' Country Waffle franchise, for example, involved a long multistep process including the respecification of packaging both physically and graphically.[28]

The outlined procedures must be separately tailored to the specifics of the beginning situation analysis, the objectives determination, and the management and subsequent coordination of the program elements, including the package design execution segment. Demand, internal constraints, competitive factors, and so on will all have effects on the augmentation or reduction of the suggested phases and steps.

Nevertheless, the two models for package design research disclose similarities that allow a phasing and steps definition. Similarities of marketing research objectives, the two general types of visual communications research, and their application and utility in package design strategy and execution allow this outlining of data and information needs.

Package Design Assessment Areas

Two general types of package design communications research that have been shown[29] to augment other marketing data will be included in this discussion:

1 *Perception Research* Measures such dimensions as shelf impact, recognition, legibility, conspicuity, comprehension, that is, the measurement of clarity of communication related to information processing.
2 *Attitude/Imagery Research* The assessment of product/package imagery either existing or planned, the evaluation of predilections to behavior, that is, the consumers' evaluative criteria applied within an expectancy model.

Various "nonreactive" research tech-niques—eye movement tracking, GSR, voice stress analysis, and so on, many of which are discussed in Webb et al.[30]—will not be detailed here. Somewhat controversial and not as useful for design diagnostics when used alone, they are mostly related to the reaction and behavior related measurements of area 2.

The balance of these two areas—whether they are labeled legibility and acceptance, perception and attitude, objective and subjective concinnity—will form major inputs into design strategy. Although it is not the purpose of this chapter to review all of the theoretical and validation studies of these two areas, it is important to understand that they must be balanced within the design strategy.

Del Coates[31] has covered the concept of concinnity in his discussion of automotive design. The same applications and balance are necessary in any design development. Objective concinnity relates to the information processing function: form, harmony, and simplicity. Subjective concinnity, as delineated by Coates, is quite different, relating more to attitudinal bases, "the stuff of market segments." Similar to attitude, which relates to behavioral intention, subjective concinnity "is formed from stored information and experience and evaluative criteria."

The balance of the two developmental/assessment areas will relate, as a result, to the existing target market segments and the life cycle position of the product. In a new product, the balance may be slightly toward shelf impact so that it may be located among competing existing products. With a package design strategy for revitalization of an existing product, the balance for certain elements—"violators" or "interrupters," design elements that announce promotional deals, cents off, premium offers, and so on—may also be toward shelf impact but with emphasis on maintaining image and acceptance equities or improving various product appeals.

In any event, these two functions of visual communication—one quite stimulus bound and the other quite value and expectancy bound—will be reflected in much of the design strategy. While the functioning of the design in one area does not preclude functioning in the other, pertinent data will indicate strategies for both. In much of the strategy, research will indicate the parameters, the referents (usually competitive), and consumers' reaction to the balance.

Whatever the techniques—qualitative or quantitative, psychometric and psychophysical, verbal or so-called nonreactive, design element or configuration oriented—basic package design functioning research is necessary to augment the market research available to the marketing manager. All are essential for package design strategy and execution planning for new product work and for existing products.

New Product Development Packaging Research Inputs

There are a number of strategies that face a company in new product development. Zaltman and Burger[32] list reformulation strategies, differentiation strategies, extension strategies, and diversification strategies, among others. Whatever the strategies and their cause, and whatever the technological positioning of the firm, new products can be classified in the following categories: flanker, product improvement, new positioning, unique product, and breakthrough.

These new products result from, and sometimes cause, a firm's strategic planning. Whatever the new product classification, either by the firm or by the target market consumers themselves, eventually certain relationships with the product's packaging must be determined.

Product and Packaging Concept Studies

The initial ongoing studies—those searching for unfulfilled needs, the market segmentation studies, demographic market descriptions, psychographic profiling, and so on—may have additional probes and investigation areas related to packaging formats and features. At a very early stage of new product development—as the product concepts are being screened and evaluated for potentials in meeting company or corporate objectives—data regarding product positioning are collected and analyzed. While their main objectives remain in relation to market segment understanding, package variables can be included. These initial studies—using appropriate sampling techniques in their quantitative orientations—define the markets and their potential demand. If packaging variables are a factor in that demand, they can then be probed further for their potential contribution.

Qualitative research may be scheduled for additional investigation and for focused direction. Using the ubiquitous focus group procedures interspersed with role playing, projective tests, verbal concept ratings, and so on, additional information—on a directional basis—can be generated. This exploratory research may prove helpful in product and packaging alternatives clarification. Product form preferences, package sizes, and so on can be probed and preliminary reactions can be related to segment needs and perceptions. This is especially useful when earlier quantitative research directs the ''sampling'' or recruiting of respondents for the concept explorations.

Preliminary Package Design Strategy

It is also at this early stage that production and marketing factors converge as part of the preliminary package design strategy. Early situation analysis notwithstanding, a variety of packaging materials will of course be available. Opportunities for their use exist even if production constraints exist. Since the new product will be positioned differently, the packaging material or combination of materials may need to be more exotic than those for existing products. Plastic films, combinations of films and paperboard, structural innovations, and other new solutions for product positioning and product quality control are in a constant development process. When these relate to emerging markets strategies—such as the growing microwave oven user segment or the retortable pouch/convenience oriented segment, two current examples—their impression and contribution in marketing strategy terms is unmistakable.

The preliminary strategy is an amalgam, as the preceding indicates. While this is early in the product development, the concept research results and the various situation analysis factors can still direct us. While we are not prepared to complete a full strategy or platform statement—some examples will be developed in the following—there is an increased focus on the new product, its positioning, and how the package design may formulate part of the strategy.

Prototype Tests

With a more detailed definition of the reactions of consumers to the product and package concepts, prototype products and their varied packaging forms can be developed for further scrutiny. Graphic concepts may be developed if structures are known and available for use.

Prototype product testing should also be related to packaging approaches. Although this could be a major packaging variable study—with a range of differences in packaging media or graphic

concepts—it could also include product variations as the major variables with only minor packaging features (i.e. opening features, dispensing structures, etc.) as part of the experimental design. Product R&D, competitive products and packaging, earlier research as well as internal constraints will all indicate the major variables and the objectives of the studies.

Nomenclature Development and Evaluation

Nomenclature development and research can move along on a separate track of the new product critical path. In some cases this is a preconcept step. In others it occurs during concept studies. In either event, if a new brand or new product designation is considered necessary, one of the most crucial elements in the design strategy, nomenclature, is involved. Eventually the chosen nomenclature variation(s) will be designed as a major design element: the brand's logotype. This brand logotype and/or product designation, also in a form of a logotype or specialized typographic system, will comprise one of the main perceptual elements.

There are a number of sources for brand names: the clients' own internal management sources, various creative sources, even the purchase of names from other firms. A variety of name generation approaches and techniques can be used.[33]

Following a preliminary legal screening is the evaluation of the names to aid selection decisions. Usually, this evaluation includes measurements of association to record initial meanings of the verbal symbols. The evaluation also involves evaluatory and preference ratings on the salient criteria discovered earlier to determine the appropriateness of the nomenclature variables. Recall and playback tests, finally, relate to the complexity and related memorability of the names.

Many of the specific findings of such nomenclature evaluations augment the earlier research. As such, they aid the preliminary development of packaging design strategy and packaging communications content. Since product concept statements are used as stimuli in certain steps of such research—to further insure the connection of the verbal constructs with the product concept—the nomenclature can give additional dimension to the product appeals, the gathering brand image, and the future resultant final execution of the graphics for the package.

Competitive Definition

To discover where competitive brands/packages are positioned in the market, continuing and complete reviews of the market are necessary. The packaging audit will be covered in more detail in the section "Revitalization of Existing Products." New product packaging development, however, also requires an audit, either as part of the situation analysis or as a separate input step. Most of the categories of new product introductions relate indirectly or directly to existing products and therefore existing brands and their packaging. While packaging audits are or should be periodic steps, the reanalysis of the data from a new product positioning vantage can be very helpful. Gaps in product mapping, investigation for areas of unsatisfied consumer needs, behavioral knowledge and similar audits may all indicate specific design strategy facets.

This competitive review may also include some of the graphic analysis techniques that the designer applies. While these are intuitive and judgmental, they serve as hypotheses generation procedures for directional creative design inputs, helping to complete the judgmental aspects of the strategy.

Development of a Strategy Statement for Package Design Development

With the data and information of the preceding phases, the marketing manager, designer, and researcher team can begin to outline portions of the profile and strategy for the new product. Check list forms have been developed to facilitate the preliminary listing of the basic requirements. While the unique aspects of each product development and positioning may not totally relate to an established check list, such guidelines as the following have proved helpful as guideline documents. (Also see the following outline for revitalization statements.)

An abridged and collapsed form would include such questions as:

What is the product category?

What is or are the brand name(s) and other required nomenclature?

What are the various legal requirements?

What materials, sizes, and configurations of packaging will be used?

What flavor or range of flavors will comprise the line?

What market segments are we trying to reach?

Where are they? (national, regional, geographical, urban/rural, etc.)

How will the product be related to other products in the line?

What basic communications requirements do we have?

- Imagery (appeals, moods, etc.)
- Perceptual (impact, brand, and product identity)

Specific retail environmental check list questions can be raised in separate sections of the planning statement:

Where and how will the new product be displayed?

Will the packaged product require refrigeration, for example?

Will it be located with other similar products?

Will it have a block by brand display, with all of one brand together, or will there probably be a mixed brand display?

Where do existing and projected planograms of product organization indicate the product will be placed?

Where do existing space management programs locate similar products, old and new? Top shelf, lower shelf?

What other store stocking conditions can affect the design's effectiveness?

Such partial product placement questions should have objective research answers since such practices as lower shelf placement will put additional emphasis on the top panel of a carton or the closure or lid of a jar. If products are stacked with only one full face panel, for example, the side and/or bottom panels of a package may be seen more frequently and will have more dominance than the front panel or label.

All of the preceding information—the concept study directions, the competitive evaluations, the production and intermediate channel constraints, and so on—will require more detailed platforms or strategy statements. In most instances, the research results will form the key references in well documented strategies. The designer interested in consumer behavior and design by objectives will have an interest in his audience, and the message content and research will offer data and information for him and the package design strategy developers.

Package Design Creation

Package structure and shape development and new package construction creations may be both possible and necessary, given specific strategies. The creative function of the graphic designer is augmented by a creative approach to structural factors since, in many directed instances, these factors will enhance the graphics and become the integral package. The use of a neck label to restrict pilferage, the development of a new opening or dispensing feature to improve product use convenience, or the positioning of a tear strip, all may be integrated into the graphics format and in fact become unique separating elements in the graphic/structure configuration.

There will be no single correct configuration. It is rare that the singular and thoroughly appropriate design strategy execution immediately emerges. Instead, a range of graphic/structural prototypes is reviewed in light of the package design strategy statement and its references. The assembled information that forms the direction in the statement—the predesign information regarding the target market and its responses to current products and packaging—aids in preliminary screening.

The best of the design solutions emanating from the strategy can be either screened judgmentally—often the case when literally dozens of package design executions are reviewed—or they may be screened by consumer panels. The latter more objective and systematic approach—using the most salient of the consumer criteria—is a less risk laden approach, especially considering potentially different perceptions among the creative team.

Postdesign Development Package Evaluation

After the pivotal creative development and screening activity, postdesign research is a further fine tuning of the strategy, a recheck of the platform and a test of all of the aspects of the strategy and its range of executions. This evaluation—which includes a review of the graphic approaches in a competitive referent frame as well as behavioral package use tests where appropriate —will have diagnostic results indicating modifications for certain elements and helping to complete the development.

While configurational responses by the consumer are the most realistic, certain perceptual measurements may indicate changes in areas such as typographical treatment or figure/ground contrast values. Various instruments such as the tachistoscope[34] are available to measure certain aspects of visual communication: the levels of brand identification or product differentiation or the effects of other design elements such as color and form and their effect on shelf impact. These can be conducted with simulated mass displays or with single package facings.

Attitude and/or Image Assessment

Using a selection of a range of techniques, scaling approaches such as the Semantic Differential[35,36] or a variety of other psychometrics, buying games, and so on are employed to measure the acceptance of the concepts. Here, the measurement is not merely of the reactions to the graphics but what the graphics convey, the reactions to the product/package. Certain measurements that compare package profiles to "ideal" profiles help in saliency analysis. Comparisons can also be made with product concept statements, with other brands and similar products. All of these will be helpful in determining the relative acceptance and potential success of the product. Product trial tests —in relation to the package promise/positioning—can also be helpful for design modification direction.

Other laboratory simulations of purchase intent or behavior, controlled store tests, and similar "sales" related measurements may also be applied at the latter stages of product/package testing for new products. However, for a full understanding of the package design communications and possible reorientations of the strategy, these approaches should allow probes for consumer reactions and responses other than purchase behavior alone. The "additional" diagnostic information is often the most useful for design strategy modification and subsequent package design modification.

Existing Product Revitalizations and Redesign Strategy Research

Many of the methods and procedures for gathering data and information on new product introductions are of course applicable to the review and reevaluation of existing products. Reliability and validity needs remain the same. There is, however, generally more information available on the shifting market segments, demographic changes in the market, market shares, tracking of new brand introductions and competitive products, and so on when a product is in the later stages of its life cycle.

In competent brand management, packaging audits are part of the continuous tracking process as the brand moves through its cycle. As the brand/product itself matures, maintenance marketing strategies become necessary. If demand drops or shares decline —possibly due to competitive new product growth—demand must be revitalized, and remarketing is the manager's task.[11,26]

The perceptions of the packaging and its design through this life cycle may serve as a major indicator of necessary changes. Values, design fads, and attitudes shift as the environment changes. Reactions to the package may serve as early indicators of the shifting trends. In some instances, package design changes may be causal factors: other brands may be into promotional pricing, the use of violator flags for cents off deals, and so on, which may affect some brand switching. The brand switchers, further, may be moving to other newly upgraded and revitalized package/products. Equities of prior years may no longer exist.

Designers' inputs may be part of the first encounter with the realization for revitalization. Reviews and analyses of the changing conditions in the retail outlets, introductions of new brands and packaging, recognition of upgraded graphics on competitive packaging, new retail environments or changes in procedures in mass merchandising may all be noticed if there are systems or procedures for this review. If the marketer's own design personnel are not part of this process, consultant design groups may be considered for the source of this periodic review. These graphic market reviews and their findings and hypotheses may then be augmented with the other types of research.

Exploratory Research

Many proponents of exploratory research fully understand the contributions of qualitative research. Such approaches as focus group discussion research have utility in offering insight into potential problems, in generating hypotheses for more structured research, and in uncovering changing attitudes toward the product and package.

Exploratory research—planned with specific sections for unaided and aided discussions of in-store conditions, display and packaging variables—can begin the assessment of the brands positioning and the consumers' views of the package's function. Such qualitative investigation may indicate the need for more quantitatively oriented investigation.

Definition of the Package/Product within the Current Market

More structured studies of the in-store conditions could be planned since they can document the competitive environ-

ment and, at the very least, help to communicate some of the changed or changing conditions to management.

However, the detailed and structured perceptual and attitudinal studies are the more definitive at this point. They establish the contemporary needs with relative precision.

While there may be strong hypotheses regarding relative loss of shelf impact or product identity, certain laboratory tests—for example, those utilizing tachistoscopic procedures—may be helpful in defining areas for design change or improvement. New competitive design displays may be dominating the shelf space; relative visibility of the existing design may be reduced. One study,[34] indicates that "there appears to be promising agreement between tachistoscopic recognition scores conducted under laboratory conditions with find-time scores from consumers in supermarkets." These tachistoscope tests and similar instrument applications, properly planned, can further help in yielding diagnostic information that the find-time tests, actual or simulated, do not afford.

Consumer attitudes and expectancies toward the package, the brand, and the product—as we have seen—can be measured through a variety of approaches. These are specifically designed to measure the package communications on various salient attributes and criteria. A variety of available measurement and analysis techniques allow an assessment of attitude shifts, the reduction of positive imagery of the past, and the mitigation of the brand's positioning. Content analysis—if scaling or similar procedures are used—may indicate the need for product attribute reemphasis or a more contemporary quality/value statement. One of the general areas that should be probed is whether the current package design supports, reduces, or elevates the brand imagery. Macdonald[37] states, "The marketing researcher should keep in mind this simple axiom: the image of a package design should be equal to or better than the image of its brand." Differential comparisons can be planned to help achieve this research objective.

The image profiles and other attitude test results may show modified reactions to symbols and props that had earlier acceptance but no longer function to enhance the brand. These findings help determine design strategy and direction.

Design Strategy Statement and Related Design Development

As in the new product development outline, the earlier data will be used by the marketing team in attending to the needs for, and extent of, package design change. All the preceding research will have given indications and/or strong direction for a redesign. Since the history of the brand/product will have been well documented and since the packaging audit approaches will augment the ongoing tracking, the management team can outline the strategy and indicate the execution:

General:
1   Should there be flavor reformulations?
2   Has the market segment shifted? Changed in size?
3   What plans for other promotional activity have been made?
4   What are the competitive positionings?
   ● Price.
   ● Promotional activity, for example, increased TV.
   ● Distribution factors.
   ● Market shares.

Production/Distribution:
1   Should physical package changes or packaging media changes be consid-

ered? (e.g., infestation control changes, increased pilferage, other changed conditions).
2   Have distribution conditions changed? (e.g., longer channels, increased need for product protection).
3   What are the existing shelving, space management, retail procedures for the product(s)?

Design Communications:

1   Have changed media uses indicated restrictions or improvement of the graphics? (e.g., television needs, changed printing capabilities).
2   Have competitive brands improved or changed their elements of communication, improving visual communications and related shelf impact?
3   What image/attribute/attitudinal shifts have occurred? (e.g., appetite appeal, product connotations, general aesthetic and appeal factors, purchase inclination measurements, etc.).

These and many more similar inputs will form the strategy statement. (See also the new product outline in the preceding.) This strategy—with its resultant directions for the executions—will attend to the factors considered most useful to the revitalization. Research will offer information so that priorities can be determined.

Creative Design Execution

Design creativity will react to these strategy priorities. In many instances, the design strategy will relate to objectively established existing equities and franchises in target markets and in graphics. A radical departure from the current graphics may not be indicated. A transitional program may be indicated instead and this will be reflected as a major direction of the strategy. Prior research will certainly be helpful in determining the range of modification or change for the execution alternatives. The graphic elements considered important to the defined existing equities (e.g., the franchise recognition elements such as logotypes) could be modified within a continuum of alternatives.

The designer begins his work within the strategy guidelines to meet the specific objectives they entail. Not only does he work with existing elements, he creates a configuration—or range of configurations—with which he tries to fulfill the specified needs. While he works with the elements of typography or specially developed logotypes, color, design forms and illustration, his objective is a total package to meet the needs and specifications.

The following is a partial listing of a design strategy statement, edited to include some of the research findings on which the strategy statement was based. The statement also indicates some of the challenges the designer faces:

Improve brand identity (confusion with other brands on shelf)

Improve product differentiation (difficult to locate specific flavor from large group)

Improve appetite appeal (package/container is used as a dispenser)

Increase quality connotations (competition has stronger quality imagery and blind product tests indicate no perceived differences)

Strengthen shelf impact (competition has more shelf dominance)

Develop tamper proof packaging (trade requirements, legal requirements, and consumer behavior)

Employ feature in unique manner (consumer needs indicate specific benefits perceived)

Postdesign Evaluation

Final design evaluation, either the replication of earlier studies or with the addition of specialized measurements, is conducted on the designers' creative output. Whether or not these studies are conducted with follow-up behavioral studies (e.g., laboratory simulations involving the manipulation of a number of variables), they serve to determine the relative effectiveness of the designs in assisting the repositioning and/or revitalization of the product.

While the main objective of these studies relates to the collection of data and information helpful to the marketing manager's decision in the packaging variables portion of the management of program elements, to determine the execution's fit with the strategy, the information may also be applied in the final modification and finalization of the graphic design.

DECISION SEQUENCE COMPLETION AND FOLLOW-UP

The application of the design within the new product or product revitalization program is part of the decision process listed in the marketing managers decision sequence. The first four general steps were discussed above and the phasing and steps for package design strategy development were included in the management of program elements portion.

After the decisions on the program elements, the marketing manager coordinates and integrates the various components. He also provides for feedback that allows the designer to assess the results and effects of his program. This allows his further follow-up since the process is continuous, a series of "adaptive feedback loops."[21] The various research inputs for package design strategy discussed throughout this chapter are very important in this adaptive feedback system.

The research phases specifically for packaging design exploration, and the various research phases related to the various tasks and responsibilities within the product management function, serve to indicate the effectiveness of the strategies and how well these strategies are executed. The marketing communications program relies heavily on these inputs since they are the connections with the target market, determining the target market's response to the product offering and the message content we have developed based on a strategy to meet general and specific objectives.

CONCLUSION

All the strategy planning, decision sequence inputs and variables have related to the controllables of the marketing mix. Within this controllable mix, packaging was given the important position it deserves. For the marketing management objectives—which may be the development of a new product to obtain a new market share, or the revitalization of an existing product for volume protection or development—research has played its creative direction and decision-related roles.

The designer's role has also been defined. He, or a group of designers or the creative agency, serves as the interpreter and facilitator of the research-based strategy. His overall aesthetic sense, his approaches to the balance of objective and subjective concinnity, his knowledge of package production feasibilities are all important contributions to the team.

But packaging design remains a team effort—much as marketing is a total company effort—and the designer must be helped in his creative efforts. The strategy developed with objective information, related to the product positioning objectives and the marketing management's decision sequence for the product within his firm's needs, is the result of coordination of the management and creative and research responsibilities.

With the properly developed strategies, based on properly developed research inputs, created within contemporary structures of creative effort and execution, the package can make just about any statement that other forms of communication can. While mass selling efforts have validity, salience and effectiveness in the total communications strategy, packaging must function on these levels to support the personal selling efforts as well as fulfilling its product protection, production, and distribution requirements.

A well documented, well planned, and well executed strategy is ultimately necessary to have packaging work most effectively in all its roles. Packaging research is the main catalyst for the planning, execution, and control of successful packaging.

REFERENCES

1 R. F. Vizza, *Paper Film and Foil Converter*, 43, No. 9 (September 1969), 72–77.

2 F. P. Tobolski, *American Paper Industry*, 50, No. 6 (June 1968), 41–44.

3 E. J. McCarthy, *Basic Marketing, A Managerial Approach*, 6th ed., Irwin, Homewood, IL, 1978, p. 261.

4 J. F. Engel, H. G. Wales and M. R. Warshaw, *Promotional Strategy*, Irwin, Homewood, IL, 1975, p. 488.

5 Ibid., p. 503.

6 Committee on Definitions of the American Marketing Association, R. S. Alexander, Chairman, *Marketing Definitions, A Glossary*

of Marketing Terms, American Marketing Association, Chicago, 1960.

7   L. Burnett, *Package Engineering,* 16, No. 4 (April 1971), 16a–16c.

8   A. Kner, *Paperboard Packaging,* 53, No. 10 (October 1968), 55.

9   W. McQuade, *Fortune,* 99, No. 9 (May 7, 1979), 186.

10 F. P. Tobolski, Contemporary Packaging Development, The Development of Packaging for Today's Market Segments, Speech to Packaging Institute, U.S.A., Chicago Chapter, December 1978.

11 P. Kotler, *Journal of Marketing,* 37 (October 1973), pp. 42–49.

12 G. Zaltman and P. C. Burger, *Marketing Research: Fundamentals and Dynamics,* The Dryden Press, Hinsdale, IL, 1975, p. 8.

13 Ibid., p. 9.

14 P. E. Green and R. E. Frank, *A Manager's Guide to Marketing Research, Survey of Recent Developments,* Wiley, New York/London/Sydney, 1967, p. 6.

15 Tobolski, *American Paper Industry,* 41.

16 M. L. Blum and V. Appel, *Journal of Applied Psychology,* 45, No. 4 (August 1961), 222–224.

17 W. Stephenson, *The Study of Behavior,* University of Chicago, Chicago, 1953.

18 R. Parcels, *Product Management,* 5, No. 6 (June 1976), 35–37.

19 F. P. Tobolski, unpublished case study in preparation, 1979.

20 Good Packaging Editorial Staff, *Good Packaging,* 35, No. 3 (March 1974), 14–17.

21 M. Ray, "A Decision Sequence Analysis of Developments in Marketing Communication," in E. J. McCarthy, J. F. Grashof, and A. A. Brogowicz, Eds., *Readings in Basic Marketing,* Homewood, IL, 1975, pp. 228–237.

22 Engel, Wales, and Warshaw, pp. 35–40.

23 Ray, p. 230.

24 McCarthy, p. 263.

25 F. P. Tobolski, Phasing and Steps in New Product Research and Research on Existing Products and Brands, unpublished manuscript, Container Corporation of America, Chicago, 1968 and modified in 1973.

26 C. R. Wasson, *Dynamic Competitive Strategy and Product Life Cycles,* Austin Press, Lone Star Publishers, Austin, TX, 1978.

27 Neilsen Marketing Service, *The Life Cycle of Grocery Brands,* The Nielsen Researcher, No. 1., A. C. Neilsen Company, Chicago, 1968, p. 12.

28 J. McCausland, *Quick Frozen Foods,* 32, No. 6 (January 1970), 103–104.

29 F. P. Tobolski, "Brightface Toothpaste," in R. J. Schultz, G. Zaltman and P. C. Burger, Eds., *Cases in Marketing Research,* Dryden Press, Hinsdale, IL, 1975, p. 128.

30 E. Webb, D. T. Campbell, R. D. Schwartz, and L. Sechrest, *Unobtrusive Measures: Nonreactive Research in the Social Sciences,* Rand McNally, Chicago, 1966.

31 F. D. Coates, *Industrial Design,* 25, No. 5 (September/October, 1978), 33–41.

32 Zaltman and Burger, p. 569.

33 F. P. Tobolski, Phasing and Steps in New Product Research. unpublished.

34 E. W. J. Faison, "Validating Recognition Speed as an Indicator of Package Design Effectiveness," in Proceedings of the Division 23 Program, Eighty-fourth Annual Convention of the American Psychological Association, Washington, D.C., September 1976, p. 30.

35 C. E. Osgood, G. J. Suci, and P. H. Tannenbaum, *The Measurement of Meaning,* University of Illinois, Chicago, 1957, p. 5.

36 J. G. Snider and C. E. Osgood, *Semantic Differential Techniques,* Aldine, Chicago, 1969, p. 50.

37 W. Macdonald, *Marketing News,* 11, No. 23 (May 18, 1979), 5.

CHAPTER FOUR

The Package and the Consumer

Ernest Dichter

With increasing discussions of trade deficits caused by import and export problems it is time that we consider the fact that a package is, in some ways, the ambassador of a country. In a recent study for Japan we found that there is considerable difference between the meticulous care the Japanese manufacturer applies to the outward appearance of his products and the lesser concern of his American counterpart. Going shopping in a Japanese department store can be a revealing experience. The package of the products and even more surprisingly the way it is wrapped in beautiful paper makes you hesitate to throw it away. The Japanese in turn often make negative remarks in connection with the way we present our merchandise imported into their country. Just as we are being blamed for being energy wasters, Japanese feel that we do not show enough respect for them. While American exporters can respond rightly that this is only an excuse, and that the Japanese government makes it very complicated to import American products successfully into Japan, there is still an important lesson to be learned.

A package is an expression of the respect we have for the consumer. We put a diamond ring on a velvet tray, we surround it with dignity and glamor, enhancing its intrinsic value.

The modern consumer, being warned almost daily how manufacturers are trying to cheat him, is suspicious. He reads the label much more carefully than before. He expects the packaging to be pleasing and to show respect for him. On the other hand, he or she is also aware of the tricks of the trade. An overly elaborate label can create even more suspicion. The consumer feels that the advertiser and packager are "trying to pull the wool over his eyes."

Since trademarks and their packaged form have broken through national barriers, the consumer has to be considered international and American at the same time.

While the United Nations has increased its membership many times, and new nations seem to spring up every few months, the "Commercial United Nations" seems to be functioning much more effectively. The Coca-Cola bottle and logotype have become much better known national emblems than the United Nations flag. More and more companies are breaking down national barriers. They have to—for economic reasons. They are *true* internationalizers— whether politicians like it or not.

A number of rules, recommendations, and pitfalls in packaging on a national-international basis will be discussed here. There are actually two kinds of packaging: visible and invisible. It is the invisible packaging that is truly important. In order to arrive at successful marketing, the common denominator on the more psychological, invisible, and often subconscious level will have to be determined.

More and more international packages and products are being sold: Volkswagens, Renault and many other car companies, Wilkinson Blades, Gillette, Dash, Kraft, Nestle, Panteen, General Electric, Ferrero candy in Europe, Cerruti in the clothing field in Europe, Maxwell House, Wrigley, Kodak, Agfa, Philips, Cinzano, Hoover, Electrolux—all these and many more have become familiar names the world over. While many of these products have been very

successful in the different countries of the world, there are others that have not achieved comparable acceptance. Based on our international experience, the following sections discuss a number of the points that we feel packagers should watch out for.

DOES YOUR PACKAGE HAVE AN INVISIBLE STAR-SPANGLED BANNER PRINTED ON IT?

While many countries begin to think more internationally, this does not necessarily mean that they are going to replace their own national characteristics with American characteristics. To think that they will is *not* a proper understanding of international marketing; yet it is a mistake made very frequently. An American hair spray manufacturer suffered a marketing failure because he simply took the American-type can, rather brassy and tinny-looking, and tried to introduce it in Europe. It so happens that in most European countries package design for everyday products shows greater solidity and more elegance than is customary in the United States. By using cheaper materials and gaudy designs a major question was raised regarding the quality of this American manufacturer's product.

Mistakes can be made in other ways, too. While a quality promise on toilet tissues has become unnecessary in the United States and many other countries, reassurance that the paper is not made of soiled rags is still necessary in some areas.

Or, in South Africa, for example, we found that reference to "labor-saving" is not as effective by far as it is in other countries either in packaging or in advertising, because native labor is cheap and abundant, and many household products are really not used by the white housewife but by her maids.

This, by the way, is an interesting factor in itself. Although an aesthetically more pleasing, more subdued type of package might be the one likely to appeal to the mistress of the house in some countries, in other countries products like floor wax, shoe polish, detergents, and furniture polish, are often bought by maids and, therefore, have to take their taste into consideration.

Also in South Africa we found that selling laxatives to natives necessitated package design of a rather brash nature, since the type of color and logotype used was to these consumers a promise of the remedy's effectiveness. Anything too mild, too subdued, and too sophisticated would represent the product as a weakly performing one and would thus be unsuccessful.

SATISFACTION OF HIDDEN DESIRES

In Norway we were told that it was considered bad taste to paint one's house too brightly, and that the colors most liked by Norwegians were white and red, in different combinations. Our motivational research study, however, revealed that Norwegians suffered from a hidden color hunger, but that the erroneous assumption of their rejection of color had been accepted in products and packaging in many areas. This went so far that a paint manufacturer was at first inclined to offer only basic, elementary colors when introducing new types of paint.

Studying folklore, using depth interviewing, and thus penetrating a little further into the *real* attitudes of Norwegians toward color convinced us very quickly that not only were they *not* really inclined to reject colors, but actually *loved* them and wanted to get as much of them as possible in their packaging.

This shows that when designing international packaging it is important to penetrate the surface of people's feelings and *not* to rely on what has been found by direct observation and immediate reaction. *Time* magazine recently talked about London as a "swinging" town. No superficial observation of Englishmen, a few years ago, would have led one to prophesy that the beatnik hairdo would originate in London, or rather Liverpool. In a way, one can consider this hair style to be just another form of packaging. Apparently, Englishmen suffered from a hunger for unusual approaches—contrary to the prevalent stereotype of their staid and conservative nature.

PACKAGING MUST REFLECT THE SOUL OF THE PRODUCT

In Germany, we recently worked on a new brand of wine actually made up of twenty different wines. It was a new concept in marketing, and the problem was to develop the proper packaging for it. Wine, to many Germans, is sacrosanct. It has to come from a specified vineyard; it has to have a vintage year, and it has to connote "connoisseurdom." Yet this new wine, called Goldener Oktober, had none of these elements. What we tried to do, then, was to find out the underlying, deeper meaning that wine has for the Germans. We discovered that purity was one such factor. We then translated this "purity" concept into advertising, packaging, and label approaches. At the same time, as we wanted to convey through all the elements of the wine's promotion and merchandising that here was a *different* concept and idea, we eliminated the cork, used a screw cap and introduced a different bottle shape. The reaction was very positive, and so a new type of wine has been firmly established in the German market. This was made possible only by our trying to get beneath the surface and understanding more fully what wine *really* meant to Germans.

In France, we faced a similar task for Fromageries Bel, one of their largest cheese manufacturers. (You may be familiar with one of their brands, La Vache Qui Rit.) Our problem was to investigate their Bon Bel brand. This is a cheese packaged in film in wedge shape. Similar to wine in Germany or in France, cheese is almost a sacred and religious thing to the Frenchman. It must have a rind; it must have "life." Wrapping it in film and in a mass production kind of shape robbed it of its individuality and life. Following our suggestions, different shapes were introduced, giving the cheese a feeling of life and modernity. We decided to address ourselves, for this particular cheese, to the young French family who had become familiar with other more practical forms of marketing and packaging in many other areas, and therefore was much less resentful when "their" cheese appeared in a new and different form.

In studies in England, we found that British youth was beginning to be more interested in the Continent than ever before but, at the same time, was also inclined to search for the lost greatness of the British Empire. Out of both these concepts, we developed a number of ideas for brand names and packaging for cigarettes and other products.

The point, then, that I am trying to make is that international packaging has to try to understand what products really mean in their respective countries and then find a common denominator on which to base practical applications.

People travel a lot and will do so more and more. Discovering known products and recognizable names and packages in foreign countries will probably represent a major element in bringing about greater unification in Europe and on a world-wide basis. At first this may seem to rob us of variety and the discovery of strange things; yet *true* understanding of the international role of packaging does not require equalization and uniformization, but almost the exact opposite. In a sense Eve's fig leaf was one of the earliest and apparently quite successful forms of packaging. It embellished and aroused curiosity. Modern packaging has the same function, that of acting as a bridge or barrier between product and consumer.

Simply showing what is in a package, even if this is done in a romanticized fashion, is only half the job. Likewise, thinking of the public and how a package might affect it, without due regard to the interaction taking place between the consumer and the product via the package, is only half the task.

Modern package design, and research testing its effectiveness, should try to assess the imaginary arch linking the two aspects of the product and the consumer and how they are related to each other. A package is not a static piece of art and design. It is dynamic in itself inasmuch as it has a beginning and an end. People start looking at it somewhere, turn it around, follow the design, and abandon it again at some other point. There is another dynamic element involved in the bridge and barrrier function mentioned before between the product and the consumer.

Testing a package design in a correct psychological way involves much more than simply a physiological test involving visibility, attention value, and the power to stand out on the shelf. Appropriate research goes also beyond purely aesthetic considerations. While it is important to know why a package is pleasing and well designed from an aesthetic viewpoint, it is even more important to measure an entirely different and new psychologial dimension. This is whether or not the package does a proper job of communication and whether its role as a three-dimensional tool between consumer and product is fulfilled. A package can have tremendous visual impact. It can be considered beautiful by all art experts and win many awards and still not do a proper selling job. There are several factors that determine the successful role of the package and that have to be measured through psychological testing.

A PACKAGE HAS TO SELECT ITS AUDIENCE

In a study for a West coast sugar company we found that because of improper packaging this company reached only a part of its potential public. Our study revealed that in the sugar consumption field there are sugar puritans and sugar hedonists. In other words, some people were afraid of eating too much sugar and still considered sugar an almost sinful kind of product. On the other hand there are people who enjoy eating sugar and have no inhibitions in their physiological or psychological needs. By ranking the various sugar packages of different brands along a line ranging from sobriety to the other extreme of pleasure and fun we could determine through systematic testing that, first of all, people indeed did have subconscious associations with the design on the package. Packages could be grouped from rather sober and stern types to the other extreme, which was much more flowery and gay using modern colors and being pleasure accentuated.

By changing, on the one hand, the package design in such a way as to move it further toward the pleasure-accentuated goal, we could increase sales. We suggested further the creation of a second line that would be competing very

definitely with the other brands with the hedonistic fun note.

In the deodorant field we found that all available competitive packages can be classified from permissive to authoritarian and from feminine to masculine. The respondents are capable of doing this very successfully. This showed that on the shelf customers unconsciously ranked packages, not so much from an aesthetic viewpoint, but rather from the vantage point of the personality of the products that it conveys. This classification of packaging along psychological lines is very important.

We are dealing more and more with a psychological segmentation of the market. It will be almost impossible in the near future for a particular brand to control more than 30 to 35% of the market. What we are witnessing is a rediscovery of individuality. The mass market is dying out very rapidly. This has important implications for packaging. Packages will have to be designed in the future so that they signal their own personality. The determination of the relevant categories is the job of motivational research. In the deodorant field there are some people who want to be warned in an authoritative fashion. Products like Veto and Ban are the ones they respond to. These people belong to a very special psychological type. On the other hand, there are people who take the fear of perspiration much less seriously and consider body odor, at least to some extent, part of a healthy human body. These people are more likely to respond to products like Fresh. This sorting out and segmentation, which is taking place more and more in our modern markets, has to be matched by the package itself. A considerable part of our psychological packaging testing, therefore, concerns itself with a determination of the role being played by a package along this scale of psychological characteristics.

NEW AESTHETIC SOPHISTICATION

There are many products that supposedly have purely utilitarian value. Among them are detergents, matches, and paper tissues. In a recent study in Germany we found that by redesigning a very utilitarian detergent package to add flowers and to make it beautiful, sales could be considerably increased. What we discovered is an increasingly emerging phenomenon. The modern consumer is becoming aesthetically more sophisticated. He no longer wants the commercial message on his everyday product carried over into his living room and his kitchen. We are beginning to discover that luxury and beauty are necessities for everyday life and not sinful extravagances. We can ascribe this not only to a development of cultural insights but also to the fact that packaging design itself has been an important and successful teacher. In the paper tissue field it has been demonstrated that by promising the housewife, first of all, the choice between various styles of design and, second, that the commercial message could be taken off with the cellophane wrapper, that she was definitely flattered and bought the product. On the other hand, the competitive product insisting on its utilitarian value and still relying on pop-up, wet-strength, and similar features suffered in comparison. In the future, package designers and manufacturers will have to count on the fact that the housewife of today has become a connoisseur and expects to be flattered by styling and aesthetic smartness.

THE INSIDE LOOK COUNTS, TOO

In some recent studies conducted in our Grocery Research Workshop for *The New York Times* we found that housewives often judge the quality of a product not only from outside of the package, but from the first look the merchandise itself offers after the package has been opened. In a study for Snowdrift shortening, we found, for example, that one of our tests was spoiled because housewives could recognize the various brands of shortening by the type of swirl and finishing that had been put on the upper layer of the shortening on the inside of the can. The more individual this swirl looked, the more it gave the housewife the impression that the can had been almost individually filled—the more a feeling of attention and homemade quality was conveyed to her. It was recommended, therefore, that a very definite kind of swirl of a unique nature be put on the inside of the Snowdrift can, and this was carried even into the "S" of the brandname itself on the label.

In studies in the frozen food field we found that much can be done to improve the inside appearance—the package of the future will have to pay much more attention to the inside story. We have to realize that once the package is opened, particularly when we are dealing with products that are being used again and again, the consumer pays relatively little attention to the outside and concentrates to a very large extent on what he sees after he opens the container. The number of liners, the material used to protect the merchandise—all these are important considerations that have, up to now, been overlooked to a large extent.

INVENTIVE CONSIDERATENESS

Again and again we find in our packaging and motivational studies enraged customers who cannot understand that with all the talk about modern packaging nobody has discovered how to package merchandise in such a way that it is really convenient for the consumer. In a study on bleaches for the Purex Corporation we found that many housewives were afraid of breaking the glass bottle and burning themselves. With the help of the designer and the company, we then designed a handle which would permit the consumer to get a good grip on the container, providing a comfortable groove for each one of her fingers. This is a relatively simple packaging improvement. Many hundreds of them are needed.

Why, for example, hasn't anybody designed a more practical can of sardines? In a study on evaporated milk we found that many people considered the typical rimless evaporated milk can as almost obscene, unpleasant, and certainly impractical.

The designer of the future will take a good hard look at all the practical considerations in packaging. Has anybody taken a good look lately at the designs of coffee cans? If aroma is indeed such a valuable aspect of coffee, as it most likely is, why expose it to the risk of evaporating each time the can is opened? Why not design it in such a way that only the right amount is being dispensed, without having to open the whole can? Why is there no sock dispenser to be mounted in the bedroom and sold by sock manufacturers?

In a study conducted on vitamins we found that sales could be considerably increased by developing an apothecary jar that permitted people to put the vitamins on the breakfast table without having to go to the refrigerator each time they wanted vitamins. Vitamins thus became a regular habit and were tied in with the practicality of the breakfast table.

Sometimes it is considerate on the part of the manufacturer not to make his product too visible. It was, for instance, a mistake for a spaghetti maker to package his spaghetti and macaroni in a box with a see-through window. There is nothing appetizing about raw spaghetti, it is even slightly repellent. This manufacturer, through his packaging, seemed to be agreeing with the idea that spaghetti was something pasty and starchy. We advised him to use the package surface to exhibit the delicious spaghetti dinner instead.

THE NEW ANGLE OF CONSUMER VISION

Today's consumer doesn't see the product on the shelf the way he did 10 or 15 years ago. He will see it differently again 5 or 10 years from now. In the past the product was directly in the line of his vision, straight across the grocer's counter. When he viewed the product in the package his eyes met it head on, and he could take in all the essentials at a glance. In today's supermarket, passing among palisades of parallel shelves filled with hundreds of products, the consumer mainly sees a gigantic blur of boxes and cans in which there are no outstanding elements. In this chaos a package must have prominent features of design that catch the eye, hold the attention, and initiate the dramatic interplay between package and patron. That makes the package come alive in the customer's hand. Nowadays the consumer's eyes meet the shelf at an angle and cannot perceive a clear-cut front packet panel. One end of the box and part of one side intrude on the area of vision, breaking and weakening the impact. In examining the packaging problem arising from this new angle of vision, designers and manufacturers might also take into account the phenomenon of peripheral vision. Here, in addition to new angles of perception, there is also present the factor of diminished visibility. In peripheral vision certain colors and designs are barely perceptible. Certain forms and shapes, effective when seen directly by the consumer, become distorted, confused, and weak in impact in the dim light and the normal angles of peripheral vision. Our modern research techniques have concentrated on more realistic elements of packaging and consumer vision. In many studies we tried to carry these findings over into package design.

THE INVISIBLE PACKAGE— A PACKAGING PARADOX

We found the following paradox: The really effective package must virtually disappear at some point in the purchase and use process. It must disappear in the psychological sense of fading into the background while the product itself comes forward to become the figure that is seen, related to, and remembered by the consumer. Sometimes, but not always, this means an effective package should be made of transparent materials. Often the reverse is true. When buying sausages at a meat department, we want the package to "fade into" the finished product, the hot, crisp aromatic meat, and not into the cold and greasy food we see through the plastic envelope. In the case of butter, for instance, we found that the matter of design and art for the package assumes paramount importance. People said: "I wish they would show butter in boxes that do not suggest just butter alone. Butter needs to go with something like pancakes and butter, or baked potatoes and butter." What the consumer is saying here is that simply illustrating butter in pats or in quarter pound sticks does not provide any appetite appeal. Butter is a combination product and goes well with other foods. We are dealing here with the need of using the package to make more of the product than it is in itself. In a sense the package is psychologically invisible, and pushes the product into the foreground.

Some time ago I was asked to testify in Washington in connection with the U.S. Senate Hearing on Packaging and Labeling Practices. I was shocked to discover how much real belief there existed in the basic dishonesty of advertisers and people connected with packaging and marketing. Some of the senators who interviewed me were convinced that designing packages in such a way as to cheat the consumer was a widespread practice. It may be important to repeat here what I stated in Washington. Packaging and labeling practices that presume that the consumer is a gullible moron can end up proving that the advertiser is the moron. There is no doubt that whenever dishonesty occurs in packaging, the guilty manufacturer should have his fingers slapped with the hard ruler of reality. I feel, however, that with increasing sophistication of the consumer, a sophistication that is developing with galloping and often unsuspected speed, deceitful and fraudulent activities (which exist much less frequently than is normally assumed) will rapidly become unprofitable. The modern consumer is a highly efficient judge of advertising and packaging practices. Any advertiser who has not yet become aware that the modern American consumer and the world consumer often have a much higher IQ than the advertiser credits him with, is due for a rude awakening.

Packaging has become even more important than ever before and will increase in importance because of an interesting development which I call the *Psychoeconomic Age.* We have reached a point where in many areas quality is taken for granted to an unprecedented degree. When people buy soap, they know that they're going to get good soap. They don't have to worry any longer about its quality. They will not get a piece of chalk. Our technological development has been so good and so fast that almost all our products are accepted by the consumer as being more or less alike, more or less uniformly good. As long as he spends the same amount of money, he knows he gets about the same quality regardless of brand.

What people then actually spend the money on today and what secures the free enterprise system, are psychological differences and the brand images that permit the consumer to express his own individuality. Most cigarettes, most soaps, most detergents could probably be grouped in two or three categories. They could be grade labeled and packaged accordingly. Instead, there are hundreds of different cigarette brands, many soap brands, hundreds of cosmetic brands. In blindfold tests most people cannot distinguish between one brand and another. This does not mean, however, that we simply buy illusions, or that the consumer is naive.

In hundreds of thousands of interviews conducted by us, one of the most frequent findings is the widespread conviction of the consumer that most of the products that he uses are almost indistinguishable from one another. Yet it is he, the consumer himself, who would probably scream loudest if an attempt were made to deprive him of this rather irrational naive opportunity to express his individuality. The American consumer wants to have a maximum of choice. He does not want to be limited to five grades of cigarettes or three sizes of a government-produced refrigerator. The businessman deals with a consumer who, while rational, is at the same time governed by emotions. The shopper for goods is not any different from the voter, (the shopper for political candidates), from the lover who picks a bride or from the job seeker who decides on the right kind of position or career. The bride should probably tell the lover what she will look like in ten year's time and that she snores. The politician should tell his voters that he's not sure that he really

has all the qualifications for his office, or that he can really do a better job than his predecessor. The one would not get married; the other, not elected.

Some time ago we were retained to help the Cooperative League and the Greenbelt Consumer Co-ops to attract more customers. Our study showed that emphasis on the good quality of the merchandise, and references to the principle of sharing profits by eliminating the middleman, offered little attraction to the modern housewife. What she really was interested in were lower prices, attractive stores, and attractive labels. When Grand Union stores called their peaches "gorgeous" and designed a corresponding label and the Co-ops called (and designed) theirs as "reasonably good," the housewife chose those sold by Grand Union. She had been trained in several decades to apply the same psychological disbelief to both statements. In other words, she felt that the Grand Union peaches were probably fairly good and that the peaches sold in the Co-op stores were probably poor.

Even in the pharmaceutical industry we found that packages of drugs have to be designed in such a way as to be more attractive to the physician in order to suggest greater scientific care.

Thus embellishment of products and packaging is universally practiced. It will become, as I pointed out before, even more important in the future. Packaging of the future will be more practical, more beautiful, more oriented toward individuality and psychological segmentation of the market. It is hoped that it will also be more honest.

Designing better packaging does not mean designing new ways of fooling the public. Nor does it mean developing packaging that only gives technical information and merely describes the content accurately. The package is basically an emotional tool. Apparently Eve knew this when she donned the fig leaf. Maybe this emotional power of packaging, its role as embellisher, as romanticizer, and its ability to hide the naked product are all responsible for the flop of the topless bathing suits. It seems that even there we prefer the packaged product to the bare facts.

EACH NATION GIVES ITS BEST TO THE WORLD

Volkswagen could have tried to translate its name into different languages and to adapt itself to national customs in its design and logotype. The fact is that they did the direct opposite. The name Volkswagen is quite difficult to pronounce for people not accustomed to the German language, yet people managed somehow. The world is, in a way, richer because of that kind of product.

As blasphemous as it may sound, the same thing is true of Coca-Cola. Though it is not as much of a tongue twister as Volkswagen, its whole appearance and the easily recognized shape of the bottle are typically American—yet it has conquered the world.

The General Electric logo, the Du-Pont logo, and many of their package designs have followed similar destinies.

In studying Tyrolean Loden some time ago, our advice was *not* to make "European" or "international" loden out of it, but to leave it with this rather narrowly localized regional designation, and yet to let it be known that it is available all over the world. Swiss Cheese, from Switzerland, of course, demonstrates a similar phenomenon.

Gas stations, such as the Total stations in Europe and other parts of the world, Shell stations, and others, indicate another important aspect. They represent the "wayside inn" with familiar features that make you feel at home even though you don't speak the language and are in a strange country. It has become evident in some studies we have done in connection with the unification of Europe that "nationalism" must be considered a disease, based to a large extent on *fear*. We still experience somewhat of a shock when we enter a foreign country and see foreign uniforms. The American would probably feel (no matter how sophisticated he may be) if he encountered Chinese or Russian soldiers, a ripple of unpleasantness, and all his logic would not help him very much nor would telling himself that all he is dealing with is a different form of "packaging" of a well-known phenomenon that in his own country is a rather familiar sight. Maybe it is a silly thought, but if all over the world all soldiers wore the same uniform, one of the major reasons for war may be eliminated—preposterous as this may sound.

While the rest of the world is still intent on raising different flags and emphasizing differences that are not always delightful, the international commercial "conspiracy," luckily, is moving in the opposite direction. International packaging is only *one* of the factors (but it could be an important one) involved in combating the fear of strangeness and with making people feel at home all over the world.

Packaging directed at the modern consumer has to consider him as being more and more internationally minded. The American consumer wants his products to be representative of his "way of life" as one way of overcoming the inferiority feeling that he often has now.

Good package design can tell the world that despite all its difficulties, the United States still does give its best even in this important field. "Made in the U.S.A." used to be a proud logo; good packaging of quality products can be of great help in reestablishing that prominence.

CHAPTER FIVE

Selecting Appropriate Packaging Research Methods

Lorna Opatow

All marketing research involves the systematic collection of information organized and used to answer a marketing question. It is the intelligence arm of business and its primary goals are to reduce risk while increasing the probability of success.

When applied to packaging, research can pay off handsomely—if you know what options are available and how they might be used. It can also be a hindrance and even detrimental when the wrong information-gathering options are applied.

This chapter will provide some general guidelines for determining when research might help and for evaluating particular methods. It is intended to help you develop your own point of view.

Design parameters:

In general, most business decisions are based on the combination of knowledge and experience we call judgment. Experience can be gained only with time. Knowledge is immediately accessible, and the sources of information are varied.

The need for information may occur at several stages during the packaging development process.

Prior to the start of a packaging program, research may be used to identify the target market and determine optimum brand positioning, suggest shortcomings in competitive products or packages that will give the product a unique claim or benefit, and help set the criteria or goals the package is to achieve.

Even in the most unsophisticated companies undirected packaging creativity is a thing of the past. If the company is not dictating creative boundaries, the government is. Both graphic and physical design limitations often start with the product itself. To these product-oriented limitations are added those of availability and cost of packaging materials, special shipping and storage problems, retailer requirements, government regulations, and the nature and extent of competion and consumer needs.

The person responsible for packaging development must resolve these often conflicting requirements while achieving all of the company's marketing goals related to the package. For consumer goods packaging the conflicts can be reduced by looking at the package through the customer's eyes.

For the customer, the package exists only in relation to the product it encloses. The following are the first questions

to be answered: Who uses this product? Who might use it? Why do they use it? How do they use it? Why should they use this particular brand? The answers to these and other questions are used to develop packaging criteria, including the designation of product, brand, and user attributes that the package is to convey. Through a skillful combination of color, shape, graphics, and copy, the designer is charged with the responsiblity of developing a package that will attract customers to the brand and ensure that, in use, the contents measure up to all implied promises.

To achieve this delicate balance, the designer should know what the specific packaging goals for this particular brand are and why they were set.

The more designers know about this the better, not only because development can focus on the specific job to be done, but also because if a decision is made to evaluate the results, researchers must at least consider doing so on the basis of a series of mutually agreed on goals.

Here are examples of some fairly universal packaging goals or criteria:

General	In-Home	In-Store
Protect Contents	Be Functional (Easy To Open, Close, Store, Use)	Gain Attention On Shelf
Satisfy Distributor Requirements	Reinforce Product Satisfaction	Identify Product
Satisfy Retailer Requirements	Remind Customer When To Repurchase	Identify Brand and Differentiate It From Competition
Satisfy Legal Requirements	Remind Customer To Buy Same Brand	Say Something Good About The Product
Be Adaptable to Illustration In Both Color And Black and White		Induce The Consumer To Buy

These criteria appear to be quite reasonable but they assume a considerable body of knowledge. For example, the in-store criteria assume that the package developer knows what kinds of stores the product is in, where it is located in the store, and what products are normally displayed near it. The in-store

environment usually differs from area to area and, in many cases, from store to store.

The in-store criteria also assume a knowledge of customer shopping habits, factors influencing in-store buying decisions, and the nature of the target market. Just consider that last piece of information, the definition of the target market or customer. At the start of a package development program designers may be given this kind of customer description: young women 17 to 24; users of the product; lower to middle-income groups.

This is the entire description that was recently given for a toiletries product, and a package was to be developed to appeal to this group. When the design program was almost completed we were asked to submit a research plan to measure the effectiveness of three packaging alternatives. In the course of discussing possible research approaches, a number of questions were raised. After searching its own files, the company revised the target market definition as follows:

Product Users: 14 years of age and over; purchase this type of product in mass merchandise/variety stores; light make-up users; more likely to admire the "All American Girl" look; less likely to spend more than $2.00 for this type of item.

Obviously, age and income alone were not the key customer identifiers for this brand.

In general, questions asked before and during the early stages of package development are critical to the success of the program.

Companies often have the results of market studies, concept tests, product tests, copy tests, distribution audits, and general background studies. If a design budget does not include field trips to check the in-store environment, shelf display pictures should be provided, as well as samples of all competitive products in all sizes plus any noncompetitive products that are ordinarily displayed in that section of the store. (The pictures should *not* be those released by the merchandising department to the sales staff, because they usually have little relation to reality.)

All of this information helps provide a realistic context for the design assignment. If possible, it should include identification of potential substitutes for the brand, stop-points in the buying process that create an opportunity for the consumer to consider other alternatives, the level and nature of brand switching within the product category, and so on.

The sources of information are varied. Any interested person can find evidence of change by following the news. In fact there is a specific technique for doing this on a systematic basis called "content analysis." It is one of the tools used to identify emerging public issues and to forecast the future importance of concerns such as those about energy, the environment, or inflation.

Surveys are a primary source of information for packaging development and evaluation. While it is sometimes difficult to determine when research should be used, there are very clear indications of when it should be avoided. No matter how carefully planned and executed, a survey cannot be justified under these circumstances:

1   No action can be taken on the results. This can happen because in reality the decision has already been made (by the government, the president of the company, or because materials are not available) or because company policy precludes any alternative action.

2   The potential dollar loss resulting from a wrong decision is lower than the cost of the research. (This is often true for line extensions where there is a clear-cut position for another flavor, size, etc.)

3   The information already exists, or the answer to the question is self-evident.

In addition to providing background information to guide the packaging design program, research is increasingly being used to judge packages. Designers may or may not be consulted in regard to this evaluation. They *should* be involved, if only to try to assure that any measurement is based on determining the extent to which the package meets the goals set for it.

In other words, if specific goals have not been set, there is no research evaluation to be made. If a package is redesigned in order to increase sales and that is in fact the *only* goal, the measurement of effectiveness becomes sales volume or a sales test, and nothing else is applicable.

Usually the redesign goals are more complex. They may involve enhancing brand identification, coordinating a line, differentiating product flavors or types, updating the product, brand repositioning, and so on.

In many cases a designer is asked to modify proposed packages or develop new designs based on the results of evaluative research. Designers and other nonresearchers placed in that position should be sure that they understand how and why the study was conducted and what the results were. They should not be afraid to question research conclusions where they cannot follow the reasoning that led to those conclusions or where their interpretation of the data differs.

Of all the criteria established for a package, the ones most often questioned by management and subjected to some kind of research evaluation are functional effectiveness, visual impact (visibility, shelf attention), and communications effectiveness (brand and product identification and imagery).

Of these, functional effectiveness is easiest to evaluate. Some aspects can be handled in the laboratory; others through trial or in-home use tests; others through focus group sessions, depending on the nature of the product and the package.

With few exceptions, the user of a product can be questioned directly in regard to whether the package is easy to handle, open, close, store, and use and whether it might protect the contents over a period of time. When measuring functional packaging effectiveness, the package can be treated as a product, and normal product-testing procedures apply.

For example, my company handled some of the early studies of retortable pouches. The technology was in the development stages and packages had not received FDA approval. Few filled pouches were available for test purposes, and those that were had a shelf life of about 1 week. Enough people had to be interviewed to determine the kinds of products that might have the best chance to succeed, the kinds of people who might be attracted to such a product, what they were now using, their ideas of the strengths and weaknesses of the proposed pouch, and so on.

The first problem was to develop a description of this new package in language that would leave no room for confusion in the respondent's mind.

To do this, a series of four focus group sessions was conducted. After each session, the concept statement to be shown to the next group was revised and clarified. The question guide was also modified to include new subject matter and to delete unproductive questioning.

Based on the results, a telephone questionnaire was developed to obtain the quantitative information needed. All respondents were asked if they would be willing to try a product in the new package and be interviewed about it at home. A follow-up study was conducted among clusters of respondents in areas close to processing facilities. (The geographic limitation was necessary because of the short shelf life of the samples.)

During the initial in-home interviews, reactions were obtained to package appearance, anticipated benefits and drawbacks, and so on. The product was left for use that evening along with an evaluation sheet to be completed by the respondent immediately after use. The follow-up interview was conducted by telephone the next day.

In general, standard research techniques were combined and adapted to suit a particular set of circumstances related to functional effectiveness.

Of all the packaging aspects that might be subject to research evaluation, visual impact is the one most often tested, but the one least often measured. A variety of instruments have been developed that purport to measure the relative visibility of various packages and the design elements for a single package. The instruments tend to lend an aura of scientific objectivity to studies. For this reason and because this kind of study can be conducted inexpensively and within short periods of time, it is often the only measurement used.

Unfortunately, the information obtained is extremely limited in application. Even more important, the need for in-store visibility is greatest when the product is first introduced and customers must seek the unknown. It diminishes in importance as advertising and in-store promotion increase familiarity with the product. Also, the package with the highest attention value may also be the poorest communicator of product and brand attributes.

If used other than as a stimulus for discussion, most instrument tests are speed tests. Which element of the package is seen first? How long does it take for someone to read the brand name? If the brand is noticed on one package in $3/10$ second and on another in $7/10$ second,

what marketing significance can we attach to the difference?

Another limitation of instrument testing is that it ordinarily takes place in a laboratory, which cannot approximate the confusion of shopping situations in the normal store environment.

It has been demonstrated that people are selective regarding what they see. They "see" things they need or want, or expect to see, and tend to "screen out" irrelevant elements. Similarly, they are more likely to see that which is familiar. This aspect of "familiarity" is the reason that current packages normally score higher on visual tests than proposed designs, especially among users of the brand.

Perhaps more than others, the designer is aware that the process of seeing involves a combination of mental and physical activities with the mind, rather than the eye, providing the bulk of visual perception. The limitations of the physical mechanism by which we see provide the basis for most optical illusions. Designers who understand how vision functions are able to consistently "design for the test" and to predict accurately the results of visual testing.

In spite of these limitations, instrument tests can be helpful in cases where test packages cannot be produced and pictures must be used, for example:

1 To develop indications of product differentiation within a line. Where a line of products is involved, mockups must be prepared for every flavor or type to be included, and the preparation of the samples is expensive. Instrument tests can be used to get an indication of the degree to which consumers are identifying all of the items in the line or are considering some or all to be additional facings of the same product.

2 Where several alternative models for

‎signed container are
‎..dered and there is a ques-
‎.e perception (e.g., whether
‎.ne container looks larger than
‎..r) or of the degree to which
‎shape enhances overall visibility.
‎. this case the models are usually
plastic or wood and cannot be shown
to respondents because they are obviously not real containers.

3 To provide indications of recognition
elements. Often, a package is recognized by its shape, configuration of
lettering, color, or overall graphics,
and there is a question about the
contribution of each element to recognition of the brand. This is sometimes the case when redesign is being
considered for a major brand.

In attempting to measure visual impact, it is more fruitful to use a simulated
product display, which helps to create
some of the visual confusion that occurs
in a store. There is no distortion of
packages due to photographic problems,
and no framing effect. Respondents are
exposed to the packages at an angle and
while in motion, just as they would be in
a store.

They may stop and look at whatever
interests them and ignore whatever does
not.

Depending on the package design included in the display, responses will differ. These differences provide a general
measurement of shelf vitality within a
somewhat confusing environment, but
without the pretense of a scientific endeavor.

Most designers can provide some
valid prejudgment of the functional and
visual effectiveness of a package. In regard to packaging communication (ideas
the package conveys about the product
and brand), designers know what they
are trying to achieve. The degree to

which each proposed package succeeds
is the question research seeks to answer.

Because the research method and
testing materials used must be appropriate to the product and the particular
design problem, there is no standard
technique or series of techniques for
packaging communication evaluation.
The particular method used depends on
the packaging goals to be measured,
economic feasibility, and the extent and
nature of the product and brand information that is already available.

When appropriate to a given product
or packaging problem, consumer research can produce actionable results if
the following requirements are met:

1 The study objectives must be clearly
stated, in writing, and there must be
no misunderstanding about the kind
of information to be obtained or
how it will be used.
2 The study plan or design must be
capable of yielding the information
the objectives call for.
3 Action standards must be established in advance. Action standards
are guidelines for judging the results. They help to specify ''how
high is up.'' It is very rare for a
package to meet all of the objectives
set for it. Management must establish priorities in advance and agree
to meaningful definitions of which
results will be considered positive
and which will be considered negative. If this is not done, there is a
good chance that different sections
of the data can be used to support
the selection of different packages,
and the value of the research is lost.
4 The people surveyed must be capable of providing meaningful information. They should be current and/
or potential customers for the product or brand so they are qualified to
judge. Someone who knows nothing

about the product or brand, and does not care, is not in a position to provide valid information about it. Many aspects of packaging are influenced by past experience with, and opinions of, a brand. For example, questions related to whether this is the "real" brand or an imitation, or whether the product formulation has been changed, must be answered by present users. As previously mentioned, respondents may be representative of the target market as defined by the manufacturer, but the manufacturer's definition may be inaccurate.

5   All the packaging alternatives to be tested must be equally acceptable. If for any reason one of the packages cannot be adopted, it should not be included in the test. The only exception is a current package that is included as a standard against which others will be evaluated.

6   Differences among alternative packages must be large enough to be measurable. Changes in copy are measurable. Changes in background color are measurable. Minor changes within a design may not be measurable.

7   If the communications value of a package is being measured, each person must be asked about only *one* packaging alternative. They may be asked about any number of brands, but each must appear in only one package. Researchers refer to this as an experimental study design. All other variables are held constant in order to measure the one of interest—in this case the package and the ways it influences people's ideas about the brand and product.

8   Questioning must be appropriate. The order in which the questions are asked must not bias the answers

to subsequent questions. At the same time there must be a logical flow of subjects. In addition, the wording of each question must be neutral in order to avoid "asking an answer". When information is sought about packaging communications, questioning must focus on product, brand, and user perceptions first, and these answers obtained before any reactions to the package are elicited.

In general, the interviewer should be able to understand and follow study specifications. The interview should be interesting, nonrepetitive, to the point. Questions should be worded in spoken rather than written English.

All questions should be structured according to how the results will be analyzed. There is a wide range of statistical techniques that will increase the usefulness of the data. Many of them, however, require particular types of questions or questioning procedures.

9   Sound package testing requires that exhibit materials shown to respondents are appropriate. When trying to determine whether or not a particular design identifies the product as a margarine and/or a particular brand of margarine, pictures may be acceptable. Pictures are not desirable for use in group interviews or most studies of communications effectiveness. After preliminary testing to screen design directions has been conducted, final studies should be based on finished-looking packages.

10   In general, the analysis or interpretation of results should take into account both patterns in the data and absolute numbers. The findings should be related to information from other sources so results can be

viewed in context, rather than in a vacuum.

For a study to be successful, the results must be used. The report must be written so that the reader can understand the data and follow the reasoning that led to the conclusions and recommendations.

One of the problems of consumer research is that there is no one best way of measuring anything. A good survey is developed by selecting from all of the technical options the combination that offers the best chance of meeting study objectives within a reasonable schedule and budget. There is usually at least one other equally valid approach. The objective is to strike a reasonable balance that will yield the required information within an acceptable amount of time and at a cost commensurate with the risk of getting a wrong answer.

One of the major limitations of consumer research is that it is far more likely to identify a loser than a winner, primarily because people are more articulate about what they do *not* like than what pleases them.

Despite its limitations, research can serve as an important source of information for packaging development. It is increasingly used to help evaluate design results because it decreases the risk of loss and enhances profit opportunities.

CHAPTER SIX

Analyzing Mechanical Package Design Criteria

Glenn R. Sontag

Every package and new product requires standards or criteria to make it an effective part of the overall packaging program. All of the design research and testing cannot overcome any shortcomings in the total design process caused by failure to analyze all of the design criteria that will be the basis for the complete package program.

This chapter will not tell you how to design a package but will delve into the most critical functions a package must perform if it is to be a successful one.

Once they are spelled out with all the alternatives, design criteria must be analyzed in detail to make sure they "fit" the basic objectives set forth in any packaging program. Each requirement by itself could have a number of feasible solutions. When coupled with other required factors, it may well be that the avenues of approach might be limited to one or two at the most. Let's take a look at some of the basic requirements that must be determined as part of package design research to show how they must dictate the ensuing work in the design of the package. Basic materials (protection and packaging shape), packaging methods, packaging machinery, printing methods, and distribution are those standards that are the most difficult and most expensive to alter once the design project is under way.

How many times have you heard about the beautiful package project that never got off the ground because it couldn't be printed or contained, because there were too many colors or because new packaging equipment had to be purchased in order to even get the product to the marketplace?

While this type of conceptual thinking is important, it must be remembered that the manufacturer of the product must be able to afford any additional costs of packaging the product if the new concept is chosen. Many times it is easier to fall back on existing packaging shapes than to buy new machinery. This is where the completeness of packaging research can reduce a costly gamble.

BASIC MATERIALS

Before a package form is selected, take a good look at the product. What is really needed to protect that product from the time it is packaged until it is opened by the consumer? Overpackaging becomes obvious to the consumer who does not want to pay for it, and underpackaging is a threat to the product itself.

Determine the geometry of the package and choose the standards based on the position of the package in the store, its relation to other like products, and how long it might have to stay there and still do a selling job. A square or rectangular package might be just the conventional form, but will a triangle package stack as well or will it keep falling down, resulting in a damaged product?

In food packaging particularly, will the constant handling before purchase effect the integrity of the product? Bacon, cheese, and other refrigerated perishables, for example, can become "leakers" and may look shopworn as a result of handling only to discourage the sale of the entire product line. Overpackaging, on the other hand, by using more costly materials could easily drive the price higher than that of reasonable competition.

Suppliers of packaging materials are more than willing to make material recommendations for containing a given product. A "low to high" range of materials and their cost relationship as compared to the barrier function can provide you with the working knowledge of the best material or material combinations to consider for your product. You will find that each structure has its positive and negative qualities, generally related to the cost of the material itself. Substitution can always be made along the line as long as you know what will be lost or gained in the overall protection of the product.

Going one step further, especially with the majority of flexible materials available for packaging, the material itself, or its look, can relate to or affect the sale or emotional appeal of the product inside. Foil, for example, has the underlying connotation of sealed-in freshness. It is a familiar sight in the kitchen and has many uses. The housewife cooks in it, covers dishes with it, gets ready made dinners in it, and wraps leftovers in it. Foil definitely means something to the consumer, but make sure foil packaging is associated with freshness rather than leftovers. One additional advantage of foil or a foil containing package is that it is one of the few packaging materials that offers deadfold characteristics. A fold or crease in the foil will remain there, so make sure you want it there.

Many packages use foil as one of the barrier materials in a multiple structure, such as dry food mixes and soup mixes, but it goes unnoticed as the foil is sandwiched between other materials. Only when you have opened and used the content do you see the foil, and then it proves to the consumer that the product they bought is truly fresh.

Polyethylenes and poly-coated

boards have a soft feel and are ideal for bakery products.

Cellophanes and polyesters have a shiny, sharp look and feel, and the consumer knows they keep crackers fresh, not stale and soggy. The polyester materials and nylon offer additional puncture and tear-proof qualities ideal for dry noodles and plastic toys that could pierce and ruin the effect of an otherwise good package.

Two packaging materials more familiar to most designers are paper and board. Depending on the cost, a coated or enameled stock offers a great deal more class and finish to the product inside, unless you are planning to package a soft, furry product, which relates better to a textured surface. In any event, make sure you consider the basic materials and/or outside surface of the package as one of the visual and "feel-able" elements when developing packaging criteria.

Once the basic materials are analyzed, the next logical criterion that is directly related to the material is the method of packaging.

PACKAGING METHODS

How you plan to close the package will determine how the consumer will have to open the package. Although the ideal package is one of the "easy open" styles that are yet completely sealed and protected until they are used, existing technology has not developed the optimum container, unless the consumer has a knife or scissors handy at all times. Bags, pouches, overwraps, cartons, bag-inabox, sleeves, trays and lids, cans, bottles are the more standard packaging methods for most standard products.

With today's advanced technology and sophisticated materials, we see motor oil in plastic bottles and composite cans rather than metal cans, bacon in metal cans rather than in flexible and board materials, stockings in plastic containers rather than in pouches; and juice and wine in bags and lidded cups rather than in glass or metal. An innovative package is not necessarily a new package form but might be a container that was previously a standard packaging method for an entirely different product line. By the same token, you would not want to consider the use of a gabled milk type carton for an insecticide (even though it might be the best container) because consumers and their children associate a container of this type with food. Graphics and advertising could offset the adverse comments, but the risks would still be there.

Another important facet of the total packaging concept is the appearance that the particular package has as the consumer sees it in the marketplace. Does it remind the consumer of something he or she is looking for? Is the consumer looking for a product in a carton when you have it packaged in a bag?

Advertising and promotion can go a long way to familiarize a consumer with specially packaged products, but what about all those impulse sales? When freeze-dried foods first appeared on the market they offered many advantages to the consumer: frozen freshness, less weight with the moisture removed, and prolonged shelf life. A metal surface was used to stress the freshness, or a can was used to keep out the moisture. When the products appeared on a grocery shelf, however, nothing could take the place of the light weight of the package of freeze-dried corn as compared to a regular can of corn. The consumers mind could not relate the same or greater cost for the two side-by-side products when the freeze-dried product showed 5 or 6 ounces less. Naturally, after reconstitution (which means additional effort), the two products looked the same and the freeze-dried product probably tasted better. In this case, without a thorough educational process, the packaging method was a deterrrent to the sale of the product. Analyzing the packaging criteria completely would indicate a new product of this type should appear in an entirely new shape or format so comparisons cannot be made until after the product is consumed.

We are now in the area of the retort package, which offers all sorts of advantages to the food processor as well as the consumer. In no way would you want an expensive product of this type to have any implications of the Army "K" rations that became part of many servicemen's diets whether they liked it or not. As this whole new form of superior food products reaches the marketplace, consideration must be given to the relationship the consumer establishes with the visible package, whether it be a foil pouch or a foil tray. Maybe a clear pouch with the same barrier properties would set the entire line of shelf stable prepared foods apart from conventional canned goods and "K" rations alike.

There are many packaging methods and each type has some sort of association with a particular food or class of food. The look of the package structure will definitely be an influence in the purchase of the product even though the graphics and related promotion can do a great deal to overcome any negative reaction.

In trying to point out specific packaging criteria to consider in a package research development project, it is difficult to keep them separate, as each one seems to have a direct relationship to another, just as the graphics related one element to another.

PACKAGING MACHINERY

When one starts analyzing these criteria, the scope becomes extremely limited unless you consider a new company that is ready to buy all new packaging equipment. Probably the most expensive part of any new product or package development is the equipment (of course the product could always be packaged by hand, but only if you don't plan to sell many packages).

If automated equipment is the only answer, the cost is great, the lead time long, and test stages time-consuming. Existing equipment within a manufacturing facility is usually the best way to go. Many of today's machines can be modified to offer a slightly different package without too great an alteration cost. Blueprints, machinery specifications, and modification possibilities considered in the early stages of a package development can greatly reduce headaches at a later date. Try to figure out ways to work around a piece of packaging machinery so that it can provide the package and geometry you want to consider.

Another approach to consider is offered by the many contract packagers available throughout the country. Almost every package shape can be handled by some contract packager. In this way a new form could be considered; the contract packaging could be part of the total development in market testing.

If the product and package are both successful, a good rationale is provided to purchase this type of equipment, using the contract packager until the installation takes place. As important as the package machinery is the configuration of the package coming off the production line. Eyespots, clear channels, foldovers, seal areas, bottom folds, blue lines, die cut requirements, perforations, length to width to height relationship, thin seals, cavity sizes, and oversized limitations are just some of the requirements to be considered. Each of these, depending on the package you are planning for your project, could be a deterring factor if not taken into consideration. A straight tuck versus a reversed tuck, an economy flap, or sealed end, could help make the difference between a successful or unsuccessful package development. These are all things that the consumer will be involved in when using the initial package, and they will help determine repeat sale potential.

Packaging machinery is by far the most inflexible of the design criteria to be analyzed, and it relates directly to the basic materials and packaging method to be considered.

The next packaging criterion, printing method, is not as critical to the package itself, but is more directly related to the cost of the packaging material and the proposed graphics. With the appropriate determination, the printing method can provide the proper atmosphere and a selling influence in keeping with the product to be sold. While one would not consider a full color illustration as a proper way to sell a package of standard screws or bolts, a special type of decorative screw or bolt might just require that colorful illustration to promote its virtues.

PRINTING METHODS

When analyzing the various available printing methods, it should be understood that creativity can do wonders with any printing method. Award winning packages are not limited to any particular printing method. The key is how well you make use of the printing method, how well the chosen printing method projects and sells the package and the product.

Lithography, letter press, rotogravure, flexography, silk screen, hot stamping, and thermography all have their place in packaging and while they are not interchangeable in all forms of packaging, there is always a multiple choice. Basic knowledge of these printing methods must be considered a de-

signer's prerequisite and will not be discussed in this chapter. Determining which type of printing offers the most beneficial relationship between the consumer and the packaged product is the important criterion to be considered.

Will a full color illustration or processed color package do a better selling job to a consumer? Is a special color or a number of special colors a prerequisite of the package graphics? If so, make sure you select the printing method that can offer color consistency throughout the run. Technically, it is possible to achieve any color using the four process colors in litho, rotogravure, or letter press, and therefore a muted color or flat shade using up to three screens of these colors can be considered. If you are counting on a particular shade or background color as a tie-in with the logotype, or to be considered as part of the total design concept, it should definitely be counted and printed as a separate color. In this way, when stacked or in a display, the color image of the display will remain consistent.

When selecting the printing method, don't let the cost totally dictate the media. It is obvious that you cannot overlook the cost relationship of printing in the design criteria, but you must decide early in the planning stage what mood and emotional effect are to be achieved by the package so that the printing will add to and complement the total package and create the proper image in the mind of the consumer.

Just walk through any supermarket and you can find countless packages that fall short in their attempt to win the consumer, particularly in the area of appetite appeal in food products. A pie carton or bag of vegetables with poorly reproduced illustrations is worse than a package without any illustration, and neither of them has the mouth-watering taste effect of a package with a beautiful illustration.

Depending on the type of illustration required in the graphic approach, a less expensive color reproduction in letter press or flexography should be weighed against the cost differential of rotogravure. For example, short runs and combination runs of many items, or the implication of constant changes in the artwork, might require the consideration of a less desirable printing method. If this is the case, consider an illustration that will adapt to these conditions, such as a tone illustration rather than process. By this I mean illustrating the appetite appeal, for instance, by showing only the strawberries with the shades of red and a drawing color rather than showing the strawberries with the green leaves, whipped cream, and other elements that would require full process colors.

Another consideration when printing costs could become a real problem is to let the product itself do its appetite appeal selling. A window carton could show the pie, or a transparent film or lamination could show and indicate the volume of strawberries in a given package. As a designer you can then use these strengths as one of the elements of the design.

While we are on the subject of printing methods I would like to point out one of the most common problems that arises once the design has been developed if proper consideration has not been given to printing requirements. Package designs often are planned to print in just one more color than is available on a printing press. While two passes through the press can satisfy your graphics, the cost will not satisfy your customer. The more colors you require, the more you limit the number of printers your customer has to choose from.

The cost of printing becomes part of the cost of the individual package, but generally the preparatory work, plates, and/or cylinders are a one-time charge and can become a "hard nut to crack" when considering the total packaging costs. Keeping the number of colors down without sacrificing the graphics will definitely decrease the upfront costs of printing preparation.

It is best to work closely with the potential printing suppliers considered by your client so that you are "in the ballpark" on costs and printing limitations. You then have a workable design criterion in the area of printing. At the same time you can incorporate into your criteria other printing limitations such as eyespots, clear channels, glue channels, fold-over areas, nonprintable surfaces, line screens, and color requirements for the matching of colors.

Whatever you design must be printed and, more than likely, mass produced. Doing this economically without sacrificing quality must be one of your key criteria.

In-depth design criteria and analysis as part of the design research is critical to the success of any package design program. Make sure you expand on the design criteria to fit the product. Not to be overlooked are the distribution and shipping requirements, product claims as well as competitive product claims, and UPC requirements.

The estimated cost and profit should also be considered. Advertising and promotional considerations for a new product will definitely affect and alter the design criteria to be established. Heavily advertised products would tend to place more emphasis on the promotion and alleviate some of the selling pressures placed on the package itself. However, the package must communicate and involve the consumer in an emotional relationship. Thorough consideration and examination of mechanical design criteria is one of the basic steps in total packaging research and design.

CHAPTER SEVEN

Qualitative Versus Quantitative Research

Nicholas T. Nicholas

CHANGING LIFE STYLE PATTERNS

Creating and marketing a consumer product and designing its package used to be a relatively simple procedure —the population was exploding (along with the war babies), the automobile was creating expressway corridors that rapidly caused cities to sprawl into suburbia, regional shopping centers became "town centers" and modern technolgy stimulated new *life style patterns* in American society.

The Agricultural Age of the 1800s gave way to the industrial revolution of the 1900s. The transition was dramatic and unique and out of its fury came the continuing need for demographic re-

search based on the need for high production in the marketplace.

But the 1960s brought great turmoil, and the business community began to witness a change of character in the consumer never before seen or understood.

Then, in the waning years of the 1970s, business was not so simple. All the great challenges and opportunities for growth were gone; recession and inflationary factors changed everything. People changed, markets changed. The products that prospered did so because companies quickly adapted to change.

The long, successful road of the industrial revolution has ended. A new *humanistic* society has begun. A society based on *life style segmentation* has grown out of changing human needs.

Business has discovered that the essential key to controlling change is finding the most accurate information about consumers and markets with which to make marketing decisions.

Since 1970 important underlying *change factors* in the consumer marketplace have altered almost every marketing plan and marketing strategy. Every product had to be reevaluated. Every research project changed.

We have seen a dramatic increase in the number of households and decrease in the total population. What is behind the increase in the number of households?

- *The Baby Boom* Children born in the 1950s and early 1960s are becoming adults and setting up their own households. The young adult consumer has replaced the 'teen generation' of the 1960s.
- *Smaller Families* A declining birth rate; fewer marriages; more one-parent families.
- *More Singles* Young people postponing marriage, leaving parental homes at an earlier age; increasing divorce rate.
- *Longevity* People living longer, maintaining their own home.

These demographics are impressive and provide us with the *quantitative* research data necessary to assist us in developing new products, changing old products, designing new packaging, and going to market. However, they do not give us the *qualitative* research data needed to understand the consumer's attitudes, motives, needs.

The Information Age was born out of the necessity for more *qualitative* input into consumer research and out of a continuing thirst for knowledge about specific markets for specific products.

In addition, there is a continuing demand for *fact* rather than *opinion* in the endeavor to succeed in the marketplace with a new product or to sustain the life cycle of an existing product in a totally new market environment.

Perhaps there is no greater demand for product fact than in package design because it must communicate with the consumer about the product at the outset. The criterion for successful package design comes from the criteria for successful product development. The two must interrelate and communicate to the consumer through product shape and package copy information.

In essence, a product cannot remain successful unless it serves the consumer on an individual basis and satisfies genuine life style needs. Product labels often fail to speak to the needs of the consumer; rather, they speak from a company point of view.

Package design cannot create consumer needs; the product must do it. The package design can communicate these needs in human benefits and selling points that assure the consumer value for the money.

Life style techniques have been used for years in psychology and sociology. Now they play important roles in marketing, advertising, and product development. Life style concentrates on attitudes and beliefs because life style is an essential ingredient in a consumer's buying decision.

There is no simple way to use demographics from a computer print-out for a package design recommendation when life style is involved. It takes people who understand people to produce and market *life style based* products and package design.

Thus life style is the foundation of today's package design research system. The components of life style research are both *qualitative* and *quantitative* in nature, and the combination creates a more accurate profile of your target audience.

Life style, then, is psychological targeting, or qualitative psychographics. It is an extension of market segmentation by demographics (quantitative). It seeks to describe the human traits of the consumer that may have bearing on products, packaging, and advertising.

The growing importance of life style can be seen in the constant growth of women in the work force, in the increased numbers of the aged, in the young adult consumer with the discretionary income, in more liberated attitudes about life, love, government, entertainment, in new emphasis on different patterns of living throughout the society.

Life style is reflected in the media. Newspapers have new life style sections: leisure, home, garden, entertainment, gourmet, apartment dwellers, singles, and so on. Zip-code and computer mailings provide a marketer with direct, pinpointed response to specific customers. Zoned or regional editions of newspapers and magazines provide the advertiser with specific life style merchandising opportunities in a certain area of circulation within a market. Radio has programmed to life style for many years in its music, news, dramas.

LIFE STYLE SYSTEMS

Dramatic, and often undetected, changes create new challenges in traditional research strategies for new products and package design criteria.

The 1980s and 1990s require techniques that can identify the changing life styles of the consumer. The emphasis on *life style segmentation* is particularly viable now because of the *change factors* in population growth, the working woman consumer, shifts in consumer segments, and the greater emphasis on the individual.

Life style consumer segments in packaged goods generally are defined by all attitudes and behavior areas relevant to the purchase and use of the product involved.

An academic definition of *life style* is attributed to Professor William Lozer of Michigan State University: "Life style is a systems concept. If refers to the distinctive or characteristic mode of living of a whole society or segment thereof. It is concerned with those unique ingredients or qualities which describe the style of life of some cultural group and distinguish it from others."

Life style techniques have been used for years in psychology and sociology. Now they play important roles in marketing, advertising, and product development. Life style techniques are vital in analyzing package design.

Life style concentrates on attitudes and beliefs because life style is an essential ingredient in a consumer's buying decision.

In every marketplace there are segments of the population that represent the potential growth for your products.

Because of accelerating change in how these consumer segments live, marketers and researchers have an increasing need for help in identifying the changes and attitudes within a demographic system. Life style identifications help solve this need.

The people who buy your product are different from those who don't buy it. By identifying your potential market by life style segments, you can design your product and package to appeal to those people who actually buy or are likely to buy your product. If you know the composition of your market, you will better understand how your customer decides whether or not to buy your product and how you can influence that purchase decision through *life style based* package design.

For example, the shampoo market

has been segmented by many factors such as sex, condition of hair, presence or absence of dandruff, and natural versus color-treated hair.

Life style segmentation is being accepted as a strategic marketing tool to define markets and thereby allocate resources more efficiently and effectively.

CRITERIA FOR LIFE STYLE SEGMENTATION

The development of package design criteria requires that three research systems be fed into the research strategy. They include: demographic data (quantitative), psychological patterns (qualitative), and change factors.

Demographics alone can no longer be an accurate indication for marketing and product decisions. Psychographics and change factors have become as important, in some cases more important, than demographics.

Demographic data may tell us a lot about the age, income, education, and family size of prospective customers — but they do not tell us anything about attitudes and living styles. They cannot clearly differentiate between militant feminists and women with traditional values. Consumer sensitivity to environmental protection cannot be clearly detected. Psychographics is aimed at making this kind of distinction and adds a new dimension to the marketing effort. Psychographic analysis helps uncover a combination of different traits not covered by demographics. Included in these would be personal values, product awareness, role perceptions, buying habits, decision making, and communication behavior.

Intermixed between demographics and psychographic data are *change factors*. These change factors are constantly working to change the market and the consumer.

Major change factors are unpredict-

able. They can alter a marketing plan overnight. Change factors can be government oriented, consumer advocated, natural resource ignited, economically or socially responsive. These factors become major market forces when applied to marketing and product viability.

Perhaps no business is more responsive to the changing consumer attitudes based on market forces and change factors than the retailer. The retailer's most valuable asset has been the ability to react to basic and fundamental elements in the marketplace. Some of the change factors or market forces are:

1 The move toward regional selling.
2 The change in consumer age levels.
3 The change in consumer spending patterns.
4 The mobility of the consumer.
5 Faster service at the point of sale.
6 The economic realities of inflation.
7 Consumer credit.
8 Mass communications.
9 The increasing cost of energy.
10 The shortage of energy.
11 Technological innovations.

The accuracy and usefulness of research could be improved if more attention were paid to demographic, psychographic, and change factors at the same time.

Life style segmentation has created "mini-booms" or "mini-markets" throughout the country as consumers change their pattern of living.

Among the mini-booms are the growing singles market, the smaller, more carefully planned family, the increasing numbers of women who work. New life style products reflect the responsiveness of manufacturers as they create products that are attuned to current lifestyles, the mini coffee brewer for the single-person household, for instance.

A mini-boom in the skiing industry

was created when 'cross-country skiing' became a family-oriented leisure activity.

Life style segmentation systems (Fig. 1) are systems that bring together the quantitative, qualitative and change factor data required to understand totally the modern consumer in the humanistic society environment.

Figure 2 demonstrates how demographics, psychographics, and change factors depend on each other in the creation of product strategy aimed at a specific market segment.

Once data are accumulated in each of these systems, they can by computerized, interconnected, and tested against each other to help you make a design decision or when to change the design criteria. These changes may include, for instance, color or copy related to a governmental order.

Table 1 shows specifically how demographic data are extended by qualitative patterns and how change factors further clarify the consumer trend.

QUALITATIVE RESEARCH IN PACKAGE DESIGN

The 1970s have seen a resurgence of qualitative approaches in package design research.

Qualitative research has its limitations in the limitations of its interviewer/moderator. There is a lack of established norms for use in qualitative research, but this is natural, because such research deals with change in behavior and atti-

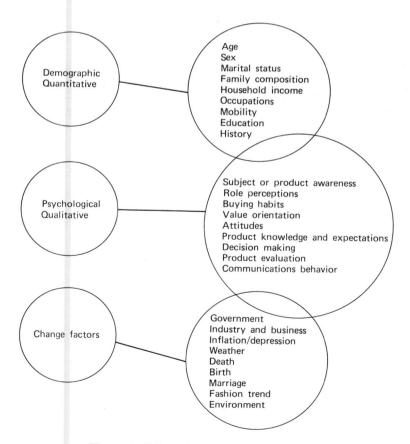

Figure 1 Life style segmentation systems.

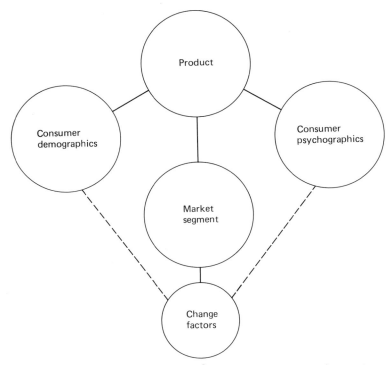

Figure 2 Interconnecting structure of life style segmentation systems.

tude. The interviewer's role is vital. Qualitative research requires skilled marketing talent to be effective. Whereas quantitative research is normally designed in a questionnaire format, computerized, and summarized for reference, qualitative research is a written analysis of the situation, product, or concept creating a *hypothesis strategy*.

A hypothesis strategy is valuable in designing additional quantitative studies. Qualitative and quantitative research must work together to be effective marketing tools.

Qualitative research is a quick and effective way to understand today's ever-changing marketplace.

Depth interviews have now extended

Table 1 Interchanging Structure of Life Style Segmentation Systems

Quantitative	Qualitative	Change Factor
25–34 age $25,000–$40,000 income Married Single homes Children	Leisure life style pattern: Racquet sports	Working women: Two-paycheck households —more discretionary monies
55+ age Fixed income Single & married Apartment dwellings	Leisure life style pattern: Reading Travel Entertainment	Inflation: Less discretionary monies.

themselves to the *focus group* concept as an effective way to assess consumer views about a product, package, service, or idea. Focus groups provide consumer feedback that is necessary to the marketing strategy and helps to prove your hypothesis.

The use of qualitative research, particularly in the form of focus groups, is proving to be the name of the game when coupled with appropriate quantitative procedures.

THE FOCUS GROUP INTERVIEW

Exhibit 1 is a case history of a new product research project that shows how Focus Group Interviewing (qualitative) is used to support quantitative data.

Exhibit 1   Consumer Market Study

Fiber Rich—Focus Report

This study was conducted by Nicholas T. Nicholas, president and marketing consultant, Retail Marketgroup, Inc. The method used was a Consumer Focus Group Session. The sessions were conducted as qualitative marketing research. The focus was on group dynamics, motivation, and product reaction to provide a systematic description in terms of everyday knowledge of the consumer in an informal (nonclinical) environment.

Two groups were selected. One group was composed of eight people *under* 40 years of age, one group of 10 people *over* 40 years of age.

The sessions were videotaped.

The participants revealed in their own words their views of the product category, The Role of Nutrition and Health in Your Life Style. They talked to each other about product-related issues.

The following segments were introduced at the session by Mr. Nicholas:

Segment 1	Subject awareness
Segment 2	Role perceptions
Segment 3	Buying habits
Segment 4	Value orientation and attitudes
Segment 5	Product knowledge and expectations
Segment 6	Decision-making process/change factors
Segment 7	Product evaluation
Segment 8	Communications behavior (television advertising recall)

Objectives

The purpose of the Consumer Focus Group Session was to find out why the product Fiber Rich failed in a Spring, 1977 test marketing program in Oklahoma City, Oklahoma and Fresno, California.

The subject category used in the session was The Role of Nutrition and Health in your Life Style. Specifically, the objectives were as follows:

1   To identify the weaknesses, if any, of Fiber Rich.
2   To identify the strengths of Fiber Rich.

Exhibit 1 Continued

3 Consumer perception of nutrition and health products.
4 Packaging perceptions.
5 Price-value relationship.
6 Consumer benefits of Fiber Rich.
7 Television advertising memory recall evaluation.

INTRODUCTION

Increasing evidence is linking the food we eat with serious medical problems among the U.S. population. Fiber and roughage in the diet is becoming increasingly significant as a way in which to prevent serious sickness.

It is, therefore, important to know the nature and extent of consumer concerns that can have practical, commercial implications for nutrition-health related products in the consumer marketplace.

Consequently, O'Connor Products Company (the marketer,) R. L. Dunn & Company (advertising) and Retail Marketgroup, Inc. utilized consumer focus group sessions to explore people's attitudes toward nutrition and health products. This investigation was designed to provide insight into why the initial test marketing of the product Fiber Rich failed.

DETAILED FINDINGS*

Under 40 Age Group	Over 40 Age Group
1. SUBJECT AWARENESS	
Consumer does not know what fiber is supposed to do.	Consumer knows about fiber and roughage and its importance to general health and longer life.
Laxative and weight-off categories are negative.	Laxative and weight-off categories are negative.
Nutrition and health important considerations, as a general rule.	Nutrition and health vital subjects for continued good health.
Vitamin and mineral category does not play a major role in life style.	Vitamin and mineral category plays a major role in life style.
Natural foods are preferable.	Natural foods are preferable.
Belief in what they eat is best guide to good health.	Belief in what they eat is best guide to good health.
Fast food restaurants do not have nutritious foods.	Fast food restaurants do not have nutritious foods.
Fiber awareness is generally through advertising and articles.	Fiber awareness is generally through advertising and articles.
2. ROLE PERCEPTION	
Against chemicals and preservatives.	Against chemicals and preservatives.

*All tables in this exhibit from Retail Marketgroup, Inc. Consumer Focus Group Session.

DETAILED FINDINGS

Under 40 Age Group	Over 40 Age Group
Concern about children's eating habits. Particularly 'junk foods.'	Concern about children's eating. Particularly 'junk foods.'
Kids play important decision making role in family's food table, particularly in breakfast cereals or fast food eating. Parents concerned.	
Single people are snackers. Eat out often. Many take vitamins as a food supplement. Some concern.	Single people are snackers. Eat out often. Many take vitamins as a food supplement. Some take Vitamin E.
Watch weight—not Weight Watchers.	Watch weight—not Weight Watchers.
Cannot afford to buy food without chemicals and additives.	Cannot afford to buy food without chemicals and additives.
Women tend to decide most often what foods will be served on the dinner table.	Men have a major role in deciding what foods will be served on the dinner table.

3. BUYING HABITS

Under 40 Age Group	Over 40 Age Group
Believes if you want to lose weight, eat less.	Believes if you want to lose weight, eat less.
Consumer forgets to take vitamins on a regular basis. Has fast-paced life style.	Consumer has good habits regarding vitamin intake on a daily basis.
Has some 'roughage' in every meal.	Has some 'roughage' in every meal.
Most products do not inform consumer about fiber or roughage diets.	Most products do not inform consumer about fiber or roughage diets.
Reject products with preservatives.	Reject products with preservatives.
Do not automatically accept what advertising says—read labels.	Do not automatically accept what advertising says—read labels.
Concern about new products on the market. What effect does long term usage have on human body?	Concern about new products on the market. What effect does long term usage have on human body?
New products need to build more confidence.	New products need to build more confidence.
Loyal purchasers of brand names in nutrition and health products.	Loyal purchasers of brand names in nutrition and health products. But will try an unknown product if price is lower and ingredients the same.
Coupon incentives are important.	Coupon incentives are important.
Bran is important in diet.	Bran is important in diet.

DETAILED FINDINGS

Under 40 Age Group	Over 40 Age Group

4. VALUE ORIENTATION AND ATTITUDES

Most people are on food budgets.	Most people are on food budgets.
Price is relative to value. Will pay more for natural food products if usage is major.	Price is relative to value.
Fiber Rich is too expensive.	Fiber Rich is too expensive.
Fiber Rich 3-A-Day plan is negative. Consumer will not remember to take before each meal.	Fiber Rich 3-A-Day plan is negative. Consumer relates to price-value relationships.
Most consumers would not consider product—at any price.	Consumer will consider product if priced around $3.50 price range.

5. PRODUCT KNOWLEDGE AND EXPECTATIONS

How much fiber can you get in a small tablet? Not believable to consumer.	Understand fiber addition to vitamin and mineral.
Do not believe in dietary plans. Believe dietary means weight watching.	Do not believe in dietary plans. Believe dietary means weight watching.
Fiber Rich dietary plan is for someone who wants to lose weight.	Fiber Rich dietary plan is for someone who wants to lose weight.
3 grams of fiber per day does not seem like very much fiber.	3 grams of fiber per day does not seem like very much fiber.
	Expect good health and longer life from nutrition and health products.
Not familiar with Fiber Rich product. Not familiar with any product with fiber in it, except All Bran.	Not familiar with Fiber Rich product. Not familiar with any product with fiber in it, except All Bran.
Dietary category is oversaturated with claims. Do not believe.	Dietary category is oversaturated with claims. Do not believe.

6. DECISION-MAKING PROCESS/CHANGE FACTORS

Personal doctor is a major influence. But do not see doctor more than twice a year.	Personal doctor is a major influence. See doctor more than twice a year.
Belief: most doctors do not know anything about nutrition.	Doctors will discuss nutrition if you ask them about it.
Food and Drug Administration lacks believability to the consumer.	Food and Drug Administration lacks believability to the consumer.

Under 40 Age Group	Over 40 Age Group
Medical evidence must be accompanied by authority.	Medical evidence must be accompanied by authority.
Media has the most influence on medical information and products. Magazine articles important.	Media has the most influence on medical information and products. Magazine articles important. Look to newspaper food sections for data. Magazines in doctor's office.
Product labeling is vital decision factor.	Product labeling is vital decision factor.
Very conscious of product ingredients.	Very conscious of product ingredients.

7. PRODUCT EVALUATION

Basic packaging is attractive.	Basic packaging is attractive.
Dietary plan copy turns off consumer from total product.	Dietary plan copy turns off consumer from total product.
Package bottle does not look medical.	Package bottle does not look medical. Should be darker color.
Packaging makes product look like 'plant food.'	Packaging makes product look like 'plant food.'
Do not believe quotation on package front " . . . increasing medical evidence . . ." Consumer wants to know who said it.	
Fiber with every meal is not believable or understood.	
If One-A-Day added fiber, they would create a new product (i.e., mineral with fiber or vitamin with fiber, etc.).	
Would not explore product usage— whatever the price.	Would explore product usage if price were lower.
33 day supply—too expensive at $6.95.	33 day supply—too expensive.
Fiber Rich story does not promise anything to consumer.	

8. COMMUNICATIONS BEHAVIOR/TELEVISION ADVERTISING RECALL (FIG. 3)

Skeptical about products on television.	Skeptical about products on television.
Offensive.	Offensive.

FIBER RICH

30 SECOND TELEVISION COMMERCIAL

Eating a lot doesn't mean...

my family gets the proper nutrition...

So I supplement our diet with Fiber Rich.™ Fiber Rich is new!

It has up to 200% of the U.S. Recommended Daily Allowance of...

essential vitamins and minerals. But, Fiber Rich has more...

10 times more fiber then the same amount of bran.

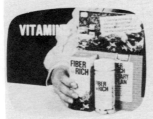

Vitamins....

Minerals.....

and the important addition of fiber!

I want all three! That's what your family wants....

and that's what they get in Fiber Rich.

BOOTH ANNCR: (Available for local dealer tag)

Figure 3

108

Figure 4

DETAILED FINDINGS

Under 40 Age Group	Over 40 Age Group
Fiber Rich television commercial is too brief. Does not sell fiber benefit.	Fiber Rich television commercial is too brief. Does not sell fiber benefit.
Up to 200% copy is misleading. Consumer has been conditioned to 100% minimum daily requirement.	Up to 200% copy is misleading. Consumer has been conditioned to 100% minimum daily requirement.
Television commercial gives more information than package front.	Television commercial gives more information than package front.
Television copy: "10 times what you need in bran" is misleading. How much bran?	Television copy: "10 times what you need in bran" is misleading. How much bran?
Magazines have most important credibility in new product information. Good Housekeeping.	Magazines have most important credibility in new product information. Good Housekeeping.
The commercial reflected a large drug company.	The commercial reflected a large drug company.
Dietary Plan on package front, shown in commercial, is a rip-off and come-on.	
Single people do not relate to people in commercial.	Do not relate to people in the commercial.

CONCLUSIONS AND MARKETING IMPLICATIONS

The consistent negative and positive responses are particularly critical in considering any marketing, distribution, packaging, sales and advertising ramifications of Fiber Rich (Fig. 2.)

Several conclusions have been suggested by this qualitative study:

1 *Plus-Forty Market*   Fiber and roughage are important to the (over) 40 age group. Fiber Rich can be a viable product if it can be positioned to the plus-forty market instead of to the young woman 24 and over. (There are over 23 million consumers over the age of 50 in the United States today.) However, marketing and communications strategy must include the following:

- A relationship between the food people eat and fiber supplement.
- Strong public relations and fiber information strategy.
- The use of newspapers and magazines as prime media sources. The use of television as a support medium.
- The creation of consumer-information point-of-purchase displays.
- Repositioning of the product itself in relationship to price structure.
- The merchandising of the enclosed 'dietary plan' booklet in a more aggressive manner.
- Develop a sampling system through doctor-patient relationships.
- Develop a 'trial offer' program for Fiber Rich.
- Determine life cycle of Fiber Rich in relationship to its positioning level in the marketplace.

2 *Fiber Rich Packaging*   Basic copy changes are required in the Fiber Rich packaging in order to make it a viable product to the (over) 40 age group.

- Remove reference to 'dietary plan.'
- Review the recipes in the booklet.
- Stress vitamins and minerals stronger.
- Use a quotation which has more authoritative influence.

3 *Advertising*   Develop advertising strategies that inform the consumer about fiber benefits. This can best be done in print media, using television and radio advertising as a reinforcement.

SUMMARY

Discussions in reference to people's feelings about nutrition and health in their life styles, and particularly high fiber, bulk, and roughage in daily diets, were conducted during a Consumer Focus Group Session.

The session was conducted using two separate groups of consumers (under 40 years of age and over 40 years of age) to determine how each group perceived nutrition and health in their life styles and how, in particular, they might react to the new product called 'Fiber Rich,' a vitamin and mineral food supplement with natural fiber concentrate.

Asked to rate the importance of three nutrition-health categories, that is, weight-off, laxative, and multiple vitamin and mineral, the consumer groups strongly identified with the vitamin and mineral category as the most consistently important one in their daily life style patterns.

Fiber Rich was identified more strongly with the vitamin and mineral group. However, there were many extreme concerns about Fiber Rich as a marketable product in its present form and content.

There is generally great concern about nutrition and health products being sold in the marketplace. Recent government findings and publicity surrounding medical findings have caused the consumer to become extremely cautious about nutrition-health products.

One of the most significant finds of this qualitative study was the discovery that Fiber Rich is more strongly identified as a viable product with the over 40 age group, men and women combined. The initial test marketing of Fiber Rich was aimed at women 25 and over.

Fiber Rich received a basically negative response from the consumer groups. The most important reasons were (1) price, (2) package copy content, (3) intake requirements (three per day).

The brand name Fiber Rich was considered

good. However, there was concern about who the manufacturer was, and the consumer seemed to seek an important identification between the product and the producer.

Some of the most consistent findings of this Consumer Focus Group Session were:

1 *Knowledge of Fiber*   There was an important lack of knowledge and information about fiber. The over-40 group identified with it, however, more clearly than the under-40 group. The common response to fiber was that the consumer was aware that certain foods, such as fruits and cereals, contained sufficient fiber for their normal requirement. Both groups believed that what they ate determined their good health, not what supplement was used daily.

Kellogg's All Bran cereal was the most commonly mentioned source of good fiber, yet the consumers did not understand how much fiber was needed or how fiber really relates to their total eating habits.

The most important sources of information for fiber, and in particular, nutrition and health products, were newspapers and magazines. Authoritative articles were mentioned as the most convincing source of information. The *Reader's Digest* articles about fiber were mentioned often. In addition, Dr. Ruben's book, *Save Your Life Diet* was referred to often in the discussions in both group sessions.

2 *Fiber Rich Dietary Plan*   Both consumer groups felt strongly that this copy was misleading. The copy related the product more to weight-off than supplement. The word 'plan' was a negative point with most consumers because it represented a strict regimentation of their daily habits.

3 *Shopping Habits*   The consumer is more likely to buy nutrition/health products in a supermarket than any other retail outlet. The supermarket provides a good point for comparison of nutrition-health products. Labeling is a prime consideration for information.

The consumer would like to purchase health food products with natural ingredients, but the high price level causes the consumer to seek more value for the money.

When fruits and vegetables are out of season, the consumer will consider vitamins and minerals to supplement diets.

4 *Price Relationships*   Most families in both age groups work with a budget. Price is relative to value. The Fiber Rich price of $7.95 was considered too high in relationship to the amount of daily intake required. The consumer would be willing to try Fiber Rich is the price were lowered to the $3.50–$3.95 level.

5 *Fiber Rich Packaging*   The total packaging concept is very good and delightful. The major problem with the packaging is the copy, particularly the words "dietary plan," and " . . . there is increasing medical opinion that dietary fiber is useful in maintaining good health." The consumer does not believe the quote. They want authoritative comments they can believe.

The booklet, which contains fiber information as well as recipes, is a good combination element to the product package. Most consumers, however, felt that the recipes contained too much fiber foods and that there would be no need for a supplement like Fiber Rich.

6 *Product Acceptance*   Fiber Rich was generally not accepted as a viable product by either age group. The most significant reasons were (1) believability, (2) price, (3) intake formula (3 per day).

The over-40 age group, however, was more willing to explore the product and had more need for it. This group would try the product if (1) the price were lower or (2) the doctor recommended it.

The under-40 age group is against any type of additives and preservatives in their diets and is therefore basically against any form of pill intake. They do not feel a need for dietary supplements. Natural foods are more valuable to this group.

7 *Television Advertising*   The Fiber Rich television commercial received a high level of general creative acceptance. It reflected the company behind the product as a large drug company. The most negative response referred to the copy: "It has up to 200% of the U.S. recommended daily allowance of essential vitamins and minerals. But, Fiber Rich has more . . ." The consumer is conditioned to believe "up to 100% of U.S. daily allowance . . ."

PART TWO

The Tools of Package Design Research

CHAPTER EIGHT

Using Focus Groups in Packaging Research

Sanford G. Lunt

The role of packaging is becoming more important in the marketing mix for a successful product whether it be old or new. The horrendous budgets needed to develop a new product or reposition an old one today have made it necessary for smart marketers to learn as much as possible about their ideas before they commit enormous amounts of development dollars from corporate funds.

In many instances the risks are so great that testing tools have to be utilized to help minimize this risk. This is one of the major reasons that people involved in packaging are becoming more research oriented and adopt research techniques from other marketing areas for their own use.

One such technique is the use of the focus group. It is a research tool that differs markedly from other forms of research, primarily because of its sample size and its motivational attributes. Focus group interviews are free flowing, relatively unstructured discussions by small groups of individuals guided by a moderator. The focus group interview is one of the most often used, most often misused, and most controversial form of consumer research. It is, however, a research tool that is capable of providing usable results both quickly and inexpensively. When used properly, focus groups can be most helpful in developing an aesthetically pleasing, communicative, and functional package.

HISTORY

There is evidence that interest in the group interview goes back over 50 years.[1] The incorporation of the focused interview into the group situation took place in the early 1940s. The focused interview was developed to meet certain problems concerning communications research and propaganda analysis.

During World War II, Dr. Herta Herzog and Paul F. Lazarsfeld were assigned by several war agencies to study the social and psychological effects of specific efforts to build morale. This work was based on research done to find out what people liked about radio programs. Groups as large as 300 people were assembled to witness a broadcast or rebroadcast of a specific program. Likes or dislikes were recorded on tally sheets at given intervals during the course of the program, or opinions were indicated on a mechanical device such as the Lazarsfeld-Stanton program analyzer.

After the broadcast, participants would be interviewed to obtain reasons for their reactions. They were then broken up into small groups, and it was in these groups that the focused interviews took place. During the course of this work, focus group interviewing was developed to a relatively standardized form.[2]

DEFINITION

Focus group interviewing is a technique for probing the thoughts, feelings, and emotions of the individuals involved. Thomas Lea Davidson, in an article in the *American Marketing Association Combined Proceedings in 1975,* very ably defined a focus group interview. He said that a focused group interview is a qualitative tool for collecting information, in which a number of respondents simultaneously discuss a given topic under the guidance of a moderator. It is not a measuring device, not a way to generate numbers of any sort. It is instead a way to learn, to gain understanding, to search for a meaning.

Focus groups are a tool to explore, to diagnose, to gain an impression of what is going on. Most important, they are a way to learn, not measure. They are a real experience with real people.[3]

VALIDITY

One of the biggest concerns voiced by critics of the focused group interview is the validity of the technique. Focus group reports often start with a disclaimer: "Qualitative research is exploratory in nature. Findings should not be considered conclusive or projectable." It is concern with the reliability of the sample size, the objectivity of the moderator, and the effective interaction of the group that has caused concern about the "validity" of focus group interviews.

There are even those who say that there is no theoretical basis for this technique in psychological literature. It was thought to be unscientific and therefore unreliable. The use of the disclaimer resulted. But such disclaimers are unfortunate because they cast aspersions on the findings of the study, and in most instances are unjustified.

In research, validity is usually defined as the ability of a technique to measure what the researcher thinks is being measured. Its value—or validity—comes from learning something that is useful about how a few people talk, think, and feel. Validity of focus groups should be based on the results of the session—whether something useful was learned.[4]

Group interviewing has survived its

critics because it has important assets that allow it to compete favorably with other forms of research. The important thing is to learn its strengths and weaknesses and to bear them in mind when using the technique.

BASIC ELEMENTS OF FOCUS GROUP INTERVIEWING

The basis for group interviewing stems from the fact that people think, act, and behave in social settings. It occurs in group situations where people can feel at home and be themselves. Here views and feelings that are closely related to actual behavior merge easily into speech. Most people are accustomed to verbalizing, and it is when we talk, and take other's reactions into account, that we get to know what we really think.

In the group situation a person is asked an opinion about something—a product, a distribution system, an advertisement or a package. In contrast to the individual interview, in which the flow of information is unidirectional (from the respondent to the interviewer), the group setting allows the opinion of each person to be considered in the discussion. In the individual interview, the input is stimulated only by the interviewer. In the group interview, each individual is exposed not only to the stimulus of the moderator but to the ideas of the other group members as well. He also has the opportunity of submitting his ideas for consideration.

There are, however, two basic elements that must be present before a group interview can be meaningful. One, and possibly the most important, is that the individuals in the group share a common interest that is relevant to the topic under consideration. The composition of the group can be comprised of individuals with divergent backgrounds. Religion, nationality, or occupation for example could have no bearing as long as a common interest in the subject matter for discussion is shared (i.e., pipe smokers, headache sufferers, or auto parts distributors).

The second basic element is that there must be a social interaction at some overt level among the members of the group. They must react to one another so that there is an interchange of ideas. If all of the remarks are directed to the moderator, then the advantages of the group setting are precluded.

ADVANTAGES

What are some of the results that can be generated by the group process but may never be achieved in the individual interview? First, the interaction among group members stimulates new ideas about the topic under discussion that might never come to light in individual interviewing. There is a chain reaction in the group, as well as an insight into ideas that an individual might never have thought of and consequently could not have commented on. The group interview also provokes an activating of forgotten details.

A second value of group interviewing is the opportunity to observe directly the group process. This is evident when individuals in the group react to each other. For example, a housewife who dislikes serving leftovers and admits that she does so under social pressure is supported in her feelings by other members of the group. Those in support "turn on" the remaining members who serve leftovers and mask them with gravies. Here the attitudes of women toward serving leftovers would be reflected in the way they behaved toward each other in the group.

A third advantage is that it provides some idea of the dynamics of attitudes and opinions. This is especially found in the flexibility or rigidity of an opinion when it is exposed to a group and there is the chance for two sides to be heard. Within the time span of a group session, an opinion that is stated with finality and apparent deep conviction can be modified a number of times by the social pressures or new information that may be provided by the group.

Discussion in a peer group can often provoke considerably greater spontaneity and candor. It is not unusual for group members to ignore the presence of the moderator completely. Candor is permitted not only because the members of the group understand and feel comfortable with one another, but also because they draw social strength from each other. Participants also try hard to contribute.

A fifth advantage is that the group setting is emotionally provocative. For example, a discussion of the weight of housewives ranging from 25 to 45 in age may be very difficult for the moderator to broach. But in a group interview, one of the younger and most slender members said "Weight is not a problem for me yet, but I imagine that for older women like yourselves it could be." This provoked a rather emotional discussion of growing old and weight gain that proved quite informative in eventually positioning a new product concept for weight watchers.[5]

Another advantage is that the technique is flexible in format, giving the moderator the opportunity to pursue pertinent avenues of information that were not in the original outline. This is an advantage over the individual interview which follows a formal, rigid questionnaire.

Focus group interviews are also timely. They can be conducted quickly and inexpensively when there is a need for basic information. The results are easy to understand as well because they are played back in consumer information and there is the opportunity to probe in depth for a complete response.

A final attribute of focus group sessions is that the marketer can participate directly in the research process. He can watch "the consumer in action" through a two-way mirror, via closed circuit television, or by sitting in the room with the group. He or she has the opportunity to ask questions directly if in the room, or by sending in notes to the moderator if watching from outside. Here questions that were not thought of prior to the group session can be answered immediately.

DISADVANTAGES

Now let us look at the other side of the coin, the disadvantages of focus groups. Here are some negative aspects that should be considered when contemplating the use of this research tool.

Possibly the biggest negative is that groups rarely constitute a satisfactory sample. The number of respondents is small, and in most cases the respondents are not selected on a probability basis. Usually twelve or more people are invited to participate. This is done to take into account dropouts, those that do not show up for one reason or another. Eight to ten is considered in most cases to be a workable number of individuals. There are even some researchers who believe that five or six is best. They feel that the smaller the group the greater the opportunity for each respondent to express his or her opinion.

Because the group is small, focus groups should never be used when the information to be collected is presented as being representative of anything— users of a product, nonusers, or a given market. Focus groups do not produce hard facts or projectable data; they are not definitive. They can provide insights or clues into the nature, but generally not the extent, of an attitude or reaction. Quantitative results are almost impossible to obtain and for this reason focus groups should not be used alone to influence business decisions.

Even if a large series of groups is conducted to build the numbers up, quantitative findings are not possible. The nature of the focus group with its unstructured question format inhibits such findings. The interaction of the individuals in the group can shift the direction of the questioning, making comparisons of a given point difficult to combine. It is only when the context of the entire group is taken into consideration that the findings are meaningful.

Furthermore, in a given group some people talk too much or are overly opinionated or domineering (the "leader syndrome"). Others are inhibited and say very little. Both have their effect on the benefit of group dynamics. As a consequence a group may yield only controversial or superficial data. A good moderator, however, can minimize this problem.

Another aspect that is considered to be negative is the inability, because of time limits, to probe in depth into an individual's motivations. If the interview lasts for 2 hours and there are 10 members in the group, for example, this leaves only 12 minutes time to spend with each individual, so that intensive exploration of each person is not feasible. This makes a case, as noted before, for smaller groups.

As a good moderator can minimize problems, a poor one can virtually eliminate the effectiveness of the group. If the moderator interjects the wrong stimulus, the net effect is response that is not relevant to the objectives set forth for the group. The moderator is one of the major elements in the success or failure of focus group interviewing, and he or she must be chosen with care.

An unusual but important disadvan-

tage could be management's involvement at the actual session. In too many instances conclusions are drawn from one person's comments. Because those in the group are real consumers or of a special interest segment, their comments are construed as being gospel. It is very easy for those listening to hear what they want to hear and, as a result, misuse the information. This reaction to the responses by group members is not the best use of the technique.

Because focus groups are relatively inexpensive, easy, and fast to do, there also is a tendency to conduct them instead of quantitative research. The cheap, "quick and dirty" focus group under the guise of "a little information is better than none" can be the most misleading information available to influence important decisions. This can lead to improper positioning, overly optimistic sales goals, and, in the case of packaging, poor direction for graphic and functional design.

USES OF FOCUS GROUPS

What then are the most productive uses of focus group sessions? These sessions can be used to:

1 Generate hypotheses, insights, or cues for confirmation at a later date with quantitative research.

2 Unearth unfulfilled needs and problems when little is known about the subject.

3 Eliminate bad ideas, or find negatives.

4 Screen for potential target audience.

5 Add to and build the knowledge base of information, and provide consumer insight.

6 Create new ideas and generate a broader perspective about them.

7 Learn consumer language or terminology.

8 Obtain information from an interaction among the participants in the group.

HOW TO PREPARE FOR A FOCUS GROUP SESSION

In order to get the most out of a focus group session it is imperative that all the participants in the project fully understand the problem that is to be discussed. This is usually done by establishing a series of goals that are clearly and concisely put into writing. They serve as guidelines and help ensure that the session stays on target. They are used to determine the purpose and objectives of the research.

The next step is to prepare a moderator's guide or a topical outline. This is done by establishing a list of questions that the research is to answer. These questions are then structured sequentially to fit into a logical framework for the moderator's use in leading the discussion. The list can be in the form of notes, questions, or an outline to ensure all predetermined points are covered in the session.

After the goals and objectives have been set it is most important to determine who should be included in the group. Obviously this depends on the purpose of the group. If the package in consideration is a revision or improvement, then the group could consist of users of the product. If the purpose is to attract new customers, then the users of competitive products, or nonusers of the category, could be used. The selection of the members of the group is critical to its success; thus a great deal of care should be used in choosing the group.

Recruiting the members of the group can be done in many ways. You can do it yourself through local clubs, organizations, and church groups, for example, or you can get professional help. Most large cities and many small ones have

market research firms that have interviewers on their staffs. Some even specialize in group sessions. They can recruit a group that will meet your specifications. Normally a screening guide is developed and given to the research firm. Let's assume that Whitehall Laboratories is considering a package design change on Anacin tablets. They are looking for users of pain relievers, specifically Anacin users. A screening questionnaire is then prepared. It is relatively simple in format so that the screening costs are kept to a minimum. A typical screening questionnaire for Anacin users, for example, could be as shown in Exhibit 1, p. 118.

Recruiting is usually done 2 to 3 weeks before the session but can be accomplished in 1 week if speed is important. Where possible, a follow-up letter with details is sent to the respondent, and a confirmation call is usually made a day or two before the group is to meet. In most cases an inducement is offered to the respondent. This can range from cash (ten to fifteen dollars for consumers and up to fifty dollars for professional people) to prizes or the promise of refreshments.

Next, where should the session take place? Usually a market research firm specializing in focus groups has facilities for such sessions, a special room with a two-way mirror or closed circuit television. In other instances someone's house, a motel room, or a business office can be used. Any of these are fine as long as the people in the group are at ease in the setting.

Another very important aspect in the group session is the selection of the moderator. He or she should have excellent credentials in conducting group sessions, as well as a thorough understanding of the reasons for the session and what is expected from it. All too often the objectives of the group session are not achieved because no agreement

Exhibit 1   Screening Questionnaire

1   Do you or any member of your family work for an advertising agency, manufacturer of drugs or toiletries, or a market research firm?

        Yes ( )   No ( )

        (If yes, terminate interview)

2   Which of the following products have you used in the past three months?

Mouthwash	( )	Hair spray	( )
Deodorant	( )	Toothpaste	( )
Pain reliever	( )	Shampoo	( )

3   Which brand of pain reliever do use most often?

Bayer	( )	Tylenol	( )
Anacin	( )	Excedrin	( )
Bufferin	( )	Other	( )_____

        (Please specify)

4   In which of these age groups do you fall?

18–24	( )	45–54	( )
25–34	( )	55 and over	( )
35–44	( )		

5   Sex?

        Male ( )   Female ( )

between moderator and client was reached prior to the start of the session.

The following are some of the qualities that should be looked for in selecting a good moderator:

1   The ability to establish a rapport with the group quickly and unobtrusively.

2   The use of nondirective approaches that permit participants the opportunity to fully express their opinions.

3   Ability to control the group in an atmosphere that allows candidness on the part of the participants to generate productive information.

4   Flexibility to depart from the topic outline to maximize the potential of the group.

5   Ability to focus on key issues and minimize the discussion of extraneous matters.

6   Proper budgeting of time to get the most out of the session.

The role of the moderator is to establish a relaxed, open atmosphere at the beginning of the session; this encourages interaction among the respondents. The participants should feel that it is *their* group and that the moderator is genuinely interested in their ideas and comments.

Groups usually start by the moderator announcing who he is, describing his company, explaining the purpose of the group, and telling why it is important to get the help of the participants. The next step is for each person to introduce and tell a little about himself or herself. Name cards are placed in front of each person so that the moderator and other members of the group can call each other by name. The moderator then goes through the session using his outline as a guide. Finally, at the end of the session, the moderator summarizes the findings and asks the group for confirmation of what he said.

A group usually runs from 1½ to 2 hours. Longer sessions are possible, if warranted. The session is usually tape recorded or a stenographer is on hand to take notes. Some groups are even videotaped so that management can get a better feel of the session. An edited version is prepared that presents a succinct format, eliminating the portions that are unproductive.

Once the session is over, it is usual for the people involved in the project who were attending the session to sit down with the moderator to discuss his impressions and reactions while the session is fresh in his mind. The moderator then usually prepares a detailed report of his findings and conclusions.

Now that the group session is completed, a logical question is: How many groups are enough? Unfortunately there is no *right* answer to this question. Geographic differences, age, sex, ethnic background, and income levels could make a difference. The budget available for this form of research will be a dominant factor in determining how many sessions can be conducted.

The results of the groups are also a factor. In general, if four or more are conducted and similar results are obtained, then enough sessions have been done. If after these sessions radical differences are found, then it is time to stop, reevaluate the procedures, and do several more, altering the format.

APPLICATION TO PACKAGE DESIGN RESEARCH

Now let us see how focus groups apply to package design research. The following are two examples of projects that were recently completed for Abbott Laboratories on Selsun Blue, an antidandruff shampoo, and on a new product called Selene, a hair conditioner especially designed for people who use an antidandruff shampoo.

An extensive series of market research studies were conducted by the Abbott Laboratories' Consumer Products Division to attempt to find ways to broaden Selsun Blue's position in the antidandruff shampoo market. The concept testing phase revealed an opportunity for the brand to become the only leading antidandruff shampoo to come in three formulations (dry, oily and normal), so that the consumer could choose the one that was right for his or her particular hair type.

It was imperative to find out whether such a change would tarnish the favorable therapeutic image of the brand, create any confusion on the part of the consumer, or possibly enhance the brand's status as a specialty product in the eyes of dandruff sufferers. Packages specifying each type were designed, and the focus group was used to determine consumer reaction. Focus group sessions involved groups of current users of the brand, as well as users of competitive products, broken down into different age brackets.

The topic outline used by the moderator was designed to learn about each participant's usage of dandruff shampoos and reaction to the new packages. (See Exhibit 2.) During the session a questionnaire was passed out to the participants to obtain unbiased reactions to the designs before the discussion by the group. These questionnaires were collected by the moderator and helped in probing for in depth reactions to the designs. They were never tabulated nor used in any quantitative manner. They were simply an aid to help make the group more meaningful. (See Exhibit 3.)

The response to the concept of three formulations was overwhelmingly favorable. It actually enhanced the image of the brand and gave further reason to buy Selsun Blue. Not only was this reaction

Exhibit 2   Selsun Blue Dry, Oily, or Normal (DON)
Focus Group Sessions

I.   Introduction
    a.   Warm-up
    b.   Explanations
II.   Respondents background information
    a.   Type of hair
    b.   How long using dandruff shampoo?
    c.   Other brands used
        1.   Cosmetic (Dry, Oily, Normal)
        2.   Dandruff
    d.   Have ever heard of Selsun Blue?
        1.   What heard about Selsun Blue?
        2.   Where learned about Selsun Blue?
    e.   How often shampoo?
    f.   Likes or dislikes with products they are currently using
    g.   Reaction to hair after shampooing
    h.   Frequency of shampooing
    i.   Number of latherings per shampoo
    j.   Conditioner or rinse usage
III.   Show Dry, normal, and oily bottles
    a.   Fill out questionnaire
    b.   Discuss DON
        1.   How perceive the three?
        2.   Which one, if any, for them?
        3.   Strengthens or weakens Selsun Blue therapeutic image?
        4.   Any one of DON formulas more or less effective?
    c.   Read description for each type
        1.   Was this how perceived?
        2.   Are there any negatives?
IV.   Show Selene package
    a.   Fill out questionnaire
    b.   Discussion of what bottle says

---

obtained from current users of Selsun Blue, but users of competitive products wanted to try the new formulations as well.

These particular focus group sessions were also used to obtain information about another new product that Abbott was considering introducing, Selene, a hair conditioner especially developed for people who use antidandruff shampoo. Earlier research revealed some confusion in regard to its association with Selsun Blue. A set of communication objectives were developed to be sure that the important elements of the package were emphasized in the session. It highlighted the brand's emphasis on quality, unique formulation, and its heritage. (See Exhibit 4, p. 122.)

The questionnaire format was also used in this portion of the focus group session to help the moderator to obtain once again each participant's reaction prior to the group discussion. (See Exhibit 5.) This phase of the group session revealed that some changes were warranted in the emphasis of the elements of the package design. These changes

Exhibit 3   Selsun Blue Questionnaire

1   Which one of these products is most interesting to you personally? (Check one.)
   _____ Selsun Blue for dry hair
   _____ Selsun Blue for oily hair
   _____ Selsun Blue for normal hair
   _____ None of these

2   If one of these Selsun Blue products is interesting to you, please indicate how you feel about buying it.
   _____ Definitely would buy it
   _____ Probably would buy it
   _____ Might or might not buy it
   _____ Probably would not buy it
   _____ Definitely would not buy it

3   Why do you feel that way?   _____
   _____
   _____

4   If none of these products is interesting to you, please say why.   _____
   _____
   _____

5   How do you think the Selsun Blue product you are most interested in is different from other dandruff shampoos?
   _____
   _____

6   How important is this difference to you?
   _____ Very important
   _____ Quite important
   _____ Not very important
   _____ Not important at all

7   Please check your hair type and age group.
   Dry ( )   Normal ( )   Oily ( )
   Under 25 ( )   25–34 ( )   35–44 ( )   45–54 ( )
   55–64 ( )   65 and over ( )

resulted in a package change prior to the brand's ultimate national introduction.

These two specific examples of uses of focus groups concerned the image projected by the package design and the problems and how they were resolved. The following are some other areas in which focus groups can be helpful in determining the effect of packaging in the overall product concept:

It is possible to learn how others see your package. Find out if consumers have the same or a different image of your package from the one held by management. You can also learn how your package compares to competitive packaging. Even in mock-up form various designs can be evaluated, and this can be done alone or in a competitive situation.

In most instances, the product concept has been established before work is done on tbe package design. Focus groups can provide the opportunity to see whether the package makes its contribution to the image sought for the

Exhibit 4	Selene Package Communication Objectives

1   *Brand Name*   Communicate brand name.
2   *Generic*   Communicate that Selene is an instant hair conditioner. This is particularly important in light of.
    a. Past confusion between Selsun Blue and Selene.
    b. The fact that we are using the same bottle as Selsun Blue.
3   *Quality*   Communicate that we are a premium product worthy of the premium price that we are going to charge.
4   *Speciality*   Communicate to dandruff shampooers that we are a special product designed for them.
5   *Heritage*   Communicate that Selene is from the makers of Selsun Blue.
6   *Unique Formulation*   Communicate those semi-unique characteristics of the product's formulation that, when combined, do indeed make us unique.
    a. Pro-vitamin enriched.
    b. Oil-free.
    c. Protein enriched.
7   *Product Description*   Communicate the product's size and type.

product. This can be done by learning about:

1   *Shape of the Package Without Graphics*   What does it communicate?
2   *Visibility of Key Elements*   Flash package or packages on screen and see what is played back. What does design mean? Does it fit into the image being created by product concept?
3   *Standing Out in a Competitive Crowd*   Run respondents through a mock display area. See what is remembered about the package.
4   *Communication of Ideas*   Words to use and what they mean.
5   *Color*   What it connotes.
6   *Typeface*   Legibility.
7   *Functional Aspect*   Does it work? Does it open? Safety cap work? Are directions clear?

These are only a few of the many things that can be learned from focus groups. They will provide clues, uncover negatives, and give direction for the next steps in finalizing the package functionally and graphically.

CONCLUSIONS

Focus group interviews can be a valid research technique in packaging research. They can provide valuable information to help make decisions to develop and refine good package design approaches. They are a tool to explore, diagnose, and gain an impression of what is going on. They can give insight into what can be expected from the user of a product or service.

Despite all this, group interviewing does have its critics. They claim that samples are too small and not selected on a probability basis. Moderators can have a biasing influence and affect the results of the study. The individuals in the group can talk too much or too little and affect the interaction of the group. Results are difficult or impossible to

Exhibit 5 Selene Questionnaire

1 Based on the bottle that you have just seen, how
   interested are you in buying this product?

Definitely buy the product.   ( )
Probably buy the product.   ( )
Might or might not buy the product.   ( )
Probably not buy the product.   ( )
Definitely not buy the product.   ( )

2 Why do you feel that way? _____
   _____
   _____

3 Please tell us in detail what you think this product
   is for? _____
   _____

4 Do you think there is anything special about the
   product?
             Yes ( )   No ( )
   If yes, which? _____
   _____

5 What if anything do you like about the product?
   _____

6 What if anything do you dislike about the
   product?
   _____

7 Do you use a conditioner or rinse?
             Yes ( )   No ( )
   If yes, which?
   _____

8 How often do you use a conditioner or rinse?

9 Please check your hair type and age group.

       Dry ( )   Under 25 ( )
    Normal ( )   25–34   ( )
      Oily ( )   35–44   ( )
                   45–54   ( )
                   55–64   ( )
                   Over 65 ( )

quantify. These are justified criticisms that must be taken into consideration when using this technique.

It is important to understand not only these weaknesses, but the strengths of focus groups as well. If used properly, they can be very productive. Start by following good research practices. Know and understand your objectives. Be sure that they are concisely and simply put into writing. Get all those involved in the project to contribute to the topic outline, so that all bases are covered in the study.

The moderator is the key element and should be chosen carefully. He or she should be thoroughly informed of your goals and objectives so that they can be implemented accurately. The recruiting of the group members is also extremely important. The site for the session must be chosen with care as well so that the group members feel at ease. Those watching the session should take part, if possible, with direct questions or notes to the moderator to stimulate reaction to new insights generated by the session. Tape record or videotape the session so that nothing is lost for the analysis of the findings.

Remember that focus groups are a qualitative research technique. They provide the opportunity to probe in depth. Their major asset is the interaction that occurs among the participants in the group, something that does not happen in other forms of research. It is this interaction that can get to the depths of the problem and provide insights that are invaluable in creating effective, functional packaging. Focus groups help solve problems quickly and inexpensively. They are a way to learn, not measure. If you understand their limitations as well as their assets, you can use them to great advantage.

REFERENCES

1   E. S. Bogardus, "The Group Interview," *Journal of Applied Sociology,* **10** 1926, 372.
2   "What Do We Really Know About Daytime Serial Listeners?" by Paul F. Lazarsfeld and Frank N. Stanton (Eds.), *Radio Research,* Duell, Sloan and Pearce, New York, 1944, pp. 1942–43.
3   Thomas Lea Davidson, "When If Ever Are Focused Groups A Valid Research Tool:" American Marketing Association, 1975 Combined Proceedings (1975), p. 141.
4   Ibid, p. 141.
5   Alfred E. Goldman, "The Group Depth Interview," *Journal of Marketing,* July 1962, p. 61.

CHAPTER NINE

Using Tachistoscope, Semantic Differential and Preference Tests in Package Design Assessment

Donald Morich

A consumer package normally goes through many tests before it gets out of the design laboratory and onto the production line. A great deal of time, energy, and money is spent in completing routine tests to insure the package physically works for the product (i.e., it holds the exact amount of product required, it protects the product itself up to acceptable standards, it withstands certain strains and stresses, it dispenses the product without difficulty, etc.). However, once the label is fixed to the package, further testing is often by-passed—

the package is technically satisfactory and it looks "finished."

It's the unusual company that takes the next step and consumer tests its finished or final stage packages. More often, a label design is judgmentally selected and the product/package is produced in quantity. Reasons generally cited for skipping the consumer test phase fall into one of these categories:

- *There's no time*. The people who have created the package have finished their job and the marketing group is anxious to get it to the marketplace.

- *There's no money.* Thousands and thousands have already been spent on package/label development.
- *The package will work.* It's mechanically well constructed and the label was designed to tight specifications concerning the precise way in which the product will be positioned to the consumer.
- *Label design is a creative endeavor.* The design will influence consumers both consciously and unconsciously and therefore its total effect will be impossible to measure.

There is, of course, truth in all of these statements. But to use them as excuses for not undertaking a consumer research program is to underestimate the contribution a well planned and executed consumer test can make to the package/label design development process. This final, consumer evaluation step should be part of this process, not an adjunct to it.

At the "moment of truth," that point at which the consumer finally decides whether or not to purchase the product, the package is the summation of the communications efforts the company has made on behalf of the product. It's at this point of sale that the package must accomplish several important things:

1 It must tell the prospective buyer what the product is called; it must communicate the brand name quickly and clearly.
2 It must tell the prospective buyer what the product is and what it does—the person looking at the package should be able to note quickly what type of product it is and what it's for (a nonaerosol spray antiperspirant that keeps you dry). It's also important to communicate clearly the ways in which this product is differ-

ent/better than others the prospective customer may be considering—the package/design must "sell" itself to the consumer. These objectives are often accomplished by graphic treatment/design elements as well as words.

3 It must inform the prospective buyer how nutritious the product is, what ingredients it contains, and, in some cases, what each ingredient's purpose is.
4 The package carries the Universal Product Code, which tells the consumer nothing, and a price sticker or stamp, which tells the consumer what she really wants to know.

Thus as a communications vehicle, the package/design must transmit a tremendous amount of information about the product to the prospective buyer, some of it spelled out in copious detail, some of it implied, but all of it consistent with the basic product positioning/level of performance built into the product and its marketing strategy.

At the same time, there are some constraints on package design development. A package is a means for a consumer to recognize the products she knows and uses. One of the axioms in package design is that a new or revised package should not be drastically different from the label/package consumers are familiar with. This is not to say packages should not be changed, rather that they should be modified on a gradual, step-by-step basis. Quaker Oats oatmeal has had a number of significant package graphics changes over the years, yet each one maintained the integrity of the previous design—a design that consumers knew well and could easily recognize in the store. Even on those occasions when the product itself has undergone significant reformulation and the marketer is anxious to have consumers recognize they have a new reason to try the brand, a drastic design change is a questionable business tactic. It seems that consumers are willing to accept reformulation of the product as an important improvement, but they consider package/label design changes unnecessary.

Also, certain package configurations and/or colors are traditionally acceptable for certain products. Catsup, for example, illustrates this point. Virtually everyone instantly recognizes the shape of a catsup package, and in controlled use-test situations nearly everyone will agree that this type of container is extremely difficult to use. But the package configuration is traditional, it is catsup, and any other shape simply would not be suitable for the product.

It's generally agreed that the package is an extremely important element in the marketing mix for most consumer products. It does have multiple functions to perform and, therefore, no single consumer research measure can do an acceptable job of evaluating the total effectiveness of a consumer package. There are at least three measures that are important considerations in evaluating consumer package/label designs:

- *Impact* How much intrusiveness does the package/label design have; does the package have the ability to "jump off" the shelf and be recognized quickly?
- *Image* What does the package communicate to consumers; what kinds of impressions and/or perceptions does the package create?
- *Preference* Does the consumer like the package; is it aesthetically pleasing to the prospective customer?

In order to be maximally effective, these three measures must be in balance. They must work in concert to produce the kind of impact that's necessary for the brand to be noticed (quickly) on the store shelf without sacrificing those design elements that help communicate the kinds of positive product impressions (image) that are important to the success of the brand. It is entirely possible to have a tremendous amount of intrusiveness, to communicate certain product characteristics very effectively, or to have people generally agree that the package is attractive. However, if any of these are accomplished at the expense of the others, the wrong balance has been struck.

It's easy to get the wrong balance. For example, it's possible to achieve a high degree of impact using certain design devices—fluorescent ink colors or unusual package configurations will invariably produce high recognition. However, the use of these overt attention getting devices may distract the consumer from the package's communications effort in other directions. Bright colors often are interpreted as harsh or too strong, and unique package shapes could prove difficult for consumers to handle. Or, as is the case with catsup, a new, different package shape might not be acceptable to a large proportion of consumers. Conversely, pastel colors and swirly, curly graphics often are interpreted as mild or gentle, but could well be so soft that the package would fail to achieve impact at the point of sale. Finally, simple consumer preference, often based on an individual's assessment of the attractiveness of the design, may have little or no relationship to which package the consumer will purchase— they want the product inside, not the package. They recognize this and so must we.

The effective package design, then, is the one that is most successful in integrating these three properties—impact, image, and preference. The consumer research test methodology must, then, be designed to measure these same properties effectively. We have developed a research methodology that measures consumer response to test package/label designs in terms of these three decision criteria.

This system is an experiment, in the sense that the environment in which the package alternatives are studied is one in which the variables (stimuli) can be controlled and manipulated. The respondent is not in her normal context (a supermarket) but in a tightly managed interviewing station or facility. Also, she is often asked to react to stimuli that *represent* the actual package(s), for example, 35mm slides or photographs, rather than the packages themselves. This does make it possible to isolate and control key variables and to study the consumers' response to these variables. Since all package alternatives included in the test are exposed to this identical treatment, the absence of a real life environment affects equally all packages being studied and should not affect an individual's assessment of one package relative to the others in the test scheme. By studying respondents' reactions as the test variables are manipulated, it's possible to make judgments concerning the extent to which these variables affect a respondent's perceptions of the packages being studied.

IMPACT MEASURE

The first measure in this consumer research system is an impact measure. How intrusive is the package design? What elements of the design do consumers notice? How quickly do they notice certain package/label features? This is the kind of information developed in the initial section of the research methodology. An impact measure is the first step in the research, not because it is the most important element to consid-

er, but because impact, in order to be correctly measured, must have a clean or uncontaminated exposure. Visual or verbal cues of any type have a tremendous influence on this measure, so much so that results tainted by prior cues are virtually meaningless.

A tachistoscope (T-scope) is used to measure the impact of salient visual elements of each test package. It is a device that precisely controls the amount of time a stimulus (package) is exposed to the respondent. By strictly controlling both the number of exposures and the length of time of each, the packages in the test are ensured equal exposure time. Thus any differences in the time required by respondents to pick out salient print or graphic elements is ascribed to the test variable, the package itself.

The T-scope used in this portion of the research is electronically controlled for increased accuracy and ease of operation. Each exposure interval is clearly marked and the control knob has a definite click stop point for each interval. Once the instrument is set up it can be operated even in total darkness. The T-scope exposes the test package to the respondent a total of eight times. The first flash exposure is at a speed of $1/150$ second and each succeeding exposure is lengthened until the maximum speed of 2 seconds is reached. The T-scope is set at the following speeds:

Exposure 1	$1/150$ second
Exposure 2	$1/100$ second
Exposure 3	$1/50$ second
Exposure 4	$1/25$ second
Exposure 5	$1/10$ second
Exposure 6	½ second
Exposure 7	1 second
Exposure 8	2 seconds

Immediately after each of the eight exposure intervals, the respondent is asked to report exactly what she saw.

Her verbatim comments are recorded, coded, and tabulated to form the basis of this portion of the analysis. The impact measure provides these types of information:

- Brand name recognition.
- Product description playback.
- Product/brand name misidentification.
- Recognition of symbols/logos.
- Identification of other salient graphics.

It's possible to calculate average recognition times or mean scores for each of the above points, and it's also possible to display the test results cumulatively to determine the pattern that respondents follow in viewing the test package(s).

Table 1 is an example of the brand name impact measure for four different test packages/brands. It shows the percentage of respondents in each test who were able to identify correctly the brand name of the package at each exposure interval. The Mean Recognition Score, the average length of time it took respondents in this test to identify correctly the brand name of the package, is also presented in the table.

The second way to examine results of the T-scope questioning series is to accumulate responses for each successive exposure interval. Table 2 shows the cumulative percentage of respondents who mention various visual elements of the package as the test progresses.

The T-scope device, administered and tabulated in this manner, provides a total recognition profile of a test package, by time period. The analyst can use the results to identify which package elements are noticed more quickly and can also track the path or pattern the respondent is following in her attempt to complete this test successfully—that is, to report exactly what she sees at each exposure interval.

Table 1 Correct Brand Name Identification

	Brand A	Brand B	Brand C	Brand D
Correctly identified package brand name at:				
$1/150$	52%	12%	10%	—%
$1/100$	20	4	24	4
$1/50$	6	12	10	—
$1/25$	10	16	8	8
$1/10$	10	8	16	6
$1/2$	6	24	20	42
1 second	—	8	4	22
2 seconds	—	—	4	8
	100	84	96	90
Not correctly identified package brand name	—	16	4	10
	100	100	100	100
Mean recognition score (In seconds)	0.05	0.26	0.26	0.67

Table 2 Visual Response*

	Exposure Time (in seconds)							
	$1/150$	$1/100$	$1/50$	$1/25$	$1/10$	$1/2$	1	2
A bottle	4%	18%	26%	28%	30%	30%	32%	32%
A bottle with handle	12	18	20	20	20	22	22	22
A blue bottle	36	40	44	48	56	56	62	64
Correct product ID	4	6	18	18	18	18	20	38
Visual item 1	2	6	6	8	10	20	24	24
Visual item 2	4	8	8	10	22	32	36	40
Incorrect brand ID	40	46	54	58	64	68	68	68
Correct brand ID	12	16	28	44	52	76	84	84
Yellow color	4	16	22	30	34	38	48	48
White color	12	20	24	36	40	42	42	42
Red color	4	8	14	20	22	24	26	28
Blue color	10	12	20	26	28	28	28	28
Pink color	10	10	10	10	12	12	16	18
Copy point 1	—	—	—	—	2	10	22	30
Copy point 2	—	—	—	—	—	4	16	24
Copy point 3	—	—	2	4	6	20	50	76
Ounces of product	4	4	4	4	4	6	6	10

*Selected responses.

The technique is especially useful in comparing test package alternatives for two important reasons—it produces consistent results in a test, retest situation and it is able to discriminate among test alternatives. If the same stimulus (package) is tested more than once, the results will, with a high degree of probability, be identical each time. Second, if the test packages are capable of producing different levels of impact or intrusiveness, this testing technique will measure the difference.

Table 3 illustrates the ability of this measurement system to replicate its findings. Brand Name Recognition Scores are shown for two consumer packages, each of which was tested on three different occasions with three different samples of respondents. The brand name recognition score is the av-

erage length of time, in seconds, it takes a sample of respondents to identify correctly the brand name of the test package. For Brand M, results differed by only 0.12 seconds in the three tests; for Brand S the scores differed by only 0.06 seconds in three successive tests. If the same packages are tested, it's likely the

Table 3 Mean Recognition Scores

Brand M Package		
Test 1		1.03 seconds
Test 2		1.11 seconds
Test 3		1.15 seconds
	Average	1.10 seconds
Brand S Package		
Test 1		0.76 seconds
Test 2		0.77 seconds
Test 3		0.82 seconds
	Average	0.78 seconds

results of the impact measure will be the same. The table also indicates that the T-scope test is able to discriminate between different package designs. Brand S was correctly identified about ⅓ second sooner than Brand M (Average Recognition Score of 0.78 versus 1.10).

Another illustration of the test's ability to discriminate between packages is shown in Table 4. In Test 1, four packages from the same product category were tested. Results indicate Brand A was correctly identified five times faster than Brands B or C and thirteen times quicker than Brand D. Test 2a included two packages with identical label graphics but different brand names; Test 2b included two packages with identical brand names but different graphics. Results clearly show significant differences in impact. These tests scores also show both brand name and/or graphics can influence recognition. With identical graphics Brand F was correctly identified about twice as quickly as Brand E; yet with identical brand names Graphic Y performed much better on the measure of correct name recognition.

The conditions of the exposure of the packages to the respondents is always a point capable of generating long discussions. There are two basic conditions under which a package may be tested by using this T-scope impact measure:

Table 4 Mean Recognition Scores

Test 1	
Brand A package	0.05 seconds
Brand B package	0.26 seconds
Brand C package	0.26 seconds
Brand D package	0.67 seconds
Test 2a	
Graphic X, brand name E	0.59 seconds
Graphic X, brand name F	0.25 seconds
Test 2b	
Graphic X, brand name F	0.25 seconds
Graphic Y, brand name F	0.07 seconds

1 An individual package, by itself, can be studied without reference to a competitive frame. The respondent is exposed to only a single package stimulus.

2 The test package can be studied as it relates to its competitive frame or environment. It can be shown to the respondent as part of a "typical" shelf array in which the test package is one of several brands/packages displayed and collectively viewed by respondents.

Results of previous tests indicate the more viable testing technique is to deal with a single package exposure rather than a multiple package situation. Exposing the respondent to only a single package generates results that go into more depth relative to the test package. In a single package test design, it is likely that information will be volunteered on the attention-getting characteristics of such design elements as:

- Colors and/or combinations of colors.
- Unusual or distinctive package shapes.
- Brand name recognition.
- Playback of product category description(s).
- Package graphic elements such as illustrations or photographs.
- Corporate symbols or logos.

It's also likely that test results from a single package testing technique could more readily identify elements of the package that are being misread or are in some way confusing to the consumers participating in the test.

In a shelf array test situation, respondents are asked to look at a number of different brands as well as a number of packages or shelf facings for each brand. This normally results in a display of 20 or more packages. When asked to concentrate on this type of stimulus, respondents tend to focus their attention on picking out the brand name of each group of packages shown. They will be sure they have correctly identified the brand name on one set of packages, then move their attention to the second set, then on to the third until they are confident they have correctly noted each separate brand in the shelf array. In a practical sense, this procedure leaves respondents little or no time to comment on any other aspects of the packages displayed.

Also, since there are a large number of packages to look at, respondents tend to be a bit cautious about relating what it is they see until they are reasonably sure of themselves—thus their recognition for correct brand name identification is usually much slower than in the single package stimulus tests. As a result, the dispersion of mean recognition scores is not as great as in the single package test mode. This point is illustrated by the following chart of Mean Recognition Scores:

		Fastest Time	Slowest Time
Single package	(24 cases)	0.04 seconds	0.73 seconds
Shelf array	(28 cases)	0.44 seconds	1.50 seconds

In the 24 Single Package T-scope tests, the fastest time recorded was just under $1/20$ of a second, nearly 20 times faster than the slowest mean recognition time recorded. In the 28 Shelf Array T-scope tests, the fastest time recorded for correct brand name identification was just under ½ second, about three times faster than the slowest score of 1½ seconds.

Another point in favor of selecting a single package mode for testing the impact of packages is that this system better handles the built-in bias that consumers have for brand names they know well. Packages that carry familiar brand names tend to achieve much higher recognition scores than packages with brand names that are less well known by respondents. Thus tests in which some of the packages carry new names will invariably indicate that the established brands achieve faster recognition scores. This will of course occur in both single package format and in shelf array test situations. The single package scheme does have greater ability, however, to discriminate between test packages and is more likely to show which of the new brand name/package alternatives includ-

ed in the testing scheme has the most impact.

Finally, the most compelling argument in favor of using a single package mode rather than a shelf array exposure is that test results are parallel for both situations. If several different brands are included in a test, the "winner" is likely to be the same brand/package in both situations, the nearest competitor would be in second position in both testing schemes, and the slowest recognition speeds would probably be the same packages. The rank order of test results, in terms of speed of brand name recognition (impact), would probably be the same.

Thus if two test designs do an equally adequate job of measuring the speed of brand name recognition of one test package relative to other test packages within one test series, but one of the testing modes also provides information concerning the impact of other package elements, the methodology that can generate the additional information is the logical choice.

Table 5 shows the similarity of results for four brand names/packages tested under both a single package mode and a

Table 5   Mean Recognition Scores

	Single Package	Shelf Array
Brand A	0.05 seconds	0.44 seconds
Brand B	0.26 seconds	0.90 seconds
Brand C	0.26 seconds	1.10 seconds
Brand D	0.67 seconds	1.35 seconds

shelf array situation. Clearly, Brand A "wins" in both methodologies, Brands B and C are distant seconds and Brand D does the least effective job of gaining correct brand identification.

The next point to consider in completing this type of package design research is the actual physical form the stimuli (packages) used in the test should be. There are three basic considerations. (1) In what form is it practical to produce the package/design alternatives for testing purposes? (2) Will the materials used to execute the test allow for geographical flexibility? (3) Can the test be easily administered in the field or does it require specialized training and/or equipment to function properly?

A testing methodology that utilizes 35mm slides to represent the test packages is usually a good choice. It's portable, takes little mechanical aptitude to operate, and requires no special equipment other than a T-scope device and a 35mm projector/viewing screen. Most important, 35mm slides can represent accurately the actual test packages in terms of color reproduction and are much less expensive to produce than tight, mock-up packages. For example, if a new product package is under consideration, only one of each test packages need be produced. That model can be photographed and the test can be completed in several markets simultaneously without fear of destroying the prototype model.

Yet the testing system must be flexible enough to vary from this methodology when it's appropriate. If, for example, one or more of the test packages

in the research use fluorescent ink colors, it would not be possible to use 35mm slides because these inks cannot be reproduced adequately in full color photography. In fact, the only way to represent this type of test package accurately would be the test package itself. In one testing situation in which these types of packages were included as alternatives a shadow box was built, and the T-scope procedure was modified so that the actual package inside the shadow box was illuminated in a controlled fashion.

A final consideration prior to conducting this type of impact/image/preference test relates to sample selection. A package test is not unlike any other form of market research investigation in the sense that sampling methods and selection criteria are important to the successful completion of the study. For the majority of consumer package tests the sample should be structured to include a high percentage of product category users. Establishing a subquota of users of certain brands is also a good plan. It was noted that brand familiarity influences how quickly respondents are able to recognize brand names. Familiarity/experience with the brand also influences a consumer's image perceptions of the product. Thus it is important that this sampling variable be controlled from test cell to test cell. If one-half the respondents in Test Cell A are users of the test brand, one-half the respondents in Test Cell B should also be users. Since the package test is essentially an experimental design, the researcher can tightly control the sampling procedures to be more certain that important differences that are noted in the test results are traceable to the package variations, and not the result of sampling variation.

The physical place in which the T-scope portion of this testing system is completed is almost always a small conference room setting in a central location interviewing facility. The room is arranged so that the 35mm projector is set up to flash approximately "life size" sequential exposures of the test package on the viewing screen. Respondents who meet the sampling eligibility requirements and agree to participate in the test are brought into the room and seated about 6 feet in front of the screen. Each participant is told she will see a picture flashed on the screen very rapidly and will be asked to report exactly what she saw each time a picture appears on the screen. If she normally wears glasses or contacts while shopping she is asked to wear them during this test. The interviewer in the room with the respondent controls the T-scope mechanism and also records the respondent's verbatim commentary after each package exposure.

This type of impact measure is a purely physical measure of how quickly respondents can pick out or notice characteristics of the test package shown to them. Thus mechanical variations in the test setting can greatly influence the test results. For this reason, this portion of the test must be closely controlled and monitored. As much as possible, the physical characteristics of the test environment should be identical for each location or city used in the test. Instructions given the respondents are quite detailed and are read, word for word, to each participant.

In this type of T-scope methodology, respondents begin to "learn" the technique as the subsequent exposures are shown to them. They try to win the game in the sense that the first few exposures have taught them what to look for on the screen. For this reason, exposing any single respondent to more than one test package (or shelf array) in this type of testing system is not an acceptable procedure. To do so would only introduce a bias into the test that could well blur the results. It is much

wiser to deal only with respondents who have not been preconditioned to the mechanics of the test procedures.

A series of eight T-scope exposures seems to be the optimal number for this testing procedure. The fastest speed, $1/150$ second, is the starting place because it is the point at which a few people can actually pick out certain package elements—speeds quicker than this are just a flash to respondents. The slowest speed, 2 full seconds, is long enough for all salient package graphics to be noticed—slowing it down further only results in redundancies. Dividing this $1/150$ to 2 full seconds range into eight intervals provides a reasonably high level of discrimination between the points.

The absolute scores generated via this T-scope methodology are only useful when a number of packages are tested and their scores are directly compared. The emphasis of this measurement is on how well each test package performed relative to others included in this or previous tests. Knowing a test package achieved a Mean Brand Name Recognition Score of 0.39 seconds doesn't mean much unless norms or other directly comparable impact scores are available.

After the respondent has seen the eight exposures of the test stimulus the T-scope impact measure is complete. Each participant is then asked to move to a second interviewing station, and an entirely different set of questions are administered.

IMAGERY ANALYSIS

The objective of this section of the testing system is to determine what types of images and/or impressions are communicated by the package/label designs being tested. This is done by having each respondent rate the test package on a long series of attributes or dimensions that might be used to describe the package/product. Normally 25–30 of these attribute statements are included in this semantic differential rating scale technique.

These attributes are designed to reflect the opinions consumers may have about the test package/brand. The list may be generated from prior consumer market research studies such as focus group sessions on the brand or product being tested. New product concept studies or product positioning studies are also a useful source for constructing this attribute list. Ultimately, however, brand management and the research analyst have the responsibility to anticipate the consumer's response to the brand/package and to be certain the final list reflects these possibilities as well as those dimensions known to be important to consumers in deciding to buy/not buy the test brand/product. The importance of the attribute list cannot be stressed enough. It forms the basis of the imagery analysis portion of the package research study. If the "right" attributes are not included in the list, the "right" consumer response pattern will never be measured.

The attribute list should cover three broad dimensions:

1 Product efficacy dimensions. These attributes measure the extent to which respondents believe the product inside the package will live up to performance expectations. Examples of these types of attributes are:
- Cleans pots and pans without rubbing.
- Makes silverware sparkle.
- Rinses off easily.
- Especially effective in removing grease.

Other product-related dimensions focus on such things as:
- Would be economical to use.
- It's convenient to use.
- It's a modern product.

2 Dimensions related to aesthetics assessment of the package:
- It's an attractive package.
- The colors are cheerful.
- It's an eye-catching design.
- Sprays on easily.

3 Statements of a self-referral nature that reflect the respondent's personal interest in the product:
- It's my kind of product.
- I'd use this product every day.

In this section of the test the actual package makes an ideal stimulus. Respondents can hold it, shake it, and examine it, front and back, straight up and upside down. However, it is often not practical to work with actual packages or package prototypes. In these instances, full-color 8 × 10 inches photographs of the test package can be used. This is an inexpensive means of reproducing original designs. It can be done quickly, and respondents can generally accept photography as a reasonable substitute for the real thing. The only proviso is that the photographs must reflect the original package/label design accurately. Colors must be very close to the original design specifications, and the size of the single package in the photograph should come close to the actual size of the package.

The impact measure (T-scope) portion of this testing system places emphasis on purely physical measurement. The imagery measure emphasizes consumer opinions/perceptions—to what extent does the test package communicate certain attributes or dimensions. In the first instance, sampling error, in a statistical sense, tends to stabilize relatively quickly and a sample size of 50–60 respondents is adequate. The imagery measure requires a larger sample base before the test results begin to stabilize. A base of 150 respondents is the minimum sample requirement.

Unlike the impact measure, where each respondent can view only one test package, this section of the test can accommodate up to three package exposures. Previous studies have shown that a consumer's opinions of one package do not strongly influence her opinion of a second or third package. Thus if each respondent is asked to rate three packages in the imagery section of the study, it's possible to keep the two sections of the package test in balance in terms of sample base requirements. Reliable test results will still be generated for both the impact and imagery measures. The test design takes this form:

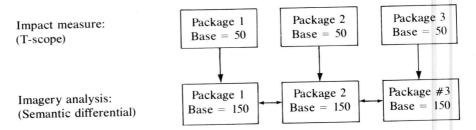

| Impact measure: (T-scope) | Package 1 Base = 50 | Package 2 Base = 50 | Package 3 Base = 50 |
| Imagery analysis: (Semantic differential) | Package 1 Base = 150 | Package 2 Base = 150 | Package #3 Base = 150 |

If more than three test packages are included in the test design, rotations can be established so that each alternative gets equal exposure. This approach is preferable to having each respondent rate four or more packages in the imagery section. The latter approach is too time-consuming, and respondents could easily lose interest in the process. If that happens, results are suspect.

The actual form of the scale that is administered in the imagery analysis section of the questionnaire is the next consideration that must be addressed. The options are almost limitless. Many companies already have strong preferences about the exact form/wording of rating scales used on their consumer research studies. As a consequence, several scale variations have been used in this section of the study—all with good results. This simply means that consumers understood that the scale was designed to place a direction and inten-

| Cleans pots and pans without rubbing | O | o | . | . | o | O | Have to rub to get pots and pans clean |

	Definitely Agree	Probably Agree	Probably Do Not Agree	Definitely Do Not Agree
Rinses off easily	4	3	2	1

	Absolutely True	Mostly True	Somewhat True	Somewhat False	Mostly False
A modern product	( )	( )	( )	( )	( )

| Sprays on easily | True 10 | 9 | 8 | 7 | 6 | 5 | 4 | 3 | 2 | Not True 1 |

My kind of product

sity to their opinions, and these scales were sensitive enough to pick up differences in consumer perceptions between package design alternatives. The five examples below are semantic differential scales that effectively measure consumer attitude in terms of both direction and intensity.

Thus the key considerations in choosing a scale are:

- Do respondents understand the scale?
- Is the scale sensitive enough to measure different opinions?
- Does the company have a track record/norms for the scale?

Once a scale rating device has been selected, the administration of that device in the field is essentially the same for all scales. The interviewer reads the list of statements and has the participant respond with her rating for each item on the list. It's a good idea to have the actual scaling device printed on a separate card so the respondent may refer to it as each statement is read to her. Having the field interviewer administer these scale ratings, as opposed to having them self-administered, has several advantages:

- It's faster.
- Rotations and/or starting points can be used.
- Respondents are less likely to mark the same rating point for all attributes; the presence of an interviewer seems to force participants to be more thoughtful and consider their responses more carefully.
- Much fewer "No Answers" are recorded.
- It avoids misinterpretation and misunderstanding.

After the respondent has rated each of the 25–30 attributes for one package

alternative, the process is repeated until a maximum of three packages are rated on this attribute list. These ratings form the basis of the imagery analysis portion of this package test system. Mean scores are calculated for each attribute and these scores are profiled, or compared, among the test alternatives included in the study. Variations in these rating scores are indicative of each package's ability to convey different impressions to those consumers seeing it.

There are occasions in which the use of mean scores will not tell the entire story. In a situation where the distribution of a particular response is skewed (i.e., a normal frequency distribution curve would not properly describe the response pattern of consumers on a particular attribute or statement), the analyst can use a "top-box" score rather than the mean score in reporting and/or analyzing test results. For example, if the scale ratings for many of the attributes included in the study show a bilateral distribution pattern (concentrations of responses at both ends of the scale) a mean score would be misleading. In this instance, the reporting of the percentage of respondents marking the attributes at the highest scale rating point (the "top-box" score) provides a more accurate picture of the consumers' response to the package being studied.

The scale rating results developed in this section of the study provide two important pieces of information:

1 They pinpoint those attributes that are highly consistent with respondents' attitude toward the product/package design. Attributes that show a high degree of agreement are indicative of those properties of the product/package that consumers will accept as truthful and realistic product claims. Conversely, those attributes that are scored low by consumers are dimensions that respondents find hard to believe about the package/

product being tested. It would, for example, make little sense to position the product as a high quality or expensive entry if the package for the product clearly conveys a low cost/ low quality image.

2 They provide a profile of one test package versus the other test packages included in the test. By examining the scores on a side-by-side basis the analyst can identify the areas on which consumer perceptions of the test packages differ. The decision as to which package is the best fit for the product positioning has to be made on the basis of a thorough examination and understanding of these profile scores.

There are, of course, no right or wrong answers in intrepreting these imagery analysis scores. The data must be ana-lyzed in concert with what the stated objectives of the package test are. What have the packages/labels been designed to communicate? How well does each package perform in light of these criteria? These are the key questions addressed by the imagery analysis section of the package test system.

Table 6 is an example of the tabular detail generated via this questioning sequence. Brand A is the leading seller in the product category, Brands B-1 and B-2 are design variations of one of the other newer brands in the category. Notice two things: first, there is a reasonably wide range of agreement/disagreement with the attributes listed on the chart (the "top-box" scores range downward from 72 to 7%), an indication that the scale rating device is measuring differing consumer opinions/perceptions relative to these attributes. Next, the scaling technique is able to differentiate

Table 6 Image Profile Scores*
Percent marking "top-box"

	Brand A	Brand B-1	Brand B-2
Reduces static cling	72%	19%	15%
Softens clothes	69	31	25
Clothes have a clean scent	72	17	17
Has a pleasant fragrance	66	13	11
Convenient to use	57	52	44
Works in any cycle	28	41	43
Has a unique advantage	27	42	42
A versatile product	25	47	41
Economical	43	13	21
Lasts a long time	42	14	15
Worth the money	46	7	7
Attractive package	69	46	47
Modern	64	51	50
For my family	69	23	32
Necessary for my home	54	37	44

*Selected responses.

between packages tested—Brand A clearly has a different profile than either Brand B-1 or B-2, and even among these two alternatives a somewhat different attitude profile is evident.

Throughout the test, respondents are asked to deal with the test stimulus (packages) on a monadic basis. In the impact portion of the test they are exposed to only one package. In the imagery analysis portion they do handle three alternatives, but only one at a time. At no time are they asked to directly compare the relative merits of the test packages.

Careful study of the impact and imagery measures will often supply sufficient consumer feedback to make solid judgments concerning the effectiveness of the test package(s).

One useful analytical device is to review the results for each package tested in terms of the four quadrants of this grid:

Strong Impact Measure Scores Correct Imagery Perceptions **No Changes Necessary**	Weak Impact Measure Scores Correct Imagery Perceptions **Reappraise Impact**
Strong Impact Measure Scores Incorrect Imagery Perceptions **Reappraise Imagery**	Weak Impact Measure Scores Incorrect Imagery Perceptions **Significant Changes Needed**

On occasion there are situations when the test scores for particular packages are very similar. In these instances, it is useful to try to exaggerate whatever small differences do exist in the minds of consumers. This is accomplished by asking respondents to "pick a winner" from among the alternatives presented.

DIRECT COMPARISON PREFERENCE

In this section of the study, each respondent is asked to compare directly the test packages for the first time. A limited number of statements or dimensions related to her impressions of the "image" communicated by each package design is read to her, and she is asked to select the one package/label design alternative she prefers for each statement. This list often includes such factors as: most economical product, most effective product, most attractive package, highest quality product.

It's important to keep in mind that this questioning technique tends to force respondents into choices that they might not otherwise be in a position to make. It's not likely a consumer will ever see three different packages for the same brand on sale at the same time. The intention of these questions is not to determine a "best choice," but to magnify the differences in consumer perceptions that may exist. When read in conjunction with the other information collected in the testing system, the discrimination produced by this questioning technique can produce meaningful information.

A final preference question is normally included in this section of the study. Respondents are asked to distribute a total of eleven votes among the packages in the test. Votes are distributed on the basis of how strongly each respondent feels her preferences are— it's a measure of the intensity of preference. It's possible one package is so well liked it gets all eleven votes, or votes may be more evenly distributed across the alternatives, an indication that preference for one package versus another is very slight.

Table 7 Package Preference

	Package Preferred (%)			
	Package 1	Package 2	Package 3	Package 4
Safest product	39	18	23	20
Most effective product	33	19	24	24
Easiest to use	37	20	26	17
Best overall product	36	20	23	21
Most expensive product	17	12	38	33
Most attractive package	24	15	33	28
% Constant sum scale votes	28	20	28	24

Table 7 is an example of this question sequence. Four alternatives of the same brand were tested. Package 1 clearly is preferred on those dimensions related to product performance (safer, more effective, easier to use, and best overall), but it is not the most attractive package nor is it the most expensive package in the test. Though the individual preference scores for each dimension are quite strong, the constant sum scale question shows the actual intensity of preference to be quite weak. Package 1 and Package 3 are, in fact, equal in terms of consumer preference, with the other two alternatives not far behind.

SUMMARY

This chapter has discussed a model for testing consumer packages or label de-signs. It is a reliable, proven methodology for evaluating consumer response to different packages using three separate criteria:

1 Impact (T-scope technique).
2 Imagery (semantic differential).
3 Preference (forced choice).

Incorporating these three discrete measures into a single testing system offers the opportunity to examine and analyze the test results in an integrated, systematic way. The methodology is flexible, both in terms of geography and in terms of types of packages that can be studied, and the time and financial commitment on the part of the sponsor is modest. The test design is summarized in the following chart.

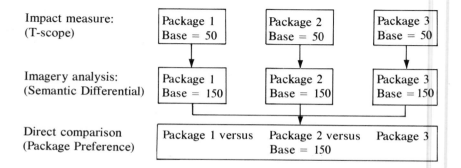

When read in concert, these measures provide a maximum amount of informa-tion on which to base important packaging decisions for consumer brands.

CHAPTER TEN

Visiometric Testing:
How a Package Communicates

Lee Swope

The role of the American consumer over the last 200 years has become increasingly intricate. This role has been complicated by the tremendous number of products and services offered the American consumer. For the most part packaging has not helped the situation; it has made shopping a complex decision for the consumer. Today's supermarkets, discount stores, and retail outlets display thousands of products packaged in a bewildering array of communicating devices. Unless a consumer is actively searching for an individual product or brand, the overall result is chaos rather than communication. In effect, the consumer is being acted upon by an overabundance of communicating devices that tend to blend together and to confuse. Communication and motivation, the prime essentials of effective packaging, are lost.

Like it or not, this is today's marketplace, a marketplace fashioned by marketers and packagers. This marketplace puts tremendous burdens on the package designer because he must not only design packages that motivate and communicate to prospective consumers, but he must design packages to compete successfully in an arena that is already saturated by well designed communicating and motivating packages. The designer's task thus becomes one of designing the better mousetrap, a difficult assignment in today's marketplace, made even harder because it becomes increasingly difficult to recognize when the better mousetrap has been attained.

Over the years designers have used three basic methods to decide when the best communicating and most appropriate package has been attained: success or failure, gut feel, and experience. All three methods have merits, but their drawbacks far outweigh their merits. For example, in the success and failure method any product that fails in the marketplace is assumed to have been packaged inappropriately and ineffectively. On the other hand, if a product is a great success then the package presumably is a winner. Not a bad philosophy except it doesn't take into account that the package and product are inseparable for the initial sale only. After the first purchase the product is on its own. If a product is less than adequate sales will die, and if a package is less than adequate, sales will die. Consequently the success and failure method tells the designer very little because it fails to take into account all the other factors that influence the sales success of products and packages, that is, advertising, product quality, display, and the marketplace.

The second method is the "gut feel" method. This method employs an almost mystical affinity for the marketplace and what the consumer sees in a package. Few of us are gifted with anything approaching an accurate "gut feel." We can say what we like and why, or we can tell you what we think consumers will purchase and why, but we are so close to packaging, and packaging hinges so much on subtle nuances, that designers tend to develop a biased eye. In other words, designers think like designers, not consumers.

The third method is called experience. There is no substitute for experience, and it will produce effective packaging more often than ineffective

packaging. Unfortunately, when a designer relies too heavily on experience he begins to design packages that look like other packages. He relies on the proven packages and is reluctant to try new approaches. He has few failures and few smashing successes; what he does have is a lot of in-between packages that are neither failures nor successes.

There is a fourth method seldom used by designers and used only occasionally by marketers. It is called Visiometric Testing. Visiometric testing is a series of tests that measure the relative effectiveness of a package. The series of tests is designed to pinpoint and measure how effectively packages communicate through what the test subject, and ultimately the consumer, perceives through his sense of sight.

In order to discuss visiometric testing, or any market and package testing for that matter, we have to realize that packages appeal to only two senses: our sense of touch and our sense of sight. Packages appeal to these two senses in a variety of ways, of which four are basic.

The first method of package communication is color. Color has the advantage of being able to set a mood immediately, to scream and shout or to be subtle. Consumers have been preconditioned to react to color as a stimulus. Color is probably the fastest and most important visual communicator available.

The second communicator is shape. Shape is a strong communicator because we as consumers identify certain products merely by the shape of their package. Package shape can lend an air of femininity or masculinity; shape can communicate a feeling of solidarity and quality.

The third communicator is typography and lettering style. Typography is an extremely important communicator because it conveys product information in a way that is different from color or shape; typography communicates through language. Typography becomes the advertiser as well as an identifier. Typography can be the prime package identification element or it can be a supportive element; in many cases it fills both roles.

The fourth communicator, most subtle of all, is texture. Texture, real or imagined, communicates from two aspects: one, it reinforces our preconceived ideas of how packages for certain products should feel and two, it reinforces each of the other three communicators.

In order to determine how effectively a package makes use of the four communicators, a series of tests can be employed. Each test is structured to tell the designer and marketer about the effectiveness of each communicator.

One such test is the eye movement test, which indicates where the eye is traveling on a package, and the sequence in which the different graphic package elements are being observed. An eye movement test is important in determining how well the overall graphic layout and typography attract and hold the eye.

The test itself can take many forms, from a simple movie camera fitted with a special close-up lens taking pictures of the eye, to a highly sophisticated computer-controlled unit, which bounces a beam of infrared light off the retina. Both systems track eye movement physically as the test subject views the test package. The test indicates what the test subject saw and that the test subject looked at certain elements for a given amount of time and in a certain sequence. The test is completely objective and is designed not to give subjective perceptions of a test package.

To be aware of the viewing sequence of the different elements on a package is extremely important, because in order for a graphic element to be effective it must first be seen. If that graphic element is the most important element on the package, then the element had better be seen at the beginning of the viewing sequence, and the eye should return to that element or linger long enough to ensure recognition. The test is essential in pinpointing unimportant graphic elements that may be stealing the "lion's share" of the prospective consumer's viewing time while the really important graphic elements, those essential to package communication, may not be viewed for an amount of time sufficient for recognition and may therefore never be seen by the consumer.

The eye movement test can be likened to a family on vacation traveling from New York City to Miami, Florida. The test tells us the family indeed traveled from New York to Miami and that they went by way of Richmond, Virginia; Charleston, South Carolina and Orlando, Florida. The test also tells us that the family spent three days in Orlando and two days in Richmond. The test tells us nothing about the quality of the vacation or what the family noticed en route. In order to determine what the family noticed on vacation, a second test must be employed; this test is a tachistoscope test.

The concept of the tachistoscope had its beginning during World War II. The armed services used the idea of a test that could allow a test subject to view an aircraft silhouette for a given amount of time in order to train antiaircraft gunners to quickly differentiate friend from foe. After the war, market and package research took over the test and developed it to measure the time required by a test subject to identify a package, or certain elements of a package.

The test is simple. A test subject is shown a package (or picture of a package) for a measured increment of time, usually a fraction of a second, and asked a series of questions. After the test subject answers the questions (being unable to answer a question is just as important as answering), he is again shown the package for a measured increment of time, only this time the viewing time is increased. Once again the same set of questions are asked. A single test usually consists of three to four viewings ranging in time from ¼ to 3 seconds.

The physical test equipment usually consists of two pieces of equipment, a stimulus box or screen and an electronics unit. The test packages (or pictures of the test packages) are placed in the stimulus box or are projected on the stimulus screen. The electronics unit controls the amount of time the test subject is permitted to view the package. For example, the stimulus box would consist of a large box in which a package is placed. The test subject would face the box in a darkened room. The front of the box would be cut away but covered with a glass panel set at an angle to prevent the test subject from seeing anything in the box until the stimulus box lights are turned on. The electronics unit contains the timer to control the lights in the stimulus box.

A typical tachistoscope test would work in this manner: A test subject would be seated in front of the stimulus box and asked to look at the glass panel and answer the test moderator's question after each flash of light. The test would then begin by the moderator activating the electronic component to flash the lights in the stimulus box for ¼ second and then asking the test subject a set of questions.

What did you see?

What colors did you see?

What product did you see?

Did you see any brand names?

Did you see anything else?

Normally, at ¼ second the test sub-ject would see very little and could answer few of the test questions. Typical answers would include, "I saw something reddish," "I think it's square," "It might be cereal," or "I didn't see anything but a flash of light."

The test continues. The same package is viewed a second time, but this time the viewing time is increased to ½ second. Again the same questions are asked. At this point the test subjects begin to recognize products and brands. The test continues to a viewing time of 1 second and again the subject is asked the same questions. If the subject is able to answer all of the questions and completely identify the package and product, the test is concluded and a new package placed in the stimulus box. If after 3 seconds of viewing the product and package are not identified, the test is concluded and a new package placed in the stimulus box; the procedure of viewing and questioning continues for each package in the same sequence.

The result of a tachistoscope test is to identify those packages that are slow communicators. If a package is telling a test subject what it contains and the brand name very quickly, then it becomes easier for the shopping consumer to identify the product and brand. The tachistoscope is also important in discovering cases of mistaken identity. If a package or product is being mistakenly identified as a completely different product with a completely different end use, the package is never given the chance to communicate. On the other hand, some products and packages rely on mistaken identity to attract consumers. These packages want to look like the package used by the market leader.

When employing a tachistoscope test, there are a few points to take into consideration. The first point is that test subjects become more proficient as the test progresses. If a series of 10 packages were tested, the first package will be the poorest performer and the later packages better performers because the test subject learns what to look for and generally becomes more accustomed and comfortable with the test. Consequently the viewing sequence must be varied from test subject to subject to eliminate the test subject's learning ability. For instance, a series of eight packages A–B–C–D–E–F–G–H would be tested in that order for the first test subject. The second test subject would view the order as B–C–D–E–F–G–H–A; the third test subject would view the sequence as C–D–E–F–G–H–A–B, and so on. Another method is to use a "ringer" package at the beginning of the test. A "ringer" is a practice package not part of the test, which is used to acquaint the test subject with the test. Even though a "ringer" is used it is a good idea to mix the sequence of packages because test subjects become better at a tachistoscope test with practice.

Another variable is individual eyesight. Subjects with excellent eyesight tend to do better on a test that relies on eyesight as the key ingredient than do test subjects with poor eyesight. To try to structure a test to include only those subjects with excellent eyesight would allow the test to indicate how a package performs in an ideal situation, but to use a test group with varying eyesight would tend to simulate the real marketplace. It is preferable to simulate actual market conditions whenever possible.

If the tachistoscope test can be likened to the family on vacation, the test may indicate that the family, while passing through Richmond, Virginia, fell asleep and missed the sights, but were wide awake and saw Disney World near Orlando, Florida. In brief, the eye movement test tells where the family went, and the tachistoscope test tells what the

family noticed during their trip. Neither one of the tests tells about the quality of the trip or how the family enjoyed the trip. To determine how a test subject feels about a package, an additional test would be employed, a sensation transference test.

The sensation transference test is the simplest of the three tests because it requires no sophisticated equipment, only a product, a package, a questionnaire, a test subject, and a test moderator. However, the test demands a great deal of preparation. Thought must go into the preparation of the questionnaire, packages, and products.

The test is based on the assumption that two identical products packaged in two packages that are different in graphics or shape will evoke different feelings towards those products. If the products are identical and a test subject has certain feelings or impressions towards one and different feelings or impressions towards the other, then something other than the product is causing those feelings. A transfer of a sensation from the package to the product has taken place, thus the name of the test: sensation transference.

The effectiveness of the test is determined by how well the test is prepared. For example, if three package concepts for cookies were being tested, the ideal would be to have three different packages prepared and to have the same cookies loaded into the three packages. Each test subject is then allowed to open the packages and taste the cookies. If the test group consists of 100 subjects, then 300 test packages must be made up. The cookies must be inspected to insure they look alike and have no cracks, chips, or breaks. The cookies must appear as identical as possible.

The second important preparation is the questionnaire. The questions must only relate to the product and never mention the package. The test subject must believe he is giving his opinion on the test products. If cookie packages were being tested, a typical questionnaire may have these types of questions:

Which cookie tasted best—A, B, or C?
Which cookie is the freshest—A, B, or C?
Which cookie is the highest quality—A, B, or C?
Which cookie would you purchase for your family—A, B, or C?
Which cookie is the most expensive—A, B, or C?

A second part to the questionnaire would be used to rate the cookies on a scale of 1 to 10. A third part of the questionnaire would ask general questions not necessarily related to the test (background marketing questions). For example:

How many dozen cookies do you buy a month?
What brands or brand do you usually purchase and why?
Who usually consumes the cookies in your household?

A very important part of the test is the opening statement given by the moderator to the test subject. In the case of the cookies test the moderator would begin by explaining that the three packages of cookies represent the cookie manufacturer's three formulations of the same cookie. The manufacturer would like you to taste the cookies and fill out the questionnaire. The moderator must be careful never to mention the package as being part of the test.

I am constantly amazed at the wide range of answers and impressions given during a sensation transference test. Test subjects normally find a great deal of

difference between products and will swear they can taste or see differences. A test subject should never be informed about "how the test works" because then he can no longer be used as a test subject.

Unlike the eye movement or tachistoscope tests, which give objective results, the sensation transference test deals in subjective impressions. Typical reactions from test subjects indicate views on quality, taste, product expectations, color, and appropriateness of packaging.

To draw once again on the analogy of the family on vacation, sensation transference indicates the family had a great vacation. The food was excellent en route, but the motel rooms were uncomfortable and everyone thought Disney World was fabulous. In short, the three tests indicate the road map of the vacation (eye movement), the vacation snap shots (tachistoscope), and vacation quality (sensation transference).

Any program of visiometric testing or any market testing program can have pitfalls so severe that erroneous test results may doom a product and package to market failure. The number one pitfall is the test subjects. They are human beings and respond as humans with all their prejudices and psychological defenses. In many instances they want to please the test moderator and will subconsciously react to his or her voice inflections or body movements. When confronted by direct questions which might threaten the ego or self-image the test subject will bring all his defense mechanisms into play and tell the researcher half truths or outright lies. A good example is a researcher asking a test subject whether he drank expensive wine or cheap wine? Very few subjects would admit to drinking cheap wines because of the social connotations.

A second pitfall is the test group. A random test group would come from widely differing socioeconomic environments. There are few times when a random sample is required because most packaging and products are aimed at specific markets, and to include test subjects who are not a part of those markets may produce invalid results. If a product is sold mainly to women in upper income levels, any test including women not in that economic group is misleading because the test subject in a lower income group may love the product and indicate she would purchase it when in fact she can't afford it. Identify to what market the product and package must appeal and use a test subject representative of that market.

The third pitfall is to assume consumers are interested in (and know about) packaging. Consumers purchase products. They can tell you what products they like and why, but they can't tell you much about packaging. A consumer can tell you, "I like that package," but a statement of that nature has no value to the researcher. Never ask a test subject to evaluate your package.

The fourth pitfall is to assume test results are conclusive evidence that a given package or product will be a success in the marketplace. Test results are indicators and should be used before, during, and after a design program to measure how well packages communicate. Test results plus market testing plus experience, however, will very nearly ensure a successful package.

In summary, the key to effective packaging is:

1   A product that is at least as good as your competitor's, appropriately priced.

2   A sound marketing plan that takes into account packaging, advertising, and the marketplace.

3   A package that makes full use of the four basic communicators.

4   Objectivity rather than subjectivity when viewing your own package and market.

If the above four conditions are met, visiometric test results are effective in developing a package that communicates, motivates, and sells the product.

CHAPTER ELEVEN

Paktest: An Approach to Package Graphics Evaluation

Larry S. Krucoff

When assessing a package design, a marketer often asks how well the package identifies the brand and the product, how easily the package can be located when shelved among competition, what the package communicates about the product it contains, and what the packaging contributes to overall product appeal. The designer keeps these same questions in mind when he creates his suggested solutions for the marketer's products.

The marketing researcher is the mediator who helps evaluate the designer's answers to the marketer's questions. The marketer presents the designer with a set of specific design objectives (brand identification, product appeal, etc.) and limitations (package size, shape, existing graphics, etc.). The designer presents the marketer with one or more recommended solutions, and the marketer must decide which solution best meets his objectives. One way is to base the decision solely on judgment. To insure enough precise data for an intelligent, effective decision, however, the marketer often turns to the researcher.

One of the tools available to the researcher is Paktest, a system for testing package and label designs. The Paktest system was developed to address specifically both marketing and design problems. Integral to the Paktest system are the design objectives of product and brand identification, shelf recognition, and affective communications. By establishing objective measures for identification, recognition, and affective communications, Paktest permits decision makers to gauge the degree to which proposed package designs meet his needs.

Identification, recognition, and affective communications are independent of one another. High or low scores on one of these performance dimensions do not predict performance levels on the others. This independence of performance measures means that a specific package design always embodies compromise — a balance of sometimes reinforcing, sometimes conflicting design components. Packages, for example, that are unique, easy to recognize and easy to locate on a supermarket shelf may be low in brand or product identification, or may have low consumer appeal. Thus independent measures for each design performance dimension are needed to permit designers and marketers to examine more precisely the impact and interplay of discrete design components.

Paktest offers an approach to package testing, not a standardized service. All projects are custom research. They vary by product category, number of items to be tested, and specific focus. Some projects, for example, may involve multiple label facings (as in canned soft drinks) or design variations within an overall format (such as flavors within the line). Typically, however, research will evaluate single package alternatives, and the Paktest approach examines these single package alternatives for the *visual comprehension* measures of in-store recognition, visual impact, and shelf impact and for the *affective communication* measures of image mediation and overall appeal. (See Fig. 1.)

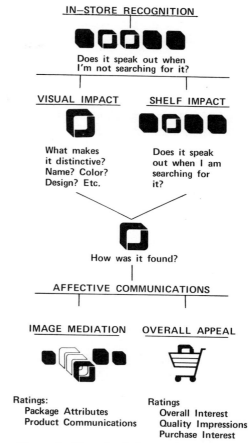

Figure 1   How the Paktest system works.

IN-STORE RECOGNITION

This performance dimension indicates a package's ability to be seen on the store shelf. To accomplish this, respondents are shown a series of in-store scenes on a slide projector, and the slides simulate a walk through a store. As the various aisles of the store come into view, respondents are asked to name all the products and brands they see.

When the shelf that contains the test design appears, the interviewer records all the brands named and the order in which they are named. The percentage of respondents who mention the test brand in total and the percentage who mention the brand first or second form the basis of the in-store recognition performance dimension.

VISUAL IMPACT

This performance dimension indicates a package's ability to communicate basic information about itself. Basic information includes product, brand, supporting copy, and the visual salience of supporting color, art, and design elements.

Using a tachistoscope, respondents see a slide of the test package for a brief period, usually 0.5 seconds or faster, after which they are asked to describe everything they remember seeing—words, pictures, colors, designs, and so forth. Respondents are then exposed to the same package for a second viewing, this time for a somewhat longer period of time, usually between 0.75 and 1.5 seconds. They are again asked to describe what they saw. Visual impact is the percentage of respondents who mention each of the design's visual components.

Two exposures to the design are routinely employed in order to isolate primary visual impressions from secondary ones. The speeds at which the packages are shown are pretested and selected on the basis that some but not all of the design components will be seen by the average respondent. The Paktest method uses two exposures rather than three or more since experience has shown that two exposures are sufficient to discriminate among package alternatives for the same components (e.g., brand and product identification) and to discriminate among components within the same package (e.g., background colors and copy line treatments).

SHELF IMPACT

For this, the second Paktest measure of recognition, respondents are asked to find a specific package. The faster they are able to locate correctly the test package, the greater shelf impact the package has.

The performance dimension is the length of time taken to find the design in question. After seeing the package during the visual impact measure and describing what they saw, respondents are told to "find the package you just saw." A shelf display containing the test package and other products is then shown and the amount of time respondents take to find the package is recorded by a reaction timer, an electronic clock designed specifically for Paktest, and accurate to 0.01 seconds. To determine if respondents have located the correct package, a schematic of the shelf display is shown, and respondents must indicate the correct location of the test design for their answers to be counted.

Shelf impact is the average of all find-times of all respondents who correctly locate the test package. In addition to respondents who mislocate the test design, everyone who takes longer than 5 seconds is eliminated from the count. This elimination institutes a control for visual acuity, reducing sample variance. (See Fig. 2.)

Using Paktest's systematic procedures, the shelf impact recognition measure builds from the visual impact identification measure. The interviewing sequence of the visual impact measure immediately followed by the shelf impact measure is critical. The sequence allows respondents to form their own visual identifiers. We do not tell respondents that their cue is the brand name or the color or the shape or anything else. Instead, respondents are simply told to "find the package you just saw."

The two store shelf measures, shelf impact and in-store recognition, are different aspects of visual performance. In-store recognition emphasizes brand comprehension in a shelf environment.

Figure 2 The Tachistoscope (T-scope) and reaction timer (electronic clock) in use. This apparatus can be set two ways. In one mode, the picture will flash on to the screen for a specified interval. In the second mode, the electronic clock is engaged and the image appears on the screen until the clock is stopped.

It simulates a shopping situation where a shopper is casually surveying store shelves on which there are a number of brands, and it is the brand name or some other equally strong visual element that strikes a responsive chord within the shopper. Shelf impact, on the other hand, assumes that the shopper is already familiar with the package. In real life this familiarity is developed through advertising and product experience. For test purposes, this experience is created through being exposed to the graphics during the visual impact test. This exposure familiarizes respondents with the package, and we reinforce impressions of what was seen when we ask respondents to describe in detail what they remember seeing.

What respondents tell us they saw on the package (how well it identified itself) thus precedes and is directly related to how quickly the package can be found when it is surrounded by competition during the shelf impact measure. One other advantage of this particular testing sequence is that it reduces some of the advantage a current package naturally has over a new design when trying to locate the product on a shelf.

IMAGE MEDIATION

The image mediation performance dimension indicates a package's ability to affect perception of and attitudes toward the product. With the package design in view, respondents rate the product using a battery of bipolar word scales. These scales have proven to be particularly sensitive to the impressions packages project about the product they contain. Both package attributes and product characteristics are rated. Package attributes are often represented by scales such as attractive versus unattractive, modern versus old-fashioned, plain versus fancy, and simple versus complex. Product attributes might include word pairs such as nutritious versus empty calories, for everyday use versus for special occasions only, expensive versus inexpensive. Particular bipolar word scales used for any project are determined by the communications goals set by the marketer.

The absolute levels of response to each of the attribute scales as well as the rank order profile of all attributes are carefully analyzed. While Paktest would examine, for example, the relative levels of attractiveness for each design being considered, the analysis would also investigate the relative position of attractiveness within the total brand profile. Attractiveness, it should be remembered, may not always be the ultimate goal. For a laundry detergent, too pretty a package may leave the impression that the product is for delicate items, not at all appropriate if a marketer desires to communicate the feeling of heavy duty.

OVERALL APPEAL

This performance dimension is concerned with product attributes that indicate personalization: for me versus not for me, high quality versus low quality, would buy versus would not buy, and so on. Once a package establishes its personality (image mediation), this measure determines the degree to which that personality is appealing to the potential customer. Again, the order of interviewing is important. Acceptance is built on product impressions. The product impressions are thus forced to the surface before the overall appeal measures are asked.

SUMMARY

The Paktest approach of integrated measures is produced in five successive steps:

STEP 1   *In-Store Recognition*   This first step tests the ability of a design's brand name to be recognized in a shopping situation.

STEP 2   *Visual Impact*   This is a measure of visual identity. By showing the design in isolation during two successive tachistoscopic exposures, this measure establishes the salience of the separate design components and reveals the immediate visual impressions a package leaves with the potential consumer. Moreover, exposure here familiarizes the respondent with the design and thus sets the stage for the next visual test, shelf impact.

STEP 3   *Shelf Impact*   This third measure of recognition is based on find-time, that is, how long it takes to locate a design when it is surrounded by competition. This measure assumes some familiarity with the package being sought and thus tends to simulate the typical shopping situation.

STEP 4   *Image Mediation*   This dimension of package performance reveals the ability of design alternatives to mediate attitudes toward, and perceptions of, the test product. For this measure the package is shown and, while in view, it is rated on bipolar word scales for packaging and product attributes. Ratings are then combined to form a profile of the brand, thus revealing its personality as filtered through package graphics.

STEP 5   *Overall Appeal*   This is a measure of personalization, or overall acceptance of the product. This measure builds on product impressions. For this reason, the overall appeal measures follow the image ratings.

By following these steps in the study of graphic design performance, Paktest establishes specific and reasoned measures for the critical goals of identification, recognition, and affective communications. Through the integration of these measures, Paktest provides clear guidance to both marketers and designers regarding the effectiveness and interplay of alternative package designs.

CHAPTER TWELVE

The Depth Interview

Clifford V. Levy

DEPTH INTERVIEWS: BACKGROUND, DEFINITION, EXAMPLES

Everyone who looks at packaged merchandise today is consciously and subliminally influenced by what he or she sees. For the past 30 years or so the package itself has assumed a share of the selling process along with merchandising and advertising. Because of the fierce competition between packaged products and between brands of the same products, manufacturers have learned that motivating the consumer to select their product and their brand is both an extremely important factor and a continual procedure.

Motivational Research is used to determine how to motivate the consumer to want a product or brand enough to buy it. The depth interview is the data collecting method most frequently used in motivational research. I want to briefly review the social, economic, and marketing history leading to the need for motivational research in packaging so that you will have a better perspective of this whole topic.

Background

A depressed people and a depressed economy in the United States was reawakened and rejuvenated by the war economy that began in 1939–1940. Jobs were available in shipyards and munitions factories, but these were far from the most highly depressed regions of the country. Becoming a mobile people rather than a settled people was the first important social change. The wages paid to shipyard and munitions factory workers were much higher than most people had ever earned in their lives. They could buy many of the things they had dreamed of having. This was both a social and an economic change. With our entry into World War II much merchandise became scarce or not available at all, so a great deal of money was saved in banks.

Four million or so servicemen were sent to every part of the world to fight the war, and although they spent most of their earnings, a good deal of it was saved. These men were exposed for the first time to other cultures; this was an important social change that would reflect on their behavior for a long time after the war.

At the end of World War II the servicemen streamed home; they had hopes of good jobs, college educations, and, in general, the fulfillment of the American Dream of a home of their own and a family of their own. Most of the war workers decided to stay where they were, in the cities, and and not to move back to the small towns and farms from which they came. The cities now bulged with servicemen looking for jobs and war workers needing new jobs, and both groups had a fair amount of money saved. More important, they had dreams to fulfill.

Factories retooled for a peacetime economy. The long awaited new cars began to roll off the assembly line and people lined up in automobile showrooms to buy them. Thousands of new families were starting and the demand for all household goods swelled; back ordering and waiting for stoves, refrigerators, beds, and so on was part of the buying procedure.

The federal government and many state governments opened the most important valve to enable the economy to expand to its fullest and to allow the American Dream to come true for millions of people. Home loans to servicemen were guaranteed; now home ownership, formerly reserved for the moderately rich and the very rich, would be shared by most segments of the middle class.

Land developers and builders chose to build homes in the suburbs, a natural choice because of more and cheaper land. This meant that schools would be built there, utilities would be extended, and all the consumer's needs would be provided for within his own community.

The supermarket was a relatively new and untried idea at that time, but not for long. The corner grocer in the city could handle one customer at a time, and often only the grocer himself was allowed to remove the packages from his shelves. This wasn't fast enough, not only for the consumer, but for the store owner, and the manufacturer as well.

We can see that the postwar prosperity caused a marketing and merchandising revolution not only in the supermarkets, but in the department stores, drug stores, and later in the discount-variety stores as well. The most important factor here relative to product packing is that the consumer was now relating more closely to the merchandise he or she was buying. Add to this marketing phenomenon the emergence of television as an entertainment and advertising medium. From its very beginning television advertising increased the sales of thousands of products. The packages had to be attractive on TV commercials so the viewer would recognize them when she saw them in the stores. The term "ask for" relating to most supermarket products had to be eliminated from advertising because the customer bought by what she saw. She selected the merchandise directly from the shelf.

Technological improvements in printing, photography, and packaging machinery have all contributed to the greater and greater importance of package design as a sales tool and have caused motivational research to become an important aid to the selling process.

Qualitative (Depth) Interview/ Quantitative Interview

About 90% of all marketing research studies have one or more of four basis objectives. They are concerned with consumer awareness, consumer opinion, consumer attitude, and consumer behavior. The 90% includes marketing studies for product concepts, products in test markets, established products, changes in established products, product advertising (both pretest and recall studies), and package design studies.

The quantitative study uses these objectives to measure the number, or quantitative measurements, of the replies of respondents to direct questions. These numbers are converted to percentages for: (1) all the respondents in the study and (2) segments of the respondents determined by their demographics—their sex, age, marital status, education, income, frequency of product use, and so on.

The qualitative study, or depth interview, uses the same objectives, consumer awareness, opinion, attitude, and buyer behavior. The purpose of the depth interview, however, is *not* to determine how many people are aware of or have various opinions about a product or how consumer's attitudes and buying behavior differ, but *why* they differ, and *how* these various characteristics are formulated.

Depth interviews go beyond the respondent's immediate verbal response to a question. For example, the quantitative questionnaire may have two questions about toothpaste. The first question may be "Which brand of tooth paste do you buy most often?" followed by a list of brand names for the interviewer to check. Below this will be an open-ended question: "Why do you buy that brand most often?" below which will be a blank space on the questionnaire for the interviewer to write the verbatim response, and answers to the probe questions.

Qualitative interviews substitute a question guide or interview outline for the questionnaire; the interviewer formulates her own questions based on the guide, and she records the respondent's replies on lined paper, or in some cases on a tape recorder. If toothpaste brand preference (consumer behavior) is the objective of the study the interviewer will open with an invitation to "discuss" it. She may say something like: "Mouth hygiene is talked about and written about a great deal these days. What are your thoughts about it?" You can observe several points about the opening statement. (1) The interviewer has mentioned 'mouth hygiene' as a topical subject. (2) She has not quoted any authority or opinion maker. (3) She has invited the respondent to 'give her thoughts,' that is, to discuss it.

The respondent's initial reply, her first statement in reply to this invitation-question, will give the interviewer the clue she needs to probe intelligently in depth, for more ideas and expressions of feeling from the respondent. (Incidentally, the respondent's first reply to this question will be extremely important to the research analyst of this study also.)

I am not going to suggest what the respondent's first reply to this question might be, because there are infinite possibilities. This example of the 'invitation-question' is meant to illustrate how quickly the respondent can become involved with the topic, and the many possibilities for follow-up or probe questions.

As an exercise the reader may wish to give his or her own reply to the invitation-question and may want to ask this question of 10 or more different people. If you try it, keep a record of the kinds of replies you get.

Summary

- Qualitative (depth) interviews and quantitative interviews use the same four basic objectives.
- Quantitative studies are primarily concerned with the number and percentage of replies to the questions; the 'what,' 'who,' and 'how many.'
- Qualitative studies are concerned with the differences in response, that is 'why' responses differ and 'how' opinions, attitudes, and behavior are formulated.
- The opening question for a depth interview must be (1) broad in scope, (2) objective, and (3) within the understanding of the respondent.

Depth Interview: Before, During, and After

The data collection portion of the depth interview is conducted by a trained interviewer who asks questions of a respondent, records the replies, probes for more (i.e. deeper) information, which she records. After the interviewer has covered all the areas of her question guide to her satisfaction she concludes the interview; then she will type her interview and deliver it to her supervisor or client. This is the physical aspect of the depth interview, but a great deal of thinking, planning, conversation and discussion, worrying, and background research is completed before the interviewer faces the respondent.

Purpose and Question Guide

As in all types of research, careful decisions must be made regarding the problem, and plans must be made for the study designed to solve the problem.

Nothing is more important in planning the depth study than deciding explicitly and describing exactly "What is the purpose of this study?" This is the responsibility of the researcher and his client, and quite often the purpose will be very elusive. Many problems may need solutions, but only the core problem can become the purpose. Different researchers have different techniques for isolating the purpose. I use this one: I write the words "What do you want to know" on a piece of paper. Below this I list my answers to this question. I ask my client to give his answers to the question also. Sometimes I'll have 20 things that I want to know. To put the list in proper order I first join those ideas that are similar or close, and I'll keep the rest.

I will examine the list and test each to determine how each is related to the other questions. This is a search to see which questions are dependent on other questions for their solution. After I have determined the order of importance I will rewrite the list putting the most important questions at the top and the others in descending order. This will usually solve the problem because the number one question will be the purpose of the study and the other questions will be the objectives. If we find answers to the objectives, we will have fulfilled the purpose.

This list serves another need: the objectives become the question guide. Since they are now in order of importance they need only be rearranged into topic areas and reexamined to see if *everything* we want to know is fully covered. We can examine this organized list and add additional questions.

A typical question guide will contain about seven topics to be covered in a 90 minute to two hour depth interview. Each of the topics will have between one and six subtopics, and each subtopic will have about three to five questions or discussion areas. This is a lot of data to gather in less than two hours. Not all of the questions, however, require deep probing; several of the questions guide the interviewer to determine what to ask and what questions to skip based on the respondent's buying behavior, attitude, and awareness. Since some of the questions are somewhat loose, the respondent's reply may give her response to more than one question.

The Respondent

Compared to the sample size for a quantitative study, the respondent in a depth interview study is part of a very small sample. She may be one of 20 to 100 people selected to be interviewed. Because of this small sample each respondent must fit several characteristics that the researcher feels will cover the 'market' for the problem being studied.

The researcher will list the characteristics of the respondents and the number of interviews to be conducted with each characteristic, or combination of characteristics. This is called "quota sampling," in contrast to the "random sampling" method used for quantitative studies.

Obviously the sample will include those people who can contribute the most to the enlightenment of the researcher and his client. The question that would be asked at some point prior to the depth interview is "Who is most likely to buy this product?" or "Who are the heaviest, or most frequent, buyers of this product?"

Here are two hypothetical examples:

1 A depth study is being conducted to determine (a) the best sales points for a new sports car and (b) a motivational advertising approach for this car. The car will sell for $18,000 to $25,000.

2 What are the characteristics of the car buyers in this price range?
- *Sex* Men more than women, but many women will be interested and can afford to buy the car.

- *Age* Primarily young people, although a few older people buy these cars. The term young at this point means under 30 or 35; older people are over 30 or 35, but more likely they would be 40 or older.
- *Income* The buyers would have to earn at least $20,000 a year, probably closer to $25,000 a year.
- *Occupation* In order to have a $20,000–$25,000 income and still be under 30 years old the person would have to be a professional, an executive, have his or her own business that is prospering, or be an energetic middle management person.
- *Life Style* The prospective respondent/sports car buyer would have to be "into sports cars." The car would have to be vital to his life; it must satisfy some important need.
- *Buyer Behavior* Most of the respondents would be owners of sports cars because they would fulfill several of the characteristics, primarily the life style characteristic.

3 The sample quota for this hypothetical study would approximate the number of respondents and characteristics shown below:

Total = 20

Sex: Male =	14	Female = 6
Age: 18–25	1	1
26–30	4	2
31–35	5	2
36–45	4	1
Income: $20–25,000	8	3
25–30,000	3	2
30,000 and over	3	1
Occupation: Professional	4	2
Executive	6	2
Self-employed	2	1
Other	2	1

Life Style: The researcher would provide the interviewer with 'screening' questions to determine if the prospective respondent truly fits this characteristic. Obviously owning a car of this type would automatically fulfill it.

Buyer Behavior: The researcher would also decide what make of sports car would qualify the person to be a respondent. These qualifications would be in the form of screening questions; the cars would be designated by make, size, foreign versus American, price paid, and how the car is used, (i.e., street and road driving, rallies, racing).

4 It is the interviewer's task to locate and qualify the respondent; if you refer to the quota chart you will see that the interviewer's task is not simple. For example, she must find a man between 18 and 24 earning between 20,000 and 25,000 dollars a year, employed or self-employed in one of the selected occupations, who owns a qualified sports car. Keep in mind that every respondent must fulfill all of the characteristics but that the combination of characteristics varies because she has four age groups, three income groups, and four occupation groups.

5 A respondent characteristic that interviewers hope for but have no control over is his or her ability to communicate. The lucid respondent is much easier to interview, but the interviewer's main task is fulfilling the quota by characteristics, so that she can only hope that the qualified person is also a good and willing talker.

The Interviewer's Task

The depth interviewer has little more than her interviewing skill, her question outline, and personal charisma to hold a respondent on a single topic for 90 minutes to two hours. She will have a small gift or cash gratuity for the respondent at the end of the interview, but never enough to compensate for what can be a grueling experience for him or her. The experienced depth interviewer will carefully plan for the interview so that it is most conducive to her task.

She will make sure that:

- The respondent is willing to be interviewed.
- The respondent has set aside the time for it.
- They will not be overheard or interrupted during the interview.
- The place of the interview, usually the respondent's home, is quiet and comfortable.

She will prepare herself by:

- Reviewing her outline and briefly rehearsing her opening questions.
- Having her materials and several pencils with her.
- "Psyching herself" to a point of confidence.

At the beginning of the interview she will:

- Establish rapport with the respondent as quickly as possible.
- Avoid small talk.
- Not discuss the interview outside the context of the question guide.
- Be businesslike but pleasant.

During the interview the depth interviewer must:

- Ask or paraphrase the questions from the question guide.
- Probe for the fullest possible reply.

- Determine if the reply to a question covers any other topic or subtopic.
- Continually record the respondent's answers verbatim.

Obviously she has her hands and her mind full. In addition to the above the interviewer must be aware that she wants 'depth replies,' and she must continually evaluate the depth of the response she is getting to a topic or question and decide if it is deep enough or if she should probe for more detail. The experienced depth interviewer has certain nonverbal probes that she will use.

- The pause or wait probe will often encourage the respondent to think and to fill the silence with another statement.
- The repeated word probe (i.e., Comfortable? Stylish? Exciting?). The one word with a verbal question mark will signal the respondent that he or she 'must' have more to say about that subject.
- Sometimes a friendly but quizzical look at the respondent will be used as an effective probe.

Near the conclusion of the interview the interviewer will look over her question guide to see if every topic and subtopic has been covered and also to jog her memory concerning any area that she feels might warrant an additional question or probe. When she is satisfied that the interview is completed she will thank the respondent, deliver the gift, and leave.

Handwritten depth interviews are rarely acceptable, most often they are typewritten by the interviewer. Shortly after the interview she will assemble her question guide and verbatim replies and begin typing. The researcher will have decided on one of two styles for the typed interview:

- One style is 'first person narrative.' For this the topics are written and the response is typed as though the respondent is talking. Probe marks (P) are omitted. A new subtopic is shown by parentheses within the typed paragraphs. This style is the easiest to read and analyze, but it does not show what was "off top of the head" and what was probed.
- The other style shows the topic, subtopic, and question and is typed in the same manner in which it was written by the interviewer. Probe marks are shown; pauses are indicated; fragments of speech are shown; probe questions are written out to show exactly what the interviewer said that prompted a given reply.

The typed questionnaire is returned to the researcher who must analyze the content and draw conclusions from it.

The Analyst's Task

It is the analyst's task to read the depth interviews, summarize the findings in each of them, and, further, to summarize all the findings collectively. From his summary he draws conclusions and makes recommendations to his client.

Unlike the quantitative study, the depth study does not rely on statistics for its summary. Each depth interview is an entity unto itself. The analyst looks at each and identifies the characteristics (i.e., demographics) of the respondent; as he reads he must be continually aware that these are the words, opinions, attitudes, depth of awareness, behavior, images, of an individual. For example, take an unmarried woman lawyer, age 32, earning $30,000 a year, who owns a Datsun Z model car. The analyst must think of *this* woman's life style, her hopes, dreams, worries, and frustrations as they relate to the purpose of this depth study.

He asks himself, "What has she said in this interview that might be a motivating force for the sports car advertising campaign?" He knows that she represents the 'market' for the product, and he wants to know what advertising approach and theme will move her enough to go to an auto showroom and look at this car.

The analyst works slowly and carefully. He or she has a routine that includes the examination of every reply to every question; during this examination the analyst must weigh each response to determine how emotionally important each is.

His first reading of the interview will enable the analyst to see the respondent's style: the length of her replies, the kinds of words she uses, and scope she covers in replying to a question, the level of her replies—are they straightforward or does the respondent relate her reply to things other than the question?

The analyst will look at the probed questions (if they are indicated in the typed interview) and note the length and depth of the probed replies. He wants to know if the respondent held back any feelings at any time during the interview, and if so, is this significant?

The analyst looks for patterns in the replies to various questions: Are all the replies the same length or do they vary? What topic areas or subtopics appear to be most important to this respondent? Why? Does the respondent use the same words or a wide variety of words to express herself?

All these preliminary steps help the analyst to get to "know" the respondent. He or she must have some idea of the respondent's personality so that proper evaluations of her replies can be made. If a respondent frequently uses superlatives such as "greatest," "finest," and "excellent," the value of these

words is diminished, and the analyst must be cautious when rating or coding responses that include these types of words.

If a respondent gives an expansive reply to a certain topic when her normal style is short, succint replies, it tells the analyst that the respondent is deeply concerned by *this* topic.

The analyst's task is to convert the 10 or so pages of the interview into a summary and then to convert these 20 or 30 summaries into a larger summary with conclusions. The analyst will use some method for coding the responses to the questions. He does this in order to determine the frequency of similiar types of response, and frequency is the basis for his analysis and his summary.

The larger summary and, more important, the conclusions, will be directed to the specific purpose for the study, in this case what advertising sales points for this automobile will be most effective.

DEPTH INTERVIEWS FOR HEALTH FOOD PRODUCTS PACKAGE DESIGN

One of the most dynamic package design competitions in the 1980s will be waged for health food products. Market forecasters feel that health foods are gaining momentum at a rapid pace and that hundreds of these products will leave the health food store in favor of the much wider exposure available to them in the supermarket. Because of this forecast I have chosen health food products to illustrate how depth interviews can be used in a "real-hypothetical" package design project.

Background on Health Foods

As of this writing, the health food market is supported primarily by two divergent groups. The first group is represented by the young consumer who is willing to pay a premium price for natural foods (i.e., those foods that are unencumbered by preservatives and those grown without chemicals). This group buys a wide array of products, including vitamins, bulk yeast powder, raw milk, soy bean milk, cookies, cake, bread, and even pizza made from natural products. Currently these products are found in "health food stores" located in most large cities, or they are bought by mail from companies that specialize in these products.

Older people or people on special diets are the other major supporters of health foods. Low cholesterol dieters form the base for this group. It also includes consumers who need certain foods for various allergies and diet supplements such as special vitamins produced in small quantities. Except for the vitamins, the health foods for this group are sold in supermarkets and smaller grocery stores. Imitation eggs, Egg Beaters, imitation bacon and hamburgers by Grillers, and chemical milk substitute products for coffee and cereal are examples of the products found in supermarkets.

Health foods for the weight conscious are found in many supermarket sections. Examples are low fat or imitation milk, cottage cheese, and ice cream, yogurt, and liquid meal supplements; also a full line of low calorie soft drinks and "lite beer" are being added almost daily to the health foods list. Many varieties of health food candy and candy bars are making headway into the market at a rapid clip. Formerly confined to the health food store, they are now found in drug and discount stores and supermarkets. Cosmetics and toiletries, soap, toothpaste, and shampoo are making their appearance in the health products arena. These products are tradition-

ally health oriented because of their "youth giving" qualities; their move toward "health emphasis" is easy to observe in the marketplace.

Those products presently found in the supermarkets show that some, but very little, research was conducted for the package design. Most of the supermarket products are there to fulfill the needs of a minority of the consumers. Many researchers feel that this market is not taken too seriously at this time because it is a tiny segment when compared with the total market.

One example of health emphasis packaging found in today's market, and a probable forecast of what the package designer of the 1980s will compete with is the Fabergé organic shampoo: its label says, "With Pure Wheat Germ Oil and Honey—Instant Conditioner." These words appear above a drawing of wheat stalks. This label addresses itself directly to people who know about and respect the health giving properties of wheat germ oil and honey; therefore, they are sales appeals directed toward health food buyers.

How will depth interviewing be used to assist package designers for these types of products? At what points during the package design assignment will it be used? What motivations will be tapped? Who will be the best qualified respondents? How will the depth interviews be translated into sales points and graphics?

The Client and the Product

Dr. Rudolph Webster Health Foods, established in 1919, was recently acquired by the National Foods Company. From the beginning Webster's had a highly respected line of natural foods that were widely distributed in health food stores only. Although it is financially successful, National wants to take this line of products into the supermarket for wider distribution. After a series of focused groups and a quantitative research study, National has decided that the natural food products must have a new package design before entering the supermarket competition.

The Depth Research Assignment

Depth interviewing is only one phase of the research that will go into this project. A great deal of secondary or library research will be conducted; intensive research on food product packages now on the market will be conducted. Packaging engineers will be consulted; color consultants, dieticians, and other professionals will add their input.

Our first project is to determine the consumer's *concept of the product(s)*. We will determine this from four areas of questioning that explore public awareness, opinion, attitude, and buying behavior of health foods, specifically natural mixed-grain cereal. This is our most important task. After we have completed it and submitted our report, and after three prototype packages have been approved, we will conduct a second depth study. The purpose of that study will be to determine which of the three prototypes best fits the potential buyer's concept of the product. The study plan for the first assignment will be the following.

Sample: We will conduct 100 depth interviews divided between four groups of consumers: (1) young, natural food oriented people, (2) older people who use specific products for health problems, (3) the diet conscious but not natural food oriented, and (4) those who currently use little or no natural, health, or diet food products. The sample will be further defined by age groups:

Group 1, between age 18 and 29.
Group 2, between age 45 and 70.
Group 3, between age 15 and 39.
Group 4, between age 18 and 65.

Next, the sample will be defined by the respondent's usage qualification and sex; we will give a quota of interviews to each group and subgroup.

Group 1 Must be a regular buyer of natural food products purchased from a health food store for at least 1 year. Must have a good knowledge of nutrition. Must use at least three "comparable" food products—products made from natural foods, for which a comparable product can be found in grocery and supermarkets. Respondent must use at least one supplement product such as yeast, wheat germ, high grade yogurt, or vitamins. *Quota:* Total = 20, Male 8, Female 12, can be single or married as long as the respondent buys the food for the household. Age categories: 18–24 = 10; 25–29 = 10.

Group 2 Must use at least one low cholesterol food product and must have been advised by a doctor to cut down on cholesterol intake. The product *user* for this group will usually be the man of the house, but the food buyer for the household will usually be the woman of the house. *Quota:* Total = 20, Males 5, Female 15. Age categories: 45–54 = 10; 55–65 = 8; 66–70 = 2. Males that buy food for the household: 3. Males with cholesterol problem where wife buys the food: 2. Females who buy the food for husbands with cholesterol problems: 12. Females who have cholesterol problems: 3.

Group 3 Qualified respondent is person who buys food for the household, or who specifically buys liquid or solid diet meal supplements for her/himself. They must use these products to replace at least three meals per week; they must be at least 10 pounds overweight by their own estimates; they must have awareness of diets and must have been on one or more diets within the last year. *Quota:* Total = 15, Male 3, Female 12. Age categories: 15–18 = 4; 19–24 = 6; 25–39 = 5.

Group 4 A qualified respondent is the person who does most of the food shopping for the household, whether male or female, the head of the house or not. There are no food buying or diet qualifications for this group. This is called the "target market" for Websters products when they appear on the supermarket shelves, because this group will require the greatest amount of motivation to be converted to natural (that is, health food) products.

In addition to the age and sex quotas included for the other groups we will include a quota for household income and ethnic types so that this sample will be somewhat representative of the "average" supermarket shopper. *Quota:* Total = 45, Male 10, Female 35.

Income: $10,000–13,999 = 5
         14,000–17,999 = 10
         18,000–21,999 = 10
         22,000–24,999 = 10
         25,000–29,999 = 15
         39,000 and over = 15
Ethnic: Caucasian = 35
         Black = 5
         Hispanic = 3
         Asian = 2

Notes about the Sample: Detailed sampling as illustrated above is absolutely essential for depth interviewing studies. The 100 respondents quota used here is probably larger than most clients would allow, but for launching a product into a new distribution system it would be necessary. The need for the four groups is understandable: The first three groups have knowledge and awareness of health, diet, or natural foods, the fourth group does not. Sex quotas are based on the sex of people currently buying these products; those who use fad diets, and those who are told to cut down on cholesterol. The age quotas are needed to get opinions that vary with age and experience.

The quotas are mutually exclusive; the interviewer must find a person who's experience fits him/her into one of the groups, then she checks off her quota for age, sex, income, and ethnic group. At the beginning the quotas are fairly simple to fill. Toward the end of her assignment the interviewer must find that one individual who fits all of the quota cells; this is difficult and time-consuming.

QUESTION GUIDE OUTLINE

What do we want the respondents to tell us about food in general, and health foods, and natural mixed-grain cereal? We want to know how they perceive the products; what they mean, what they symbolize; if, why, and how they use these products, and what "feelings" they have for the products, whether they use them or not. How did they first become aware of the products? Did a doctor, friend, magazine article, radio or television program introduce them to the product, and what influence did this introduction have on their eating habits? In short we want:

- Answers to the questions in the four question areas—awareness, opinion, attitude, and buying behavior.
- We want to know *why* these are their answers and *how* these answers were formulated.

We will prepare the question guide outline with these two ideas in mind. Also, we must keep in mind that the levels of awareness and interest of the four quota groups on these products are diverse, and we must not waste the respondent's time on deep probing in any area where the respondent has little or no knowledge.

Question Guide Outline—Natural Food Product Concept Depth Study

Interviewer: Respondent screening for qualifications and quota groups is conducted prior to the interview. Make note of the respondent's group and ask only those questions and probes that pertain to her/him.

Food History

Different people have different attitudes toward food. Some people are "into" foods, the gourmets for example, others can take it or leave it, and think of food merely as fuel. How do you regard food? (Probe, Probe, Probe)

Are your attitudes toward food different from what they once were? When did you change, why did you change, what changed them?

Was food important in your family when you were a child? Types of meals? Special occasions; Thanksgiving, Christmas, birthdays, for company?

Are you familiar with nutrition? What does it mean? Are you interested in it, low, medium, high interest? Have you read about it? Sources?

Current Food Usage

What is your meal eating pattern like? Three a day; less, more. What for breakfast usually, for lunch, for dinner? Between meals what do you eat? Why that? At parties, picnics, camping, vacation, what do you eat? What do you give up?

Special foods that you need? What? Why? For how long? Why did you start; who or what got you started? How important are these to *you,* to your health?

Ideal Foods

What would be your ideal dinner if you could eat anything you wanted? (For each item ask: Why that? What does that do for you)?

Food and Health Attitudes

To what extent is good health dependent upon food? (Probe, Probe, Probe).

Are your own ideas and decisions more important than the ideas of someone else: food expert, doctor, other? Why?

Are you careful or cautious about what you eat? When you eat? How much you eat? Why?

Do you ever intentionally skip your own eating plan? What occasions? Why? For how long? What is your favorite "diet escape food?" Why that? (Even though you have screened the respondents and know their quota place, ask the following of everyone, and skip those questions that are not appropriate.)

Special Food, History

What 'natural' foods have you used in the past two years or so? (For each ask) Why did you start using it? Who recommended it to you? What was your reaction to it? What did you like about it? What didn't you like about it? What has it done for your health, your weight, your feeling of well-being?

What 'health' foods have you used in the last two years or so? (For each, ask the same questions that you asked for 'natural' foods in the preceding paragraph).

What 'diet' foods have you used in the past two years or so? (For each question ask the same questions that you asked for 'natural' and 'health' foods.)

Attitude Toward Natural Foods

How do you feel about including natural foods in your diet or replacing some of the foods you have been using with natural foods? (Important probes!) If the price of natural foods was the same as foods you now use? If you could buy natural foods in a supermarket when you do your regular food shopping? What does it mean to you that "no preservatives are added" to a food? What does "pure food" mean to you? What does "organically grown food" mean to you? In what way would the use of natural foods improve your health? Would you consider using some (additional) natural foods that you found you liked if you felt they were beneficial? Why do you say that?

Mixed-Grain Cereal

When I say, "mixed-grain cereal," what do you think of? What does this mean to you? What would it be made from? Why do you say that? Who would use this product? (Probe, probe) What kind of people would they be—young, old, rich, poor, intelligent, not too intelligent; fast living or slow living people; people who worry a lot, or don't worry very much; most likely men or most likely women; would they be very much like you or not very much like you?

How would you use mixed-grain cereal if you used it? (Probe) Would you use it for breakfast in place of what you now use? Would you use it every day or once in a while? Why? Without knowing too much about it please tell me what you think mixed-grain cereal would taste like? Probe: Dry, bland, sweet, exciting, uninteresting. (Ask "Why" for each adjective.)

Sentence Completion Questions

The researcher will include 10 or so sentence completion questions at this point in the depth interview. The purpose of 'sentence completion' questions is to determine to what extent the respondent is oriented toward (a) natural foods and (b) mixed-grain cereals.

Is his/her orientation toward health food its imagery, its use, the benefits he might get from it, availability, price, or is he partially or totally turned off from any orientation toward it?

ANALYSIS, INTERPRETATION, AND REPORT

It is not possible to hypothesize what the researcher will find when he looks at the 100 completed depth interviews from this study. Needless to say the length, depth, and value of the interviews will vary widely. The interviewer's skill in 'drawing out' the respondent is a very important factor in the success of a depth interview study also. The respondent's orientation toward natural, health and diet foods, and his/her buying behavior will determine the contribution of each interview toward a meaningful report.

Although the analyst looks for emotional involvement between the respondent and the product, neither the length of the interview nor intelligent replies will necessarily render this emotional affinity. One or two words, a phrase, or a simple sentence that truly expresses the respondent's feelings can direct the researcher to uncharted analytical territory.

The questionnaires will be read and analyzed by group and, separately, by sex, by income category, and by ethnic group. The researcher will code the questionnaires as I explained previously.

The researcher wants to present a report to his client that tells how people relate to the product and what ideas, symbols, or colors will have the highest emotional appeal. He particularly wants to do this for the four groups.

In addition he must be able to back up his analysis and interpretation with as many examples as he can find in the questionnaires.

THE PROTOTYPE DEPTH STUDY

The prototype study that I described previously is the second assignment in which depth interviews will be used.

Those semifinal designs that the client, the designers, and the advertising agency feel have all the qualities needed by a natural food product entering supermarket competition are subjected to a depth interview study.

This study is far more structured than the first study. The purpose is to determine which of the two or three prototypes best fits the potential buyer's concept of the product and which will do the best job of *selling* the product.

The study plan will include 30 to 50 interviews; at least 60% of these will involve people who rarely if ever buy natural foods, and the balance will be divided between long time users, health food buyers, and dieters.

The procedure for this study will include a brief orientation of the respondent regarding what the interviewer wants him/her to do. There will be a few questions on eating habits and food buying patterns. The packages will be coded A,B,C, or 1,2,3 so that they can be shown in a different order to the various respondents.

The questioning will begin when the interviewer places one of the packages on a table, keeping the other packages covered. Her questions will include the following:

- Please look at this package and tell me what it means to you. (She will ask the respondent *not* to describe the package.)
- How do you feel about the size, the color, the words? What do they say to you about the product inside? (Probe for the fullest possible response for every item.)
- What do you feel about the package that makes you want to buy it? (Probe)
- What is there about it that turns you off? (Probe)
- How does this package compare to

packages for products you now buy and use? (Probe)

The same questions, with slight variations, will be asked about each of the prototype packages.

After questioning about the prototypes the interviewer will show all the packages for her final questions:

- Which package has the strongest appeal for you? Why do you say that?
- What is the one thing about that package that appeals to you most?
- Which package has the second highest appeal for you? Why do you say that?
- What is its strongest point?

These questions will conclude the interview.

The researcher/analyst has one purpose when he reviews these interviews:

- Which of the two or three packages has the most appeal? For what reasons?
- The emotional quality of the appeal is a factor that he must weigh in making his decision.
- It is possible, in a three package runoff, for a package to have one extremely strong point, but still not be considered the most appealing overall. The researcher may suggest that this point be incorporated into the design of the most emotionally appealing package.

The depth interviewing technique is a unique, somewhat exhaustive, research methodology. It is more expensive then other techniques, but it is also more rewarding. Used properly it can gather data not available by other methods. One who plans to use depth interviewing for package design research should study it carefully, and when the decision is made to use it, the user must have faith in what depth interviewing tells him.

CHAPTER THIRTEEN

Package Testing in Miniature Stores

Roy Roberts

A miniature store gives the package tester the opportunity of simulating a real life situation.

Since a package is, in effect, a salesman in a store, a store environment is an ideal place in which to test a package. Unfortunately, few stores will permit someone to alter their usual shelf setups and interrupt their traffic flows in order to conduct a package test (though I often wonder whether store owners wouldn't be receptive to having their stores used after hours if the incentive were adequate.) The next best solution is to have your own miniature store.

In order to accommodate the widest variety of packaging it is important that various kinds of store fixtures be used in the miniature store.

I regard the following equipment as necessary (See Fig. 1, p. 168.):

- Adjustable shelving.
- A store freezer.
- An open refrigerator case (often called "dairy" case or "meat and dairy" case).
- A "check-out counter."

The quote marks indicate that one does not have to acquire a manufactured check-out counter, which can be too elaborate and costly (and perhaps too large) for a test store. A simple formica top with an adequate substructure is all that is necessary to give the appearance of a check-out counter.

Figure 1   This miniature store fits into an area only 7 feet wide but contains 7 feet of adjustable shelving, a 7 foot display freezer and a 4 foot display refrigerator case. There is a small check-out counter at one end (not shown). Small as it is, respondents feel they are in a store because of the equipment and display of products.

What is most important is to give the respondent the feeling that he is in a store. If we are to bridge the gap from package testing as an abstract process to package *buying* or *selecting* as a real life process, then the respondent must be in an environment that simulates the feeling of being in a real store.

Accordingly, the store must be stocked with real packages. If the freezer case is not being used in the study, it must nevertheless be stocked with empty packages, and the refrigerator case can be filled with all sorts of empty packages of products normally found in such a case: dairy and juice cartons, bacon and link sausage cartons, cheese jars, and so on.

The shelves where the test packages are to be displayed must have the look of a "section." So a dog food package test must be conducted on shelves filled with a variety of the types of packages normally found in this section of a store,

preferably with all of the various varieties represented. In the case of dog food, for example, this would include not only canned, semimoist, and dry dog food, but various sizes as well.

Depending on the size of the miniature store there may be room for only one section, such as dog food. A better arrangement is to have enough additional shelving so that other products can be displayed, too, in order to give a more realistic atmosphere to the store.

MINIATURE STORE TESTING COMPARED TO PRIOR TESTING PHASES

With most of the package testing techniques leading up to miniature store testing the respondent is quite aware that he is testing a package, whether one is obtaining physical measurements such as in T-scope tests or attitudinal information.

The respondent therefore focuses an abnormal amount of attention on the package in general and often on its separate elements.

In the types of testing needed to develop a package it is almost impossible to avoid the respondent's awareness that he is testing a package.

In real life, however, people do not buy a package, they buy a product. The package's function is, of course, to present that product in its best light, to communicate its benefits, or to be recognized as the package of a product that one has seen in television, print advertising, or some other form of advertising that bears the brunt of communicating the benefits of the product.

In any case, the package is normally surrounded by competitors with whom it is fighting for shelf space or recognition. Since each package is shouting "buy me!" this should be taken into account in testing the package. In a miniature store it can be.

In an abstract situation the package standing alone might score high on many criteria, yet when standing side by side with competitors it may do poorly. Even a simulated situation in a miniature store does not necessarily compare to a real one if we are merely obtaining perceptions or attitudes from a respondent.

If, on the other hand, we can place the respondent under the same type of pressure he is usually under when making a purchase decision, we are much closer to a normal situation.

Accordingly, we should try to set up a situation in which he may actually purchase the product, or at least believe he is purchasing the product.

MINIATURE STORE TESTING TECHNIQUES

There are several techniques that can be used depending on the development stage of the package or label.

Real Purchasing of Real Products

The ideal situation is when the manufacturer can provide a quantity of the package to be tested. This usually means a small production run, which may be costly, but provides the tester with an opportunity to come as close to reality as is possible in a testing situation. The package should of course be filled with the product. Given these circumstances, we can now put the respondent in an actual buying situation.

At this point we must discuss various methods of putting the respondent in a buying mood.

Remember, in the real life world people usually go to a store in order to fill a need that they have at that exact moment: "I need laundry detergent" or "I need something for dessert for the kids" and so on. Of course, they also buy many products on impulse, over and above the shopping list they prepared.

In a miniature store, however, with its limited displays, this type of normal shopping is not possible. So what can we do to approach reality? The following are some guidelines.

Be Sure that the Respondent Is a User of the Product Category

This is assured by asking a number of screening questions. I feel it is important to disguise the actual product category we are dealing with by asking for product usage in a number of areas (e.g., "Which of the following products do you buy regularly?)".

Product Category
( )Dry breakfast cereal
( )Frozen meat pies
( )Canned dog food
( )Bottled spaghetti sauce
( )Peanut butter
Frequency of Purchase?

The package we are testing may be for

dog food, but when the respondent comes into the store he is not sure which of the above products we are interested in.

The main reason for the disguised screening is that often respondents are screened days ahead of time and invited to the miniature store. If they know that it is to obtain reactions toward dog food or dog food packaging they will become, I feel, overly conscious of the product category and perhaps its packaging. Since we don't want people to come into the store as dog food "experts," it is better if they do not know exactly what they are going to do when they visit the miniature store.

The frequency of purchase question is to enable us to weed out infrequent purchasers in favor of frequent ones. This question can be revised to include not only frequency of purchase but also quantity of purchase. (Someone who may buy dog food by the case only once every 30 days would be a better dog food customer than one who buys a few cans a week.)

Give the Respondent an Incentive To Buy

The most realistic package test is one in which the respondent lays out his own cash to buy the product. If we are to measure how well a test package does compared to a standard, we must have product movement. In order to get the maximum movement we must provide a stimulus. This usually takes one of several forms:

- *Seed Money* This is cash in a little envelope that is given to each respondent as he enters the store. The amount should have some resemblance to the amount of incentive given in new product introductions such as "cash off" incentives. Usually it is 25¢ to 50¢ depending on the selling price of the products to be bought. (Of course, if the product

sells for less than 50¢ then seed money should be 10¢ or 15¢.) In some cases the respondent is told that he can keep the money whether he purchases or not. In other cases he must return the money if he makes no purchases. There is probably more pressure to make a purchase if the latter method is used. Either way the respondent is paying for the major portion of the purchase with his own money.

- *Cents-off Coupon* If a coupon is given instead of cash then it must specify that it is valid for that day only in order to stimulate purchasing.

- *Discount or Sale* Discounts may be communicated in several ways. A list of the products for sale may be given the respondent showing the regular price and the discounted price. The savings may also be indicated in an adjacent column. Another effective stimulus is to have a sale on the products in the test category. This should be indicated by price markers on the shelves themselves rather than price stickers on the packages which might interfere with the design of the package being tested.

All the above have been stimuli designed to give the respondent the feeling that he will get a bargain if he buys today. Of course, none of these should replace the incentives respondents usually receive for visiting testing centers to "cover their expenses." These will vary from place to place depending on how easy or difficult it is to recruit the sample.

Advertising as an Additional Stimulus

If the product category is one in which most brands use TV or magazine advertising, you can heighten interest in the product category immediately prior to the time respondents enter the store.

This is a technique used in Yankelovich, Skelly & White's highly effective Laboratory Test Market program in which shopping in a test store is combined with gathering extensive product usage data from consumers, focus groups to obtain in-depth attitudes after purchase, and telephone call-backs to check on satisfaction with the purchase and future purchase intentions. (Purchasers are also given the opportunity to repurchase.)

After giving product usage and demographic information, respondents are shown a half-hour TV show, usually a situation comedy in which commercials for brands in the test category as well as other commercials are spliced.

If the product category is one in which TV advertising is rarely or never used then slides of magazine ads can be shown at various points in the program. This may be too elaborate or not even possible for many testing situations, but it is one way of providing an additional stimulus, which in turn should encourage more people to buy in the product category and thus to provide enough sales for a statistical base.

Build a Realistic Display

In order to build a realistic display one has to visit a number of representative stores in the area from which the respondents are coming and make a plan of just how the particular product category is set up in each store. In other words, if the test is being conducted in Chicago, then the stores to be checked should be in Chicago since there may be local brands in the product category and there may even be local idiosyncrasies with regard to placement of products by sizes or brand. The store display information should include:

1 Which brands are on display.
2 Relative positions of brands (Is Brand "X" usually to the left or right of Brand "Y"?).
3 Number of facings per brand for each size.
4 Where products are displayed from top to bottom of rack.
5 Store prices for each brand per size.

A minimum of three stores should be visited and each should represent a different chain. If the three stores differ as to how they set up displays in the product category, then follow the one from the most important chain in the community.

Test Each Package Alternative Monadically

In a test where people can actually buy the product, do not ever put both a current package and an alternative package for the same product on display at the same time. This is not the way things are in the real world and so *this type* of package test should avoid it. For each package alternative to be tested there must be a separate sample of respondents. Everything in the display remains constant except the package alternatives to be tested.

Note that it is extremely important that the cells of respondents for each packaging alternative be matched as closely as possible on all demographic and use characteristics. This is to insure that the only significant variable in each cell is the package being tested.

Let us say, for example, that the package being tested is for a family deodorant used by both men and women. Two new package designs are being tested, with the third variable being the current package. This gives us three cells, one for each variable. If Cell A is composed of 60% men and 40% women, Cell B has 50% men and 50% women, and Cell C has 40% men and 60% women, it is entirely possible that differences in scores on the effectiveness of each package could be influenced as strongly by the differences in the balance between men and women as the difference between the packages themselves.

Recruit a Large Enough Sample

How large should the sample be? I know of no hard and fast rules on this. However, be guided by one very' important fact: unless enough people actually buy a package in this type of test, the numbers will not be large enough to be meaningful.

Since there is no guarantee that *anyone* will buy, it is a good idea to start out with perhaps 100 respondents per cell and monitor the data closely as the test progresses. Initial results may be so lopsided in favor of one package that a larger sample may not be necessary.

For example, let us say that there are four package design variables and three constant competitive packages. Since we now have four cells, 400 people should be recruited. Let us also say that after the 400 people have participated the sales look like this:

Cell A

Test design 1 (new)	12 units
Competitor A	6 units
Competitor B	19 units
Competitor C	35 units
No purchase	28 people

Cell B

Test design 2 (current package)	24 units
Competitor A	7 units
Competitor B	15 units
Competitor C	30 units
No purchase	24 people

Cell C

Test design 3 (new)	30 units
Competitor A	5 units
Competitor B	17 units
Competitor C	30 units
No purchase	18 people

Cell D

Test design 4 (new)	34 units
Competitor A	9 units
Competitor B	20 units
Competitor C	25 units
No purchase	12 people

It is clear at this point that test designs 3 and 4 are doing much better than test design 1 and are also outselling the current package. Cell A can now be dropped and the test can be continued with 100 respondents each in the remaining three cells.

Let us say, that in the next phase test designs 3 and 4 continue to do a better job than the current package, but that there is still no clear-cut superiority of one design over the other. You can continue the test dropping Cell B, and test another 200 people and then examine the results to see if there is a clear-cut winner. If there is not, then other considerations may come into play:

1 Does one package do better against the major competitor?
2 Is one package cheaper to produce than the other?
3 Does one package do a better selling job in the home so that it will produce better repeat sales?

This last factor may also be tested. It is related to promise and performance.

It is possible that one of the test designs outsells the other because it implies that the product inside will be more satisfactory than the product inside the other test package, therefore heightening expectations of product performance. If the consumer then is disappointed, he may be less inclined to repurchase the product than the consumer who has bought the package that didn't do as well.

How can we test the propensity to repurchase?

One method used by some of the leading research companies is to follow up with telephone interviews to determine not only consumer satisfaction with the product but also to give the consumer the opportunity of repurchasing. Usually home delivery of the product is arranged.

While on the subject of a package doing a better job of selling in the home, I am reminded of an article that *Modern Packaging* magazine featured many years ago (June 1958) about the importance of side panels. I was interviewed for the article and remarked at the time that many package designers striving for a cleaner look wanted very little product information on the side panels.

Indeed, in some cases not even the brand name appeared on the side panel. Yet it is my feeling that in kitchen cabinets and pantries many products in boxes are stored with only the side panels exposed. What better place is there to keep the brand name in front of the consumer, as well as product information that could reinforce the consumer's positive feelings about the product?

In a package test involving buying in a miniature store the customer may pick up the package and study all panels before buying. Since there are many brands on display, however, the chances are small that the consumer spends much time examining the side panels. At home, however, this competition between packages usually does not exist and the consumer may want to read the side panels to help reassure him that he made a wise choice in buying the product.

It is therefore possible that a test design that does better in a miniature store, where many consumers may be responding impulsively to colors or a photograph or some other design element, may not do as well on repurchase for reasons given above. Thus if it is within your budget to conduct a repurchase phase you may discover that what seemed at first to be the best design will ultimately not produce the sales of a second-best design.

Ask Questions To Gain Insights

One important advantage of testing in a miniature store is the opportunity of determining why a respondent chose one product over the others. Whether a self-administered questionnaire is used or a personal interview is conducted, you should try to obtain this information. If your store is equipped with a one-way mirror and an observation room with sound then you can gain valuable insights by having a short personal interview conducted immediately following the purchase while the respondent can still look at the display and refer to it. The client can then observe this procedure without intruding. Of course, it is not necessary to have a one-way mirror to obtain the information. It may be recorded on videotape or audio tape, or the interviewer may simply fill out the questionnaire. However, the feelings are best communicated by actually hearing and seeing the interview.

The questioning can be as simple as: "Please tell me why you bought this particular product today rather than any of the others?" Or it may ask also for comments on the products that were *not* purchased. This may prove particularly useful if the test package was not bought.

What about those people who bought nothing? They, too, can be questioned. Since we know that they use the type of product on display and that they have been given an incentive to buy, why didn't they buy?

Usually there are three possible reasons:

- They have a large enough supply at home.
- They are not going directly home from the test store and may not want to carry packages with them.
- They didn't bring enough money with them.

It is possible in some cases to turn nonbuyers into buyers by making accommodations. For example, if you are testing a frozen product, particularly in warm weather, people may not want to carry it around if they are not going directly home, or if home is so far that there is danger of the package thawing. In this case you can supply insulated bags. You can also assure the respondent that if she wants to leave the testing center and do other shopping you will keep her package in the freezer until she returns and can take it directly home. If the respondent doesn't buy because she doesn't have enough money with her, you always have the option of extending

credit and sending her a bill for the amount of the purchase.

If they still do not buy you can still interview them but weight their answers less than those of the buyers. For example, this kind of questioning can be used: "Although you did not buy today, the next time you are in your regular store and you need this type of product, which of the products on display will you buy?"

You can also find out which one they would be least likely to buy. In both cases, ask for their reasons.

Real Selling With Promised Delivery

The technique we have been discussing can be used where a small production run of proposed packages enables you to have on hand enough stock so that the consumer can actually take the product home.

What do you do if you have only a few prototypes of each package? One technique is to give one of the buying incentives discussed above and take the money, but tell the respondent that the shipment has not arrived yet and that as soon as it does you will send or deliver the product. (A frozen product would need delivery by a car or truck equipped with an insulated container filled with product and dry ice.) The respondent still feels that she is buying, and she is not unaccustomed to buying some products this way since we often have purchases delivered.

You now have several options, some more costly than others:

1 *Deliver the Product in Its Current Package* (if it is already on the market this is how those who saw only the current package will expect the product to arrive anyway.) To those who were in a different cell and will be expecting a different package you

can enclose a note saying that the product is the same as the one she bought, but that the new package is not ready yet. Few people will ask for their money back.

2 *Mail a Refund By Check* This will save delivery costs. You can enclose a note saying that the shipment has not arrived and rather than hold on to the money you are returning it.

Since neither option 1 nor 2 permits you to follow up with a telephone interview to check on satisfaction with the product as discussed previously, you may choose to wait until the most effective new design is determined.

If a production run can then be made and the product can be delivered in a matter of weeks, those who bought can be contacted by mail or phone and advised that there will be a short delay. Those who do not want to wait can then ask for a refund. The majority, however, will probably wait and can then be interviewed by phone.

Real Selling, But Those Who Buy Take Home Only the Current Package

This will work only when the product is already on the market and the test is to determine if a redesigned package will be more effective. The respondents who choose one of the new package alternatives are told that the supply of new packages has not arrived, but they can take home the same product in the current package. Since they were given an incentive to buy, most people will be glad to take home the product in the current package. This does not permit you to follow up with a telephone interview to compare consumer satisfaction derived from one package alternative compared to another, but you can run

the in-store test nevertheless. This again is a technique to be used when there are only a few prototypes of the new package alternatives.

WHAT TO DO WHEN THE BUDGET DOES NOT ALLOW USE OF THE ABOVE TECHNIQUES

What If Too Few People Buy the Test Product? It is possible that none of the package design alternatives excite the respondents and that so few buy that there is not enough of a statistical base to draw conclusions.

Or perhaps a competitor so dominates the product category that most people choose that brand in the test and again you do not have large enough numbers on the test brand to draw any conclusions.

The obvious but most expensive solution is to expand the sample until enough people have bought the test brand so that you can determine which of its package design alternatives is most effective. This, however, could be prohibitively expensive. If too small a number of respondents buys the test product this particular technique will have to be abandoned in favor of less realistic ones, some of which are described later on.

Hold a Drawing

This is a technique that simulates reality to the extent that the respondent is asked to choose the product he most wants to win a supply of. Depending on the nature of the product it can be a 3 month supply or a 6 month supply or perhaps a 1 year supply. This will place most respondents under a pressure similar to that of choosing a product for which they are paying money. They are still trying to determine which product will give them the most satisfaction. Assuming that there is not a significant, perceivable price differential between the various brands on display, holding a drawing is one technique to determine which package promises the most benefits or attracts the respondent the most for whatever reason.

All the other elements of the test as described under the selling techniques remain the same, including the postselection interview. After the test is concluded a drawing is held and the winner is sent the supply. Note that if you do not want to wait until the winning package is in production you can send the current package if the product is already on the market. If it is not and you do not want to wait, send the winner a check for the value of the merchandise.

Behavior Prediction

Up to now we have not asked respondents what they would do in a given situation. We have placed them in a situation and they have behaved. We have exposed them to a realistic display in a miniature store and they have either purchased or indicated what product they want delivered to their home if they win a drawing. We have placed them under the stress of making a selection of a product they will keep and use.

If we do not want to make a commitment to deliver anything, then we must take a step back from reality and ask the consumer to predict his behavior. We can still keep the same cell design, but instead of giving the respondent an opportunity to buy, we merely ask him: "The next time you need this type of product and assuming your store has the same products for sale as you see here, which one will you buy?"

Of course you ask for reasons. You can also ask what the second choice would be and also what he would be least likely to buy and why. With any of the methods described above it is possible to administer as simple or comprehensive a questionnaire as you wish,

depending on budget, time considerations, and so on.

I feel, however, that a miniature store testing situation where we have deliberately avoided even referring to the package or using terms like "package design" is not the place to obtain the type of in-depth attitudinal information that should be acquired in the more preliminary phases of package testing, and that is described elsewhere in this book.

Providing a Realistic Environment for an Attitudinal Study

Another use for a miniature store is to provide a realistic setting for obtaining attitudes toward a package in the developmental stages of design. While this type of study can be done on almost any table or shelf, it may be useful to place the package in the environment in which it will be sold, whether it is a meat and dairy case, a freezer, or shelving for canned or packaged goods. Some products, for example, often are placed at the same level in stores. Ice cream is more often than not placed in a freezer that you reach down into for the package. In fact, in some freezer cases all you see is the top of the ice cream package. Inasmuch as this affects package design, it would be appropriate to obtain attitudes on this package as it sits in a freezer case rather than at eye level on a shelf.

Shelf Impact Studies

A miniature store test can be used to supplement a T-scope test to measure the impact a package has on the shelf when surrounded by the "clutter" of competitive packages.

Recognition Test

One method is to give the respondent a shopping list of perhaps five or six products and a basket to put them in. The shopping list might look like this:

Cookies	Nabisco Butter Cookies
Canned spaghetti	Franco-American
Soup	Lipton Chicken Noodle
Facial tissues	Scott (white)
Dog food	Ken-L Ration Beef Flavor

In the miniature store there must then be a cookie section, dry soup section, canned pasta section, facial tissue section, and canned dog food section. The respondent is then timed with a stopwatch. Let us assume that the test packaging is a redesign of the Franco-American spaghetti can and that there are three alternatives plus the current can. The sample will then be divided into four cells, with each cell shopping for a different Franco-American can. Since everything else in the store remains constant, and assuming that each cell's sample is matched demographically, then the only variable is the spaghetti can design.

Should one cell be able to complete its shopping trip faster than the others, we would conclude that the can design is thus making a stronger impact, so it is recognized faster and picked up faster.

Recall Test

In this type of test the respondent has no advance idea of what she should be looking for.

Only one section is needed for this test. Let us assume that the test package is for children's chewable vitamins. The section would contain at eye level the various brands and sizes of children's chewable vitamins, surrounded by some adult vitamin packages, and on other shelving would be other health care products, such as aspirin and cold tablets. The respondent would be told by the interviewer that she is going to be taken to the health care section of the store and once there would be asked to look at the products, standing at a set point.

She would then be given a certain time to look at the display. This can be varied depending on the complexity of the display. Usually 60 seconds would be the maximum time allowed. She would then be taken to an interviewing room. There would first be unaided recall where she would name everything she remembered seeing in the display by product category and brand name. Then an aided recall phase is conducted in which she would be given the product category that is being tested (assuming that she did not mention it in the unaided recall section) and she would list all of the brands she remembered seeing in that category. You can go further, of course, and ask for detailed information on the test package.

The criteria of effectiveness would then depend on how many people correctly recalled seeing the test package and identified the brand and the extent of information these people could give you about the package. Any number of alternative designs can be tested this way, as long as the sample cells are matched demographically and by usership characteristics.

There are probably other methodologies that can be carried out using a miniature store.

Keep in mind that while the illusion of shopping in a real store is important, it does not preclude the possibility of setting up temporary miniature stores in almost any location, even church basements, if there is no other place in a particular locality to set up such a store.

The atmosphere can be enhanced by point-of-sale posters or even cardboard displays that are often found at the ends of aisles filled with products.

Much store equipment can be rented for the project, so that if you need a refrigerator case or freezer it can probably be arranged.

Regular adjustable shelving, while available as a complete manufactured unit in metal, can also be built inexpensively as a wall-hung unit using standards, brackets, and lumber.

Lighting is of course very important, and any miniature store should be well lit with fluorescent lighting. Whether to use cool-white or warm-white bulbs may depend on what is being tested. The best advice is to conduct your own local survey of stores to see what color of bulb is being used the most in the section where the test product category is being sold.

The important thing is that you tell the respondent she is going into a store where she will be able to buy or select real products and (in some cases) spend real money. Once she understands that she can really buy you have bridged the gap from asking her to *predict her behavior* to *watching her behave*.

Most important is the fact that the results of the miniature store testing can now give you a more reliable prediction of how effective a package will be under actual selling conditions than the physical and attitudinal studies leading up to miniature store testing.

CHAPTER FOURTEEN

Package Testing in Shopping Malls

Doris G. Warmouth

OVERVIEW

Marketing research companies are doing more package testing in their shopping mall facilities today than they did a few years ago for several reasons. First, manufacturers are ever more aware of the importance of packaging in selling consumer goods in an increasingly competitive market in which consumers can literally choose from dozens of competitive products on the supermarket or drug store shelf. The increase of mass marketing, self-service merchandisers means there are fewer sales people to sell products. In this environment, the package must do its own selling.

Further, new products are continually being introduced into the marketplace and the packaging of a new product is critical to its success. If the consumer cannot easily and quickly identify the new product on the shelf, the chance of that product being a success is greatly diminished.

Additionally, marketing research facilities in shopping malls are ideal locations in which to conduct packaging testing. Consumers are in a shopping mood, and in an environment that simulates actual purchase situations and conditions. Further, shopping malls provide a ready supply of consumers of varying ages, income levels, socioeconomic groups and product usage habits from which to select target market consumers to interview. Other factors that make shopping malls particularly suitable for

179

package testing include the ability to control conditions under which packages are tested and the fact that most of the techniques currently used to test packages can be readily administered in a shopping mall interview facility. One last, important factor is that such studies can be conducted in shopping malls at reasonable costs.

SOME PACKAGING TESTING TECHNIQUES SUITABLE FOR USE IN SHOPPING MALLS

Almost all the currently used techniques for package testing can be conducted with ease in a shopping mall interviewing facility. Included in the most frequently used techniques are:

1   Focus group discussions.
2   Tachistoscope tests.
3   Semantic differential tests.
4   Other connotation and denotation studies.
5   Simulated purchase situations.

Each of these techniques, as used in shopping mall interviewing, is discussed in detail in the following.

Focus Group Discussions

Focus group discussions are used most commonly to test new package concepts—a squeeze tube for jelly, for example—or to gain insight into consumers' reactions to a dramatic shift in packaging, such as the change in packaging of sugar from boxes to plastic bags or of frozen pies from a cardboard box to a clear plastic container.

Most marketing research interviewing services located in shopping malls have one-way mirror rooms that allow clients to observe these discussions as they take place. These rooms are wired

for sound so that observers can both hear and see the discussion in process. The discussion is also recorded, with two separate sound systems generally used as a safeguard measure in the event that one system fails.

The focus group room itself is set up either as a living room or in conference room style. The living room style has the advantage of being homelike and informal. The conference room, while more formal, is convenient for displaying packages, for passing materials from one person to another, and for writing or filling in questionnaires.

Consumers are recruited to attend these sessions based on predetermined demographics (e.g., age, income, household composition, or marital status) that reflect the target market for the product, and on the basis of product usage. All screening for product or brand use is blind to prevent respondent bias. The particular brand or product being tested is always concealed in a group of products or brands during the screening process.

Some groups consist of users and nonusers of the product category, while other groups are composed solely of product category users. In the latter case, users of the particular brand being tested and users of competitive brands are generally recruited to attend the sessions. Other focus groups are composed of consumers with varying levels (light, medium and heavy) of use of the particular brand or product being tested.

As in all package testing, it is important to get reactions to packages not only from present users of the brand/product but also from potential users. A package that appeals to present users but is unappealing to potential users does not serve its purpose any more than does a package that appeals to potential users but that current users of the product dislike.

All prospective participants are also

given a "security" screen to eliminate consumers who are (or with family members who are) employed in marketing research, advertising, or the industry of the test product. Such consumers are atypical of consumers as a whole and would distort findings of the focus group discussion. Prospective participants are also screened for previous participation in focus group discussions to eliminate "professional respondents." At minimum, no one who has attended such a discussion in the past year is permitted to attend; under strictest conditions, all participants must be virgin, never having attended a focus group discussion. Another requirement for focus group discussions is that none, or at most two, of the participants know each other. This restriction allows for frank, individual opinions being given, rather than having participants being overly conscious of what they are saying and the effect of their opinions and comments on their friends.

The topic of the focus group discussion is described to prospective participants only in general terms, for example, "about beverages," never specifically, for example, "about some new packages for tea," to prevent consumers from coming to the groups with prepared, preconceived ideas and opinions. Participants in group discussions are given a cooperation fee, presently about 10 to 20 dollars, for attending the session. The number of participants in the group ranges from 6 to 12 depending on moderator preference. Groups are limited to this size so that each participant has enough time to express his or her opinions and to keep the group to a manageable size for the moderator to handle.

The actual recruiting for discussion participants can be done in several ways. One is to recruit in the mall itself by having interviewers approach mall shoppers and screen them for eligibility to attend the focus group. Another commonly used way to recruit group participants is to screen by calling randomly from local telephone directories. Both of these recruiting methods result in good cross sections of consumers attending a focus group discussion.

Another, less costly, way to recruit participants for these discussions is to use organizations, contributing the co-op money to the organization instead of paying it to individuals. The use of organizations, however, has two potential problems associated with it. One is that participants may not have been properly screened and the other is that they are likely to know each other.

The groups themselves, led by trained moderators, are informal and free flowing. Participants are identified by first name only and light refreshments are served to help create a pleasant, relaxed environment. In this atmosphere, the moderator introduces the topic of product packaging, frequently beginning by asking for consumers' opinions of current packages, problems they have encountered with them, packages liked and disliked, and so on. The conversation is then led to a discussion of the particular packages being tested. At this point the moderator may simply give a verbal description of the test package(s) or have mock-ups or prototypes to show members of the discussion group. The remainder of the session is then devoted to a discussion about the test packaging.

While results of these sessions are obviously not projectable to the total market, they do provide valuable, in depth insights into consumers' perceptions about packaging, reactions to the new packaging (either positive or negative) or intended use of the new package. Frequently, potential problems with a particular package, which can be corrected before any further testing is done, are uncovered at this stage of the research.

Tachistoscopic Tests

This test, commonly called the T-scope test, is one of the most frequently used techniques for package testing conducted in shopping malls. This technique is used to measure the impact of the package and to evaluate package graphics and design.

Respondents are taken through the T-scope test individually. In this procedure, consumers are shown slides of test packages at very fast speeds such as $1/100$, $1/50$, or $1/10$ second. The actual speeds used vary with the package being tested and must be determined experimentally. Generally, the same slide is shown to the respondent several times, with each successive exposure being at a slower speed. After each exposure the consumer is asked to report what he or she saw on the screen—any words, colors, pictures, shapes, anything at all. In these quick exposures, which approximate the amount of time a consumer has to look at a package while walking down a supermarket aisle, the most salient elements of the package design are seen. The impact of the package by itself and in comparison with competitive packages can then be determined from the degree to which each component of the package design is recalled.

Consumers are recruited to take part in a T-scope study from mall shoppers on the basis of predetermined demographic and product use characteristics. The recruiting and screening process is similar to that described in the section on focus group discussion recruiting. The T-scope sample generally consists of users of the product category, split between users of the test brand and competitive brand users. This screening takes place on the mall itself, and those consumers who qualify are invited into the interviewing office to take part in the T-scope study.

It is critical that the screening for eligible respondents be done carefully so that the study is conducted with the desired target market consumers. For example, if a cereal package were being tested, respondents would be almost certain to include women who had young children living at home since this is an important segment of the cereal market; if the test package were designed for a one- or two-serving frozen entrée, the sample would include persons living in small households. Sample requirements are of course determined by the client, but it is the responsibility of the research company to ensure that the study is being conducted with the proper respondents.

There are a number of technical considerations that require careful attention when a T-scope study is being conducted. Because of the visual nature of T-scope testing, respondents who normally wear their glasses for watching television or movies must wear their glasses while participating in the T-scope test. Obviously a person who needs glasses to watch television or movies needs glasses to see slides flashed on a screen for extremely short periods of time. Therefore, during the screening process all potential respondents are asked if they normally wear glasses and, if so, do they have their glasses with them. Consumers who normally wear glasses but do not have them available at the time of the interview cannot participate in a T-scope study.

Because of the technical nature of a T-scope study, the test is conducted in the interviewing office or an enclosed area rather than in the mall itself. For example, it is important to control the amount of light in the room where the T-scope is being administered and to keep the amount of light constant for all respondents. If the methodology calls for the room to be dark, it must be dark for all respondents. If the room is to be lighted, which is possible if the proper type of screen is used, the amount of light in the room must remain constant for all respondents.

Another consideration for a T-scope study is that there should be nothing to distract the respondent's attention during the test. Because the packages appear on the screen so briefly, the slightest distraction can cause respondents to miss seeing the package entirely and then report seeing nothing when, in fact, they were not watching.

It is also particularly important to put the respondent at ease for the T-scope procedure. First, it is an unfamiliar situation for most consumers, so they may feel a certain amount of apprehension about the test. A good interviewer can minimize the amount of anxiety a respondent feels. Second, unlike almost all other research, there *are* right and wrong answers to T-scope questions and it is difficult to keep the respondent from knowing this. The use of practice slides, to accustom the respondent to the procedure, helps to mitigate the anxiety and "test" problems. By showing the practice slide at relatively slow speeds the respondent is likely to be able to identify the product correctly and therefore feel comfortable and satisfied with his or her accomplishment.

Another consideration is that every respondent should see the same size packages when they are flashed onto the screen. The projection size is easily controlled by maintaining constant distances between the T-scope equipment, the screen, and the respondent's chair. This distance, obviously, must be fixed so that the package is projected to a size that can be read from the viewing location but is not so large as to be unrealistic. Control of the image size across the sample of consumers included in the study is essential for comparability of results.

A further technical consideration is that the slides should be projected on the center of the screen to reduce distortion around the perimeter of the projected image. (Putting the slide into a metal frame between two pieces of glass also helps to control distortion.)

Finally, it is important to check the T-scope equipment itself during the course of the study. Particularly at the fastest speeds it is impossible to determine by observation whether the equipment is working properly or not. (Try distinguishing between $1/100$ and $1/50$ second and you will quickly appreciate this problem.)

Each of these conditions or potential problems can be controlled in a shopping mall interviewing facility that is carefully and correctly supervised. The amount of control needed explains why T-scope studies must be conducted in the interviewing office or an enclosed area as opposed to being conducted on the floor of the mall.

Semantic Differential Tests

The semantic differential technique is also a standard package testing technique that is well suited for use in shopping mall interviewing. The semantic differential, which measures connotations of the package and the product contained therein, is frequently used in conjunction with other package testing techniques such as the T-scope test procedure.

Consumers are given pairs of opposite words with a series of spaces between the polar words and asked to indicate the space on the scale that best describes the package or product. The conventional number of points between the polar words is seven, but other odd numbers of points are used. This use of a multipoint scale allows the measurement of both direction and intensity of feeling or belief.

This technique measures three factors—evaluative, potency, and activity—with respect to packaging. An example of the evaluative factor is attractive/unattractive; strong/weak is a potency measure, and active/passive is an example of the activity factor. The actual choice of polar words used in a semantic differential package test depends on the connotations and characteristics that the package is designed to convey about the product it contains. Different sets of scales are generally used for the package and the product itself. Package scales might include dimensions such as modern/old-fashioned, attractive/unattractive, easy-to-read/hard-to-read, interesting/boring, and colorful/colorless. Scales used for the product might include high quality/low quality, expensive/inexpensive, and strong/weak. Clearly, dimensions used for a frozen dessert pie would differ from those used for a shampoo. The important thing is that the scales measure the extent to which the package communicates the attributes that the manufacturer wants to convey about the product.

Consumers are screened for eligibility to participate in a semantic differential study on the basis of predetermined demographic and product use characteristics. Some studies also require the respondent to have certain attitudinal characteristics, such as being a trier of new products. Screening takes place on the mall, with the interview itself being conducted either on the mall or in the interviewing office. The choice of interview location is dependent on the use of other package testing techniques in conjunction with the semantic differential. If the semantic differential is used with a T-scope test, the interview will almost always be conducted in the interviewing office or an enclosed area. If, on the other hand, it is used alone or in conjunction with some other technique, the semantic differential can be administered on the mall itself.

The test package is on display while the consumer is completing the semantic differential. If the semantic differential is being used with a T-scope test, the test package is usually shown on the screen the entire time the respondent is completing the semantic differential. In other cases a sample package may be on display while the respondent completes the semantic differential.

The interviewer has the important job of explaining the semantic differential to the respondent without in any way influencing the respondent's answer. Since many people have limited ability to think abstractly, it is especially important that the semantic differential be explained carefully and completely to the respondent. This lack of ability to think abstractly is one reason why the semantic differential is often presented to the consumer in words—very attractive, somewhat attractive, and so on.

The semantic differential scales are explained to the respondent by telling him or her that there are no right or wrong answers, that we simply want to know which point on the scale best describes the package or product to the consumer. Respondents are further instructed that the more like the word on the left they think the package is, the closer to that word they should mark the scale; the more like the word on the right they think the package is, the closer to that word should they mark the scale. The semantic differential is a self-administered technique. The respondent is given a sheet of paper with the test scales and asked to X the place on each scale that best describes the package to him or her. The positions of the positive and negative words are rotated so that both positive and negative words appear on each side of the scale. This is done in order to prevent column checking and to encourage the respondent to consider each scale individually.

While the consumer is completing the semantic differential, the interviewer monitors the respondent's answers to check for completeness and lack of comprehension on the part of the respondent. One indication of this is straight line checking—every scale marked in the extreme 7 position, for example. While this does not necessarily mean that the respondent doesn't understand the procedure, in the majority of cases further probing uncovers the respondent's lack of comprehension of how the procedure works.

If used in conjunction with other package testing techniques, the semantic differential is usually administered after the other portion of the interview because it is an aided evaluation of the package. The consumer is judging the package on the basis of preselected attributes rather than freely reporting the attributes he or she believes the package has. In order not to bias answers given to unaided portions of the interview, then, the semantic differential has to be administered at the end of the interview.

Other Package Connotation and Denotation Studies

This type of package testing has as its goal the determination of what the package conveys or suggests to the consumer, that is, the connotations of the package in addition to the direct or specific meaning the package denotes to consumers.

There are several ways, in addition to the semantic differential, to measure package connotations and denotations in a shopping mall environment. Consumers can be shown actual test packages and asked open-ended questions about them, such as:

- What comes to mind when you see this package?
- What type of product would you expect to find in it?
- How much do you like or dislike this package?

- What do you like about it?
- What do you dislike about it?
- How would you describe a company that would use this package?

This method of package testing requires excellent interviewers who are capable of probing respondents' answers to produce detailed information on how consumers perceive a product and its package. The quality of the results of this type of package research is heavily dependent on the interviewers' ability to conduct open-ended interviewing and requires strict control over the interviewers by the research company conducting the interviewing.

Alternatively, packages can be rated on selected attributes such as legibility, attractiveness, quality of the product inside, nutritiousness of the product (in the case of foods), strength of the product (in the case of pharmaceuticals), or suitability of the product for particular uses.

This more structured type of interviewing offers the advantages that it requires less dependence on the interviewer's skill and that it provides an evaluation of the package on all attributes of interest by every respondent. The disadvantage is that the package is rated on preselected attributes, which may not always include all those that consumers would freely mention in unstructured questioning.

Simulated Purchase Situations

This package testing technique is a behavioral measure of how consumers react to test packages. Conditions that a consumer experiences in a retail store are simulated by displaying the test product on a shelf along with competitive products. The arrangement and facings are carefully designed to replicate conditions that exist in retail stores.

Consumers are shown the display and asked which product they would like to buy with money that was given to them at the beginning of the interview. This technique is often used to test new designs against a present package design, with results being measured in terms of the relative portion of purchases in which consumers select each of the test brand packages. It is not meant to test the absolute percentage of the time the test package brand is selected instead of the competitive brands.

Consumers are screened to participate in this type of package test on the basis of appropriate demographic and product use characteristics, with the sample including both buyers of the test brand and buyers of competitive brands (unless, of course, it is a new product in which case there are no buyers of the test brand). Individual studies may have additional screening requirements.

When several packages for the test brand are being evaluated, each consumer is exposed to only one test package. The samples of consumers seeing each version are carefully matched on demographics and brand usage so that any difference in ''purchase'' rate is a function of the package and not of the characteristics of the respondents.

It is also important that the position of the test product on the shelf be comparable for each of the test brand packages. The test product may always be placed in the center of the display or it may be rotated with competitive brands. In the latter case, care must be taken in the administration of the research that comparable rotations are used for each version of the package being tested so that position bias does not affect the results of the research.

Before consumers are shown the packages, they are given a fixed amount of money to ''spend'' on the products displayed. This amount may be calculated to allow the respondent to buy only one package or several packages depending on how the study has been designed. In either case, the consumer is shown the products and asked which one(s) he or she would like to buy with the money given to them. (In fact, respondents are almost always permitted to keep the money after they have made a purchase since test packages are rarely available in a sufficient quantity to allow consumers to take them home.)

Differences in ''purchase'' rate between the test products, then, are due to the differences in the packages since all other conditions are held constant. Here again, rigid control of field conditions by the market research company is essential to produce accurate results.

This type of package test can easily be used with other package testing techniques such as connotative and denotative measurements. If other measures are used, they are administered to the respondent after the simulated purchase so that the purchase measure is pure. The other measures can provide valuable diagnostic information about the packages such as the strengths and weaknesses of each package or the reasons one test package is superior to another, which can be used to improve the winning package further.

SOME OTHER ELEMENTS OF PACKAGE TESTING

In addition to testing packages for their impact, design, and connotations, there are other elements of the package that require research. These include testing packages for:

1   Size and shape perception.
2   Product line identification.
3   Mechanical features.
4   Information content.

Each of these measures, which require

special consideration, is discussed in the following.

Size and Shape Perception Studies

Package testing can be conducted to determine how consumers perceive packages in terms of their size and shape. Two packages of different shapes may hold an identical amount of product but one may be perceived to hold more than the other. The package that appears to be larger would surely be the one a manufacturer would prefer to use.

This kind of package testing may be conducted on the mall itself with consumers being recruited to participate at the interviewing station. This involves screening the potential respondent for eligibility in terms of product use and demographic characteristics specified by the client to conform to the target market for the product. Consumers may have to be of a certain age, have young children in the household, or be married, depending on the particular package being tested, or they may have to be users of a particular product category such as beer. If the package(s) being tested are for a new type of product, the screen frequently involves describing the product to the consumer and asking him/her how likely he or she would be to buy it. In this case, eligible respondents are generally restricted to those who indicate a positive intention to buy the product if it were available.

The packages are displayed to the respondent, who is then asked which one holds the larger (or largest) amount or, more indirectly, which one he or she would buy if the packages were all priced the same.

In addition to affecting consumers' perceptions of relative amounts of product contained therein, the shape of the package may affect consumers' perceptions of the product itself. This might include dimensions such as quality, purity, nutritional benefits, and strength. The shape of the package may even affect perceptions of the product's price. To measure the effect of shape on these attributes, consumers are also asked to rate or evaluate each of the different shaped packages on appropriate attributes. The resulting data are then analyzed in conjunction with size perceptions to select the optimal package shape.

This package testing requires that care be taken to ensure that the position of the packages in the display is rotated so that each test package appears in each position an equal number of times. The order in which each package alternative is asked about is also rotated to allow for position bias.

Package Line Testing

Packages designed for a line of products have common elements of design in order to identify the product line, with individual variations to identify specific products, for example, shampoos for dry, normal, or oily hair, or different flavors of frozen dessert pies. This package testing, then, is conducted to determine if consumers recognize individual products as being part of a line and if, having done this, they are able to identify the individual products within the line.

This research can be done by showing respondents photographs of the products for a controlled (usually brief) period of time, by showing a display of actual packages for a controlled time, or by using the T-scope procedure discussed previously.

In any case, the objective is to ascertain whether consumers recognize the products as being variations in the line and whether they are able to identify the individual products within the line correctly.

The displays or photographs are designed to look like what the consumer sees in a retail store. A shampoo display would have the same arrangement and multiple facings that a consumer would encounter in a supermarket or drug store. A frozen food would be arranged as the consumer sees it in the supermarket freezer. In the latter case, a decision has to be made about showing the sides of the packages as they appear in shelf freezers versus showing the tops of the packages as they appear in a chest freezer. This decision is generally based on the most common type of freezers used in the market areas of the product's distribution. If both types of displays are commonly used, both will have to be used for the package testing.

The interviewer in these studies has the critical task of probing fully for everything the consumer can recall about the test packages, particularly with respect to product differentiation, without in any way leading the respondent into reporting differences that he or she did not really observe before prompting.

Individual product variations or formulations are frequently distinguished only by color—a different color cap or different color printing on the packages. This, too, presents particular problems, those associated with color blindness, which must be addressed in the research procedure. In this case, respondents must be given the additional screen for color blindness before they are permitted to participate in the study to prevent a distortion of the results of the research.

Testing the Mechanics of a Package

The introduction of a new product or changes in a present package design can involve the way a package is opened, how the product comes out of the package (a pump versus an aerosol spray, for example), how the package can be re-

closed, how it is stored, and how easy or difficult it is to handle or use.

It is as important to test these mechanical features of the package design as it is to test graphics or package information communication. This testing of the package mechanics can be done efficiently in a shopping mall.

After being screened for appropriate demographic and product use characteristics, consumers are given the test package and asked to use it. They may, for example, be asked to open a present and proposed package. Then they are questioned about what they like and dislike about each package, their preferred type of package, which package is easier to open, and so on.

Interviewers must be careful to present the alternative packages objectively, describing them as "different packages" or "two packages" rather than revealing to the respondent that one is a new package (which to many people implies better or superior) and the other an old package. Rotation of the order in which the packages are tried is also critical to the success of the study.

The interviewer has the additional responsibility here of reporting nonverbal responses to the test, for example, a struggle to open a package because the packaging material is too strong or ill-fitting, a lack of ability to open it on the first try because the consumer was not using the proper method of opening, or the need for the consumer to read the instructions carefully before being able to open the package. These nonverbal responses can be as or even more important than the consumer's answers to direct questions.

Package Information Content Testing

In addition to serving as silent salesmen of a product, packages can provide the user with salient information about the product. Every package contains information about the quantity it contains and other mandatory requirements such as ingredients and nutritional value (in the case of foods) or use cautions or warnings. However, packages can also contain optional information such as preparation suggestions, instructions for opening and closing the package, storage or serving suggestions.

Package information testing is done to determine whether the information on the package is clear and understandable to consumers. Consumers are asked to study the package carefully and are told that they will be asked questions about it. They are then asked what they can recall, if there was anything confusing or hard to understand, and how they evaluate the information on the package. This type of package research sometimes uncovers the fact that the package communicates unintended messages or misinformation, which, of course, has to be corrected.

It is also important, however, to ascertain if there is sufficient information on the package, that is, to find out if there is additional information consumers need or want to be able to use the product correctly to obtain maximum benefits. Sometimes preparation instructions are not clear to consumers, for example, "defrost 2 hours," without specifying if this means at room temperature or in the refrigerator. The interviewer in such cases has the delicate task of ferreting out the missing information from consumers without leading or suggesting possible missing information to them. New products in particular require packages that give the consumer detailed information on how to use the product.

Packages can also help to increase product use by suggesting different ways to use the product. This can be in the form of menus, occasions for serving, use of the product as an ingredient in a recipe rather than as a dish by itself, and so on. Various alternative ways to present these uses are tested in shopping malls with the aim of determining the most effective presentations. Generally this is done by measuring interest in the suggested uses, intention to use the product in the suggested ways, and buying intention.

SUMMARY

There are many types of research objectives to be met in package testing, many methods of conducting package testing in shopping malls, and many different kinds of packages tested. Whatever the research objective, the methodology, or the type of package involved, the research company has certain basic responsibilities to the client, including:

1 The need for strict control of the study by adhering to the prescribed procedure.

2 Accurate screening of consumers to ensure that they are eligible to be included in the study because they meet product use and demographic requirements as well as the security screen or other requirements of the study.

3 Professional interviewing where the questions are asked exactly as written and probing is done correctly to provide the maximum information without leading or influencing the respondent.

4 Proper rotation of product order so that each package appears in each position with equal frequency.

5 Complete reporting to the client of additional information, comments, or questions from respondents that did not result directly from answers to

specific questions on the questionnaire.

6 Thorough initial training of interviewers and subsequent checking of the interviewers' work by the supervisor to ensure that all interviewers are handling the study in the same way and that the quality of the work is high.

By assuming these responsibilities, the shopping mall interviewing company can fulfill its obligation to carry out the package testing research correctly and accurately. Decisions about the test packages can then be made with confidence on the basis of the research results.

CHAPTER FIFTEEN

Color Research in Package Design

Bonnie Lynn

Brands that have favorable psychological connotations to consumers succeed. Those that do not have favorable psychological effects often fail.

Product quality is undeniably the make-it-or-break-it of any consumer product. If the product fails to provide the satisfactions the consumer expects from it, repeat sales will cease and the brand will fail.

Actually, most marketing failures are not failures of new product or of inferior products in existing categories; most marketing failures are failures of brands.

The package is the visual identity of the brand. The package represents the product at the point of sale; the representation must be effective, motivating. The package is the visual brand image and silent salesman.

Unquestionably, advertising plays a major role in motivating the consumer. Television, printed and radio communications are motivating factors. A "good" ad campaign motivates consumers to buy. A "poor" one does not promote interest or discourages consumers. But the best product quality and "great" advertising can be defeated by an ineffective "brand image" at the point of sale or a package that does not sell, that has an unfavorable psychological effect.

To assure marketing success, it must be understood that at the point of sale the package is the product. The package is the final stimulus before product use to which the consumer must, and will, react when making her purchase. If the package presents the brand in a favorable light, that is, if the package conveys more favorable associations and desirable connotations than competition does, the brand will be put into the shopping cart. If the package is not motivating, the brand will almost definitely fail, no matter how fine the product is or how effectively it is advertised.

In self-service shopping, which most present-day shopping is, the package must first of all attract the consumer's attention. It must then hold attention and inform the consumer clearly about the product. It must convey that it contains the highest quality product. It must accomplish all this better than the competition, or at least as well.

Thus ensues the battle of the boxes. Competing for attention on the store shelves, the boxes, bottles, cans, and jars fight frantically to win the consumer over.

What does all this have to do with color?

In light of the importance of the package in achieving marketing success, package design—including all components involved in it (logo-symbol, typefaces, color)—has become a specialized, highly competitive, and very important field. Taken separately and together as a unit, the packages must be effective both optically and psychologically.

The consumer reacts to the package on a conscious level to a limited degree. She identifies the product that the package contains; she may recall (be conscious of) advertising and/or past personal experience with the brand. In some rare cases, she may react to the appearance of the package on a conscious level. Consumers' unconscious reactions to the package, however, have an even greater role in determining if the package will be put into the shopping cart.

Many of conducted studies have shown that "image communication" is superior to, or more effective than, linear communication, and that semantics are not the most effective kind of communication. In other words, lines (words) are not as effective as forms and images. Forms and images have much greater motivating power than words.

It should be understood that people have built-in defense mechanisms against words. People are not defensive in relation to forms (images, symbols) or to colors, because people are not aware that they are affected by forms and colors. The effect colors and forms have on consumers is largely unconscious.

No one says, or even thinks, "This store has this sign, in these colors, with this imagery, because they want me to buy here." To consumers the purpose of the sign is merely to identify, to inform. The consumer is conscious only of the literal communication.

To consumers the purpose of a package is to serve as a container for the product, and the purpose of the label is to identify the product. However, two different packages with the same product inside may have a totally different psychological impact on the consumer. One package may create acceptance by many more consumers than the other.

It is in the psychological realm where color plays a major role. Color, like imagery, is a vital factor in package design.

To be effective, a brand name that is appropriate for the product must have attention-getting power and psychologically appealing imagery combined with psychologically effective color (or colors). The package as a whole must have a dominant identifying color, and the other colors on the package must be in harmony with it in order for the brand to present a total favorable image.

I am stressing the role of color combined with imagery because color cannot be independent of space or form.

Even in the highly specialized field of package design, color is generally used haphazardly, subjectively, without realization of the power it has in communicating and motivating. The subjective color selection of the package designer, or preference of the brand manager, is too often the deciding factor in the creation of package graphics. The assumption is that "if I like it, others will."

The selection of words and images is frequently made in the same way. But the greatest absence of knowledge is revealed in color selection.

Package designers, as those who are in a position to make package design decisions, generally are not concerned with measuring the optical and psychological effect. "Good taste" does not result in a package that contributes to marketing success.

Basic physical and optical principles must be implemented and psychological effect must be given consideration. The psychological effect has to be determined by measuring with the right tools, with special devices that are designed for discovering psychological effect. It cannot be determined subjectively, on the basis of personal preference.

In addition to an optical and a psychological role, color also has a legal role. It is usually not under government control. Marketing and advertising executives cannot use psychologically potent words that are not factual. They cannot make exaggerated claims or even factual claims which might be misunderstood. But they are free to use color symbolism. The FTC rarely dictates the kinds of colors, images, and designs that can be used.

Most marketing programs that fail do so because few executives making marketing decisions recognize they were made on a subjective basis. Color selection is generally made subjectively.

Then there are marketing executives who have research conducted on the assumption that people can and will reveal their true feelings and actual attitudes in response to direct questions in interviews about package designs and about colors.

Because the motivating factors involved in consumer purchasing occur largely on the unconscious level, only research methodology that is indirect can reveal the unconscious factors that affect the consumer at the point of sale. Only indirect methodology can reveal the true feelings and actual attitudes of consumers.

It was brought out earlier in this chapter that color cannot exist without space or form. A package incorporates forms, space, and color. The arrangement of these components constitutes the package design.

In predetermining the effectiveness of a package design, the design components must be tested one by one. First of all, the effectiveness of the brand name must be determined. Second, the effectiveness of the logo and/or symbol must be studied. Third, the dominant package color should be determined on the basis of research. Finally, the effect of the entire package design should be measured. Only by this step by step procedure can a package that will motivate consumers be predetermined.

There are thousands of colors; a number of colors have to be selected as a starting point. Prior to the step by step packaging research, knowledge about color (and symbols), attained through previous research studies, can and should be applied toward a way to stack the odds in your favor.

From many indirect method research studies much knowledge has been gained about the powers of colors. Over the past 35 years we have been collecting and updating information relative to the powers of colors: preference ratings, retention ratings, visibility ratings, and the associations (favorable and unfavorable) conveyed by colors as they are associated with a variety of household, personal, and recreational products and services. The data from this research are applied to package designing to reduce the number of experimental designs.

Through consumer attitude testing, quantified qualitative tests with samples of consumers conducted by indirect method, it is determined which of the three, four, or five package designs, created with the aid of color information, is the most effective. Testing this reduced number of proposed designs cuts both design and research costs greatly.

Basic physical and optical principles about color can be of great value to the designer in creating an effective package. For example, the designer should be aware that color attracts and is remembered in accordance with its degree of visibility. In other words, the color that can be seen at the greatest distance is the color that attracts the eyes the quickest and is remembered most easily (color retention). This is true even when the color is seen at close range.

A pure hue (a color that has no part black, white, or gray) has greater visibility (attraction power) than any of the tints, tones, or shades derived from that hue. Some pure hues, however, have much greater visibility than other pure hues. Also, a tint derived from a hue of high visibility has greater attraction power than a tint derived from a hue of low visibility. This is known to most designers, yet it is widely disregarded in package designing.

It should be clear that colors that attract attention first and are most easily remembered are not necessarily colors that have great preference or appeal. Color preference does not generally correlate with color retention. Although a color of high preference and great appeal is an important factor for an effective package, high visibility, high retention color must generally be used as well. Since the attributes of wide preference and great visibility can rarely be achieved through one color, the domi-

nant color should be determined on the basis of the product category. In food packaging it is not difficult to combine colors of high preference and great visibility and retention. In packages for cosmetics, the problem is more difficult.

Direct color symbolism is an important factor in effective package design. Most package designers are aware that green represents green vegetables, red symbolizes meat, and white communicates pure. It was learned, however, that many designers were not aware that some greens symbolize "poison." Indirect color symbolism that reaches the subconscious mind is not generally considered in design studios. Such information can be derived only from research conducted with consumers by indirect method.

When shopping in a supermarket a consumer may suddenly decide she would like to have another package of the beef gravy mix that proved to be so delicious. She has only purchased the product once and cannot remember the brand name. The shopper will go to the area in the store where that type of product is on display and look for the color scheme of the package. If the package has high retention (memorable) colors, the product will be recalled easily.

The trademark (logo/symbol) has an important role in identifying and recalling a package, in addition to having a psychological role, favorable or unfavorable. The trademark is recalled much more easily if it always appears in the same color or colors.

Colors are generally divided into two psychological groups, cool and warm. The warm colors are those that are predominantly yellow or red. They rate much higher in visibility and retention than the cool colors, which are predominately blue or violet.

Colors of high preference, which are those that have favorable associations in the unconscious or subconscious mind, play important roles through the phenomenon of sensation transference. The colors on the package affect the consumer's perception of the product in the package. The effect of the colors, and of other elements on the package, is transferred to the product in the unconscious mind of the consumer.

Tests conducted on the basis of what people want (not what they say they like) reveal that there are geographic, national, ethnic, cultural, and economic factors in color preference.

For example, a specific red received a much higher preference rating with Italians and Mexicans than with Scandinavians and New Englanders. Also it had a much higher preference rating with Italians in lower income groups than with upper income Italians. A grass green had a low preference in rural communities, but a very high rating in a steel mill community.

For practical purposes, research showed that persons who had many emotional outlets and/or the ability to purchase emotional satisfactions showed preference for diluted and neutralized colors. Those who had only a limited variety of emotional outlets (either because of lack of education or because of low income) showed a distinct preference for pure hues in large doses, particularly those that were warm, such as yellows, oranges, and reds.

Our studies have shown that cool colors have a sedative effect and have proved to rate highest in preference with less outgoing or inhibited people. On the other hand, warm colors are more appealing to those who are outgoing and uninhibited.

Controlled studies with colors have shown that color sensations produce physical reactions. People often feel cold in a blue room and warm in a red room, without realizing that color, not temperature, causes this feeling.

There is a strong tendency for most people to seek, unconsciously, a balanced diet of calming (cool) and stimulating (red) colors. This explains why complementary pairs of colors produce psychologically pleasurable sensations. Complementary colors are cool-warm color pairs.

The most important color factor to be considered in package design is the symbolic power of color. Motivation research studies conducted over the past 35 years have proven color symbolism to be a very powerful marketing tool.

Color symbolism plays an important role in our daily lives. From the beginning of history, color symbolism has had a place in human affairs. It is involved in our spiritual values as well as our physical well-being.

Color exerts a strong symbolic force that has, through tradition, influenced us. In Western civilizations we are conditioned to associate red with festivity, blue with distinction, purple with dignity, green with nature, yellow with sunshine, and so on.

Colors express qualities of sex, strength, durability, reliability, cleanness, freshness, efficacy, professionalism, culture, and nationality. Color by its nature is highly symbolic. Also, through the ages different cultures have assigned different symbolic meanings to specific colors. For example, in Western cultures black symbolizes mourning and red festivity. In some Eastern cultures, white symbolizes mourning, but red (as in Western cultures) means festivity.

Color symbolism is relative to the object with which the color is associated and to the context in which the color is seen. Color symbolism can be applied to packaging in two ways: to suggest the product and to suggest favorable characteristics (attributes) of the product.

Yellow may symbolize sunshine and heat, but when a yellow package is seen in a refrigerated dairy display, it is likely to symbolize (suggest) butter or margarine.

Using color to symbolize the contents of the package is very advantageous to the manufacturer; it aids the consumer in identifying the product.

Symbolizing the contents of the package is effective in such categories as fruit products since one commonly associates such colors as orange with oranges, violet with grapes, and yellow with lemon, grapefruits, and so on. Green is used extensively in canned vegetables as it is commonly associated with plants. Consumers are conscious of this kind of symbolism. Designers should not disregard this.

There is, as I pointed out, color symbolism of which consumers are not conscious. For example, colors of the peach-pink family have high preference and favorable associations ratings when associated with cosmetics, but drop radically in rating when linked with hardware.

A large percentage of people have favorable associations with, and show preference for, a certain green when that green is associated with a vacation. The same green drops in preference when associated with most food products. Another green has a high rating when associated with foods and a low rating when associated with clothing.

In addition to the color that is symbolic of, or identifies, the product, a package should have an accent or dominant color that identifies the brand. Ideally, it should be a high visibility, great retention, color that is complementary or related to the product color.

Often, the product identity color has to be left out because it identifies a competitive brand. High preference, great visibility colors have to be used with the assumption that consumers know the color of the product.

Cans of coffee can be found in bright reds, oranges, and blues. Milk is available in similarly bright colored cartons. Manufacturers of chocolate bars use red, white, orange, yellow, and other brightly colored wrappers for their brown bars. The bright hues have great appeal to the young.

For cleansing products, detergents, and similar impersonal products, there is no need for product identity. It is the brand that needs specific identity. The symbolism need not be direct or affect the conscious mind of the consumer. The color symbolism should be that of performance. Strong visibility is also very important.

The specific satisfactions that a detergent might provide include cleaning power, whitening power, ease of use, gentleness to clothes, and saving time. A detergent package can suggest most of these intangible satisfactions (product attributes) with color symbolism.

Color symbolism may be overused, if a number of brands of the same type of product use practically the same package color or color scheme to suggest their attributes. At one time, for example, over half the detergents employed blue as their dominant package color. (It was pointed out that blue symbolizes cleanliness.) The blue detergent cartons, however, when displayed side by side, may have conveyed that they were all similar, old-fashioned products. Therefore the shopper's eyes would be induced to roam to a detergent package with a more contemporary color scheme, in which blue was only one color. Today, detergent brands are in boxes of yellows, oranges, and reds (high visibility, attention-getting colors), with some blue.

The competitive factor must, of course, always be considered in using color in packaging. If a competitor uses a color, you may not want it. You will want a color combination that will give your brand specific identity. Your package should also be identified by a symbol—a design that serves as a focal point. The focal point should be in an effective color or colors.

Many problems have to be resolved in determining package colors. How much of which hue should be used because of its attention-getting power? How much of which hue should be used because of its high preference rating? How much of which hue should be used because of its appropriate symbolism?

Making these decisions should not be a purely subjective matter. Based on extensive data from package research, general guidelines with regard to selection of package colors can be made.

In selecting package colors, the powers of visibility and retention, preference, and favorable symbolism must be given consideration. Color of competitive packages is always a factor.

First, the package must attract attention. Then it must have a favorable psychological effect—favorable associations that are transferred to the product it contains.

It is important to be aware that studies have shown that color visibility and retention generally do not correlate with color preference. It is necessary, therefore, to have both a high visibility color and a high preference color on a package. The type of product should be the basis for determining which should be dominant—the strong visibility or the high preference color.

The physical and optical color principles and color information derived from studies with thousands of consumers can serve as a general guide in producing package designs for the purpose of communicating to and motivating consumers to buy the products the packages contain.

The creative process, however, plays a major role. Within the limits of these guidelines, the designer has to endow the package with specific identity and unique character. The package has to be distinctive and have an easily recognizable, easily recalled image, with all components having roles.

We generally recommend that a package designer produce several proposed designs (from three to five) for a product. After the designer has produced several package renderings, the designs (packages) are put through two kinds of research: (1) display tests—visibility, readability, and eye flow—that reveal the involuntary reactions to the package and, (2) indirect method consumer attitude tests that reveal which of several proposed package designs is most motivating and why.

People reveal their attitudes in spontaneous expressions and in unconscious reactions. Consumers express themselves freely in situations in which there are no ego involvements or prestige identification factors and in which no defense mechanism plays a part. A direct question puts one on guard, but an impersonal, simple questionnaire does not arouse defense mechanisms; responses are uninhibited. A "control" is also used to gain cooperation and promote interest in the test.

Consumer attitude tests are structured to provide quantified qualitative information. For each test there is a specially designed one page questionnaire. Personal interviews are conducted with a large enough sample of consumers, representative of the market, to be statistically significant. Because the methodology is indirect, the respondents are not aware of the specific information being sought. The respondents reveal their reactions by checking polar attitude terms (favorable and unfavorable) in response to each unit in the test.

Consumer attitude tests have proven that effective use of color will heighten the appeal and effectiveness of packaging.

CHAPTER SIXTEEN

Consumer Research as a Design Guide

Milton I. Brand

In the last analysis, the function of design is to create the perception of product desirability in the mind of the ultimate consumer. Actually, product desirability is "necessary" but not "sufficient" to fulfill design's real job—that of creating the perception of product superiority in the mind of the consumer. The operating idea in these statements is "in the mind of the consumer." As simple as this idea is, it is surprising how many manufacturing companies and how many designers use another criterion. "I *know* what is the best product for the consumer; I *know* what is the best design for the consumer." Obviously the consumer is an untrained individual. He or she may not know what is "in" in contemporary design or what the elements

for good design are. Is not one of the functions for design then that of presenting "rightness" to the consumer, of educating the consumer to understand and (it is hoped) to react positively to good design?

The answer to that statement is an unequivocal—no!

Everything then depends on the consumer's mind. It depends on knowing precisely what it takes to create the perceptions of desirability and superiority. Unfortunately, however, this kind of insight is not there just for the taking. One cannot ask a consumer what he would like either in terms of a product or the design associated with it. The basic rule in product and design strategy is a simple one. The consumer does not know what he needs or desires until he knows what he can get. There is only one completely "pure" mechanism to take advantage of that situation. Make up a number of different products (or versions of a product), imbue them with a number of alternative design interpretations, put them out for sale in a number of different (matched) markets. Then all one has to do is observe which version(s) sell best against competition.

Unfortunately, this kind of approach is very expensive, very time-consuming, and never quite solves the problem. Maybe another product or design version would have done better than the winner of this very expensive test. Yet in one form or another this approach is utilized by many companies in achieving their goal of the "ultimate" product or design.

The best of all worlds is one that changes the rules of the game somewhat. Instead of testing to see how good a job we have done in product and package design, perhaps we can create a new kind of research that will give us product and package design direction that in essence would ensure that our efforts are proceeding into potentially productive areas.

The new approach to be recommended is based on the psychology of the buying situation. We must ask ourselves precisely what it is that a consumer buys. Does he buy that physical, three-dimensional product he sees sitting on a shelf? Interestingly enough, that is not what he buys. That product on the shelf is sensed by sight, touch, and any of the other senses at a human being's disposal. This sensory information is interpreted by the brain, processed by the brain, and stored in the memory of the brain. This stored information is then itself processed. The processed information is compared with "stored" information on personal needs and desires. It is compared with stored information about

competitively available products. On the basis of this processing, a decision is made as to whether this "class" of product should be considered for purchase. If the answer is "yes," then a decision is made as to whether this new information (received from the product on the shelf) is as good, better than, or less good than information about the competitive products. On this basis a decision is made to buy, not buy, or look for more information. The "sell" is made in the mind, not on the shelf.

The reason that different people buy different products after they have decided to buy a class of product is often based on the simple fact that different people process the information about the product on the shelf differently. What precisely is the best "set" of mentally stored information about that product on the shelf that will generate the highest probability of positive purchase decision?

It is in the answer to this question that the strategy of predesign guidance lies. To solve the problem we must know how to get inside those consumer minds. Predesign research is based on a basically simple idea—somewhere in the minds of consumers lie the answers to all marketing, product, and design problems.

The consumer knows what product he buys; he knows what TV shows he watches; he knows what magazines or papers he reads; he knows what ads he has seen; he knows what other people have told him about products. He even knows why he buys what he buys; why he reads what he reads; why he watches what he watches; why he behaves the way he behaves.

To become "heir" to all this vital and rich information, one need only ask people in the market to "dump" their memory; ask questions—receive answers; ask them with depth intensity and get intimate understanding; ask them more cursorily but in great numbers and obtain highly reliable statistics about market behavior.

Unfortunately, as with many "ointments," this one has several "flies" in it.

1 People do not always remember what they did.
2 People may lie (or exaggerate) if a truthful answer might prove embarrassing.
3 People seldom know "why" in truth they have behaved as they did—and so people invent or rationalize the "why's" of their life "ex post facto."

The indictments above need not be 100% true to affect the validity and reliability of research findings. Even if they are only a "few percent" true, research findings may have doubt cast on them. Worse, however, than the idea that these dangers may only be a few percent true is the idea that we never know really how many percent true the problem is.

How much more pleasant life would be for the researcher or problem-haunted executive if there were some way to get at the information within the individual in a way that protects against the inadequacies of questioning protocol. The quest for this "royal road to knowledge" has led to the "black box" theory of obtaining consumer information. For the designer to understand this theory, it is necessary to digress and to explain a classic problem in "network theory"— the complex field of electricity, electronics, and mathematics.

One of the unique problems in network theory involves the famous "black box." The professor places on a table in front of his students a steel box welded shut on all sides and corners and painted black. He informs his class that within this black box there is a complex "passive" electric network. That is, it is a circuit composed of resistance, capacitance, and inductance but which does not contain diodes, transistors or vacuum tubes. The professor indicates that the job of the class is to determine precisely what that network is, without opening up the box and looking.

The student is shown that there is an electrical input terminal sticking out of the left-hand side of the box and an electrical output terminal sticking out of the right-hand side of the box.

With that, the professor places on the table three electrical instruments— two oscilloscopes (devices that display electrical wave shapes on a cathode ray tube) and one signal generator (a device that can create a variety of different electrical wave shapes). He connects both the signal generator and one of the oscilloscopes to the input terminal. He connects the other oscilloscope to the output terminal.

"Now gentlemen," the professor says, "observe that as I turn the appropriate knobs on the signal generator a perfect sine wave appears on the input oscilloscope. Thus we know that a perfect sine wave is going into the circuit in this black box. But look at the interesting wave shape that now shows up on the output. Please make copies of this input and output wave shape in your notebooks."

Thereafter the professor moves some more knobs and creates a perfect square wave on the input. This creates a still different wave shape on the output. The students copy these two shapes as well.

Finally, the professor twists the knobs once again and a "pulse" wave (a spike-shape electrical wave) appears in the input and once again a unique and different wave appears on the output. The students copy these shapes.

"Now gentlemen, we have observed

that the electrical circuit in the black box has caused these rather pure wave shapes going into it to be changed into the unique wave shapes coming out. Obviously, there is a unique electrical circuit in the box that is capable of performing these transfers that you have observed. Your job, based on your observations, is to come up with a hypothesis of precisely what electronic circuit must be in that black box in order that this transfer should have taken place."

"Once you have such an hypothesis, your job is to make it work for you. I would like you to predict for me, therefore, what the output picture will look like if I put a pure sawtooth wave into the input. Come back one week from today with your prediction."

One week later, the students return and the professor reassembles the equipment. He "twists in" a sawtooth wave that appears on the input oscilloscope and immediately a unique output wave appears on the output oscilloscope. "How many people correctly predicted the shape of the output wave?" Only a few students answer affirmatively.

"Well, you students are through with the experiment. Now, as for the rest of you, you still have the same problem—coming up with a hypothesis of what's inside the box. You are, however, in better luck this week than you were last week. Last week you only had three bits of information to work with—this week, you have four bits of information. When you come up with your new hypothesis, use it to prodict what the output would look like if I put a triangular wave shape into the black box."

A week later, the process is repeated and this time a few more students indicate their prediction had been correct. "For those of you who still haven't gotten it right, you are in great shape. Now, you have five pieces of data to work with. Please don't worry about

being able to do this job. I have enough wave shapes tucked away in my back pocket to keep us going until the end of the semester."

Within 5 weeks, every student had "solved the problem" of what was in the black box.

In a sense, the market (or, more precisely, a market segment) can be viewed as a black box. All the information is indeed locked up in their collective minds. It only awaits an effective device for getting through the protective facade surrounding everybody's mind and put there by nature to keep strangers out (and even to keep themselves out). Over the past few years, there has been some interesting experimentation in attempting to exploit a technique to do precisely this. For the lack of a more esoteric name it can be called the "Stimulus/Reaction" approach to research.

It is clear that stimulus/reaction has at least one major advantage. Cause and effect relationships are clear. If one takes a 3 inch long woman's hat pin and jabs it strategically into the hind quarters of another individual, it is quite likely that individual will jump at least one foot. There is little possibility that the relationship between the reaction (the jump) and the stimulus (the jab) cannot be made.

Psychologists have used stimulus/reaction approaches for years in word association tests. "Black," they say, and the respondent answers "white." "Mother," they say, and the respondent answers "father." "Love" yields "hate." "Up" yields "down." "Life" yields "death."

There are certain expected norms for reaction to the various word stimuli. To whatever degree an individual answers in unique ways "out of the norm," the psychologist has a basis for analysis. In a sense, Rorschach tests and Thematic Apperception Tests and other such mechanisms are also stimulus/reaction devices to get insights about individuals otherwise not readily obtainable.

The stimulus/reaction approach for marketing research is based upon the same principle. Let's consider the following example. The problem is to understand behavior, attitudes, perceptions, and brand predispositions with regard to the purchase and consumption of crème de menthe. There are literally "scores" of idea stimuli that can be associated with crème de menthe. A number of these are the following:

1 *Color* Deep green, green, pale green, other colors and shades.
2 *Imported/Domestic* Manufacturered in France, manufactured in the United States with French implications, manufactured in the United States or other countries with non-French orientation.
3 *Viscosity* Thick/syrupy, normal, thin/watery.

Other possible idea stimuli are aging, bottle shape, design motifs, brand names, and so on.

If each of the many ideas indicated in the proceeding were considered a stimulus and assigned a code S_1, S_2, S_3, ... S_i, we would have generated quite an array of idea stimuli.

If it were possible to expose an individual clearly and cleanly to stimulus S_1 and obtain a clean and clear reaction to this stimulus, R_1, we would have a unique procedure for playing the "black box" game. After ever so many S's have yielded an equivalent number of R's, we could make our hypothesis and from this start predicting what R's would result from the next grouping of S's.

Unfortunately life is not that simple. The laboratory-like approach of word association tests does not work well with marketing problems. Respondents stop talking like consumers and either view themselves as laboratory subjects or as experts. What people do well, however, is to discuss a subject of mutual interest. The trick, then, in exposing stimulus S_1, for instance, is to include it in a seemingly aimless discussion with the respondent. When the respondent takes his turn in the discussion, the reaction R_1 to this stimulus S_1, will be somewhere in his comments. If the stimulus S_1 is carefully and strategically placed in the researcher's remarks, then a competent, perceptive researcher can ferret out from the totality of the respondent's remarks the reaction R_1.

To put it more graphically, the research exposes S_1 to the respondent in the midst of a lot of "background noise." The respondent plays back R_1 in the midst of a lot of "noise." The skill lies in constructing the researcher's remarks so that the key element without a question is S_1 and in being experienced enough to recognize R_1 in the response.

This, then, is the way the researcher plays the "black box" game. After a number of stimuli have been exposed and reactions noted, he will develop a "behavorial hypothesis," which he will use to predict what the probable reactions to still other stimuli will be. The process continues progressively until the researcher has developed a hypothesis that not only accounts for the reactions to stimuli that he has observed but also serves as the basis for meaningful prediction.

If one had to rely on individual personal interviews to utilize this technique, one would face a very large and tedious job. Fortunately one can take advantage of that unique tool of research—the depth group technique. It is ideally suited to the solution of "idea problems."

There is a uniqueness to the dynamics of a well-run group that is the basis for this approach. Properly preconditioned, a group will attempt to seek a consensus viewpoint. An experienced moderator will encourage group members to challenge each other's statements, to argue with other members and even with the moderator. When people argue, the purpose is to get another individual to think the way they do. This process is exploited in such a way that for most stimuli the moderator can discern the main consensus of group reaction (what most people ultimately agree upon) and can note minority opinions if they exist. When he repeats the process with several groups, he will reinforce his idea of what the overall group consensus is.

Progressively, then, a researcher is able to establish a set of ideas that indicate that $S_1 \rightarrow R_1, S_2 \rightarrow R_2, \ldots S_i \rightarrow R_i$.

It is interesting to note that in groups of relatively matched samples, half the reactions that one will ever get to a controlled set of stimuli seem to appear in the very first group. Although the premise is not mathematically precise, it further appears that half of the remaining possible reactions will appear by the end of the second group. The third group will contribute half of what yet remains.

Any experienced moderator knows that for a given matched sample set of groups the law of diminishing returns (actually the law of halves) works particularly fast. After three groups it appears that close to 90% of all the ideas that one can ever generate from group probing will usually have been exposed. Not only that, but extended group studies among matched samples have indicated that the remaining 10% which may yet be exposed in still more groups, rarely contains ideas of first order importance. In other words, important reactions come forth quickly.

Of course, if a number of relevant respondent cells are logically required to cover a given problem, each of these should theoretically be composed of at least three matched cell groups.

The moderator notices the kinds of consensus reactions he gets to the stim-uli he has exposed to the groups. He hypothesizes (a creative function) why people are probably reacting to these stimuli in the manner he has observed (other observers may come up with different hypotheses). He formulates key hypotheses into a set of new stimuli and, on observing the reaction to these stimuli, may draw inferences as to their validity.

In this manner a skilled researcher can develop a theory of market behavior that proceeds from his own experience, sensitivity, and intelligence. This procedure does not define market research in the classic sense since the respondents are used only as an aid to the researcher's own creativity, not as a source of cold fact. Perhaps this makes the approach less a science than an art.

In the process of executing this stimulus/reaction procedure, the researcher obtains two benefits:

1 He develops an understanding of market behavior that is derived from the marketplace creatively, is validated within the marketplace, and is not based upon the inadequacies of questions and answers.

2 He creates (for given market segments and given product or brand areas) a library of Stimulus/Reaction relationships (S/R building blocks). He may use these as ingredients in creating communication strategies.

Point 2 above is probably one of the most fascinating elements of this approach. In fact, the job of the marketer is to create a predisposition within the market to purchase a product (whether this predisposition results from intrinsic product or package design character or unique marketing argument). In the course of conducting stimulus/reaction research, the researcher generates a number of insights as to what the "perceptual sets" are for all competing brands. His job is to create a "perceptual set" for this brand (or new product) that is more predisposing than those for other brands. In a sense, we can view a "perceptual set" as a group of idea reactions that each individual in the market stores in his brain for each brand of which he is aware.

Thus it is conceivable that a researcher, now with a fairly intimate understanding of market dynamics, can design a set of reactions that, if they were in the market's mind, would create relative predisposition to buy his product. But how can he induce this set of reactions in the market's mind? *There* lies the great trick of the stimulus/reaction building block.

If he has done his job well, he will know for every reaction he desires to have in the market's mind the stimulus that will probably create it.

Thus for a given desirable "reaction set," he can define a desirable "stimulus set." He need then only "artfully" intertwine these stimuli in a literate fashion to obtain a *concept statement*.

A concept statement thus represents a set of stimuli that create a set of mental reactions tested to predispose the consumer to perceive product superiority. The job of the designer, then, is to study this concept statement and the constituent stimuli that make it up and to interpret as many of these stimuli as possible into design parameters. In essence, his great skill, his artistry, his creativity is focused on this one job. He must create a physical product and package with colors, shapes, forms, and materials which in his mind send out messages to the consumer as nearly like the messages that have been created by the individual stimuli.

How well he has performed that function is now readily testable. We already have group data on what consumer reactions are created by exposure to the concept. We now expose one or more design interpretations of that concept to similar groups of consumers and observe the reactions to those design approaches. How closely do the reactions to the design match the reactions to the concept? Has the designer effectively interpreted the concept stimuli into design form? Probing reactions to design can provide the designer with these insights.

The designer now need not worry about any other function of his design except the one and most important one—how closely do I match concept reactions? The concept and its reactions become the criteria against which his efforts are measured. If he must go "back to the drawing board" several times in order to home in on that objective, he knows that these are productive efforts that will ultimately lead to a winning product package.

The *concept* is then used as the perceptual blueprint from which to create product/package characteristics, design directions, and even advertising strategies.

The reader may find this description somewhat complex and even esoteric. In practice, however, it represents a relatively inexpensive technique. More important, it is a technique that is not time-consuming.

In summary, the *concept* is created in lieu of a test product. The concept is evaluated rather than a product. This is inexpensive and fast. All other functions are now addressed to matching the concept "in real life."

CHAPTER SEVENTEEN

Trademark and Brandmark Research

J. Roy Parcels

As corporations grow and diversify, the number and nature of publics they reach increase. Modes of identification that once delineated the company and its products and services no longer suffice. Acquisitions present new problems as management weighs the advantages of bringing the new entity under the corporate identification umbrella versus the possibility of building on its existing strengths.

At the other end of the marketing spectrum is the new company. Like an infant born of a natural mother, this entity must be given its own unique identity, establishing its place in the market and its right to mature.

For the established, growing corporation, the problems of trademark and brandmark development can become especially acute. Strong growth of particular segments of the corporation can quickly overshadow the parent's identity. New product areas may benefit substantially from a grafting of corporate marks onto brand identities, but to what extent? Too powerful an endorsement can undermine brand individuality, while abstractions or understatements may be glossed over by the publics to be reached.

This element—the publics to be reached—lies at the heart of trademark and brandmark research. The necessary first step in this research is a clear profile of these publics, obtained through meticulous study.

Once these publics have been identified, each must be explored for its attitudes toward the corporation in all of its manifestations. These attitudes, evaluated qualitatively, can not only help to define potential problem areas for the corporation, but can also suggest new marketing opportunities.

No better illustration of the problems and opportunities that await discovery through research exists than the experience of The Borden Company, a large, diversified corporation with roots reaching back nearly a century and a quarter into America's past. With its origins in the milk and dairy industry, Borden had created and employed one of the best known trade characters in twentieth century marketing: Elsie the Cow. In use for many years, she has helped to sell some two billion packages annually. No other cow ever labored so mightily for her owner.

Even so, when The Borden Company decided to reevaluate its identity, there were sound reasons to suspect that important changes needed to be made.

Figure 1 The old brand name in combination with the Elsie-Daisy mark.

First, significant diversification had taken place. Moving into paints, plastics, industrial and agricultural chemicals, and other areas with high-quality products often established under their own trademarks and brandmarks, the company had in the meantime grown from regional to world-wide marketer. The time was long past when the famous Elsie character with the associated "script Borden's" trade name could adequately describe the corporation and its functions. (Fig. 1)

Other emanations of corporate identity growing out of the original marks, the company's management concluded through preliminary research and intuitive observation, were just too closely associated with its dairy products. Even an extensive internal survey conducted by the company reflected this fragmentation. A complex organizational structure had evolved with many companies, groups, divisions, and departments. Within a single division there was diversified use of the Borden's mark, and even letterheads and business cards varied widely.

Questions and doubts about the corporate image were cropping up in the minds of upper management because of the realization that a favorable, accurate visual representation would have a tangible impact on the company's future. Internally, it would help executives and other staff to project a strong image of the great strides being made in diversification to the consumer and financial communities. Moreover, this image would come into even sharper focus because corporate plans called for an even greater change in the character of the company's business in the future.

Following a comprehensive review of the corporation's total communications system and all related parts including corporate name and nomenclature, a pilot research study was conducted. One segment was directed at the general public and consumers to measure recognition of the Borden script logotype as used primarily by the milk and dairy division. The other was designed to obtain financial analysts' opinions and degree of knowledge about the total corporate operation. This second segment consisted of in-depth interviews with security analysts at leading brokerage firms.

The findings in the two studies were quite revealing and, as it turned out in later independent research, accurate. It was determined, for example, that Borden was widely regarded as a good company, which produced high-quality products and was good to its employees. The company was thought of as being primarily in the milk and ice cream business, with less than 4% of respondents realizing that it made nondairy products.

But Borden did not have the image of a dynamic, exciting, or diversified company. To the contrary, "old-fashioned," "staid," and "a good stock for widows and orphans" were the descriptions most frequently used.

All analysts queried thought of Borden as an excellent grade security, but few showed enthusiasm. One major broker did not even have the Borden Chemical folder in his "chemicals" file.

Another element of the research segment involved extensive interviews with the general public and consumers to evaluate the style of lettering employed by the company. Six alternate lettering styles and two oval shapes with fictitious names were tested to determine the equity, if any, in the Borden's lettering style. (Fig. 2)

Through these and related investigations, it was determined that the "script Borden's" could be shortened to "Borden" without loss of recognition. It was also learned that a different style of lettering could add measurably to the company's image for leadership, progressiveness, and aggressiveness.

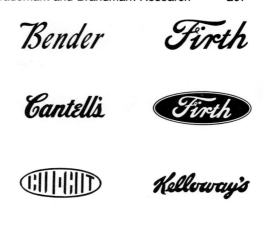

Figure 2 Some of the illustrations used in the consumer study to determine the design equity in the then current Borden's brand name.

Borden	Ford
Campbell's	Ford
Du Pont	Kellogg's
Firestone	Reader's Digest

A series of field trips, plant tours, and interviews with Borden staff, related to a complete review of corporate literature and nomenclature used throughout, suggested that a complete system approach would be needed to unify total corporate communications and to control them in the future.

As the result of information gathered here and through earlier research, it was possible to propose the single word "Borden" as the corporate designation, dropping the words "The" and "Company" from company nomenclature, with the word "Division" to be adopted for the entire system of company groups. It was also proposed that a single letter style for the Borden logotype be used for all corporate and divisional nomenclature, to reflect the desired image better, and that a second typestyle be adopted for universal use where the lettered Borden logotype might not prove

Figure 3 The present corporate mark in combination with the Elsie-Daisy mark.

feasible, as in the legal address lines of certain product packages. (Fig. 3 and 4)

While research was justified and proved invaluable in the determination of a direction for the overall corporate identity program, there was another valuable lesson to be learned from the initial studies. It was that a large, diversified corporation, understood only fragmentarily by its various publics, can gain perceptible, measurable strength by integrating more closely its corporate marks with its product brands, particularly when the corporate marks have elements of authority and universality designed into them. Acting as a kind of seal of approval, the corporate mark, when integrated with product brands, can help introduce new products more quickly and economically with greater assurance of public acceptance. At the same time, it can broaden the base of

Figure 4   The present corporate mark.

public recognition and support at a low investment level. It is not widely understood, for example, that for the typical nationally marketed product public exposure through packaging may be up to 1500 times greater than exposure created by a multimillion dollar television campaign.

The corporate mark, however, can have its limitations. This is particularly true of established companies moving into innovative new product areas. Here the focus of trademark and brandmark research shifts to the particular product package and can include the package itself.

Though product names may arise from a variety of sources, a productive source to help in narrowing down the list of names before testing is *The Trademark Register of the United States*. This book is also a good source for trademarks that may fulfill the new marketer's requirements. A trademark listed in the *Register* may be available for purchase if it is not in general use. This can be the least expensive way to obtain a mark.

Most often brand name candidates are in typewritten form when presented to consumers for evaluation, and the selection of the final name is based in part on the results of this test. A procedure often neglected is the testing of the leading name candidates in a design context. After the determination of the most effective typewritten names, these "winners" should be rendered in graphic design form and then evaluated against marketing criteria. (Fig. 5)

Some names, because of their combinations of letters, lend themselves to designs in keeping with the marketing objectives better than others. Thus the name candidate ranked fourth in the initial evaluation of typewritten names may become the number one choice when displayed in a graphic design context. Inclusion of this important step helps to assure that the selected name

Figure 5 Some typewritten brand names and their corresponding logotypes. The following registered trademarks are the property of the following companies:

Jolly Time	American Pop Corn Company
Patio	RJR Foods, Inc.
Wilbur	Wilbur Chocolate Co., Inc.
Beech-Nut	Beech-Nut Foods Corp.
Bidette	Youngs Drug Products Corp.
Pfeiffer	Pfeiffer's Foods, Div. of Hunt-Wesson Foods, Inc.
Palmer	R.M. Palmer Company
Cottonelle	Scott Paper Company

will be the most effective one for all marketing efforts.

Though the criteria for evaluating trademarks and trade names are quite subjective in many product and service industries, there is a surprising degree of uniformity in corporate procedures for nomenclature creation and follow-through search practices, as interviews with executives in the publication-broadcasting, automotive, industrial chemicals, financial, and packaged foods industries reveal. Practically all of the companies surveyed now rely to an extent on trademark computer banks, particularly for products that will be placed in international markets. These banks, however, have their limitations and prove more effective for words than symbols.

There is almost complete uniformity,

however, in trademark and trade-name search procedures. The marketer generally uses two external search firms; one to conduct the original search, and the second as a backup. These organizations are given several names and designs, in order of preference. At the same time, where it is expected the product will be marketed outside the United States, a connotation search may be conducted in one or more foreign languages. One factor is universally true: corporate or "house" marks must be searched more broadly than brandmarks.

Following the search procedure the mark is registered, generally in all countries where the product could conceivably be marketed. The number of nations most frequently mentioned is between 142 and 146 with, of course, the connotation search having included only major language groups.

Research of trademarks and brandmarks for international markets is becoming increasingly important both to established corporations and new marketing entities. Interviewed corporations agree that the most efficacious technique is to standardize designs world-wide. This, in turn, implies the development of the most effective trademark and tradename devices possible for the original U.S. market. It is generally accepted that designs that prove effective in the American marketplace will be equally accessible to overseas consumers, with modifications primarily related to linguistic and regulatory differences.

One area of growing importance to all marketers is postdesign research. In this step all elements comprising a total package design are given expert consumer and store testing. This means placing the trademarks or brandmarks in their packaging contexts with configurations, illustrations, and the overall graphic format.

Research at this point indicates any weaknesses in the marks themselves and their flexibility, or in the other package elements.

The extra time involved in this step is worthwhile, when compared with the potential losses caused by an ineffective design.

When the optimal package or other expression of the graphic system has been created, the marketer may then elect to seek international registration not only of the mark but also of the distinctive face panel design, or "trade dress" of the container, and of the descriptive material or verbiage. In some cases it is possible to register the entire package.

Research can be helpful in one final area of trademark and trade-name development: postmarket evaluation. The periodic audit can be applied to all the publics affected by the initial program. Such audits, conducted in a manner similar to the research program initiated at the outset, can not only measure the effectiveness of that program but also point out weaknesses and problem areas. These can be corrected with modifications attuned to the specific publics.

In any research related to trademarks and brandmarks, the marketer should understand that techniques must be exploratory, qualitative, and related to the objectives of the corporation or the product. Thus the validity of findings won't necessarily depend on numbers but rather on the knowledge and the perception of the interviewer and on the attitude and reactions of the respondents.

Research is in fact a screening process aimed at obtaining new, objective information rather than justifying decisions already made. The marketers that recognize it as a tool are those who stand to gain the greatest benefits from it.

CHAPTER EIGHTEEN

Research Design and Analysis in the Testing of Symbols

Louis Cheskin

Business executives are concerned with finance, product development, and production facilities. They rarely assign great importance to a corporate or brand symbol.

In the past, the corporate symbol was merely an identification, a trademark. Often a company was identified by only the name, which was generally management oriented; the company had the name of the founder or the name was selected by an executive or group of executives.

The name Sears & Roebuck is obviously that of two individuals. In more recent years, Consoweld was selected as the name for a plastic laminate by the management of Consolidated Water Power Company. Because the plastic laminate was to be produced by a division of Consolidated Water Power Company and is produced by "welding" sheets of kraft paper, the executives reasoned that Consoweld incorporated parts of consolidated and welding—the manufacturing process. Although the product was to be sold to consumers, the consumer was not given consideration in naming the product. Printing the name in some kind of type was considered satisfactory. Obviously a symbol for Consoweld did not enter the mind of the management.

In the 1960s and 1970s, large corporations began paying considerable attention to corporate symbols. There is evidence, however, that the recognition of the symbol as having a role in a corporation does not mean that the symbol serves the corporation most effectively.

The symbol should be consumer oriented, public oriented. The purpose of a symbol is to communicate to, or have an effect on, consumers, the public, not be satisfactory to management or to meet the "taste" of the chief executive.

In the late 1970s, National Broadcasting Company (NBC) spent a very large sum of money on its symbol. Of course, NBC had the means to give the new symbol great promotion at little cost. But did NBC know how effective the new symbol was? What effect did the symbol have on the viewing public? It can be assumed that NBC did not know the answer.

If research was conducted before adoption to learn how effective the new symbol would be, questions were asked about the symbol. The answers to such questions are meaningless. The questions were about art, about design, or perhaps about trademarks. But answers to such questions do not reveal how effective the symbol was in representing the character of NBC.

The public (TV viewers) cannot say how they are affected by the symbol because they do not know the answer. The effect of the symbol is on an unconscious level. People cannot give answers to questions about factors of which they are not aware.

If I suggested that the NBC symbol may have had some adverse effect on the public, I do not believe that management would take this seriously. After all, it's only a trademark. Traditionally, trademarks have not been considered psychological factors or vital communication devices.

There is a high probability that top management selected the design from a number submitted by the design studio. Such terms as "modern," "up-to-date," "clean," "simple," "effective," "I like it" were used. Terms like "psychologically effective," "motivating communication," and "specific identity" are not in the frame of reference of most designers or of corporate executives.

"Research design and analysis" in most corporations still consists of making art (design) critics of consumers and/or of management committees judging proposed designs. In some corporations the chief executive officer makes the selection.

Our educational system is fragmented, although the disciplines are more integrated than they were half a century ago. We now have socioeconomic studies, instead of sociology and economics. Business schools have courses in social studies and some rudimentary psychology. But the effect of forms, colors, and designs is not in any curriculum I know. Although my books on marketing and marketing research have been required reading in many business schools, they have not reached the majority of top corporate executives. They are unknown in design schools.

The top executives of most corporations have their frame of reference from an earlier period. Although the executives have been trained in finance, business strategies, production principles, and logistics, they received little or no orientation in consumer behavior or in how various segments of the population, consumers, are motivated.

Our problems now are very different from those of the past. Yet corporate executives want them solved as in the past. Symbols were not marketing factors half a century ago. Only in recent years have executives become aware of the importance of symbols. But most do not know how to determine their role.

This chapter may therefore be of interest to corporate executives and sufficiently brief for them to read.

EFFECT OF IMAGES ON THE SUBCONSCIOUS

The effect of a symbol is on the subconscious mind. People are not aware that a symbol has an effect on them; therefore, they are not defensive in relation to a symbol. Yet a symbol may have favorable or unfavorable associations and desirable or undesirable connotations.

Subtle differences in design may be very different in effect. For example, a bird drawing may symbolize a dove or a pigeon. A dove symbolizes delicacy, femininity, peace; a pigeon is considered a dirty bird.

A logo (lettering style of the name) may symbolize old, strong, powerful, solid associations, or it may evoke new, delicate, elegant, soft connotations.

Triangles with sharp points symbolize industrial character and have a favorable effect on most men and an unfavorable effect on most women. Circles and ovals have favorable associations in the subconscious mind of men and women.

A mere geometric image, however, has no specific identity and therefore is not an effective corporate or brand symbol.

The effects of symbols have been established by extensive research conducted on an unconscious level—the kind of research that reveals actual reactions of people, not what they say. Quantified qualitative studies reveal that people practice sensation transference; they transfer the sensation produced by the company symbol to the company it represents.

In the marketplace there are symbols that are effective marketing tools; there are symbols that are sterile; there are symbols that do harm to the companies they represent.

Generally we think of power in relation to physical energy. Few are aware of psychological power or of devices that have psychological power. Still fewer are aware that symbols—nonverbal elements—can be used as tools to affect the behavior of people.

In fact, most of us, or almost all of us, are unaware, or refuse to admit even to ourselves, that a shape or a color affects our behavior.

Not many are aware that communication has an effect on both a conscious and an unconscious level. Few are aware that subconsciously they are motivated to act or react in a certain way by symbols. In fact, almost all people attribute no vital significance to shapes or colors.

When over a thousand individuals were shown two shapes, a triangle and a circle, and were asked which shape they liked better, the responses were 50-50; half of them responded that they liked the triangle; the other half declared a preference for the circle. Almost all of them indicated that they did not see any significance to the question: "What's the difference, what value is there in the shape?" was the expressed attitude.

To those who had no orientation in the field of psychology, it meant that shapes—triangles and circles—have no significance, or have no meaning to people. To those who had orientation in the behavioral sciences, it meant that the law of probability was operative in the responses, as in tossing a coin, and that the images had no meaning to people on a conscious or rational level because they had no practical or material value. The responses did not show, however, that the images had no impact on the subconscious minds of the respondents;

they did not mean that the images had no effect on the respondents.

We had the knowledge that each individual reacts to stimuli, sometimes consciously and sometimes unconsciously and that although individuals strive to be, or give the impression that they are, rational and practical, actually they are irrational and impractical in their behavior patterns.

SENSATION TRANSFERENCE

We designed a series of experiments so that the participants in it would not be aware of any factors that were not practical or of some value.

We tested products that were identified by symbols. The first study involved a product that was presented to consumers in two kinds of packages, one identified by circles, the other by triangles. The result was that over 80% of the consumers wanted the product with the circles, because it was of better quality, they said. The products were identical.

I had difficulty believing the results of the first 200 interviews. But after 1000 interviews I had to accept the fact that the majority of consumers transferred the sensation from the circles on the container cover, which obviously was pleasurable or favorable, to the contents in the container, and they transferred the sensation from the triangles on the container cover, which obviously to the majority was unpleasant or unfavorable, to the contents in the container. This study and those that followed were conducted in two ways: (a) after using the product from both containers and (b) by merely seeing the containers, with respondents matched demographically. The results differed by less than 2%.

In the analysis of this study in 1936 I first used the statement "sensation transference". The consumers trans-

ferred the sensation produced by the container to the contents.

It meant that to the subconscious mind of the majority, the circles had favorable associations and produced pleasant sensations and the triangles had unfavorable associations and produced unpleasant sensations. It was clear that the sensations produced by the "images" were associated with the product.

The experiments (1935–1941) were conducted with various products—tea, coffee, cookies, cosmetics. These experiments, with images and colors, established the fact that people practice sensation transference. They provided the clearest evidence that symbols—images and colors—have a great impact on the subconscious minds of people, that individuals are affected by symbols. Some symbols have a favorable effect; other symbols have an unfavorable effect on most consumers.

We had conclusive evidence that by associating a product or an article with a symbol (or color) that has favorable associations, the product or article becomes endowed with favorable connotations. The same is true of corporations.

ROLES OF SYMBOLS

As a marketing device, a symbol should rate high in visibility. It should always have specific identity. It should have favorable associations and desirable connotations to the subconscious or conscious mind.

An optically and psychologically effective symbol may be the single most vital element in promoting the growth of a brand or company. The Standard Oil Co. of Indiana (also American, now Amoco) symbol was adopted in 1945–46 after our research showed that it had three facets of favorable symbolism: the oval (Fig. 1) means pleasure to the sub-

new   old

Figure 1 Standard Oil of Indiana introduced the oval with torch in red, white and blue in 1946. It rated higher than any competitive company sign at the time and in recent years.

conscious mind. (Ovals and circles are associated with the first pleasurable sensation a human being has; these are also organic, biological, and planetary forms.) "It is our pleasure to serve you" has a deeper meaning when associated with an oval or circle. The torch means oil. Red, white, and blue means American. Also, the symbol rates high in visibility—attention-getting power.

Other oil companies introduced new symbols. Very extensive research of many proposed sign designs was conducted by us for one of these companies. The research showed clearly that each sign design for Sinclair Oil with the prehistoric animal downgraded the brand (Fig. 2). The research showed that to motorists the dinosaur symbolized "sluggishness" and lacked the connotations of "speed" and "up-to-dateness." But to the chief executive officer of the Sinclair Oil Company the symbol meant oil. He had read when he was a boy that oil is as old as the dinosaur.

Our research was disregarded and the symbol was adopted. Result: The name Sinclair, many years on the mar-

ket, became worthless. The new owners of the company discarded the brand name.

In the fast-food field there are a number of successful companies. McDonald's is the brightest of the stars. Many fast-food operators have been making raids on McDonald's management. Ten or more of McDonald's executives have been recruited by competitive fast-food operators. Assuming that these executives were among those who were supporting the McDonald new look and multifaceted expansion program, they cannot bring success on a par with McDonald's to their new operations (Fig. 3, p. 216).

Our research showed that two great assets McDonald's has are the name McDonald's and the Golden Arches.

The name McDonald's symbolizes economy and frugality to a significant degree that contributes to marketing success. The Golden Arches make a still greater contribution to the success of the company.

Figure 2 Sinclair Oil was a long established company that adopted this symbol with dinosaur, disregarding the research information that the dinosaur was associated by consumers with "sluggish" and with other unfavorable terms, not with "power," "up-to-date," or "oil" as the Chairman of the Board thought. The name Sinclair is no longer in existence.

Figure 3 The McDonald logo-symbol was evolved from wall and roof supporting arches. It is a very effective logo-symbol because it has great visibility, rates high in memory retention, and has pleasurable associations in the subconscious mind, with Freudian implications.

Originally, the arches served a functional purpose—inexpensive framing and support for walls and roof. These functional arches have evolved into a symbol—a most effective symbol. The yellow arches symbol has great visibility, specific identity, and highly favorable associations with Freudian implications in the subconscious mind of the consumer.

The name McDonald's with the "Golden Arches" symbol is not easy to match in effectiveness for the fast-food market.

How To Get A Corporate Or Brand Symbol

Most executives believe you get a symbol by employing a fine designer. This is a necessary step, but it is neither the first step nor the last.

The first step is to learn the strengths and weaknesses of the old symbol and/or the "image" of the company or brand.

An old symbol may have some assets that should be retained. If an old symbol has favorable elements, they have an advantage over any new design elements

because the old elements are associated with the company or brand. They already identify it.

Research often shows that an old symbol should be improved, not changed. It should be more effective in visibility, in specific identity, and in favorable associations.

If the company or brand takes on a new character, the symbol should have a new character, in keeping with the new "image" of the company or brand.

How To Achieve An Effective Symbol

Is there a way of knowing which of a number of symbol designs is best or which will be most effective?

There is a procedure for developing an effective symbol that has been followed many years. Getting a designer to produce several designs and an executive or a committee to make a choice will not result in having an effective symbol.

Before a symbol can have a psychological effect it has to have impact—visibility and attention-getting power. There are basic principles to be followed for achieving these.

Several hundred studies conducted in 1935–1941 and thousands of studies conducted since 1945 have shown that great impact equals great memory retention. The image or color that gets attention first will be recalled easiest or remembered longest.

Great impact and great attention-getting power (or great recall), however, do not always correlate with great preference. Yellow has greatest impact of all colors; it does not have greatest preference or appeal and is not associated favorably with some products. A triangle is an effective attention-getting device; it does not have great or universal appeal and is not favorably associated with many products.

A circle or an oval rates high both as

an attention-getting device (impact) and in preference (appeal). However, by itself it generally does not have great "psychological value" as a marketing tool because it lacks specific identity.

A symbol having attraction power and impact is necessary but not sufficient. It must create favorable impression; it must have favorable associations that will be transferred to the company it represents.

In determining the psychological effect of an image it is important to have full understanding that learning that the symbol has appeal as a design is of no value to the company. It is necessary to learn the attitudes toward the company the symbol represents.

Symbols are important factors because people practice sensation transference. They judge a product by its package, a book by its cover, a person by his or her attire, and a corporation (or brand) by its symbol.

After it has been ascertained that the symbol is seen as it should be, the attitudes of representatives of the segment of the population of interest to the company are determined by obtaining their reactions. They constitute a sample that is demographically grouped and classified.

Determining effectiveness of a symbol (package or ad) is not the same as determining the effect of a work of art, which is an end in itself. A symbol has to communicate effectively about the company (or brand) with which it is associated, not about itself.

Because sensation transference from the symbol (package or ad) to the company it represents is the reason that the symbol is a major factor, the company executive cannot gain from comments that his symbol (package or ad) is good or beautiful, modern or up-to-date. He can gain only if people (consumers) who see the symbol react favorably to the products or services the company offers,

because they unconsciously transfer the favorable effect of the symbol to the products or services.

The symbol is generally integrated with the logo. The most effective logos are logo-symbols; the name and symbol are most effective when they are a single point of focus. Standard Oil (Amoco), McDonald's, and the new Hormel signature are such logo-symbols.

Generally, the components, logo and symbol, are first tested separately and then together, as one unit, to determine how effective the logo-symbol is in impact and in psychological effect. Instruments are used for measuring impact. Controlled association tests and interviews with individuals are conducted to determine the psychological effect.

Brief interviews, in an association test in which the respondents have no defense mechanisms, have to be conducted to learn how the logo-symbol rates in favorable associations and desirable connotations.

Association tests conducted with controls for determining attitudes, and ocular measurements conducted with instruments for determining impact have 35 years' validation in the marketplace.

As was pointed out at the beginning of this chapter, however, the psychological effect of images and colors is still not widely understood among corporation executives and knowledge about how to measure the psychological effectiveness of images and colors is often absent.

Executives who are not subjective use polls. Pollsters operate on the assumption that men and women can and will always reveal their real feelings and true attitudes. This assumption was proven fallacious in the studies conducted experimentally in 1935–1941 and in hundreds of studies of consumer products conducted by us since 1945.

There is now conclusive evidence that people will not give answers (to questions) that will put them in a bad light or will in any way downgrade them socially. Most people give "prestigious" answers to direct questions.

Also, marketing involves psychological elements. Most of the time the respondent (consumer) is not aware that he or she is affected or influenced by a symbol. People are not aware of psychological factors and therefore cannot tell about them, even if they want to reveal their reactions.

Some corporations, ad agencies in particular, employ depth interviewing services. Depth interviews are conducted mostly because the researchers like to use their skills, or make use of their training, in psychology. There are many schools of psychology—Freudians, behaviorists, existentialists, and so on. Each psychologist introduces his own subjectivity into the interview.

If there are 10 interviewers who are psychologists there will be 10 kinds of subjectivity in the interview results, or 10 kinds of results. This kind of information is worthless. It is not a reliable basis for determining the effectiveness of a logo-symbol or any other marketing device. Also, the cost of the worthless report is very great.

Consumers are not interested in logos or symbols, they are interested only in products and services. If consumers are asked to evaluate logos or symbols, the evaluations are based on their art "standards." The best art is not necessarily the best symbol or selling tool.

Focus group interviews have been widely used in the 1970s. In a group there is generally a leader, sometimes two or three, who influences the others in the group. The results from focus group interviews are actually equal to interviews with one, two, or three individuals. The individuals in the group behave as critics, not as consumers. Their verbalisms do not always reveal actual attitudes.

Some rely on T-scope tests. The T-scope was used in World War II to train men to identify aircraft. The T-scope is not relevant to marketing because not all individuals take the same amount of time to look at objects. People view symbols, as they do packages in a supermarket, as well as ads in a magazine, at their own pace. Some people are leisurely and take a long time; others are hurried and take a short time. Under different circumstances, people take more or less time to look at things.

There are still researchers who use judgment scales. These may be appropriate for testing products and product quality, but they are not appropriate for testing graphics or psychological effects, of which consumers are not aware.

HOW TO MEASURE SYMBOL EFFECTIVENESS

In conclusion, I am pinpointing what has to be done to measure effectiveness of graphics, research design in testing symbols, logo-symbols, as well as packages and other marketing tools.

The test design that is used has to be fit onto one page. It has to be structured so that the test can be controlled. A U.S. national sample (number of respondents) has to be large (close to 800–1000, or at least 600) so that the qualitative information is quantified and represents the segment of population of interest to the company. The interviews have to be conducted with individuals, not in groups, and they have to be brief to get spontaneous response, eliminating inhibitions and defense mechanisms of respondents. A control has to be used to ensure correct sampling. Because words do not have precise meaning or the same meaning to all individuals, synonyms have to be used to get the answer in depth, to get nuances of meaning.

STEP 1   Use research to determine the kind of "image" the company has for its segment of the population or consumers and/or to determine the strengths and weaknesses of the old symbol.

STEP 2   Get a competent designer to produce four or five designs, using the research results as guidelines.

STEP 3   Put the proposed designs through research to determine which is the most effective in visibility, attention-getting power.

STEP 4   Measure how the symbol, or each of a number of designs, rates psychologically in terms of favorable associations and connotations.

If there is an old symbol, it is also necessary to test the most effective of the symbol designs against the old symbol to determine how much more effective (if at all) the proposed symbol is in communicating about the company.

This methodology outline with the controls on sampling and ensuring uninhibited responses is reliable only if the tests are conducted by indirect method based on the principle of sensation transference.

About 20 years ago, Oscar Mayer had a recommendation from a prestigious design studio to improve their logo. Instead of accepting the recommendation, the management decided to submit their logo to research. The tests showed clearly that their logo rated very high in every aspect (Fig. 4).

Several years later another design studio submitted three logo designs that were considered "modernizing and bringing up-to-date the Oscar Mayer logo."

Because some executives at Oscar Mayer wanted to be up-to-date, it was decided by the management to submit the proposed designs to research for testing against the old logo.

The research showed that each of the proposed "up-to-date" logo designs downgraded Oscar Mayer products, that the old logo rated much higher (almost

Figure 4   The Oscar Mayer logo-symbol has served the company well for many years. There have been attempts to "modernize" it. None of the proposed designs rated as high in favorable associations and preference for the brand as this logo-symbol.

three times higher) than the best of the proposed designs in favorable associations, desirable connotations, and preference for Oscar Mayer products.

Recently, we were presented with the problem of determining the effectiveness of the Hormel Company signature (symbol-logo). On the basis of a series of tests conducted by indirect method, guidelines were established. The strengths and weaknesses of the signature (symbol-logo) were revealed in the research.

The designers thus knew exactly which element to improve without changing it radically and what to add to make the signature more effective. Specifically, the research showed that the color identity should be retained, but that the hue of the red should be

Figure 5   Hormel's new corporate and brand signature—logo-symbol—has a three dimensional effect. It rates 94% in favorable associations. It is a very effective marketing device.

changed; the logo had to be improved to increase its readability; and a design device had to be added to give the corporation specific identity with a favorable psychological effect to express distinction and high quality to the subconscious mind of the consumer.

After the designers produced several logo renderings on the basis of the research guidelines, the designs were put through research. (The orange red was changed to a deep red on the basis of accumulated data.)

The proposed logos were first analyzed through readability measurements. Then the four logo designs that passed these tests were put into a controlled association test with the old logo to determine which one had greatest psychological appeal or rated highest in favorable associations.

The most effective logo was then integrated with five proposed symbol designs. These were screened through an association test to determine which symbol-logo rated highest in favorable associations and promoted greatest preference for Hormel products.

The final steps were to test the new signature (symbol-logo) against the current one and to test the packages with the new symbol-logo against current packages. Packages of three product categories were used.

In each of the tests against the current corporate signature and three packaged product categories, the new symbol-logo rated much higher in favorable associations and preference for the products. (Fig. 5)

Thus, by predesign testing and determining the strengths and weeknesses of the old Hormel signature, logo, and packages, then by testing a variety of proposed designs, a new symbol-logo evolved that had great optical-psychological effectiveness as a corporate signature.

A recent visual effect change of the

Figure 6 Hart Skis introduced this combination of logo and symbol in 1979, in conjunction with new colors for the various kinds and prices of skis.

Hart Skis mark (Fig. 6) based on a predesign research–postdesign research program is an excellent example of the effectiveness of such an approach. First the current logo and symbol were tested against four competitive logos and symbols. Then four new designs, (logos and symbols) were tested separately against the present logo and symbol and tested together to determine which logo-symbol combination was most effective. Finally, colors were prescribed, on the basis of accumulated research data, for the entire line and for each price category. Management reported an immediate, substantial sales increase.

IMPORTANCE OF SYMBOL

A corporate or brand symbol is much more than an identity. The symbol is the visual manifestation of the company. It characterizes the company; it has psychological implications. It communicates and has an instant effect. The effect can be favorable or unfavorable, or it can be neutral. The corporate symbol may have no effect; it may, in effect, be impotent or sterile.

An advertising campaign that is not effective can be changed, often, if necessary. A more effective campaign can be created. A corporate symbol should be as effective as possible because it is (or should be) the permanent and most vital communication about the company.

PART THREE

The Role of Design Assessment in Product Development

CHAPTER NINETEEN

Positioning, Segmentation and Image Research

Lawrence E. Newman

In the 1960s and 1970s, manufacturers began to place greater and greater reliance on market research to furnish important insights into the process of marketing products to the consumer. Techniques have become standardized within consumer product manufacturing organizations for the assessment of product efficacy and consumer satisfaction, the tracking of product performance in the marketplace, the determination of the worth of new product ideas, the definition of market segments, the understanding of consumer needs, and so on. Historically, the role of the package per se in the generating of consumer appeal and purchase interest, in the conveying of the intended product image, and in differentiating or sometimes even aligning itself with competition has been sorely neglected. This situation, however, has begun to change in the past few years, with manufacturers increasingly cognizant of the importance of measuring the contribution of the package in the marketing process.

This change in the perception of the importance of the package has been dramatically noted in the types of package design research undertaken. Formerly the typical approach to package design research was either by judgment alone, with no significant expenditures for package research, or by a few questions pertaining to relative package options, appended to a research study fielded at that time, (e.g., exposure to several package variations at the end of either a concept test, a product test, or even an advertising test). More often than not these "tag" assessments constituted nothing more than a popularity contest with little consideration given to the more subtle factors at work. Now, however, manufacturers are allocating specific research funds for package design research, with these studies designed solely for the purpose of measuring the worth of various design options. With this willingness to elevate the importance of package design research, professional researchers have been able to utilize creative thinking for the understanding of the specific problem at hand and the determination of the factors underlying the interaction of the package and product acceptance. Thus researchers have been better able to seek and apply the best methodology to answer marketing questions that in some way incorporate a packaging consideration.

Herein lies, perhaps, the key to the state of the art today. Researchers have learned that it is most important to elicit the marketer's objectives for the package so that certain key questions can be raised and answered. For example:

- What does the marketer wish the package to accomplish for the product, beyond the level of stimulating purchase interest?
- Is the package intended to convey a particular image?
- Is the package intended to project a particular position within the product category?
- Is there a specific segment of the population to which the package is intended to appeal?

Only when researchers can determine the answers to these types of questions from the marketer/manufacturer can a research methodology be formulated to provide the desired information.

It is not necessary, nor should it become necessary, for a particular research design/methodology to be rigidly applied to answer a package problem as it arises. Rather, the individual complexities of a particular problem should be explored and the researcher given the freedom to apply whatever methodology is felt to answer that problem best. It is especially in the areas of determining the imagery and the positioning and the segmentation fit that an *individualized, custom tailored* questioning sequence is mostly used. Nonresearchers (i.e., marketers) are usually most comfortable with "standardized" techniques that afford consistent measurements, and usually one "key" measurement that, from test to test, provides normative data from which an easy decision can be made. When the issue is imagery, positioning, or segmentation, however, the marketer cannot be provided with easy, one-step numbers to measure against previous research; rather, the subtleties of the questioning sequence, the composition of different cell samples and analysis across different cells, the inferential questioning, and so on provide the necessary information that allows the researcher to piece together the various elements to arrive at the key analytical conclusions and recommendations. In other words, more than other types of research, this imagery package design research calls for greater reliance on the design and analytical capabilities of the researchers servicing the marketing people.

It might be worthwhile now to look at some of the varying possibilities for these more subtle package design research problems.

IMAGERY

The most common occurrences in the marketing field today are the instances when the manufacturer has a particular image he wishes to convey for his product and, in the design of the package for the product, attempts to have the packaging reflect or complement this image. Examples that could be mentioned might be the manufacturer of a gourmet frozen vegetable who wishes to convey a *high price / top-of-the-line* image, or perhaps a panty hose manufacturer whose product is *high quality* and who wishes the packaging for the hose to reflect this high quality image. Other examples might be a *fun* image for a chewable children's vitamin or an image of *therapeutic efficacy* for a dandruff shampoo. In today's marketing world the manufacturer who markets a number of products usually has an overall image which he feels it is necessary to project for his full line of products, and a secondary image that he wishes to project for any new product he is developing.

Given that the image is somewhat set in the manufacturer's mind prior to development of the package, it is relatively easy for researchers to design a study that measures whether the package is correctly projecting this intended image. These studies, either on a qualitative or quantitative basis, are more fully discussed in the following paragraphs.

Qualitative Techniques

A quite simple means of measuring the image projected by a package is to gather approximately 10 target group consumers in a room and have them participate in a focus group session led by an experienced, marketing-oriented moderator. The package would be introduced, and a wide variety of reactions obtained. The advantage to this technique is the opportunity for new issues to be raised that would not have been raised had questioning been conducted in a quantitative, structured environment. Another obvious advantage to this technique is that "likes," "dislikes," and the critical "why do you say that" probing can be effected to a far greater degree than is the case with the quantitative, questionnaire technique.

A major disadvantage of the focus group technique is the relatively small number of people involved in the study, as well as the group dynamics syndrome at work within any one room containing 10 consumers unknown to one another. Specifically, it becomes impractical to conduct a large number of these focus group sessions for any one packaging problem, with a maximum of four different sessions usually considered the outside limit for any one problem. Thus whatever observations are noted cannot be judged with the same confidence and reliability as are observations drawn from a quantitative study. Furthermore, in any one session there is the tendency for one or two panelists to be the more vocal, forceful group members, pressing their opinions on other panel members and perhaps intimidating them to the point that these more naturally reticent panel members may not truly express their opinions. Although the professional moderator can be relied on to limit this liability, it cannot be eliminated totally because of the natural tendency of many panel members not to challenge another panel member with an opposing opinion.

One further disadvantage of the group technique is the difficulty encountered when there is a relatively large number of package options available. The bias that exists because of the order in which the packages are presented necessitates a very large number of different discussions in order to provide for all combinations of "clean" reactions. Most often this ideal rotation is simply too impractical to effect. From a realistic, practical standpoint, therefore, all package options would have to be shown either together or in certain groupings, thereby diluting the depth of response to be achieved for any one package option, which was one of the original reasons for doing this qualitative work.

Thus the overall worth of qualitative techniques for package design imagery research should probably be limited to instances where a preliminary, in depth reading is required or where a more in depth reaction is felt to be necessary to complement a quantitative study.

Quantitative Techniques

The most typical quantitative technique for the measurement of package imagery calls for interviewing a group of respondents in person, either in respondents' homes or in high traffic shopping centers, exposing them to the package either by showing them a prototype of the package or a color photograph of it and asking a series of questions from a structured questionnaire.

This technique is most sensitive when a monadic design is employed, that is, when any one respondent is exposed to only one package and asked questions only about that one. A *minimum* sample size of 100 respondents seeing any one package might be used. (This 100 sample size is an accepted minimum standard, although there are many researchers who prefer the larger sample sizes of 150, 200, or 250 because of the greater confidence to be had in reading differences from cell to cell. Of course cost factors to a large extent dictate whether these greater sample sizes can actually be used.) Each new or test package would have its own cell of 100. It is always advisable to have, no matter how many test cells, one *control* cell, where either the current package or a competitive package is shown.

The preference for the monadic de-sign for this research is related to two key factors. First, were a paired technique employed, that is, where more than one package is shown to respondents, a situation is too easily created in which a "popularity contest" would result, an occurrence that would neither address nor discover the imagery issues to be researched. Furthermore a paired technique maximizes or forces differences and could overemphasize a minimal difference that actually exists between the packages. A second support for the monadic design is that it constitutes a more realistic test environment, comparable to that in which the consumer is placed when entering a store. That is, there is only one package design for the brand and not two different designs as there would be in a paired test.

The questioning sequence for a monadic image study might include many questions related to the following aspects of package imagery:

- Overall rating of the package on a 1 to 10 "excellent" to "poor" scale.
- Specific likes and dislikes about the package.
- How the product would be described to a friend.
- Rating on the package/product on various attributes and characteristics (e.g., "high quality," "good value for the money," "attractive," "suitable for displaying for guests," and "easy to use"). Each of these attributes/characteristics would be rated on the same 1 to 10, "excellent" to "poor" scale.
- Purchase intent for the product, on a five-point scale running from "definitely would buy" to "definitely would not buy."
- Reasons for the stated purchase intent.

It can thus be seen that by comparing

across cells one can easily discern differences in the projected imagery between the test package(s) and the current/control. Although this methodology is quite simplistic and easily capable of establishing the imagery of the package, it might be too easily applied to a problem without complete thought being given to additional questions that might be added. For example, if price/quality are the image factors to be projected, special questions should be employed ascertaining the perceived price of the product, the perceived price relative to competitive products, and so on. If efficacy is the issue, then detailed questions should be used to establish all the image variations of the product's efficacy. It is critical to the maximum utilization of the research that the marketer and researcher explore fully all image ramifications of the package.

Other quantitative techniques that can be used for measuring the imagery of the package are any of the "visibility" techniques, that is, tachistoscope (T-scope), eye movement, and so on. These techniques, of course, provide primary measurements of how quickly consumers recognize the product. In addition to wishing to measure the speed of recognition, marketers often also want to measure the imagery created at these different levels of recognition. Thus while a marketer may be interested in knowing what image is conveyed to a consumer who is given a maximum amount of time to study the package, this marketer may also be interested in the "real world" situation of what image is conveyed when the consumer does not have this maximum time, but rather sees the product sitting on a shelf in a supermarket as he quickly walks down the aisle. These visibility measurement techniques therefore afford the marketer the opportunity to read the image factor in a simulated real world situation.

POSITIONING

Marketers have become increasingly aware of the value to be gained by correctly "positioning" their products. Prior to entering a new product category, most marketers attempt to assess consumer satisfaction with existing products in the category. In this way, and with this information gained, a new product to be introduced can be positioned to fulfill some untapped need. Similarly, a marketer with an existing product can look at the market, examine consumer needs and satisfaction levels, and attempt to *reposition* his product so as to exploit a "gap" in the category. Regardless of the circumstances, then, the marketer faced with the task of understanding the influence of the package on positioning the product has to look to research to uncover for him either the extent to which the package supports the intended position or, with no specific position intended with the formulation of the package, how the product is perceived and thus the position it occupies in consumers' minds.

Often circumstances show that a position has to be created for the product *prior* to the finalized product and/or package. The research necessary at this initial stage is thus conceptual in nature and would not measure the impact of the package per se. For example, a manufacturer might be interested in finding some unfulfilled need, let's say in the aerosol antiperspirant deodorant market. With the knowledge that his research and development people have uncovered an extra dry formulation for an aerosol antiperspirant, and with the assumption that the antiperspirant category could conceivably support this new type product, the manufacturer commissions a study among antiperspirant users wherein levels of satisfaction with existing products are fully explored. (This quantitative study could be preceded by a series of focus groups where consumer satisfaction levels with current products are fully explored and the idea of an extra dry product is presented and considered by the panelists.) Questions consumers would answer about existing products and the brand they use most often in the quantitative study would fall into the following areas:

- On a voluntary basis, all the things they like about the brand they currently use most often.

- On a voluntary basis, all the things they do not like about the brand they use most often, with a follow-up probe on how they think their brand might be improved.

- Overall rating of the brand they use most often on a 1 to 10 scale running from "excellent" to "poor."

- Rating of the brand they use most often on a long list of specific product characteristics and attributes on the same "excellent" to "poor" scale.

- For the same list of product attributes and characteristics, a rating of the *importance* of each on a 1 to 10 scale running from "extremely important" to "not at all important."

These questions would form the basis of measuring the status of the market as it currently exists. Ideally, the sample size would include enough people to provide large enough bases for the analysis of different brand user groups representing the leading brands in the category. The differences noted in the ratings of the brands on given attributes relative to the importance these people attach to the same characteristics indicate the "gaps" that exist in their needs being filled. For the manufacturer in this example it would be ideal if the attributes of "goes on dry," "keeps me dry," "is not sticky," and "prevents perspiration for a long time" all show high levels of importance to consumers, while at the same time exhibiting a situation where the existing brands in the category are not being highly rated on these same attributes. Were this the case, the opportunity arises for the extra dry formulation to represent a viable positioning.

This study would not be concluded at this point, since it would also be worthwhile to present the respondent with a conceptual presentation of the manufacturer's proposed product, the extra dry formulation. The conceptual presentation should not present the contemplated package unless the package decision has been finalized. The questions asked would follow the same format as those discussed for the imagery research (i.e., overall rating, likes, dislikes, perceived advantages, rating on the same characteristics/attributes, etc.). Analytically, the ratings and responses to the concept are matched to the importance and "brand used most often" ratings to determine the extent to which the concept appears to be capable of narrowing the gap in key unfilled areas.

Once the manufacturer digests the information from this study, he can then proceed with product and package finalization. It's at this point that a specific research study can be conducted on the package/product to determine the extent to which the package is communicating the now solidified extra dry positioning.

The specific methodology options for this assessment are quite similar to those employed for the imagery studies. Qualitative work could be conducted again to furnish a preliminary feel for the in depth responses to the package/positioning fit. If more definitive information is required, or if there are many variations of packages being considered, the focus group/qualitative research will be insufficient for providing actionable data.

A typical piece of *quantitative* research for this problem would involve interviewing, in person, a sample of respondents (in this case perhaps 100 to 150 aerosol antiperspirant users) and exposing them at the beginning of the interview to a picture or a prototype of the new package. In addition to the "standard" questions of overall rating, likes, dislikes, purchase interest, and so on, special questions could be formulated to ascertain, on an unaided basis, the perceived positioning. For example:

- Based on what you see from this picture, what benefits do you think the product will provide you that other products on the market are not currently providing?
- What brands of antiperspirants does this seem to be most similar to? What brands does it seem to be least similar to?
- To what extent does this product appear to be an improvement over other products on the market? Does it seem to be a very big improvement, a big improvement, a small improvement, or no improvement at all?

Following this, a detailed explanation of the new product's benefits would be given along with the package picture or prototype. Questioning would be almost the same as the preexposure to the benefit explanation, with overall rating, likes, dislikes, ratings of product characteristics, similarity to other brands, and improvement obtained. Results can be analyzed in several ways—postbenefit explanation versus prebenefit explanation, prebenefit explanation by itself to see on that unaided basis what the package is conveying, postbenefit versus the earlier study's information, which indi-

cated the current brand ratings and importance ratings, and so on. If only one new package is to be tested, these analyses would show the ability of the package per se to deliver the positioning required. If more than one new package is being considered, the cross-cell comparisons, with each new package being surveyed among a separate, matched sample of respondents, will clearly point out the relative ability of each package to meet the marketer's objectives.

Positioning research for packages can also lead to some important *structural* changes in the design of the package itself. For example, a manufacturer of soft drinks might decide that he is interested in exploring some of the problems that consumers have with using any and all soft drink products. His intent would be to uncover some consumer problem that conceivably is not being satisfied with existing products and to enter the market with either a new product, line extension, or change in his present products intended to satisfy this unfulfilled need. He would commission a research study the sole intent of which would be to uncover these consumer needs.

The study would be a two-stage project with the first stage being a series of focus group discussions among target group members. The group discussions would seek to have consumers state all the problems they have with using soft drinks, with no restrictions or inhibitions placed on respondents' freely associating the most "far-out" suggestions. In these discussions consumers would be encouraged to talk about their problems with the soft drink product itself, the problems in purchasing, and the problems in using, which would easily lead to unearthing problems people have with the packages, whether bottles or cans.

In this example it is quite possible that consumers talked about many functional aspects of the soft drink package, one of which might have been that cans are a problem because they can't be resealed when only part of the can is consumed at one time and when the consumer might want to save the remaining product. Mention of this problem might easily stimulate another panel member to say that they find it to be a problem that manufacturers of soft drinks do not make a small 4 ounce can available, a packaging innovation that they feel would solve this very problem. Certainly the manufacturer, hearing these comments spoken by consumers, might commission his packaging people to construct such a can and place this new product in test market to measure its viability. The manufacturer would be remiss, however, in not understanding the *depth* and breadth of the problem in the population, since it would not be wise to go into test market from the remarks of a few group discussion panelists.

Thus a second phase of the study would present a sample of 150 to 200 consumers with a long list of the problems developed from the focus groups and have them indicate for each problem, including those pertaining to this small can serving, how great a problem it is (perhaps on a five-point scale running from "very great" to "none") and how often this problem occurs (also on a five-point scale running from "all the time" to "never"). When results are tabulated, a judgment can be made as to the extent of the problem in the population in terms of the depth and frequency of occurrence. For this example it is quite conceivable that the problem with a small can could rank relatively high in the order of category problems and would thus recommend to the manufacturer that he develop such a 4 ounce package. Armed with his research findings the manufacturer would feel confident that he is meeting a consumer need.

SEGMENTATION

Segmentation as it applies to package design research is closely allied with the positioning and imagery research already discussed. "Traditional" segmentation studies do not concern themselves with package issues, nor would a package design problem evoke the need for a separate segmentation study to be conducted. The reason for this is that "traditional" segmentation studies are extremely detailed and costly and are generally undertaken to investigate "global" problems. Examples of these global problems are:

- Understanding a product category in terms of the different types of people, both demographically and psychographically, using various products within the category.
- Segmenting the category into particular *benefit* groups (e.g., in the analgesic category where benefit segments might be developed for speed, strength, safety, etc.).
- Determining the perceptions of consumers as to which products in the category are most like one another, thus forming different *product* segment groups.

Research projects investigating and supplying this information tend to be quite extensive in time, scope, and cost and can be justified by marketers to answer the large questions, but rarely deserve justification for addressing relatively small problems such as a package design question.

There are, however, instances when new package designs are being considered, or variations for an existing package are being developed, that necessitate a research study that would take into account the different segments existing in the market. As an example, let us use the product category of facial tissues, where a manufacturer has an existing product that is available in five different current packages. His design people have developed six new packages and wish to see which of the new packages have greater consumer appeal than the current so that a new mix of the best five packages can be marketed from the eleven now available. If this were the only consideration for the research, a relatively simple package design preference test could be constructed to answer the manufacturer's question. Segmentation plays a part in this example, however, because the manufacturer has already conducted a major global segmentation study on the facial tissue category and discovered that there are three basic segments of the market—a high priced, quality segment (in which his product is included with the new packages to be tested), a medium priced, medium quality segment, and a low-priced segment. In designing the study for the package research it was therefore felt important to have three different samples or cells of consumers representing these three segments, since it was believed that it would be necessary to measure the viability of the new packages among the current franchise/segment, as well as the potential these new packages might have for making inroads in the other segments. Although it might have been argued that little was to be gained from the appeal to the low priced segment, the manufacturer felt confident that one or more of the *new* packages might be capable of garnering some purchase interest from this segment.

The test then was conducted among 300 consumers, with quotas established for 100 in each of the three segments as defined by the earlier segmentation study. The study design called for exposing the complete array of 11 packages, both old and new and asking con-

sumers to rank their top five favorites. Then these top five were placed in a shelf display with competitive products and the respondent was asked to indicate his anticipated purchase intent behavior for the next ten purchases of facial tissues. This variation of the five-point purchase intent scale is the frequently used "constant sum" technique, for which a similar measurement is obtained at the beginning of the interview, with changes noted from the preexposure to the new packages to postexposure. Results from this constant sum question display the drawing power of the various package options in each of the different segments.

It is therefore easy for manufacturers who have already conducted global segmentation research on their products to take into account these segmentation results when designing their research for package design problems. As mentioned earlier, the segmentation information is most useful when factored into the design of package research, as opposed to designing the package research within the framework of a segmentation study. Those marketers who have no experience with segmentation should first attempt to integrate segmentation theory (whether user, product, or benefit segments) into smaller scale image and positioning studies and then apply the derived information to a custom-tailored package design study addressing the specific problem at hand.

CONCLUSIONS

As has been seen, the inclusion of imagery, positioning, and segmentation considerations in package design research is quite common today. It is paramount that each new package design problem that arises be carefully investigated to discover the type of information needed to make correct marketing decisions. This careful consideration will inevitably lead to the application of current methodology to the problem, always with consideration given to exploring all avenues of inquiry including the various considerations of imagery, positioning, and segmentation, so that there is no necessity to view the final data with a feeling that something has been overlooked. This can be most successfully accomplished when the marketer places the burden of correct research design in the hands of competent market researchers on his or her staff or, when staff capabilities do not permit, of consultants who provide these services.

CHAPTER TWENTY

The Package as Communication

Burleigh B. Gardner

To begin with, we must recognize that a package performs two functions: (1) as a container for the product and (2) as a communication that will influence the purchaser.

The first area is critical and involves many considerations. The package must protect the product, must facilitate its use, must facilitate shipping, handling, and distribution, and must be feasible in terms of cost and technology. This is a field for specialists such as packaging engineers and is outside the scope of this chapter.

The second function, that of communication, has over the last 50 years become increasingly critical, especially in the broad field of consumer products. Every package that we see on the shelves of the supermarket, drug store,

the liquor store, or any retail outlet is a visual communication, a complex set of symbols that has an impact on the customer. The purpose of this communication is to make the potential customer desire the product. It is a deliberate effort to win customers, as a politician's speech is a deliberate attempt to win the votes of his audience. In a consuming society in which every customer has a multitude of choices, the importance of the package as a communication is well understood. This is so well understood by everyone (including customers) that probably every package we use has been designed by some expert, and a vast number of them have also been studied by researchers to determine consumer reaction.

In the last few decades a major field

of package design research has developed. It is composed of hundreds of firms, consultants, and research specialists, with a wide variety of techniques for studying the communications aspect of packaging. Probably every consumer research organization in the country has studied packaging in one way or another.

The various methods of this research are all based on certain assumptions about the package as a communication and what is important about it. Each has its use and each rationale has some truth to it. Each seems, however, to be only a partial truth. There is no one technique from which researchers can say that any package will overcome all the others in winning the minds and dollars of consumers. In fact, we are in a state in which what we *don't know* about the consumer's mind and how to research it is greater than what we do know. All assumptions are only partial assumptions or relevant only to certain situations, and research can at best only eliminate the gross errors and increase the odds of success.

With these limits in mind, I will discuss further some of the theories and methods of research, but will not attempt to go into detail on any one method. These are covered more adequately in other chapters by specialists thoroughly familiar with each.

WHAT DO WE MEAN BY COMMUNICATION?

I addressed this question earlier as follows:[1]

When we talk about a package as a communication, we are referring to a very complex set of reactions in the mind of the beholder, all of which contribute to a feeling about the product inside. The nature of this feeling can contribute to the anticipation of what the product will be like and the satisfaction it will provide. In the case of many food and beverage products the package may actually affect the sensory response to the product. Thus the same soft drink in different bottles may taste differently. With high-volume consumer items, where there are many competing brands, even subtle differences in package communication can have an important impact on sales.

The package becomes a symbol that penetrates the mind and triggers meanings that often are not put into words and that to some degree may vary from person to person. With us a package may have some generalizable meanings because in our culture we have all been exposed to it in similar contexts. Since as far as we know most symbolic meanings are learned, however, many package symbols will mean something different in different cultures. A package or a logo that is rich in meaning to Americans will have entirely different meanings to aborigines who have had no experience with the product. (That is not to say that there are no symbols with universal meanings due to the human condition. Phallic symbols, for example, appear widely, and they appear early in man's history.)

Within our culture, however, a package, a name, a logo, tell the consumer many things and arouse feelings or expectations that are relevant to whether he will accept or desire and use the product. The problem of package research lies in part in studying this symbolic meaning and how it will affect the sale of the product.

When we talk about a package as communication we are talking of its symbolic meaning. To quote Dr. Sidney J. Levy from his article, "Symbols for Sale:"[2]

Several years of research into the symbolic nature of products, brands, institutions, and media of communication make it amply clear that consumers are able to judge grossly and subtly the symbolic language of different objects and then to translate them into meanings for themselves.

Consumers understand that dark colors are symbolic of more 'respectable' products; that browns and yellows are manly; that reds are exciting and provocative.

Dr. Levy further describes a study of symbolic meaning:

In a recent study of two cheese advertisements for a certain cheese, one wedge of cheese was shown in a setting of a brown cutting board, dark bread and a glimpse of a chess game. The cheese wedge was pictured standing erect on its smallest base. Although no people were shown, consumers interpreted the ad as part of a masculine scene, with men playing a game, being served a snack.

The same cheese was also shown in another setting with lighter colors, a suggestion of a floral bowl, and the wedge lying flat on its longer side. This was interpreted by consumers as a feminine scene, probably with ladies lunching in the vicinity. Each ad worked to convey a symbolic impression of the cheese, modifying or enhancing established ideas about the product.

This study could easily be of two packages, each of which would change the impressions of the product.

The significance of the symbolic communication of the package and its effect on perception of the product was also shown in an experiment in which the same beer was bottled in three different types of bottles: the usual long-necked brown bottle, a short, "export" brown bottle, and a clear bottle. In a taste test, a group of beer drinkers rated each beer on a special semantic differential, and there was a clear difference perceived among the three. Clearly, the package was affecting the perceived taste.

All of this reiterates the point that the package is a symbolic communication that tells consumers something about the product it contains. The package also often communicates at a very subjective level; it produces not firm knowledge but a feeling.

When dealing with symbolic communication, we are dealing with the processes of the mind. The eye is only a link in the communication system that transmits external stimuli that the eye perceives, to the mind, which interprets these stimuli. Dr. Edwin Land of Polaroid recently reported research that shows that the eye transmits only variations in grey (not color) and that the brain translates these into form and color. He claims that the eye transmits only a chaotic series of impulses and that the brain organizes them into forms with structure and meaning.[3]

It is important to keep this in mind with some types of perceptual research such as eye movement and pupil dilation research used in advertising and package studies. In such research we are seeking measurable physiological responses that can tell something about what is going on in the mind that is relevant to our interests. To judge the meaning of such measures when studying a package involves a number of assumptions that may or may not be true.

WHAT A PACKAGE SHOULD DO

As a communication a package has several functions to fulfill. First, it must be visible in in its setting. If it is not seen, the customer will pass it by. Conventional wisdom tells us that of several competing brands the one most readily seen will get the customer's attention and hence the sales. In usual supermarket merchandising the goal of a brand is to get as much shelf facing as possible, with the assumption that a row of many packages will get more attention and sales than a single package. Also, the package that stands out perceptually will get more attention and sales than one that blends into the background.

This is one of the partial assumptions of package research that is not always true. Theoretically, it assumes that the customer looking at grocery shelves or a similar display is making a random search and that the first package to be perceived will be chosen. The housewife in the supermarket has ideas about brands and prices that enter into her search for a given product. If she can't find one she wants, she will keep on looking, or not buy, or settle for a second choice. With a new brand its product visibility may be especially important, since until it attracts attention to itself, it will not be noticed or bought. When purchasers think of all brands as being essentially the same, shelf visibility is probably very important since the shopper has an attitude approaching random search and has no influencing attitudes about brands.

A number of methods have been widely used to test for package visibility. For example, a test subject may view a test display through a tachistoscope to control rigidly the viewing time to a limited number of seconds. The subject then names the perceived brands. If the package being tested is named more often than the other brands in the test scene, then it is considered to be more attention-getting.

In these tests it is rare for a package being tested to get a dramatically higher score than competitive brands. If the experiment uses a large sample, even small differences can be considered statistically significant, but small differences are probably of no practical importance.

There is one problem with these tests: the subjects are not actually shopping with a purpose in mind. They are participating in an experiment. In test-

ing, the subjects are sometimes afterwards asked to select samples of the brands tested as gifts. The choices made often do not reflect the measurements of perception in the test situation. At best there is only partial confirmation since in an actual choice situation the factors of brand perception and choice enter in.

With packages in shelf displays at retail, the shopper actually is receiving an extraordinary volume of competing signals. These comprise a chaos of visual impressions that the brain must organize into a meaningful structure. At our present state of knowledge it is believed that every visual impulse is received and interpreted by the brain, which selects those impulses to which it will give attention. The process is analogous to tuning a radio set to receive only certain stations: the brain "tunes in on" certain things that the eye brings to it, and tunes out, that is ignores, the rest.

Thus a shopper can walk through a supermarket with 10,000 items on display and select (and pay attention to) only a few. The shopper may pass down every aisle, stopping to consider a few selected items here and there and successfully ignoring the rest. Amidst the welter of competing impressions it is doubtful if a package can get attention unless the mind-set of the shopper is ready to let the impression in.

This is not to say that the shopper will only attend to the things that are appropriate to that shopping trip. The mind has a wide range of interests that can be reached. For example, experience has shown that a jumble of items in a shopping cart with a "Sale" sign will attract attention and sell items that the shopper was not originally intending to buy. Many other merchandising techniques try to get past the barriers in the mind to attract attention.

A package should communicate in a way that will make the customer want it. All the elements in the design should be geared to that objective. Furthermore, it should try not to express conflicting ideas. A package design should be viewed with the following questions in mind:

1 What does it intend to say about the product?
2 To whom does it want to say it?
3 How well does it say it, and to what kind of people?
4 What does it say to others?

Since a package is usually speaking to a wide variety of people and may have mixed purposes, such as to change or reinforce a brand image, to retain old customers, and to gain new customers, its message can easily be confused or can be a compromise and not the best possible message for any audience or for any purpose. Communication is always simpler for a brand that is targeted to a limited audience and that has a specific purpose.

The package designer works with symbols that range from those with a commonality of meaning, such as descriptive words that all literate people understand, to elements with less agreed-on meanings or that arouse subjective reactions that are not defined in any dictionary. Some of these meanings may be detrimental to what the designer intends to communicate. To quote again from my earlier speech:

Now let's consider the elements in a package configuration which affect its communication. One of the first to be considered is the material itself. Studies show that there are distinct differences in the meanings and imagery evoked by different materials. Thus, metal arouses feelings different from plastic, paperboard, or glass. Each evokes a different set of associations ranging from descriptive realism, i.e., hard, brittle, to the non-rational, emotion-laden ideas such as purity and warmth. This means that starting with the choice of materials, the designer is beginning to create a communication which will, in its ultimate form of the completed package, reach into the mind of the consumer and influence his reactions to both package and contents.

Next is the influence of form, the actual shape of the package. Different geometric shapes have different evocative meanings. A cube is different from a sphere or a pyramid in the non-rational feelings each evokes. For example, comparing a sphere and a pyramid we find the sphere is more soft, feminine, delicate, while the pyramid is more masculine, aggressive, strong, and even harsh. Although the reactions to the basic forms can be readily established, the design often is a complex form whose meaning cannot be entirely anticipated and must be determined by research.

Color is another factor which contributes strongly to the communication. If the package (apart from the label) is to any extent visible, its color carries a wide range of meanings. Thus, white speaks of purity, cleanliness, coldness, while red is warm, exciting, harsh, and even dangerous. There are variations in the meanings from one end of the spectrum to the other and for different variations in lightness and saturation.

Finally, there is the problem of the label which usually carries the main burden of the communication. When considered first as a visual configuration (apart from the actual words), the label combines both two-dimensional form and color. Here the influence of different colors is very important both in themselves and as they interact with the forms to create a configuration.

The goal of the design of package and label should be to have all the different elements blend together to communicate desirable qualities or feelings. Since other factors may often impose basic materials or shapes, it is possible to use color and form especially in the label to set the dominant themes. This requires a high order of skill in manipulating forms and colors for their symbolic values, not merely for artistic effects.

Let us consider some examples showing what the package communicates. In one case, a design for a box of sanitary napkins used a scene of brown plants and grasses blowing in the wind. The whole was dominated by irregular forms and shades of brown. When women were asked what ideas the package brought to mind, the general feelings were of disturbance and unease. As one woman put it "I would call it a menstrual storm." In contrast, another package with a simple floral design in pink and white evoked feelings of serenity, comfort, and relaxation. Clearly, for a sanitary napkin the package should be soothing, not stimulating.

In another test, a design for a small kitchen appliance was tested in two forms. In one form it was only in stainless steel. The other was the identical form but with a small logo in gold-colored metal on the side. The model with the touch of gold was felt to be of somewhat better quality than the plain model.

Other research has shown that color affects the perception of weight.[4]

To conclude, the package is a complex bundle of symbols that interact with the complex mechanisms of the mind. It is only this interaction that gives the package any meaning to the individual, and this meaning is what is important to the producer of the product or to the merchant who sells it. Research, therefore, must deal with two very complex systems and there are no simple or even complex techniques that tell us with certainty what the effects will be.

REFERENCES

1 Burleigh B. Gardner, "The Package As A Communication," a speech prepared for the 1967 International Marketing Congress, American Marketing Association, Toronto, Ontario, Canada.

2 Sidney J. Levy, "Symbols for Sale," *Harvard Business Review,* **37,** No. 4 (July–August 1959), pp. 120, 121.

3 Edwin H. Land, "Our Polar Partnership with the World Around Us," *Harvard Magazine,* (January–February 1978), pp. 23–26.

4 Benjamin Wright, "The Influence of Hue, Lightness, and Saturation on Apparent Warmth and Weight," *The American Journal of Psychology,* **25,** (June 1962), pp. 232–241.

CHAPTER TWENTY-ONE

How Package Design Contributes to Product Positioning

Lawrence J. Wackerman

Product positioning has come to mean many things to many people in recent years. Some marketing strategists have heralded the 1970s as the "Positioning Era," inferring that positioning, as a major marketing tool, suddenly materialized some 10 years ago. Others insist that positioning is as old as the hills, and that without creativity even the best positioned products are doomed to failure. Still others warn that positioning and creativity alone are of little help without strong market planning.

For those of us in the packaging industry, such a debate, or simply the fact there is one, may seem a bit amusing. Product positioning was an important marketing strategy for packaged

goods long before the so-called "Positioning Era" began. Planning, research, creativity, and all other elements of sound marketing are incalculably important, but we've always felt that in the supermarket where the purchase is made the package is king. It is at this point, where all marketing elements converge, that the position the product holds in the consumer's mind must be confirmed by package design.

The positioning debate, however, does signify one major change in marketing. The importance of proper positioning, and, for our purposes, the need to project that position in package design, has been recognized as a marketing element that requires long term research and planning. Packaging should begin when your product first emerges from R & D and its positioning is decided on. Moreover, the designer must be privy to vast amounts of marketing information if he is to create the right package design to support the position. Too often the package has been relegated to the role of afterthought, resulting in designs that "didn't talk" and provided even less action. The enormous and growing number of new products that failed attests to this. And because of the proliferation of "me-to" products, the vying for shrinking and overcrowded shelf space and the continuing trend to quick-decision, off the shelf purchasing, there will undoubtedly be many more in the years to come.

But what is product positioning, and what part does packaging design play in it?

PRODUCT POSITIONING

In general terms, every product should be aimed at a particular audience and designed to occupy that particular niche in the marketplace. The package design communicates to consumers the qualities of the product that fill that niche. One bakery, for example, might produce three qualities of bread—no frills, budget, and premium. In each case, package design visually differentiates the products from each other. The no frills bread would come in an inexpensive, clear plain wrapper with few, if any, design elements; budget bread would require more design elements and more color to communicate its position, while the premium bread package would represent a considerable step up in the design from budget. In each case, the design communicates its level to the consumer who seeks that particular product. If a no frills bread with its low, no frills price, were packaged as if it were a premium, higher priced bread, two things are likely to occur. Either the consumer will pur-

chase it as a premium bread, find it lacking, and never buy the brand again, or the no frills consumer, the real audience, will pass it by because upscale design and packaging qualities connote higher price.

It would be so simple if each product could find its own niche and sit on the shelves alone, but of course this is not the case. In most supermarkets one no frills product will share shelf space with dozens of others, so product positioning and its communicator, the package, take on another task: positioning the product among many similarly positioned products, *on the shelf*.

WHAT CONSUMERS SEE

Package designs may contain artwork, but the final package should not be considered a work of art. At its best, it communicates vividly and clearly to the consumer the qualities of the product it contains, which, of course, have been developed to appeal to a specific audience.

Color, for example, plays a major role as a symbolic communicator that can cause psychological reactions in consumers. Walking into a bright red room will speed up the pulse rate while a green room will slow it down. Blue can connote calmness, yellow implies low expense while lush, dark colors represent wealth or prestige. Poisons are usually packaged in ominous black and yellow cans. Toothpaste, on the other hand, usually comes in clean white tubes.

STRUCTURE AND GRAPHICS

In general, there are two major design areas in packaging that can be utilized to reinforce a product's position: structure and graphics. Structure includes materials, shape, and, to a different degree, function or utility of use. Graphics include the previously mentioned color, art work, and type (copy). Remember, these design elements, when being created, must also incorporate the better known functions of packaging such as protection, containment, handling requirements, and communicating information required by law. Costs, of course, are a major concern, and compromises involving all these elements must often be made.

MATERIALS

While the product by its very nature will usually dictate the kind, form, and style of its package, the package designer has a myriad of packaging materials and styles at his disposal, such as paper, paperboard, folding cartons, rigid boxes, spiral-wound containers, corrugated containers, flexible films, foils, metal or composite cans, glass, and plastic. Each, in turn, can be divided into numerous subcategories from which one or a combination of materials can be chosen for optimum product positioning reinforcement.

For example, a high priced facial tissue positioned to appeal through softness all but invites consumers in the supermarket to squeeze for themselves. While a conventional paperboard container would suffice, it prevents the consumer from testing the softness. However, if the product is contained in a flexible film, the consumer can squeeze the product and actualize the communicated quality.

Another tactile quality involves the packaging material itself. A premium loaf of bread, for example, would use a heavier, stronger film for its bag because the tactile quality of the material connotes a premium product.

Or, if another food product is visually

appealing without its container, the packaging should allow certain visual access by using clear materials. If the product is not visually appealing, as is the case with many frozen goods, the package should be opaque. Appetite appeal would be projected through graphics.

The natural appearance of a material can also play a big part in positioning. One of our clients, a distributor of unpopped popcorn, wanted to project an image of premium quality through package design.

After selecting a material that visually connotes quality, in this case a highly reflective, opaque metallic film, the client began to enjoy a larger share of the market. The choice of material, he feels, was responsible.

NEW IDEAS ALL THE TIME

One need only reflect on the rapid development of new packaging processes, techniques, and materials to realize the importance of keeping up-to-date and understanding that packaging is in the business of solving problems. Each one is new and unique, and if immediate solutions do not appear available, that is no reason to abandon the search. New technology or material combinations are constantly being developed to meet such unique needs. One new example was recently born in the corrugated box category, a material that, to say the least, was drab and industrial in appearance. The St. Regis Company developed a corrugated process incorporating a film coating with superior color and graphic printing capabilities. The product inside, a toaster, for example, could now be appealingly represented in full color on the package itself. And when this package is stacked among dozens of others on the shelf, it stands out. Its product position, though shared by many similar

products, is distinctly and visually uplift-
ed.

SHAPE

To some degree, package shape will be
determined by material selection, but
this should not be considered a stum-
bling block. As mentioned earlier, technological developments continue to reduce the limitations of our packaging
materials, and certain shapes, once not
possible, can now be produced.

A package's shape reinforces a product's position both in the tactile and
visual sense.

In the cosmetics field, for instance,
shape plays a most important part. A
luxury perfume aimed at the older, mature woman may be packaged in a cut
glass bottle recalling classic and expensive crystal containers. If the perfume is
targeted at a younger audience, sleek,
smooth, and modern shapes, perhaps
made of plastic would be appropriate for
attracting today's active young adults.
In each case, container shape represents
a visual and tactile confirmation of who
the product is for. At home the shape
reconfirms the consumers expectations
each time it is picked up and used.

In addition, due consideration must
be given to package shape as it relates
to shelf display. If the supermarket finds
the package unwieldy or unstackable, it
can only find a place near the bottom
shelf if it finds a place at all.

FUNCTION

A package's function, and here we mean
designed-in utility of use, differs from
other structural elements because it
often becomes part of the product as
well as its container and communicator.
Hence the product's positioning could
well include the package's utility.

A refrigerated pancake batter, for
example, has been recently packaged in
a carton that allows consumers to mix
the batter without transfering it to a
bowl. This designed-in utility of use provides convenience and saves time and
energy, and the package's message
makes sure consumers know about it.

To introduce a new vegetable shortening, another company packaged its
product in a reusable "plastic" can with
a screw top. The fact that the container
is reusable is prominently featured on
the label. In addition, the screw cap
provides better and longer storage.

GRAPHICS

When most consumers talk about package design, they are really talking about
graphics, which shows how important
this facet of packaging is to product
positioning. Graphic elements are varied, versatile, and only limited by man's
imagination (though most are definitely
limited by cost). Their effects are also
unpredictable and must be handled with
care and ample consideration.

Graphics include color, artwork,
type, and copy. Each element, as previously discussed with color, produces
certain subjective reactions from targeted consumers that can to a great degree
be controlled to support a product's positioning.

For example, let's look once again at
our three loaves of bread.

Graphics on the no frills bread are
minimal. Plain sans-serif type states the
ingredients as prescribed by law, and
the brand name on the end. The brand
name in this case includes color to enhance the product's quality and to raise
it a notch above its other no frill competitors.

The budget loaf uses many more
colors on a white film and larger, somewhat more stylish type to emphasize its

market position. Rather than being limited to end labels, these graphics encompass the overall area of the package.

The premium loaf, on the other hand, represents a major departure. In this case, copy elements emphasize the bread's all-natural ingredients. The type, while contemporary, resembles the type from "the good old days." Art work and color were chosen for similar effects. Artwork (illustrations) depicts a wholesome, multicolor farm scene that represents simpler, "healthier" times. The overall package color is green, which again connotes freshness and naturalness.

In each case the product's positioning in the supermarket is supported and confirmed through graphics.

APPETITE APPEAL

Another method for supporting a food product's positioning is through appetite appeal, and here artwork is of the utmost importance. Designers and brand managers have many possible options for emphasizing appetite appeal and can select from line drawings, multicolor illustrations, stock or custom art, illustrations and color process, and many other forms depending on cost.

Graphics for frozen vegetables are a good example of appetite appeal at work. Frozen vegetables are visually unappealing before they are cooked. But brand managers aren't really selling frozen vegetables. Rather, it is the prepared, hot-off-the-stove product that is for sale. So artwork of cooked vegetables is used to enhance the appetite appeal of the product.

A CASE HISTORY

Premarket testing a package design to gauge its effectiveness is never a sure thing. Many designers will state that until the package proves itself in the supermarket you can never be completely sure of its appeal.

In the supermarket the package can nevertheless be judged quite easily. What follows is a case in point that illustrates the power of the package.

In 1968, Ore-Ida Foods, Inc., of Boise, Idaho, began marketing premium frozen french fries called "Deep Fries." The name was appropriate because each potato was coated in oil. When cooked, the product came out deep fried, but from the oven.

"Deep Fries" was a premium product that cost up to twice as much as private label products. They were also considered the easiest, quickest, and best tasting frozen french fries available.

To reinforce this product's positioning in the supermarket, an effective package was essential.

A first package design seemed to incorporate all the essential elements needed to project a premium product. The background of the package was a clean, flawless white for purity and freshness. The product illustration was a collection of crisp and golden french fries fresh out of the fryer—appetite appeal. The illustration was placed in the upper right quadrant of the package. A gold border further emphasized high quality. Package copy included the trademark, the brand name, and the description, "Deep Fries, French Fries" in the lower right hand corner and "Deep Fried flavor and crispness without deep frying" in the top left quadrant.

The product was introduced in 1970, went national in 1972 and enjoyed good (but not great) success in the next four years.

Unfortunately, the product did not meet the sales expectations and projections originally hoped for. Ore-Ida enlisted a market research firm to find out what was happening. A survey to determine who was and was not buying the product resulted in a number of revelations.

1 The package was a success because to most buyers it projected the brand as expensive, superior, and of premium quality.

2 The reason that the product was superior, however, was not understood by most consumers, and therefore the higher price was not justified in the shoppers' minds.

3 Those consumers who did purchase the product did in fact understand why the product was more expensive. The profile of this consumer was that of a working woman who habitually purchased the best products she could afford.

4 Many consumers mistook the brand name for a description of the product but purchased it anyway because of the quality packaging.

Ore-Ida chose to investigate two presentation possibilities that would attract new customers without alienating the old.

The first option was to explain more effectively why the product was superior. The second option was to maintain the premium image but to make it less intimidating and more accessible.

In either case the key was positioning the product to a specific audience through package design.

The first new package looked almost exactly like the original, but certain copy elements were changed to tell a better story. The description, "Self Sizzling" was added to "Deep Fries, French Fries." In addition, "Deep Fried flavor and crispness without deep frying" was replaced with "The Self Sizzlers, they make your oven work like a deep fryer." The product now had a new name.

To soften the premium image, the second new package gave the brand a new parent—Heinz, a familiar, trusted name and Ore-Ida's parent company. The copy in the lower right corner was changed to "Heinz, Self Sizzling Deep Fries." The illustration, now centered, was also changed to resemble the product cooking in the oven to emphasize its easy preparation. The copy, "The Self Sizzlers, they make your oven work like a deep fryer," remained the same but was repositioned below the product illustration.

In test markets both packages produced greater sales and a larger share of the market. In the end the second new package was chosen, but each demonstrates just how important the right package can be in communicating a product's position to generate sales. The change may seem inconsequential at first, but the results say something completely different.

CHAPTER TWENTY-TWO

The Role of Industrial Research in Planning and Managing New Packaging Product Development

Daniel D. Rosener

The development of successful new packages, as that of most products, is determined in large measure by the product's ability to meet the needs of the marketplace. The term "meeting the needs of the marketplace" may be considered an oversimplification because the marketplace and its needs are generally fragmented. More precisely, a product is most successful when it maximizes its opportunities, based on the company's goals, and develops the product and marketing plan to achieve these goals. In the case of packaging, these goals are often directed at specific segments of the marketplace for a specific package. For example, the marketplace for plastic milk containers may generally be described as the milk market. But, in fact, plastic milk containers serve only a segment of the marketplace, the one gallon container segment.

Because a successful new product is based on the reflected needs of the market, it is critical that the product development team's progress be based on satisfying these needs. While this statement may approach a truism, most new products do fail, and probably more because of a company's inability to understand market needs than for any other reason. Companies become too often engrossed in developing new products that happen to fit their own internal structure and stop listening to the marketplace.

The purpose of this chapter is to develop some perspective of product development in broad terms and also to develop an understanding of the role of industrial research in product development. The first part of the chapter addresses itself to the package development team. The second part deals with the industrial research effort as a part of this team.

THE PACKAGING DEVELOPMENT TEAM

As businesses are chiefly involved in day-to-day operations, new product development is a unique activity. The standards and controls of producing and marketing existing packaging products cannot be followed in the development of new packages. The uncertainties of new packaging development require that constraints placed on the handling of existing packaging products be eliminated.

Companies oriented towards the development of new packaging products recognize the importance of establishing policies and procedures specifically for new products. These typically provide for the establishment of a team with a clear set of goals, a carefully designed plan to reach these goals, and the authority to follow through on the plan. As compared to plans for ongoing businesses, these plans and goals are more flexible because of the uncertainties of new package development. This team should be comprised of a combination of technical and marketing personnel with mar-

keting concentrating on the needs of the marketplace and engineering responsible for package development.

THE FLOW OF NEW PRODUCT PLANNING

The specific steps in new package development will vary depending on the magnitude of the opportunity and degree of risk involved. Product modifications, such as changing the type of board for folding cartons to achieve cost reductions, offer a moderate risk and do not change the company to an appreciable degree. The development of such products as the retortable food pouch and ovenable tray, on the other hand, represent breakthroughs in food packaging, offering excellent potential but at very high risk.

The organization of the steps or stages of new package development should be designed to minimize the risks involved. These stages reflect the increasing investment in time and money as the development team moves from conceptualizing through production. As the risks of failure are lower at the initial stages of development, the project is usually not abandoned unless reasons for eventual failure are quite obvious. At the latter stages of development, as the risk of failure increases and as a better understanding of the product and markets develops, the go/no-go decision becomes more critical.

Figure 1 indicates the flow of product planning for a new package concept and illustrates the role of marketing and engineering in the development. Although these basic steps must be covered in product development, the procedures within each step will vary depending on the product. Each company must devise its own procedure to suit its own needs.

The objective of the basic screening stage is to develop the product idea.

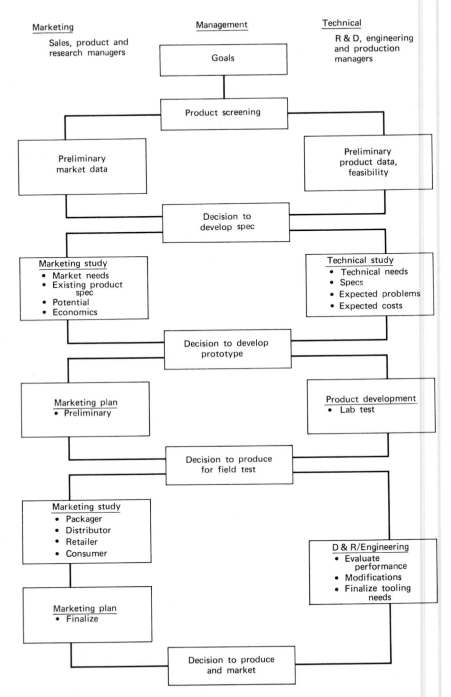

Figure 1 Product development planning steps: Packaging products company.

This does not suggest that the package be developed in its final form; rather it suggests that both marketing and engineering develop gross estimates regarding the commercial feasibility of the product. This would include some idea of market acceptance and manufacturing feasibility.

If the product survives the basic screening stage, the team begins development of the final product design. This is often the most critical stage of packaging development. Marketing must develop a clear picture of the needs of the market for the new product from both performance and economic standpoints. Engineering must determine the technical requirements for the package, including anticipated costs and production problems. This stage will be expanded in a later description of the role industrial research plays in this process.

Assuming that the product survives the second stage, it is ready for development. During this stage marketing is engaged in the preparation of preliminary marketing plans while R&D develops prototypes for in-house testing.

If lab tests are successful, the product should be given extensive field testing. There are often two levels of testing for new packaging products because the packaging supplier is typically one level removed from the ultimate consumer. Generally the packaging supplier's customer (e.g., food processor, nonfood consumer goods manufacturer) will conduct its own in-house tests and, if the results are satisfactory, move into a commercial test of its product in the new package. The packaging supplier and potential customer often work in concert at these various field testing levels. During this period, the packaging suppliers' marketing department monitors customers' reactions and measures the product against the needs that it is expected to fill.

This input, together with information previously developed, forms the basis for preparation of the recommended marketing plan. Concurrently production engineering is involved in final checks and modifications in preparation for full-scale production.

At this stage, the strategic elements of the package's development have been addressed and satisfied. Management has all the information required to reach its decision to produce and market the package.

THE INDUSTRIAL RESEARCH TEAM

Product development is essentially guided by information flowing from without (the market) inward to the development team. From the initial stage of development through to the decision to produce, decisions are based on the team's perception of the needs of the market and the company's ability to meet those needs with the new product.

It is the function of industrial research to work with the other members of the product development team in providing the information required to develop the product successfully. The industrial research team is most effective when included at the time of creating the product development plan. Although this may appear to be obvious, industrial research is employed often only as an afterthought; it is frequently introduced at the field test stage or considered only when such team members as marketing and engineering disagree on development and a "neutral third party" is required. Unfortunately much of the value of industrial research is lost when it is not used at the outset of development, for example, to provide input for product changes that could be uncovered in the earlier phases and possibly eliminate "losers" earlier in development.

Another factor that often affects the industrial research team's effectiveness is its degree of independence. In order to maintain credibility, industrial research should be responsible to the product development team manager, not to marketing or engineering. In many cases external consultants are used, at least in part because of their independence. Some companies have set up their industrial research departments to function similarly to external consultants. The department sells its time and interacts with other departments on a client/consultant basis.

OBJECTIVES OF INDUSTRIAL RESEARCH

Typically, industrial research is charged with meeting the following types of objectives:

1 Determine the current and future size of the market for the related existing packaging products.

2 Develop a profile of the competitive market, including major competitors and customers.

3 Determine the reasons for using the existing products and the strengths/weaknesses of these products.

4 Determine the relative strengths/weaknesses of the concept.

5 Based on an analysis of the above factors, estimate the potential for the new product.

6 Assist management in the development of a program to reach potential.

As the list of objectives indicates, the input from industrial research is geared to provide guidance to both marketing and engineering, to provide marketing with enough information to determine if the market represents sufficient volume, and through an analysis of competitive factors to identify the degree of effort required to enter the market.

MAJOR ELEMENTS OF INDUSTRIAL RESEARCH

The uncertainties of product development and the flexibility required through the developmental process are also reflected in the approach by industrial research in evaluating the product. Although it is important for the industrial research team to use the product concept as a frame of reference for its analysis, it is critical that the team be aware of alternatives in design in order to provide engineering with feedback on changes that would further enhance the new product's potential.

The industrial research team must be careful to reflect accurately the reasons for current product use, the advantages and disadvantages of these products, and the relative strengths and weaknesses of the new product. The pitfalls in analyzing the potential for the new product in this context are numerous and the detailed methodology will not be covered here. However, there are a few points that merit emphasis because they are crucial to successful analysis and are frequently mishandled.

Industrial research must be provided with a detailed description of the new product (or concept), including the development team's perception of the strengths and weaknesses of the product relative to existing products. In addition, the reasons for considering the new product as having potential should be communicated to industrial research. It is recognized that the development team's information may be in large measure a series of assumptions. This reference base is essential, however, so that industrial research can relate the new product to existing products.

The reasons for using a package are complex because they generally affect the differing sets of needs of several trade factors. Typically, a package must satisfy the packager, distributor, retailer, and consumer. As it is virtually impossible to completely satisfy the needs of all of the trade factors, it is critical to develop an understanding of the importance of each trade factor in the package selection process and the level of need for the various package characteristics within each of the trade factor groups. As this understanding forms the base of existing product/new product analysis, failure in understanding the reasons for product usage threatens the results of the project.

The following set of tables summarizes these particular findings in a product development study conducted for retortable food pouches.

The study covers:

1 Reasons for use of existing packages, can and frozen food packages.
2 Product development perception of retort pouch advantages relative to existing packages versus perception of these characteristics in the field.

3 Relative advantages/disadvantages of retort pouch based on analysis of reasons for use and product characteristics.

The information provided, as summarized in Table 1, sets the stage for the new product analysis. It provides the team with an understanding of the market's view towards the products and their packaging.

The importance of determining the development team's perceptions of the new product's characteristics and role in the market is illustrated in Table 2. The comparison of the development team's view of the new product relative to canned and frozen food packaging is significantly different from the views of the market factors (processors, distributors, retailers, and consumers). In this example the findings are particularly relevant to marketing as they provide insight into the problems likely to be encountered in the marketplace and allow marketing to plan accordingly.

Table 1 Reasons for Use Canned and Frozen Food

	Canned Important	Canned Critical	Frozen Important	Frozen Critical
COST				
Package		X	X	
Filling		X	X	
Processing		X	X	
Handling		X	X	
Distribution		X	X	
Storage		X	X	
Total cost		X	X	
QUALITY				
Taste	X			X
Odor	X			X
Color	X			X
Texture	X			X
CONVENIENCE				
Retailer	X		X	
Consumer	X		X	

Table 2 Retort Pouch
Development Team Perceptions Versus Reaction in Field

Development Team Perceptions	Marketplace Perceptions
A. PROCESSOR (QUALITY)	

Better quality than canned
  Taste
  Texture
  Color
Equal to, or better than frozen
  Freezer burn problem
  overcome
	Better quality than canned for
	most foods
	  Taste, texture, color
	But, improvement in many foods
	insignificant.
	  So, unless other advantages
	  would narrow potential,
	At best, equal in quality to
	frozen.
	  Equal to some foods
	  Inferior to other foods
	  Freezer burn problem not a
	  major factor

B. PROCESSOR (ECONOMICS)	

Lower cost versus canned and
frozen.
  Canned
    Material
    Inventory
    Freight
  Frozen
    Processing
    Storage
    Freight
	Currently higher cost
	  Negative
	    Slow line speeds
	    Excessive labor
	    High reject rate
	  Positive
	    Inventory and freight savings
	  Neutral
	    Material
	Potential for economic savings
	but major technological change
	required before achieved

C. DISTRIBUTOR	

Reduce storage space
Reduce freezer space (frozen)
Easier to handle
  Lighter weight (can)
  Freezer problems reduced
Lower distribution costs
  Weight reduced
  Freezer capacity reduced
Reduced damage loss
  Frozen food thawing
	Clear distribution advantages for
	retort pouch
	  Especially when retort pouch
	  food constitutes large share of
	  product line
	  Advantages less significant
	  when pouches constitute small
	  share of product line
	    May involve special handling

D. RETAILER	

Space savings (can)
  Storage and display
Lower cost of space (frozen)
  Storage and display
Easier to handle
  Lighter weight
Reduced loss through damage
	Retailers unimpressed.
	  Perceived package as specialty
	  product
	  Volume of food sold in pouch
	  too low to result in savings of
	  labor or space or lower cost
	  display (freezer)

Development Team Perceptions	Marketplace Perceptions
Dented cans Thawed frozen food	Shape of package disadvantage. Reduce number of facings Added shelf maintenance

E. CONSUMER

Economics
  Less costly than some frozen*
  More costly than can*
Quality
  Better than can food
  Equal to frozen food
Ease of storage
  No freezer
  Less shelf space than can
Ease of opening
  Tear open
Ease of preparation
  Boil-in-bag
  Reduce time versus frozen
Ease of disposing
	Consumer decision based on
	price/quality
	  Other factors minor
	Shift from canned unlikely
	  Taste
	  Cost
	  Role of can
	  Frozen food available
	Potential for penetration of frozen
	  Price/value
	Massive education program
	required
	  Quality
	  Safety

*Modified at stage two. Originally considered less costly.

The analysis summary (Table 3) places the new product in perspective by weighing the relative advantages/disadvantages of the retort pouch versus existing packages against the relative importance of the various reasons for using specific packages. As previously indicated, the selection of a package generally represents a compromise from what the market considers the ideal package. A satisfactory compromise is one where the most critical needs are satisfied.

The industrial research steps summarized in these tables generally reflect findings gathered through the several stages of product development. Frequently a brief overview study provides sufficient input for the go/no-go decision regarding the next stage. The intensity of the efforts of industrial research generally coincides with the increase in intensity of marketing and engineering through the development phases.

It is critical that the industrial re-search findings be supported by a clearly documented rationale. Although this standard should be met in any industrial research program, it is especially important in product development because the team is often comprised of members with widely diverging views on the product and its potential.

As previously mentioned, the product development process, including the industrial research steps, will vary widely depending on the situation. However, regardless of the situation, there are several basic criteria that should always be followed. Industrial research should be viewed as a natural part of product development. This includes assigning the research to professionals and not relying on the perceptions of personnel from such departments as sales and engineering. Also, the industrial research team must be accepted and respected by other members of the product development team. It is not unusual for some members of the product development team to be highly skeptical of the value of industrial research and the burden often falls on the research team to gain their respect. In addition, clear lines of communication must be set up within the group and there must be frequent contact between group members due to the uncertainties of product development. Finally, the industrial research team must be adaptable to changes in the development program. The successful development of a new product is largely dependent on the ability of the industrial research team to communicate accurately all those market factors relating to the new product to provide the necessary guidance to the development team in maximizing its opportunities for the new product.

Table 3 Product Analysis—New Product: Canned/Frozen Versus Retort Pouch

	Canned			Frozen		
	Important	Critical	Pouch*	Important	Critical	Pouch†
COST						
Packaging		X	=	X		+
Filling		X	−	X		=/−
Processing		X	−	X		+
Handling		X	+/−	X		+
Distribution		X	+/−	X		+
Storage		X	+/−	X		+
Total cost		X	−	X		+
QUALITY						
Taste	X		+		X	=/−
Odor	X		+		X	=/−
Color	X		+		X	=/−
Texture	X		+		X	=/−
CONVENIENCE						
Retailer	X		=	X		+
Consumer	X		=	X		=/+

*+/−: a plus for the pouch if volume usage is significant but a minus if the pouch is a specialty item that would require a new system.
†=/−: rating depends on type of frozen product.

CHAPTER TWENTY-THREE

Integrating Package Design Research into New Product Development:

COMPREHENSIVE PLANNING GUIDE CHARTS

Walter Stern

The development of new products is in most corporations assigned to an invididual or group of individuals whose task it is to create the product concepts and to carry them through all stages to the point where initial market tests have been successfully concluded and the product is ready to join others in the corporation's full list of brands.

Because product development is a highly complex task involving many different disciplines and virtually all divisions or departments of the corporate organization chart, it is essential at the

project's outset to have a clear-cut flow chart that delegates the various tasks involved in this process to the various participating departments at the appropriate time. In general this approach is felt to be more effective than relying either on regular meetings to brief the various groups or on the services of an independent consultant to function as the catalyst that will somehow make the different departmental contributions fall into place on schedule.

In most cases the departments most frequently involved in the product launch process would be marketing and market research, product research and testing, the legal department, manufacturing and engineering, purchasing, outside suppliers and consultants, traffic, and finally the ancillary services such as marketing and sales services and sales promotion services. Because, as we have seen in the Introduction, the product and its package are so intimately related, both are usually developed simultaneously. Table 1 demonstrates in the form of a flow chart how the above-mentioned departments interrelate in contributing to the five steps through which most product/package development moves: project concept and organization, fact gathering, tabulation and analysis of facts, creation of a number of alternative approaches, and the evaluation of these alternatives.

Tables 2–5 in turn indicate the responsibilities that marketing, technical research, manufacturing and engineering, and purchasing and traffic in turn assume in the development process. Because of the ever increasing complexity of legal considerations in the marketing field (especially in retail consumer goods), there are no stages in which legal counsel is not at least to some extent involved or consulted.

Tables 6–9 (which were contributed to this book by the Packaging Association of Canada) can serve as detailed check lists to assure that all required steps are observed at the right time and by the right department(s). In these charts the letter C (Concept) stands for the idea stage in which conceptual approaches are formulated. This occurs almost entirely in the areas of market research, package design research, product research, and is the area in which

Table 1

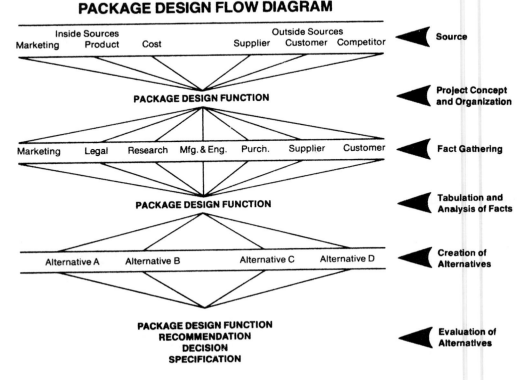

PACKAGE DESIGN FLOW DIAGRAM

Table 2	PACKAGE DESIGN MARKETING RESPONSIBILITIES				
Product Concept	**Package Concept**	**Marketing Considerations**	**Information**	**Legal**	**Market Testing**
Form	Type	Identification	Instructions	Brand Name	Customer
Arrangement	Unit Size	Competitive Practice	Recipes	Trade Marks	Consumer
	Art Work	Market Place Practice	Pricing	Net Weight	
	Copy	Product Exposure		Ingredients	
	Features	Buying Habits		Patents	
	Standardization	Use Factors		Government	
		Season Schedules			
		Distribution			

product development is most active. The asterisk stands for the suggested use of package design research at any given stage. The letter D (Development) stands for the protracted stages of concept refinement, technical investigation of all pertinent factors, the development of marketing objectives and the formulation of a marketing plan, and the first steps towards a thorough cost analysis to determine estimated cost of goods and perceived profit margins at all distribution stages.

The letter S, (Survey research) finally stands for all of the various audit and test activities that are used at most stages to determine or verify results, and it is in this general area that package design research becomes an especially important tool. Let us briefly run through these check lists to follow the course of a typical product launch.

We start (in Table 6) with the determination of possible physical form specifications and tolerances that will serve to define the product, its benefits, its

Table 3	PACKAGE DESIGN TECHNICAL RESEARCH RESPONSIBILITIES		
PRODUCT SPECIFICATIONS	**PACKAGE SPECIFICATIONS**	**INFORMATION**	**PERFORMANCE**
Protection Requirements	Material Requirements	Ingredients	Trials
From Requirements	Testing Methods	Instructions	Field Evaluation
Hazards	Storage Data	Legal Claims	Hazards
Handling Characteristics	Field Conditions	Government	

PACKAGE DESIGN
Table 4 — MANUFACTURING & ENGINEERING RESPONSIBILITIES

EQUIPMENT	PRODUCT	PACKAGE	PEOPLE	SCHEDULES
Capacity Available	Tolerances	Plant Trials	Quality Control	Supplier
New Equipment Selection	Capacitites	Measurement Specifications	Machine Operation	Plant
Performance with Product	Handling	Storage Conditions	Material Characteristics	Seasonal
Performance with Materials	Hazards	Receiving Specifications	Handling	
Modifications Required	Plant Conditions	Pallet Patterns	Storage	
Methods	Shrinkage Data	Stacking	Occupational Hazards	
Layout		Shipping Tests		
Operating Costs		Plant Hazards		
		Inspection		
		Rejection Standards		
		Adhesives		

PACKAGE DESIGN
Table 5 — PURCHASING & TRAFFIC RESPONSIBILITIES

SUPPLIERS CONTACT	TRAFFIC
Company-Supplier Liaison	Freight Classification
Materials Availability	Rates
Package Type Alternatives	Gross Weights
Graphic Arts Alternatives	Distribution Costs
Cost Estimates	Customer Practices
Prototype Samples	Shipping Hazards
Selection of Suppliers	Coding
Specification Agreements	Handling & Piling
Over-run & Under-run Tolerances	
Quality Control Standards	
Government Approvals	

Table 6 PACKAGE PLANNING—CHECK POINT INVOLVEMENT

PRODUCT CHARACTERISTICS	MARKETING M	&	MR	TECHNICAL RESEARCH	M & E	MATERIALS	SUPPLIERS & OTHERS
Physical form specifications and tolerances							
Product form: powder, granular, solids, paste, liquid, viscous, oils and greases, gaseous, fragile	C			D	D		
Product orientation	C	*		D	D		
Package content	C	*		D			
Consumer preferences		*	S				
Form change in package		*		D			
Plant trials				D	D		
Processing effects				D	D		
Raw material and packaging material storage conditions				D		D	D
Materials available		*		D		D	D
Protection required							
Shelf life required	S			D			
Home use	S	*	S	D			
Moisture equilibrium				D			
Aroma loss				D			
Aroma pickup				D			
Oil and grease effect				D			
Oxygen effect				D			
Breathing				D			
Seasonal production	S						
Storage facilities	S	*		D	S		
Hazards							
Reclosure	C	*	S	D		D	D
Pilferage	S						
Light, bacteria, mould, corrosion, rodents, insects				D			
Heat, humidity, barometric	S			D			
Sifting, leakage,				D	D	D	D

Table 6 PACKAGE PLANNING—CHECK POINT INVOLVEMENT (CONTINUED)

PRODUCT CHARACTERISTICS	MARKETING M & MR	TECHNICAL RESEARCH	M & E	MATERIALS	SUPPLIERS & OTHERS
Product, package reaction	*	D			
Shipping and handling		D	D	D (Traffic)	
Plant conditions		S	S		
Storage conditions		S & D	S	S	
Equipment available		S	S		
Seals effectiveness		D	D		
Quality control facilities		S & D	S		

C, Concept; S, Survey; D, Development.

packaging, and (in the words of the Introduction) its "perceived realities." Asterisks liberally sprinkled through the entire development process in Tables 6–9 indicate the frequency with which product/package design research can be used to firm up concepts, evaluate alternatives, and countercheck decisions. Quite obviously no realistical product development budget would utilize research in all of these steps; the decision on when to use it is critical because whenever we decide the use our own judgment (or a consensus) we tend to disregard the customer's potential perceptions, which we may find to be at variance with ours.

Probably the area in which package assessment and product assessment would be most helpful would be that beginning with Table 7. These are the steps in which the product's identity and communications tasks are defined and spelled out. They are also the steps in which traditionally personal or group judgment has played an important part in arriving at decisions—sometimes to the ultimate regret of product management. If asked to select the one time span in the development life of a product

in which package design research could make its most important contributions, we would feel that the stage outlined by Table 7 would get our vote. It is also, as shown on the check list, the stage at which legal input is most important. Incidentally, the caption "Suppliers and Others" over column 5 of our check list simply is meant to imply "suppliers and all other corporate departments participating in the development."

While Table 6 deals with the product itself, Tables 7 and 8 deal with its package and Table 9 with the total product/package entity. Table 7 will develop information and objectives that must be communicated to the package's designer in the greatest affordable detail if his creation is to produce the anticipated marketing results.

All of the separate areas mentioned under "Package Type Selection," for instance Information, Artwork and Design, Product, Market, and Size will be under investigation more or less simultaneously—no particular sequence is to be implied by their location on the check list as they are heavily interrelated. This applies also to the sequence of the tables themselves: package type selection (Ta-

Table 7   PACKAGE TYPE SELECTION

	MARKETING M	&	MR	TECHNICAL RESEARCH	M & E	MATERIALS	SUPPLIERS & OTHERS
Identity							
Features required	C	*		D		S	D
Brand name: position, style, legal	D	*					Legal
Trade name: position, style, legal	D	*					Legal
Name of manufacturer	D						
Family design quality, integrity	D	*					
Advertising tie-in	D						
Adaptable to TV	D						
Information							
Legal requirements				S			Legal
Product description	D	*		D			Legal
Instructions and uses: clarity, brevity, adequacy	D	*	D	D			
Illustration: attract, interest, instruct	D	*	S				
Price spot	D						
Code marking	C			D			
Government approval			S	D		S & D	
Artwork & Design							
Good taste: color selection appropriate for product, for retail, for consumer	D	*	S				
Competitive	D	*	S				
Consumer impression: distant, close, shelf, counter, window or display, home	D	*	S				
Self selling story	D	*	S				
Visibility— shelf location	S&D	*					
Appetite appeal	D	*	S				
Remembrance value	D	*	S				
Self sufficient	D	*	S				
Graphic arts alternatives	D	*				S & D	S & D
Product							
New, improved or regular	C			D			
Competitive	S	*	S				
Unique	C&S	*					
Market appearance	C	*		D	D	D	D

Table 7 PACKAGE TYPE SELECTION (CONTINUED)

	MARKETING			TECHNICAL RESEARCH	M & E	MATERIALS	SUPPLIERS & OTHERS
	M	&	MR				
Shelf life required	S						
Distribution method	C&D					D (Traffic)	
Maximum market penetration	D						
Special markets	D						
Recipe requirements	D		S&D	D			
Protection requirements				D			
Product physical characteristics		*		D		D	D
Product handling characteristics				D	D	D	D
Product storage characteristics				D		D	D
Processing problems				D	D		
Equipment available				S	S		
Shipping, handling, and storage facilities				S	D		
Market							
Age, income, sex, ethnic	D	*	S				
Geographic location	D						
Market place	D						
Buying Habits							
Unit of purchase	D	*					
Shelf life	D			D			
Shelf display	D	*					
Mass display	D	*					
Attractiveness	D	*					
Shelf location (eye level)	D	*					
Panel displayed	D	*					
Point-of-sale support	D	*					
Size							
Distribution methods	D				D	D (Tra	
Consumer habits	D	*	S				
Consumer convenience	D	*	S	D		D	D
Quantity purchases	D						
Reclosures	D	*		D		D	D
Recipes & instructions	C&D	*	S&D	D			
Package instructions and recipe development		*	D	D			
Prototype package	C			D	D	D	D
Possibilities with existing equipment				S	D		
New equipment available or required				S	S&D		D

Table 7 PACKAGE TYPE SELECTION (CONTINUED)

| | MARKETING | | | TECHNICAL | | | SUPPLIERS |
	M	&	MR	RESEARCH	M & E	MATERIALS	& OTHERS
Competition							
Product or package	S	*					
Economy or prestige	D	*					
Sizes, shapes, colors, designs	D	*					
Compete or be unique	C&D	*					
Functional use	C&D	*	D	D		D	D
Competitive testing	S	*	S	S			
Distribution							
Size & shape: wholesale & retail factors	D			D	D	D	D
Storage	S			D	S	S	
Stacking	C&D				D	D	D
Display	D						
Handling				D	D	D	
Price marking	D						
Check-out	D						
Self service	D						
Shelf location	D						
Retail pilferage	D						
Soilage	C						
Breakage	C						
Seasonal	S						
Tie-ins & special promotions	D						
Use	D		S				
Use factors							
Unit size	C	*	S	D	D	D	D
Product visibility	C	*		D	D	D	D
Safety	C	*		D	D	D	D
Opening and reclosure facilities	C	*		D	D	D	D
Dispensing features	C	*		D	D	D	D
Measuring facilities	C	*		D	D	D	D
Container disposal	C	*		D		D	D
Container reuse	C	*		D		D	D
Market-to-home handling	C	*				D	D
Home storage	C	*	S				D
Home shelf location	C	*					D

Table 8 PACKAGE MATERIAL SELECTION

	MARKETING M	& MR	TECHNICAL RESEARCH	M & E	MATERIALS	SUPPLIERS & OTHERS
Marketing						
Functional use	C	*	D		D	D
Neat appearance	C&D	*	D		D	D
Graphic Arts applicability	C	*			D	D
Noncontaminating inks			D		D	D
adhesives			D		D	D
solvents			D		D	D
Familiarity to consumer: form, shape, texture	S	*				
Selling features	C&D					
Opening features	C	*	D	D	D	D
Reclosure features	C	*	D	D	D	D
Structure						
Machine performance				D		
High temperature effects			D			
Low temperature effects			D			
Shipping and handling. hazards				D	D	
Loss or pick up aroma			D			
flavour			D			
volatile components			D			
color			D			
vitamin content			D			
Oxidation or other chemical reactions			D			
Sterilization requirements			D			
Alternates considered	C		D	D	D	D
Availability						
Supply & suppliers					S	D
Delivery requirements					S	
Prices & price trends					S	

ble 7) and package material selection (Table 8), for instance, would very likely occur simultaneously rather than sequentially.

Table 9 outlines the last and most important phase of the new product/package launch: a determination of whether the whole venture will be sufficiently attractive economically for the corporation to continue the previous investment in development by continued investment in manufacture, distribution, marketing, and advertising. Because product and package design research,

Table 9   ECONOMICS

	MARKETING M & MR	TECHNICAL RESEARCH	M & E	MATERIALS	SUPPLIERS & OTHERS
Minimum material design	C	D	D	D	D
Optimum price, protection, marketing design	D *	D	D	D	D
Standardization	C		D	D	
Proper cost proportion to:					
product price and cost	C *				
product concept	C *				
market design	C *				
class of product	C *				
Shrinkage comparisons			D	D	D
Machine efficiency			D	D	D
Protection for product life	C	D			
Overprotection	C	D			
Optimum weight, size, structure	C *	D		D	D
Freight classification & rates				D	

and subsequent market tests, often tend to modify the original concept and sometimes to change them to a significant degree, final economic balance sheets on a product cannot be drawn up until that final stage has been reached. It is for this reason more than for any other that design research is so attractive: while research budgets sometimes are sizable they are invariably only a small fraction of the cost of actually going to test marketing for product/package performance insights.

CHAPTER TWENTY-FOUR

Quick Test: A Program to Test Product and Package Alternatives

Richard McCullough

BACKGROUND AND PURPOSE

In 1974, Winona, Inc. developed a research approach called Quick Test. The Quick Test approach is a program to test product alternatives in a grocery store in a short period of time.

The Quick Test approach should be considered for a number of different alternatives. Some typical applications would be a pricing test where consideration of a price increase might be 10¢ versus 20¢ per unit. The objective of the Quick Test approach would be to determine the price resistance to a 20¢ increase versus a 10¢ increase. A new package or label test might consist of

two label alternatives rather than one alternative. We always recommend, however, the inclusion of the current package in the study to serve as a bench mark or norm. In this case the study would be divided into three panels—one control and two test panels.

The Quick Test approach works best for current products that are brand recognizable by shoppers. Quick Test makes no attempt to predict the ultimate market share or brand share of a product. The approach is not particularly suited for this objective.

The Quick Test approach makes no claims that it is a substitute for test marketing or extended controlled store tests. As pointed out before, the advantage of Quick Test is the speed and the flexibility of screening product and packaging alternatives prior to entering a test market situation. It also can be done to prevent detection by competitive companies.

OVERALL CONCEPT

Some specific marketing factors that can be tested using Quick Test are pricing, promotion impact, promotion alternatives, display alternatives, packaging, coupon redemption, product location, and trial purchase.

The Quick Test methodology consists of exposing shoppers in supermarkets to the product alternatives in a real shopping situation during normal store hours (Fig. 1).

Before the store opens for business, it is rearranged as necessary. Oftentimes, as many as six different supermarkets may be used in a study. If there is one alternative such as a new package that is under consideration, it is very possible that three of the stores will contain the current or control package while the other three stores have the test package on their shelves. Prior to the study all six stores are arranged to be as comparable as possible. The shelf positions for the products are arranged to have a similar number of facings per brand, and the prices in a label or packaging test are identical in all six stores. Table 1, p. 266, illustrates a typical store configuration for a package test.

On occasion a product that is normally on the shelves, for example, a private label brand, is removed from the shelves for the duration of the test. Once the

Figure 1

Table 1 Store Configurations

	CONTROL	TEST PACKAGE ONE	TEST PACKAGE TWO
Brand Y	X	X	X
Brand X	X	X	X
Brand J	X		X
Brand T (test)			
Current package	X		
Test package one		X	
Test package two			X

shelf work is completed, interviewers are stationed at the front inside entrance of each supermarket. The interviewers intercept shoppers as they arrive to do their normal grocery shopping. Shoppers are individually screened by using a portfolio book of ten photographs, each photograph consisting of several brands in a product category. Typical product category photographs consist of three to four brands of tuna fish, another three to four brands of catsup, another four to six brands of fruit drinks, and so on.

A photograph of the product category for which the Quick Test is being conducted is inserted in the book, usually in the third or fourth position. The screening consists of exposing a photograph of the test category asking a respondent a question such as "Have you purchased a product such as the one shown in the photograph in the past month?" Depending on the product category, a screening criterion is established from which to determine whether or not a potential respondent or participant is a user of that product category. When the respondent qualifies as a user of the product category, brand usage and frequency of use data are collected. Let's assume that the test category under consideration is fruit drinks. After showing tuna fish and catsup, the third photograph in the portfolio book would

consist of a photograph of different brands of fruit drinks. If the potential respondent being screened claims they have not used fruit drinks in the past month, the respondent would be thanked and the interview would be terminated at that point. For those that claim they have purchased in the past month, however, the interviewer would continue through the portfolio screening book until all ten categories have been answered. It has now been determined that a qualified shopper has been screened.

At this point, the interviewer explains the Quick Test approach to the respondent. The Quick Test approach consists of offering each shopper a 10% discount on their entire grocery purchase in the store that day. That is, if they spend $43, they will receive a rebate after shopping of $4.30 for participating in the study. The rebate is limited, however, to a $5 maximum. Therefore someone who spends $75 would receive only the $5 maximum rebate.

For their 10% discount, each shopper must agree to purchase at least one product in each of three product categories used in the portfolio screening. The three categories will consist of the test category; that is, fruit drinks, and two control categories that are simply included to disguise the overall purpose of the study. The two control categories can be varied from respondent to respondent. Some often used control categories are toilet tissue, bar soap, catsup, and tuna fish. The selection of the control categories is determined by the responses to the screening questionnaire. The respondent is not allowed to select the categories. Rather the interviewer will assign two control categories at random from the list of products the respondent indicated they do currently use or have purchased in the last month, and always the respondent will be required to purchase in the test category—fruit drinks in this example.

At this point the shopper agrees to participate and is given three lists of the brands that qualify, a separate list for each product category in which the purchase is required. The use of these lists is to avoid the confusion of a product being purchased that is not in the product category under consideration. For example, a person purchasing Orange Crush carbonated soft drink could be confused and feel that that purchase did satisfy their requirement for purchasing a fruit drink. That particular product, however, would not satisfy the requirement. Therefore the lists for shoppers include all the brands that do qualify for the fruit drink category. See, for example, the following list of brands in the fruit drink category that satisfy the purchase requirement.

FRUIT DRINKS

Gatorade
Hawaiian Punch
Hi-C
Wagner's
Welchade
Wyler's

At this point the shoppers proceed through the store and do their regular grocery shopping, which will include the purchase of the three required items, one of which will be a fruit drink.

After completing their shopping, they will go through the check-out and pay the full amount for their groceries. After paying, they will bring their register receipt to a table near the front of the store and receive their cash rebate. At this time a post-interview is conducted that will include diagnostic questions that pertain to the fruit drink category. Examples of question areas are: awareness, recognition of any new product or package, reasons for purchasing the product, reasons for brand switching if a switch occurred, and reasons for not purchasing the test product.

A Quick Test approach generally involves from 200 to 500 shoppers per product alternative. That is, if in the example we have used throughout this discussion the fruit drink has one alternative package, a total of 400 to 1,000 shoppers would be included in the study, 200 to 500 per test alternative. As many supermarkets as are required are set up to complete the number of interviews desired. These supermarkets are selected to be demographically balanced. Oftentimes during the middle of the Quick Test approach, the product alternatives are switched from store to store for balancing purposes. For example, at the midway point in a fruit drink package study, the test package will be substituted for the control package. In stores with the control package, the test package will be substituted.

RESULTS

In Table 2, p. 268, the results of a Quick Test package evaluation are shown. The study consisted of a panel of six supermarkets—two Control, two Test One, and two Test Two. The only difference in each panel was the package of the

Table 2 Quick Test Sales Results

	CONTROL	TEST PACKAGE ONE	TEST PACKAGE TWO
Brand Y	17%	19%	16%
Brand X	21%	24%	18%
Brand J	9%	12%	8%
Brand T (test)			
Current package	53%		
Test package one		45%	
Test package two			58%
Base for percentages (shoppers)	(200)	(200)	(200)

product (Brand T). The results of the Control panel shown in Table 2 represent the bench mark or norm. These data compared favorably with known market shares.

The examination of the results leads to the conclusion that Test Package Two was much better than Test Package One—and even more favorable than the Current (Control) Package.

TIMING AND COSTS

Generally speaking a Quick Test study will take from 3 to 6 days to complete. Some advantages to the Quick Test approach are its speed or the timing in which a test alternative can be evaluated. The cost is much more reasonable than a mini-market or controlled store test that requires a longer duration.

Another advantage of the Quick Test approach is its flexibility. If it is felt that a private label brand on the shelves would distort the study, arrangements can be made with the supermarket chains to remove the private label for the study duration in those stores being used in the Quick Test. In addition, if the client desires, special displays can be built and prices altered for the duration of the study. The biggest overall advantage of the Quick Test approach is that it is a real-life situation and does not include simulated shopping. The shoppers in the study are participating during their normal shopping experience; and although they are given a 10% overall discount, there is no special incentive for purchasing a particular brand or product.

In conclusion, the Quick Test method has had considerable success with utilization for testing product alternatives. The occasion for utilizing the Quick Test approach may only surface once or twice a year; when this occurs, the approach can be implemented with 2 to 3 weeks lead time. The lead time is necessary to contact participating chain stores and to request that they not feature a fruit drink in their newspaper advertising during the week of (and week prior to) the implementation of the Quick Test approach (in the example of doing a fruit drink Quick Test study).

CHAPTER TWENTY-FIVE

Developing and Testing Brand Names

Priscilla Douglas

This chapter will cover practical, how-to-do-it techniques for the development and testing of product names, illustrated liberally with specific examples. It will explore the objectives and functions of names and will give specific steps and procedures to use in the *creative development* of names. For the *selection process* measurement against five criteria is provided and procedures for the testing of names with consumers are outlined.

A century ago most goods were sold as commodities. Coffee, sugar, flour, rice were weighed out and packaged to order; cheeses, butter, soaps were cut from tubs or slabs and wrapped in brown paper. Then the packaging and labeling of products got under way providing consumers with product uniformity and the reassurance of the manufacturer's name backing the product.

Many of our old familiar brand names were family names—Smith Brothers cough drops, Ford cars, Kraft cheeses, Birdseye frozen foods (named after the founder Clarence Birdseye—not because of any association with a bird, an eye, or a bird's-eye view of anything).

Names were frequently arrived at in a casual manner. A new model car from the Morris Garage came to be referred to by mechanics and test drivers as the MG, and the name stuck.

Names with more deliberate raison d'être appeared. Listerine was named after Joseph Lister, the founder of antiseptic surgery. It still conveys an antiseptic image regardless of whether the consumer is aware of its origin and in spite of the fact that antiseptic claims or inferences are now restricted.

Lysol had early consumer acceptance because of the strong image of the disinfectant and cleaning properties of lye. Today, as a new introduction Lysol might well have negative reactions, with lye being thought of as harsh, caustic. Ivory was selected for a soap that was pure white compared with many soaps of the brown laundry type, and "Ivory" conveyed an image of quality associating itself with a costly substance. Names such as Ajax appeared—a short, distinctive, and memorable name, any association with the mythological Greek superman probably being lost on the vast majority of consumers. Life Savers is an early example of a happy marriage of name and product neither of which have become out-of-date or are likely to do so.

The case of the Edsel is a whole story in itself. This was a project for which the talented and famous poet Marianne Moore was engaged. She and others developed some 20,000 names, following which management decided to name the car after Edsel Ford. Many were the problems associated with the marketing of the car, but we can assume that the name itself did not aid in the acceptance of the product. What would one expect of an "Edsel"? Something stodgy, unexciting? What might an "Edsel" be? Perhaps a pump or a circuit breaker.

Once established, the family names such as Kraft, Campbell, Heinz, Hershey, Lipton and also the Lysols and Listerines have become household words—valuable properties. The use of these for new products—"line extensions"—is sometimes advisable, sometimes not. Because of its disinfectant image, Lysol is of great value for products that are today restricted as to the

disinfectant claims they can make, whereas in the case of Scotties, Scotkins, and so on the Scott lends the prestige of a well-known company to the product but does nothing for the individual products.

What's in a Name?
Objectives, Functions

What could/should a name accomplish?

- Describe the product or the benefit of using the product.
- Be associated with the category, use, ingredients, performance, or the benefit of these.
- Suggest the proper mood.
- Provide uniqueness, distinctiveness.
- Be memorable, "catchy."
- Be alliterative or perhaps rhyme.
- Be a play on words.
- Have kinetic value—suggest a logo, packaging, an advertising claim, a distinctive "line."
- Create a favorable impression.

Obviously a name will not fulfill *all* of the above objectives. It should accomplish *some* of them *well*. To do this a coined word can sometimes be very effective or a renowned person or place can be used. (When an *invented* personality is used it is well to have an in-the-flesh person by this name in the company or at the agency to avoid suits by individuals having the same name.)

Examples

Let us see what the following names do or don't do for the products:

Tide Short, memorable, visually associated with the frothy foam of a seaside spray, led naturally to "Tide's in, dirt's out."

Pampers   A verb turned into a noun—suggests tender, loving, pampering care.

Shake 'n Bake   Descriptive . . . rhymes . . . short, memorable.

Wish-Bone   No association with a salad dressing, not strong on raison d'être.

Seven Seas   Also not strong on raison d'être but possibly suggests flavors from the seven seas—from all over the world.

Hungry Jack   Appetizing; good product association as the "jack" rhymes with flapjack . . . also a man's name . . . a hard-working man such as a lumberjack.

Zoom 'n Groom   Descriptive of a hair dryer . . . distinctive, memorable, rhymes.

Light 'n Lively   Alliterative, memorable, creates a favorable impression as today's consumers wish to be light and lively.

Janitor In A Drum   Distinctive, memorable, lends itself to unique packaging; suggests that the janitor is going to do the work.

Crest   Not strong on raison d'être, suggestive of cleanliness, foam, riding the crest of the wave, the one on top, and so on.

Gee Your Hair Smells Terrific!   Unique, memorable, suggests use and benefit.

Tab   Short; memorable; not descriptive of a soda but suggests it helps keep tabs on calories.

Charmin   Coined name; pleasant, gentle sounding.

Duncan Hines   Use of a renowned personality.

Kool-Aid   Descriptive of the product's benefit, has become almost generic.

Hamburger Helper   Descriptive; implies benefit; leads to line "when you need a helping hand" and accompanying animated visual.

Pam   A vegetable cooking spray but could be anything. Does connect somewhat with "pan."

Hold   Short, memorable, describes benefit of a cough suppressant. Lends itself to creative execution.

L'Eggs   Very successful introduction of a new brand in a category having weak brand recognition in a new marketing outlet. Concept, name, logo, and package closely integrated.

Fresca   A coined word not descriptive of a soda but is suggesting of freshness.

Snackin' Cake   A Betty Crocker line extension that describes the product and also tells the consumer what he wants to hear, that is, go ahead and snack, it's O.K.

Aim   Not descriptive of the product but leads to the line "Take Aim against cavities."

Taster's Choice   Descriptive of something ingested—a product *preferred* over others by *experts*.

Merit   A very successful cigarette introduction. Word association favorable: the product must have something good going for it. Had high name recall.

Shout   A surprising name for a spray for stains. What to do about stuck-on stains? Shout 'em out! Playback was high.

No Jelly   A peanut candy that capitalized on the popularity of peanut butter and jelly sandwiches. Unique but somewhat roundabout. Limits revision of positioning or strategy should this be desired.

Scrunge   Coined word combining scrubber and sponge; short, memorable, not pleasant, suited to a "drudge" type of product.

1 For The Road   (Clairol portable hair dryer). Memorable; descriptive.

Handle With Care   Suitable, descriptive of a light-duty laundry detergent.

Reach   Short; memorable; descriptive of function of a toothbrush.

THE CREATIVE PROCESS: PROCEDURES FOR THE DEVELOPMENT OF NAMES

Tools, Aids

Coming up with new names is an inventive, imaginative process. There are, however, tools you should have at hand: a dictionary, a thesaurus, a Latin dictionary (particularly helpful when working on over-the-counter or proprietary drugs), a French dictionary (helpful for fashion/cosmetic/culinary products), and possibly a Spanish dictionary. Other miscellaneous sources may be helpful— Mother Goose, magazines, a book of quotations—and you will also want *The Trademark Register of the United States*. This is published annually by the Trademark Register, 454 Washington Building, Washington, D.C. 20005. You will use this in a step that is covered briefly later in this chapter.

Methodology

In starting work on the development of names exactly what are you setting out to do? What is the assignment? In delegating work or contributing to it yourself, be sure that a statement about the product is prepared. Exactly what is the product? How would you describe it? How does it function? What does it do? What is unique about it? Who will use it? When? In a competitive category, what is its distinctive positioning? If participants are vague about these points, a great deal of time will be wasted and creative people will travel down useless paths.

Should you work with others à la creative push, think-tank, or brainstorm-ing procedure? Some creative heads are stimulated this way. Frequently the experience is exciting and a lot of fun— minds are stimulated—ideas bounce back and forth, and so on. Yet sometimes, in the cold light of review, the session was not really as productive as it seemed to be at the time.

If you should find yourself as the leader or moderator of such a think-tank or brainstorming session, remember that it must be freewheeling and open. Participants must not feel inhibited in making suggestions. There should be no "put-downs." If a list is being compiled, everything should be included. It is the job of the leader or moderator to keep the group tactfully on the track.

You may prefer to work by yourself. Incidentally, you may find it productive to work on three or four name projects concurrently. While you are concentrating on one area, the subconscious often breaks forth with an idea in another. The creative process cannot be pushed. When you have reached a dead end, turn to something else.

While in the process of developing names you may have the assignment ruminating in the subconscious. In the middle of the night something may pop out of the blue. While driving along the highway—while mind-wandering in church—(you might get lucky as Harley Procter did on hearing the thirty-fifth Psalm: "Out of the *ivory* palaces whereby they have made thee glad . . .").—or during a briefing on the current share-of-market picture in centrifugal pumps, the "great idea" could flash into your consciousness.

Note that the developed names must be complete, for example, Tang (name) Breakfast Drink (generic description). As the latter is usually a simple, uncomplicated decision to make, we will devote our attentions to the "Tang" part with little attention to the generic description.

"How to" Examples

Let's consider how a few specific assignments could be handled.

Home Protective Device

CONCEPT FOR A BURGLAR ALARM

This burglar alarm is about the size and appearance of a small table radio. It is portable and battery operated. Whereas many burglar alarms require wiring around each door and window, the alarm being activated when something passes through, this alarm sends out a beam that travels to the opposite wall or other object and bounces back to the unit signaling that all is well. There are no wires that can be cut by an intruder. No installation is involved. When the beam is interrupted the alarm sounds, frightening the intruder and alerting anyone nearby. (Ob-viously the device does not *guarantee* safety.)

If you wish, stop at this point and jot down names that occur to you. Do not be selective at the outset; simply write words as they occur. While a great deal of creative work is by inspiration, much seat-of-the-pants-on-the-seat-of-the-chair is involved. Your next step may be to note the *directions—categories* in which you can work, for example, intruder, burglar; scare off; sound an alarm, make a loud noise; and guard. You can also consider the *functioning*, the *operation* of the unit.

At this point you may wish to consult your thesaurus, which can help you to think beyond the obvious "alarm" to "blitz," "banshee," and so on.

Under categories/directions in which you have worked you may have listed:

Guard	Sound Alarm, Noise	Scare
Keepwatch	Blitz Box	The Scrammer
Night & Day	Auto Blitz	Buzz Off
House Sitter	The Squark	The Routout
Stand-by Patrol	Klaxon	Exodus
Private Eye	Scream Machine	The Chaser
En Garde	Screaming Meemie	Vamoose!
Eye Spy	Banshee	Exit!
Eye Patrol	Blabbermouth	The Rouster
Argus (100 eyes, guardian)		Super Scare
Ambush	*Function*	Chaseoff
Home Warden		The Exout
The Lookout	The Bouncer	Split!
Goalie	Burglar Bouncer	Warnaway
The Interceptor	Beam Box	Blast Off
Secret Service	Boomerang	Warn Off
	Echo	

You may have parts, pieces of words that might be combined with another piece, by which system you might have put together Beam Patrol, Prowler Patrol, The Scream Beam, Rouse & Rout, Peep-Squawk, The Blastout, Blitzbeam, Sound Off For Safety, Bum's Rush, Buzz Box, Beam Bouncer, Prowler Howler, Run Robber, Blitzray, Beam Barrier, Bedlam Box, Discoveray, See and Squawk. You may have been inspired to jot down some miscellany such as Superbug, Bug Off, Hood Lam.

When working with parts and pieces a good procedure is to lay them all out on a large table and see how they fit together. While Prowler Howler may not have occurred to you by inspiration, on seeing these pieces before you they quickly connect themselves. In this way you might arrive at the Scream Machine, Eye Scream, Detectomat, Detectascope, Scat-o-Mat. If you will be working frequently on names, you may wish to collect prefixes and suffixes which would be of help in developing such names as Scat-o-Mat.

Also consider initials: A.P.B. (for All Prowlers Beware). Computers are sometimes used for name development, but with parts and pieces spread before you you can be your own computer.

Later in this chapter we cover the *testing* of names. For the present we will simply comment on a few of the burglar alarm names.

A.P.B. (for All Prowlers Beware) The name is short, it is familiar to TV watchers, and it is suggestive of the law on the hunt. It could become a household expression—before retiring or leaving the house, "Put on the A.P.B." It would have good name recall.

Argus The name is short; the association of the guardian of 100 eyes might be completely lost on the consuming public. It might be confused with the camera and indeed there might be a trademark conflict (a subject covered briefly later in this chapter).

Boomerang A unique name suggestive of the way the unit functions, that is, a signal goes out and returns. Boomerang is also used in the sense of something that failed . . . a backlash . . . "that was one burglary attempt that boomeranged." Combining boom and rang also conveys the idea of something very loud, exactly what an intruder would not wish to encounter. Also *B*oomerang *B*urglar Alarm has a ring to it.

Stampede An alarm causing intruders to stampede would be desirable for the home owner.

The Squark Short, memorable, coined word. Sounds noisy as if it would frighten off intruders.

Rouse & Rout Alliterative. Describes the benefits of the device—rouses home owners and neighbors, routs intruders.

Personal Care—Antiperspirant Product

Now let's consider a personal care product—an antiperspirant stick in a convenient, purse-size applicator. In product-placement tests it was found to have great appeal to *young girls* who liked its effectiveness and convenience. Names developed were:

Miss Manikin Keeps you neat and dry as a manikin on display.

Wonder Bar Amazingly dependable.

Shock Absorber Works when you need it. (The antiperspirant ingredient—aluminum chlorohydrate—is activated when the body starts to perspire.)

Fidelity Trust Dependability you can bank on.

Happy-Go-Dry A nonwet application—you can be happy and dry all day.

Keepsafe Dependable, reliable.

Smooth 'n Shield Gentle to smooth on and protects you from perspiration.

Totem Conveniently portable.

Jacqueline Hyde A female has two natures: one is neat, demure; the other is the "animal" nature of a human—we sweat.

Appliqué The product is applied only where you wish to place it (as opposed to an aerosol).

Confidential A personal product.

Ms. Behave You are a miss in a *body* that behaves even though *you* might misbehave.

Earlier, we mentioned the kinetic value inherent in a name. Let's consider a few of the above names in the light of this criterion.

Appliqué Logo and design elements can simulate cut-out and stitched-down designs. Feminine, neat, dainty.

Totem Cylindrical package could be a feminine, fashiony version of a totem pole.

Shock Absorber Design could suggest life's experiences: serenity with a shock about to be experienced.

Miss Manikin A manikin figure that never loses its cool even on display in the hottest store window.

Happy-Go-Dry Cool, controlled figure under a scorching sun.

Keepsafe A cameo, somewhat old-fashioned motif.

Food Products

In the realm of foods let us consider canned products. Take, for example, the concept of a complete balanced meal in a self-opening can, that can be eaten right out of the can, or warmed and eaten out of the can. Possible names:

Take-Along Dinners
Gadabout Dinners
Wanderlunch
Safari Meals
Lug-A-Lunch
Snack-In-A-Box
Knapsack Dinners

As can be seen, two of these limit perception of the product to a lunch. Snack-In-A-Box, while catchy, is light on heartiness and nutrition and would be a better name for a snack that is packed in a box. Knapsack Dinners suggest heartiness, portability, and offers graphic design possibilities.

In the case of a line of complete dinners in a can that feature an entire chicken leg (or a piece of beef shortrib or breast of lamb, etc.) with vegetables and noodles, the positioning would doubtless be one of heartiness and good nutrition. Possible names:

Mainstay Meals
Explorers Club Dinners
Minute Man Meals
Bunkhouse Dinners

Bunkhouse Dinners is a name that conveys heartiness and has masculine appeal while not limiting the market to males.

Miscellaneous

New products sometimes evolve from a waste product—a material that must be disposed of. R & D (Research and Development labs) and marketing people may be asked to find ways of using a waste product. (Unsold bread that goes back to the bakery can be made into prepared, packaged stuffing; bleaches used in paper manufacture can be reprocessed for laundry bleaches, etc.) Today we can expect increased interest in recycling and use of waste products.

Let us say that a waste product for a food processor is grease. An impregnated pad for greasing roasting pans, chickens, potatoes, and other foods could be made. These greased-on-one-side pads might be named **Grease Monkeys**. It would be important that graphics convey a clean, pure image. Testing with consumers would determine whether the image conveyed is convenient, handy, appealing, appropriate for a food-related product or whether the name and graphics are a "turnoff."

Many new products come out of a company's R & D. A window cleaner that does a better job of eliminating streaks or a floor cleaner that is superior in removing wax build-up may have been developed. Regardless of the R & D reports on the efficacy of the product, the *consumer* must perceive the difference, usually via blind product tests. In this way an aerosol foam cleaner was developed in R & D and turned over to marketing for testing. By means of placement tests in which consumers used the product in their homes it was found that the product was a superior multipurpose cleaner for all surfaces and could replace a multitude of specialty cleaning products. An additional finding of interest was that the men in the test households liked it and found it to be very effective for vinyl car tops, cleaning up in their shops, and when helping out with the household cleaning. The product was named **Arsenal** Multi-Purpose Cleaner.

This product when tested with consumers evoked comments such as "efficient"—"serious"—"a weapon against dirt"—"a store of weapons"—"very businesslike"—"gets right at the dirt"—"no-fooling product"—"a lot of cleaning ability in one."

As mentioned, the name of a product may be related to the benefit desired as a result of using the product. A diet aid named **Bikini** can conjure up visions of oneself down to a whistle-provoking sylphlike figure in a mini bikini.

Names can help solve marketing problems. In a world of huge advertising budgets the smaller manufacturer of snacks and candies can find it difficult to compete. How can he stretch a small budget—make it work overtime? Let us suppose he has a handful of snacks/candies that have been found in consumer tests to have good taste appeal—to equal or surpass known brands. Inspired by the British "elevenses" the bars might be named **The Foursy, The Eightsy, The Elevensy, The Threesy**, and so on. In this way several bars can be advertised together. As teenagers and young people are heavy consumers of such candy and snack bars and are also great radio fans,

the products could be advertised on radio, using time slots, and disc jockeys provided with fact sheets could be expected to have fun with the names: "Can I eat my Foursy now when it's only three o'clock?" and so on. Consumers would pick up the fun with a great deal of word-of-mouth advertising.

Never underestimate the value of free advertising. **Ajax** laundry detergent received millions of dollars of free advertising on prime time TV with talents such as Bob Hope trying out for the part of the White Knight, and so on. (While the subject of this chapter is the development and testing of brand names you are urged to think as an all-around advertising and marketing person in the larger dimension, hence the inclusion of material relative to situations in which you may become involved.)

EVALUATION OF NAMES: PROCEDURES FOR SELECTION AND TESTING

With the fatality rate of today's new products, it is inexcusable to go into even a test market with a weak name. The new product needs to have everything going for it.

It has been said that the selection of a name is a creative, not a scientific, process. Let us see how the process can be more objective, more scientific, and less hit-or-miss.

In considering the development of brand names we have already given some thought to *selectivity* since the creative process cannot operate in a vacuum without goals.

Preliminary Weeding

Now we will concentrate on the selection process: weeding and testing. The former involves preselection and elimination by the creative, marketing, research *insiders;* the latter, the procedures for testing with consumers—*outsiders.*

Let us say that you and other contributors have proposed a "hatful" of 100 or more names for a product. Now commences the weeding process by the *insiders.*

The SOCK-IT Criteria

Earlier in this chapter we generalized on what a name can/should do for a product. Now at the "hatful" stage of the selection process let us tighten these generalizations into specific criteria:

S *Suitability* Is there a relationship to the function, ingredients, benefit, performance, or a unique characteristic or advantage of the product?

O *Originality* Is the name distinctive, unique?

C *Creativity* Is it catchy—"with it"—alliterative; does it have rhyme, have rhythm; does it have a "ring" to it? Is it a play on words?

K *Kinetic Value* Does it lend itself to promotion, logo, and so on? (Does it take off or just sit there like a lump?) (Figs. 1–4)

I *Identity* Is it memorable; does it have recall value?

T *Tempo* Does it suggest the proper mood for its intended market (segment of the market) and create a favorable impression on this target audience?

(A name need not be strong in all these areas but should be strong in *some.*)

Note that the SOCK-IT criteria are for use in the evaluation and *weeding* process by the creative, marketing, research, and management "teams"—the *insiders—not* by consumers.

Go through your "hatful" of names and eliminate those that fulfill none of

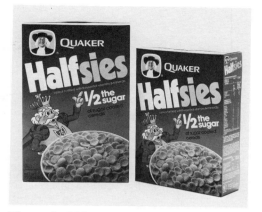

Figure 1 This logo in two colors boldly carries out the "half" theme. (Courtesy of Peterson & Blyth Associates, Inc.)

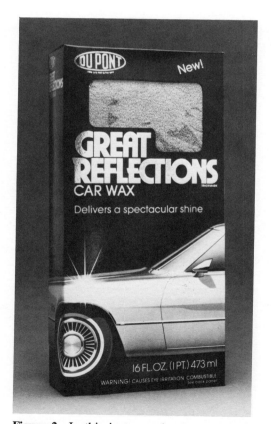

Figure 2 In this instance the name "Great Reflections" is printed in highly reflective silver on a glossy black carton conveying an image of spectacular shine. (Courtesy of Peterson & Blyth Associates, Inc.)

Figure 3 The name in bubbly, grafittilike lettering has been highly successful in attracting the attention of the youthful market for the product. (Courtesy of Peterson & Blyth Associates Inc.)

the SOCK-IT criteria, that are off target. Be careful in this elimination process. The author has experienced the hasty elimination of names that were used at a later time by others and became valuable properties. Remember that the word need not necessarily describe the product—this can be done in the generic description.

Trademark Availability

Following the preliminary weeding it is well to alphabetize the names and to check them in the Trademark Register under all related categories. Make a list of names that are particularly suitable, but that appear to be in conflict. (These can sometimes be cleared or purchased.)

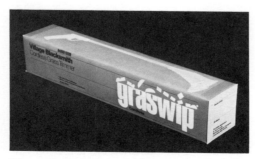

Figure 4 This eye-catching and unique logotype is an important element of the package design. (Courtesy of Anspach Grossman Portugal Inc.)

The lack of conflict in the Trademark Register by no means assures that the name is available, but this preliminary check should be a routine procedure for anyone developing or researching brand names.

You will have lost many of your most suitable, most creative names to others who got there ahead of you. Finding a good name with strong SOCK-IT value is becoming increasingly difficult although at the same time it is becoming more imperative.

Examples

Hair Dryer

Let us follow the weeding and testing procedures using as an example a hair dryer. What is the specific positioning for the product? Some possible positionings are:

- A dryer/styler with more blow power than others.
- A versatile dryer/styler with many attachments.
- A dryer/styler engineered to use less energy, yet dry fast.
- A for-all-the-family dryer/styler.
- A portable, take-it-with-you dryer.

It was marketing's decision to proceed with the dryer/styler with *more blow power* than others. The following names survived the preliminary weeding:

Blowmaster	Airblitz
Whiz-a-Matic	P.D. Cue
Zoom 'n Groom	Vortex
Huff 'n Fluff	Actionair
Chief Big Wind	Hurrycane
Windry	Dryphoon
Airmaster	Blitz Dri
Dry Up!	Windstant
The Blowhard	Power Play

At this point aging is of great value. Let the names sit and come back to them after another project or two. It is difficult for idea people to be objective about their creations but aging helps. Very often time does not allow for this as was the case in this instance.

Four names survived the second weeding process. They were Power Play, Hurrycane, The Blowhard, Zoom 'n Groom. To these were added Avanti and Fast 'n Light and testing with consumers (the *outsiders*) proceeded.

The names were tested monadically in a four-step procedure. The respondent was first shown the word on a card in a word association test. The very quantity of responses is significant. If a word elicits few associations, it is not very exciting. The *quality* of responses tells you if you are in the right ball park. Second, the respondents were asked what the product with such a name might be.

Avanti was thought of as a car—something Italian or Spanish.

The Blowhard was most frequently thought to be a snow blower or a hair dryer with other mentions of vacuum cleaner, fan, air conditioner, balloon. Some remarks about "hot air" were voiced.

Fast 'n Light suggested light bulbs, sleds, cars, boats, cake mixes, puddings, and diet foods.

Hurrycane suggested wind, storm, something moving fast, motorcycles, snow blowers, cars, airplanes, speed boats, cleaners (because of association with White Tornado), electrical appliances, hair dryers.

Power Play suggested power, strength, electricity, energy, games, power mowers, sport games, stereos, and record players.

Zoom 'n Groom was associated with speed, cleanliness, and neatness, particularly associated with hair. It was most frequently thought to be a hair dryer.

In the third step respondents were shown a visual of the hair dryer/styler with name and asked what performance they would expect from it, what advantages it would have for them, and so on. Zoom 'n Groom, Hurrycane, and Blowhard were the strongest scorers. When asked purchase interest (step four) names scored in order: Blowhard, Power Play, Hurrycane, Zoom 'n Groom, Avanti, Fast 'n Light.

In overall scores Zoom 'n Groom and The Blowhard rated highest in a close tie.

Laundry Detergent

Now let us consider an example in another product area: laundry detergents. The positionings under consideration are:

- A detergent specially for whites—to keep them "new white."
- For cold-water washing, containing an antibacteria ingredient.
- For children's colored school dresses, blouses, and so on for which no "whitening" is desired.

Of course, positioning should be established before name development is started. (This can be done via concept testing). If it is decided to pursue more than one positioning, for instance, in the three preceding positionings, each should be handled as a separate product. Steps would be:

STEP 1   Creative development of three lists of names.

STEP 2   Preliminary screening via SOCK-IT criteria. Note that in this instance the "unique characteristic or advantage"—the U.S.P.—would be of great importance in setting each of these potential products apart from the competition. The kinetic value would also be of importance. What can be done with the name in logo and package development? (For instance, **Colorfast** might be a part of a name. The letters c o l o r could be in a brilliant variety of colors on solid black or dark navy—a problem combination in laundering—and the f a s t in a speedy motif or better yet in lettering that suggests that the colors are holding fast—tightly—won't let themselves be faded away.)

Each element in the SOCK-IT Scale can be judged on a point system: Very Strong +2; Strong +1; Somewhat weak −1; Weak −2. Consider the SOCK-IT criteria from the point of view of the product category and objectives in which you are working. In the case of the detergent products we have said that S and K would be more important than other criteria. Thus we can double the value of these two factors. (For a soft drink or a candy T and C might be the most important elements, in which case these factors would be doubled.)

STEP 3   Consumer testing.

Procedure for Consumer Testing of Names

In the testing of names with consumers a fundamental principle is to avoid at all times asking a consumer to give an opinion about a name, to make a choice between names, or to state what names he or she "likes". The consumer is not qualified to judge. The following four-step procedure has been devised to avoid this.

STEP 1   In individual interviews words alone are tested monadically (each respondent sees only one). The word on a card is shown to the consumer who is asked to play a word association game.

What springs to mind when seeing this word? How does it make the respondent *feel*? What meaning does it have? What kind of a picture does it create in the respondent's mind? What does it make one think of?

In-depth interviewing can reveal that while **Chaste** might have favorable connotations for a skin cream, as a feminine hygiene deodorant its mid-Victorian aura has little appeal to today's swinging singles or to any of today's females for that matter.

STEP 2   The respondent is asked if this were the name of a product, what kind of a product might it be? What else might it be? What kind of a product would it *not* be?

STEP 3   The respondent is then handed a product concept with the complete name (such as Tang Breakfast Drink). The interviewer reads the concept as the respondent follows it. The respondent is asked: What is your overall reaction to the idea of this product? What would you expect of the (complete name)? What would be its advantages for you? Does it fill a need for you? What need? How interested would you be in using this product? In what ways do you feel you might prefer it to your usual brand? Assuming the price to be reasonable, how likely would you be to buy it? On what occasions would you use it? The respondent's usual brand and comments would be noted.

STEP 4   24 hour recall. The interviewer explains that the respondent was good enough to give an opinion about a new product and that with so many products on the market today it is difficult to remember all the names one hears. The respondent is asked for the name of the product that was discussed. Aided recall is used following this.

After interviews have been analyzed and reviewed, some names may be dropped as not being strong enough— not conveying the function and advantages of the product in a way that creates an interest in purchasing.

In the case of the detergent concepts, let us assume that following Step 4 the "new white" concept is found to be weak—does not offer enough promise— and it is decided to drop it, whereas the colored clothes concept scored well and three names look promising.

Exploratory Graphics

At this point, if they have not begun previously, the graphic designers should work with the names, roughing up preliminary logos and package designs. At this preliminary stage of rough graphics the objective is to explore the kinetic value of the names. Perhaps a dozen rough package designs seem strong after weeding and comparison with competition.

Consumer Testing of Names with Exploratory Graphics

Now the temptation arises to expose these to consumers in group sessions. This can have value but must be handled with care. It should be explained to the respondents that it is desired to get their *feelings* about some new products. Each product is identified by number—not referred to by name. Before discussion each respondent should consider each package and via a self-administered questionnaire make note of his/her *feelings* about each one—advantages—disadvantages perceived—the product he/she would be most interested in, least interested in, and so on with comments about the package. (This prevents a remark at the outset: "Oh, that horrible purple!") Then in the following discussion respondents are aware they have committed themselves and will not join

the bandwagon of a "loud-mouth." Such sessions can be helpful in a diagnostic way and can eliminate some duds, but they should not be used for decision making.

The package designs with rough logos and product description should be tested monadically in individual interviews asking respondents, as in Step 3, for their feelings about the new product and their purchase interest. This would be followed by a name-recall test, as in Step 4.

Note that the scores from testing rough logos and packages should equal or exceed scores received on the test of the name with the concept only. If they do not, more work should be done on the graphics.

Where a new product such as the colored clothes detergent is positioned against an existing brand such as Woolite, the two can be shown to consumers in a paired comparison test. In individual interviews respondents who had been screened to be at least occasional users of Woolite would be asked to consider the two products, which would be identical in degree of finish—rough package graphics or mock-ups. The new product would have to meet a predetermined score (equal, close to, or exceeding the competitive product) to be considered viable. An important consideration for management is always: how will the product fare against the competition?

Testing of Names with Concepts

Needless to say there is not just one way to test brand names. While the word association test is exceedingly important, it is sometimes necessary to go right to the testing of the concept with name. In this procedure the concepts are identical; only the *names* are different and they are, of course, tested monadically. No questions are asked about the name; if comments are volunteered they are simply noted. The 24 hour name recall is used.

Importance of Brand Names

It is sometimes argued that name recall is not of great importance, that with heavy advertising support the name will become established. This may be, but as we've said, with today's competition a new product should have everything going for it. The kinetic value of the name, what can be done with it, is important.

Regarding kinetic value, some experienced new-product managers feel they do not have a new product until they have a :30 TV spot to sell it. It is well to consider logos, package designs, ad "lines," and TV executions, along with names, early in the new-product development process. It is easy to get off target, off strategy. **Carpet Brite** was launched against **Glory** with a strategy that did not feature *brightness,* and after considerable exposure on TV, registration of the name was still poor in the test markets.

Micrin mouthwash, on which much money was spent, dropped out of the competition. The brand was regarded as somewhat dull and we can assume that the name was not of particular help. What would one expect of a product named Micrin? Something small? Perhaps had the product been in capsule form and dissolved and foamed with a sip of water, cleaning the mouth and breath, it would have had more relationship with the name.

Alka-Seltzer Plus had great difficulty positioned as a cold remedy. What would you expect of Alka-Seltzer Plus? You would expect the Alka-Seltzer part of it to relieve your stomach upset and your headache and, with its bubbly action, to help your hangover. What's the Plus? Make you feel you'd had a full night's sleep instead of a night's partying? Make you feel healthy as if you'd just been jogging? Instead of taking business from Dristan and Contac it took sales from Alka-Seltzer.

In this chapter we have covered some general procedures for developing brand names. There is very little *training* that will make you a good creator of names. Experience may. Start working at it, seat-of-the-pants-on-the-seat-of-the-chair. While you are considering names, *notice* names—in newspapers, magazines, on TV—every time you shop. What is good about them? What is bad about them?

At the start of a project analyze your product, its uniqueness, its positioning, its competition. Alone or with others develop your "hatful" of names and weed out the obvious misses via the SOCK-IT criteria.

Each project will require a testing procedure designed specifically for it. This chapter has outlined general testing techniques.

Test your names with consumers but do not let them tell you what they like or don't like.

Name Selection by the "Insiders"

As a last resort, should your team be forced to select a name, use the SOCK-IT criteria, assigning values to the points of most importance to your specific product. Then each person on the team can rate each name objectively by these criteria. You can then add the scores together and arrive at a total score for each one of the names in question.

Although it is judgmental, this procedure is more scientific than making this all-important decision based on what the president's wife and the cleaning lady like.

BIBLIOGRAPHY

Thomas C. Collins, "Selecting and Establishing Brand Names" in Victor P. Buell and Carl Heyel, Eds., *Handbook of Modern Marketing*, McGraw-Hill, New York, 1970.

Donald M. Dible, Ed., *What Everybody Should Know About Patents, Trademarks and Copyrights*, The Entrepreneur Press, Fairfield, CA, 1978.

Leon G. Schiffman and Leslie Lazar Kanuk, *Consumer Behavior*, Prentice-Hall, Englewood Cliffs, NJ, 1978.

The Trademark Register of the United States, The Trademark Register, Washington, D.C.

Trademark Rules of Practice with Forms and Statutes, U.S. Government Printing Office, Washington, D.C.

PART FOUR

Package Design Research in the Corporate Structure

CHAPTER TWENTY-SIX

Design Research in Package Redesign

Robert A. Roden

Package redesign decisions are not made easily. There is always an economic cost. There is often an individual or corporate emotional attachment to a time-honored package design. "Is that package really so bad we're losing sales?" That disturbing question swirls in the air when the suggestion is made to redesign the package. Design research should supply an answer.

The right kind of design research will provide substantive direction to a designer. It will tell which of several designs will outperform the others. At the bottom line, it should tell marketing management whether the product is losing a brand choice battle to competition on the retail shelf.

Although many industrial marketers might view the package as a vehicle for transporting the product from manufacture to user and as a medium through which to transmit certain essential information (quantity, ingredients, instructions, etc.), the reality is that in the eyes of many consumers, for many products the package *is* the product. Does the buyer who purchases a cake mix visualize the 18-or-so ounces of brown powder in the box? Of course not. She sees a delicious chocolate cake with its oh-so-moist texture and deep chocolatey frosting, just as pictured on the box and described in the copy.

In fact, the decision the buyer makes at the *point of purchase* regarding which product will do a better job may be a more powerful determinant in long run brand loyalty than the evaluation made following use of the product. People tend to perceive things according to their expectations; therefore, given substantially equal products, the one with the best packaging will be perceived as superior.

Obtaining consumer reaction to the product rather than the package is an important concept in packaging research. Consumers are not experts on packaging and package design. If they are asked to judge designs, they may or may not do it well, but they will certainly be performing an activity that is unfamiliar to them. But judging a product is another matter. Consumers certainly know what is good for them. All other things being equal, the product the consumer chooses will invariably be the one with more appealing packaging. Since that is what the marketing manager hopes to achieve—that competitive edge—consumer choice of the product over others should be the ultimate design research criterion.

Advancement in statistical techniques for data analysis have elevated some forms of marketing research to near-science. When consumer attitudes, preferences, and choices can be scaled and measured, marketing research can divulge the "why" and the "how" of product choice and whether there are segments of the market for whom the basis of choice is different. Such a wealth of information is seldom available to the marketing manager faced with one of the more important decisions for any product: how to package it.

Many technologically sophisticated techniques are available to evaluate important visual aspects of the package; however, science can't provide answers to some of the basic marketing questions, such as what to communicate about the product and to whom. What

packaging research *can* do is to tell the marketing manager whether a package is communicating effectively, and if not, why not.

Is accomplishing a package redesign a simpler, less ambiguous process than creating a package for a new product? Like many answers in the art/science of statistical research, the answer is a clear yes and no. Obviously, package design can be critical to a new product's successful introduction, but the situation is different from a design research standpoint. New product "A" confronts the buyer with a totally new choice: Product "A," or that other product you've been using, satisfied or not. The new product fights to establish an advantageously positioned image.

The established product, however, already means something to somebody. Package redesign should not progress very far without a determination of what the product means to all important categories of buyers. Have you a "franchise" with the consumer? What is it? Here we touch on the risk element in redesign: that your product will no longer be what it was to its loyal buyers, and that it hasn't really captured the fancy of some new, higher sales potential market target. If the redesign will alienate loyal users, you must ensure that the appeal to a new segment is solid and likely to pay off in a net sales gain. Such a shift in emphasis to engender trial within a new group may call for a comprehensive reintroduction strategy including heavy promotional effort.

Caution should be exercised before making drastic changes in the package unless the product itself has been changed to deliver some new benefit or to deliver the same benefits in a dramatically superior fashion. Moreover, extensive change must be accompanied by substantial supportive promotion. A total package change without product change may attract a few buyers who hadn't previously used the product. It may also confuse the product's loyal following and possibly cause them to rethink a choice they were previously comfortable with. No simple research will give an answer to the question of whether "repositioning" the product to appeal to a new segment will result in a net gain or loss. This is not a plea for added life for poor package design. Simple marketing wisdom discourages ventures into the unknown whose outcomes are no more certain than a "crapshoot."

Another critical aspect is whether the package is the only communication between manufacturer and buyer about the product or whether it will be supported by promotional and advertising activity. If the package is the only "selling" going on, it had better draw attention with magnet force.

EMBARKING ON DESIGN RESEARCH—THE STEPS TO FOLLOW

A revision of the packaging design of an existing product should be carefully managed; considerable time and expense are nearly always involved. Acrimony can be expected from the management if the process is ragged and uncontrolled, or if the result does not present the product in a better light to prospective buyers. Who is responsible for the orderly progress of this process from start to conclusion? If that question is not already answered in the internal structure and position descriptions of the marketing department, there could be a fast shuffle before the buck finally stops with someone.

A packaging specialist in the marketing communications section is a logical choice. The importance of coordination between the functions of marketing, research, and design suggests that the focal point is not in any of those three areas. In the long run, who is in charge will depend on whom the management would like to hold responsible. The marketing research director is certainly responsible for the design research phase of the process; at one time or another, however, representatives of different areas may assume a leadership role. It would not be presumptuous to compare proper execution of this type of project to a team of surgeons collaborating in a delicate operation. One person may open, another go for the organ, and another close. Yet one of the masked figures around the table knows he or she is ultimately responsible for a successful operation.

We can summarize in relatively few words the critical steps in the process. Some of the steps can be skirted, but not without courting the risk of ending up with a less-than-satisfying result.

STEP 1   Assess the marketing situation for the product. Is user interest in the product lagging? Is competition closing in? Can you legitimately say something new about the product? Spell out clearly the marketing objective of making a change. Why are you doing it, and what is intended to be the end result? This is marketing management's responsibility.

STEP 2   Critically evaluate the present package. Are there obvious defects? Do you have any feedback on the package? Challenge marketing to specify what they want the package to accomplish for them. Critique it for obvious flaws. This is the responsibility of marketing communications.

STEP 3   Map out the critical activities and events that must be covered in the entire redesign process. You can stop short of a full network diagram, but there is a necessary sequence of events, from assessing the product and evaluating the package through selection of research supplier and designer, assembly of test materials, test and test evaluation, to final design implementation with packaging engineers. This should be the responsibility of a packaging coordinator in marketing communications. Otherwise, the marketing research director should get it done.

STEP 4 Interview package research groups. Let's assume this is your first time selecting a package research group. Most of the major market research houses include package research among their wares. Ask a fellow practitioner. Check with some designers, if you're not afraid of their biases. Consult a publication such as *Industrial Packaging*. Contact research suppliers to let them know what your situation is and what the marketing objective is. If they're interested in the assignment, carefully describe it in writing so they can provide you with a written proposal that explains methodology, details, time, and cost. If they want to quote you both research and design, if they have the capability, insist they break the two pieces apart in their cost projection. Avoid being boxed into an all-or-nothing situation. Verify whether the price is all inclusive, or what is excluded. This is the marketing research director's responsibility.

STEP 5 Interview designers. Does your company have an internal design group whose use is mandatory? Is there a designer on retainer? Get an idea of what your budget can afford. The big names might be out of your price range. On the other hand, if their staff is underutilized, they may bid lower to cover overhead. Some innovative designers, just starting on their own, can often offer talent at a reasonable price; make it clear, however, that this may not be their opportunity for unbridled creativity. Give them a general idea of what kind of research you expect to make available to them. Make sure they are willing to work with

a research company and will use the input in the design process. Make sure that their price is all inclusive or detail what is excluded. Travel and materials are usually add-ons. If you haven't used a designer recently, you can find designers through the Package Designers Council or the Yellow Pages or the Black Book, but the best tack may be to ask around. You will be quickly told who is good and who isn't. Also be sure to specify package design (not all designers are equally good in all fields). This should be the responsibility of a packaging coordinator in marketing communications.

STEP 6 Prepare a recommendation to management. Eventually the time comes when capabilities of research groups and designers must be evaluated and a proposal made to the management. Management may be represented by an individual with budgetary control, a committee, or several levels in the organizational chart jointly. The marketing objective, costs, and a time schedule form the heart of such a proposal. Credentials of the selected research supplier and designer should be included, along with a rationale for your recommended choice. Your criteria should include not only professional qualifications but whether or not designer and researcher blend well together. This should be done jointly by the market research director and the packaging coordinator.

STEP 7 Review your internal administrative procedures. You'll be receiving a series of invoices that my be unfamiliar to your internal processing system. Make sure the groundwork is laid for payment with properly documented approval from the executive with the necessary approval limit. Find out in advance what kind of proof of work performed or vouchering of expenses is necessary to satisfy the internal control system. Not only will your suppliers appreciate getting paid more promptly and efficiently, but you'll spend less time expediting and explaining. If your project is expected to lap over into the following budget year but its costs are all against this year's budget, take the time to review costs to date with your suppliers and ask them to bill "up to work performed." Last year's hero for a budget surplus is this year's scape goat if this year's budget is devoured by last year's cost. If there is a time crunch on obtaining documentation, accounting can arrange an accrual against the budget year's expense accounts.

STEP 8   Control the process. Decision points will occur at various stages of the process. Some are appropriately made by individuals, others by management groups or specialists in a particular area. Be sure you've enlisted the right people to make each decision. Many individuals within the organization may want to be informed of package redesign progress, but that doesn't necessarily place them in a decision-making role. As tactfully as possible, declare who has decision-making responsibility by virtue of position or qualifications, and let them decide. The rationale for all decisions should be carefully explained.

STEP 9   Review the process critically. When it's over, you may or may not be satisfied with the result and how it came about. You owe it to everyone to review your plan—was it adequate?—and the sequence of events that actually occurred. Your plan may have been fine, but perhaps a particular decision should have been handled differently. You or someone else should learn from the experience.

AN EXAMPLE OF PACKAGE REDESIGN—SCOTCHGARD FABRIC PROTECTOR

The process just described and other package redesign research issues might best be examined in the context of a package redesign that 3M Company recently completed with "Scotchgard" fabric protector, one of its best-known consumer products. Scotchgard protects fabric by chemically coating fibers so soil and moisture will not adhere.

The product had undergone several packaging changes since its introduction as a self-applied fabric protection in 1965. Consumer awareness of Scotchgard is high—when the consumer thinks of fabric-protection, he or she thinks of Scotchgard. Over time, various elements of communication (the brand, the benefit statement, application instructions, the warranty, weight etc.) began to be crowded on the limited space on the container. The revelation of the detrimental effect of certain fluorocarbons on the ozone layer of the earth's atmosphere led to a chemical change in the propellant and a new statement on the container. In depth consumer segmentation research was undertaken to provide stronger marketing direction for the now mature product. This research showed that for certain consumer segments the package was not a positive inducement to purchase and use. Since necessary corrections in copy would not have a major impact on the package as viewed on the shelf by the consumer, it was decided that more extensive packaging research should be performed in order to parlay the newly found consumer research information into a more effective package communication.

Many packaging redesign projects begin with a review of the product in its natural environment, in this case the retailer's shelf. Scotchgard is a product comfortably at home in many channels of distribution. The shopper may find it in the nonfood section of the grocery, as well as in a drug, discount, department, variety, or hardware store. There are many other possible distribution outlets. Display conditions vary from one or two facings on a lower shelf of the laundry products section of the corner grocery, to a gigantic, hundred-case display in a center aisle of a mass merchandiser. The product's water-repellancy feature lends itself to spring and fall promotions. The trade recognizes Scotchgard as a consumer drawing card, as a promotion. In a featured display, the package's brilliant red plaid design commanded attention; on its own in reduced display space, however, the product tended to fade into the visually noisy background.

Visual inspection of the product on the shelf showed that the plaid and the red container cap were the only clearly distinguishable features of the package. Shoppers in the market for the product could locate it with some search. Consumer research had disclosed that many Scotchgard purchases were associated with the prior purchase of a high-priced fabric item, such as an expensive piece of furniture or a garment. A person aware of the protection and preservation benefits of Scotchgard might seek it out as a way to safeguard that major expenditure. Consumer confidence in the product's performance was high. For relatively few dollars, the user received peace of mind: in case of spill or stain, the expensive fabric could be wiped clean simply and easily. In a marketing sense, the product package could and should play a more dynamic role in stimulating consumer sales. Scotchgard protection lasts, but not forever. Protection of aerosol Scotchgard for wearing apparel ends with dry cleaning or washing. Depending on wear, protection for upholstery fabric eventually wears off. A large percentage of consumers had applied Scotchgard at one time or knew that it had been mill-applied to fabric on furniture they had purchased. Marketing's objectives, in this case, were to remind previous users through the package of the many uses of the product, and, further, to stimulate interest in the product and to induce possible trial by those who had not previously used the product. Any newly designed container

must have the ability to command attention; draw the consumer to the product; communicate a message. The message was that Scotchgard frees the user from worry about spills and stains. Purchase needn't only follow a major recent expenditure for an item containing fine fabric. Periodic reapplication guarantees that the protection lasts.

This was a renovation project. Much about the Scotchgard container certainly had value or "equity" in the minds of the consumer. The buyer did not pick up 12 or 16 ounces of chemical; he or she picked up an aerosol can colorfully decorated with red plaid, branded as Scotchgard, manufactured by 3M Company. Were these the most important graphic elements on the container? If so, which of them communicated most strongly to the consumer the qualities of the product that were essential to consumer acceptance of the product?

With the marketing objective clarified, the next stage in our design research task was to determine those equities of the present package. We reviewed available research techniques to select one that would identify those graphic elements that were most central—our strengths—as opposed to those that were only incidental to the consumer's concept of the product. Where was our message strong, so we could benefit from even greater emphasis? What space or graphic concessions could we make as a trade-off for strengthening the important element? Several considerations entered into the thinking that ultimately became our decision rule for selection of a package research supplier.

It is said that in real estate the three most important factors in determining the value of a piece of property are location, location, location. I personally believe that in design research, if there are three such factors, the first two, at least, are experience and experience. Let the inexperienced get their experience at someone else's expense. The greatest risk on your part may not be the possibility of poorly conceived and executed research so much as the chance that the results will be inconclusive and not provide clear direction to marketing management and the designer. A rarer, but more serious, risk is that they will be conclusive but erroneous.

The payoff of experience is perhaps greatest in interpretation of test results. Since a major objective of the research is to provide direction to the designer, that direction must be sure. While the chance of ambiguous results may not be great, if sound methodology does produce inconsistent or incongruous results, the judgment of experience is invaluable. Our Scotchgard design research relied heavily on the experience of the research firm in interpreting an ambiguous test result.

No one could presume to tell you which design research firm to enlist. Talk to several. Ask for proposals. Look at the steps in their process. Is it really going to get answers to the questions you need answered? Does it really understand your marketing problem? Hopefully, its process does not rely on a gimmick to provide the entire answer to your question. Ask its representative to show you examples of how it has resolved a packaging problem similar to yours. (The name of the product can be blinded easily enough.) You must feel assured that their methods and their staff thinking are sensitive enough to bring definitive answers. Ask for references. Find out who will be working with you. Is it a principal in the firm or a project director? Who will interpret the results for you?

Cost certainly cannot be the only basis to choose a design research supplier, but it would be unrealistic to pretend it isn't important. There are reasons that a supplier may quote higher than expected: they may think your company is very profitable and can afford it; they may not really want the job too badly. If you think they are a little high, but you like their approach, tell them and maybe they'll find a way to reduce the quote without letting quality suffer, or a way to add a little more service without raising the price. There are reasons why a supplier may quote unexpectedly low: they may simply want you for a client so badly they'll sacrifice profit margins; on the other hand, they might not really be giving you the full cost picture (you may get hit with an extra charge later). If the quote appears low, a careful probing of their method and costs for each stage should give you an idea of whether a supplier wants to begin a lasting relationship or is giving you an unrealistic cost projection, as in "You mean the steering wheel and tires are optional extras?"

There is a certain face validity or surface honesty to a proposal whose steps can be priced out at standard rates. If those steps can be relied on to do the job, you know very clearly where you stand on price. If the task of redesigning a package is viewed as strictly a design project, it might not seem illogical to transfer total responsibility to the designer. Resist the temptation. You, the client, are at a disadvantage if the design firm controls the research process. Depending on the design firm you engage, the designer may or may not incorporate research into the redesign process. If the design firm is in charge of the research process, the designer can easily be the sole interpreter of design research.

While the designer may rely heavily on results of research for design direction, he should not be in control of that research. Design is an artistic endeavor. Judgment about the relative merits of one design versus another are highly subjective. A designer who has been successful and believes he is good at the craft also believes he has excellent judgment. Artistic judgment, certainly. But not necessarily unerring judgment regarding which designs the consumer will or won't react positively to and why. No one, least of all a designer concerned primarily with aesthetics, can predict consumer behavior in advance with any certainty. Worse yet, a designer in control of the research is free to tailor that research to serve his ends and to interpret the results to back up his own bias or to suit his own convenience. One can be skeptical about whether any designer would accept research results that contradict his long-ingrained theories of packaging design. Moreover, you cannot be sure he would always disclose research results fully to you, the client. And if research found too late in the game that the design direction was not effective, can you be confident the designer would have the humility to tell you, the client, that it was necessary to start over? Designers who rise to prominence are both good and proud of being good. To some designers, good research supports their judgment. If it doesn't, it was a waste of time and money.

Maintaining control of design research is simply a matter of risk reduction. You, the client, are the loser if the redesign is not as good as it might be. Research can be valuable input to the design process. More importantly, you must rely on research to tell you if you are succeeding in the all-important task of reaching the consumer.

THE DESIGN RESEARCH PROCESS FOR THE SCOTCHGARD REDESIGN

The rationale of the design research proposal accepted for Scotchgard consisted of the following three steps, each with its own decision point:

1 Evaluate elements of the present package.

2 Evaluate four or five alternative designs incorporating the strengths of the product disclosed by the previous step.

3 Evaluate the strongest design of the previous evaluation step, with whatever modifications might be appropriate, against the present package.

An alternative proposal suggested obtaining consumer reactions to the current and three new designs monadically, then pairing the current design against each of the three new designs in consumer tests.

We felt quite confident in the research capability of both firms. In this instance, several factors swung us toward our ultimate choice. Experience certainly was one factor. More critical to us, however, was the apparent ability of the methodology to measure the subtleties of the equity in the current package. Respondents chose the product for themselves from a set of differently designed packages. They were also asked to describe the product through bipolar modifiers, either adjectives or adverbs. Differences in description could then be related to choice preference. In the Scotchgard test, respondents reacted more or less favorably to containers visually emphasizing different key graphic elements, either 3M, the word Scotchgard, or the red plaid. This test disclosed that the word "Scotchgard" featured more prominently on the container signified to the respondent a product of higher quality, better for fine fabrics, and one that would leave fabric more moisture-resistant. The first step of the research was completed. We were then able to offer this guidance to the designer, whose task it was to provide (in the second step in the research) four to six designs to test against each other, each of them featuring the word Scotchgard prominently.

Consumer survey research is a much

maligned art, often with good cause. There are as many ways to construct a questionnaire poorly as there are questionnaires. Interviewer problems can range from leading respondents to inventing the data. Two problem areas that are particularly relevant to design research are the experimental design for evaluating various package design options, and the sample sizes desirable to create confidence in the results.

In design research respondents typically are expected to react to a variety of stimuli, expressing favorable or unfavorable opinions, recalling what they are able, interpreting visual and verbal messages, and stating their preference. A difficulty the researcher may easily face in this situation is that subjects feel "forced" to give reactions they don't really have (i.e., if in day-to-day life a housewife does not really have an emotional reaction to her detergent box, reactions elicited in a test may be unrealistic). Moreover, when responses are forced, answers may not be what the respondent feels, but merely any answers to keep a probing interviewer at bay.

Another potential flaw in research design is the demand for more response from a respondent than the human creature can reasonably be expected to produce, both quantitatively and qualitatively. Quantitatively, there is certainly a limit to what the person can process and give back to the researcher in the form of opinions and preferences. A research proposal once submitted to us for a scouring pad package redesign centered around requiring respondents to evaluate six groups of product designs and six individual product designs in about 30 minutes. My sympathies would have gone out to those harried people. That is a lot of work in half an hour.

Research can suffer qualitatively if the respondent is expected to draw dis-

tinctions, and express them verbally, regarding differences that are so subtle that they may not in fact be real differences. If you were interviewed about several cake mix packages, what would you say differently about the one with the flavor at the bottom of the box rather than the top? Many important marketing decisions may be based on a mental coin-flip by several hundred housewives. In such a situation it might be presumed that a large enough sample size may minimize the degree of sampling error. This is a valid presumption if the sampling process is the source of the apparent variability rather than an error in design methodology. The size of a sample for a given research procedure is always a perplexing question to which few of us will ever get a definitive answer. One reason is that the people most able to persuade a company to invest in design are not always schooled in the nuances of statistical sampling. They will cleverly dodge the sample size question, or assume it away. Even a statistician may dodge it, knowing the precision of the measurement depends on various factors, and the "ideal" sample may be either unaffordable or unattainable. The sample size decision must be made before incidence or frequency is known; theoretically, precision and confidence level must be adjusted if incidence is other than expected.

Practically speaking, a sample size of 150 is not a bad rule of thumb for measurement of most variables. But you'd better hold your breath and pray the measured difference is not narrow. If it's a 55–45% preference, you're dead in the water unless you're willing to go with a less than 90% confidence level. You may find it necessary to test again with a larger sample. There are some research procedures in which responses can reasonably be expected to stabilize after a sample as small as 30. For example, the human body and its motor responses are such that ocular measurement or eye movement tests can provide quite stable results with a relatively small sample size.

In most cases, however, the buyer of design research should be much more concerned with *what* the design researcher is expecting the respondents to do than with *how many* of them will do it. Methodology problems have the potential to affect results much more than sample size problems.

The second step in our design research called for testing four to six container design variations against each other. The designer provided us with a dozen designs, from one similar to the present container to one that was radically different in color and design. Knowing the single graphic element that is of primary importance and should be emphasized is an aid to the designer, but that knowledge could also lead to a reluctance to stray too far from the existing design. The value of consumer research at this point is to give the designer and marketing management an idea of how different the new design should be, depending on whether newness is attractive to the consumer.

A decision-making committee from marketing, marketing communications, and marketing research reviewed the designs submitted to select six designs to test against each other to find the single best alternative to pit against the present container. Three of the six continued to emphasize the basic water-beaded red plaid cloth of the then-current container, with variations in treatment. Another design depicted items to which Scotchgard could be applied, without plaid. The other two designs were variations showing plaid cloth resisting stains. The six designs were presented to 400 respondents, soliciting their opinion of the product within the four containers and

asking which product they would want for themselves. The test produced a not illogical outcome, but one which left us perplexed. A good deal of the positive response centered on the design showing potential applications of the product; most of the remaining positive response was diffused among the three designs with treatment similar to the existing container design. The designer's judgment pointed toward the strongest of those three designs. Marketing had a particular interest in the design showing several specific uses of Scotchgard, believing it might serve as a more powerful reminder and lead users to additional uses new to them. The issue of which of the new designs to test against the then current design could only be resolved by a further test. Minor modifications were made to improve each design and they were then tested against each other.

In the retest, the design most closely resembling the then-current design made the strongest showing. We then tested the winner of the "face-off" against the Scotchgard container then on the market. The new design was preferred by a wide majority of respondents. Significantly, the new design also had the strongest appeal to previous users of the product. Since our primary objective was to strengthen the package's ability to communicate a reminder message, and secondarily to stimulate new trial, the test results seemed to support conclusively a decision in favor of a redesign which, though similar to the present design, corrected obvious weaknesses and several weaknesses that were more subtle. Electronically controlled display effectiveness tests were used to divulge those subtle flaws.

Tests described thus far were telling us about the overall appeal of one design versus another, and, qualitatively in quantified form, why the respondent perceived the product in a particular container as more desirable than another. We needed to know more, however, about the strictly visual aspects of the design. First, the present Scotchgard container was tested for display effectiveness. Did the container "stand out?" Was it easily seen? Electronic devices were used to test its visibility with a small number of subjects (in this type of test, variability is minimal from subject to subject.)

How readable was the copy to the passing shopper? Again, an electronic device was used to measure how easily various copy points could be read.

Finally, eye flow was recorded electronically. Were important elements seen or missed? Were they attractive or interesting enough to hold attention? The movement of eyes across the container was measured, noting where attention was held and where it was not held.

The then-current container scored rather well in visibility. Its splotch of plaid and bright red cap were easily noticed. Unfortunately, the typical viewer could read little, if anything, of what was being said about the product on the container. Eyes moved quickly across the container; attention was not held long at any point.

The "winner" of the second research step was then subjected to display effectiveness tests. The proposed design had slightly lower visibility, a not unreasonable result since the area devoted to plaid had been reduced. This problem was solved by using a brighter red in the plaid. Readability and eye flow results improved dramatically. The shopper could now read what Scotchgard was and what it would do for her. The new design caught the attention of the shopper and held it. The tests indicated it was desirable to relocate the 3M logo to place the corporate symbol in the flow of the shopper's gaze over the container.

Nearly 1½ years passed from the time we began design research to redesign the Scotchgard container until the product was shipped in the new container. Acceptance of the final design triggered an internal process in which packaging engineering worked with an internal design group to produce keylines which were then approved by marketing for use on the container. The new design reached the can supplier for use on the next printing. 3M's aerosol Scotchgard fabric protector had a new container design that design research showed would do a more effective selling job.

One school of thought might suggest validation of the design research by a sales test. We chose not to perform a sales test. Such a test situation could be set up, but it would require many stores in many cities to soften the effects of the many other variables that could influence the sales rate. A long test might be necessary for an item of relatively infrequent purchase such as Scotchgard. A higher sales rate could be expected and would only reinforce the consumer research that led us to the new design. A lower sales rate could be attributable to factors other than the design, which we may be unable to measure.

Any redesign should be accompanied by a promotional campaign to call the attention of shoppers to the product and stimulate renewed interest. The newly repackaged product could quietly slip onto the shelf to replace the old product, but why not get maximum impact from all the fire power in the marketing arsenal? Supporting promotion will bring back many more buyers than a redesign by itself could.

The cost of delaying a package redesign is reflected in lost sales opportunities. Such an opportunity cost is difficult to measure. If market share slips, or unit sales volume declines, it would make good sense to test your brand against competitive brands to see whether your brand is at a disadvantage on the retail shelf. In this way, design research would be an input to a redesign decision, rather than simply input for the designer after a decision has been made. Research must be a source of direction in the decision-making process the marketer goes through to satisfy himself he has accomplished his objectives for the product. Over the long run, design research could make a richer contribution by flashing a signal that a redesign should be considered.

CHAPTER TWENTY-SEVEN

Evaluating the Finished Package

Jack Richardson

The designer has recommended it. It's not quite what you expected, but you like it. Your boss isn't so sure. How do you determine if that new package will really help your brand?

My objective in this chapter is to demonstrate that packages can be evaluated, that the methods are not particularly difficult, and that *you* can do it successfully if you'll follow a few "simple" suggestions. *FIRST, KNOW WHAT YOU'VE SET OUT TO DO.*

We've all heard the one about the airline pilot who reported to his passengers that he had some good news and some bad news. The good news was that they were setting a new all time speed record. The bad was that the guidance system had broken down and he didn't know where they were going.

The most important single requirement in package evaluation is a clear understanding of what a package is supposed to do and where it is supposed to be going. Three "tools" can be very helpful in providing this understanding and direction.

One of these is a *Philosophy of Packaging*. A clear-cut, well-thought-out, generally-agreed-to philosophy about what a package really *should do* can be of tremendous value in determining what questions will be asked, what methods will be used, and how successful you will be when results are presented to management and the consumer. My guess is that almost all package testing disasters—the ones that killed off "good" packages or approved "bad" ones—can be traced either to someone who did not think through to the basics of what he was doing or to someone who didn't reach real agreement with other members of the "team" about the process.

There are undoubtedly many philosophies of packaging, and each can be expressed in many ways. Table 1 is one.

Table 1 Philosophy of Packaging

A package has four basic functions:
- It contains, protects, and dispenses the product.
- It provides information.
- It helps the product become visible.
- It promises physical or emotional benefits.

It is usually *not* the function of a package to be liked for its own sake.

A firm that believes in this philosophy will almost inevitably provide testing direction that leads to measures of product performance, communication, visibility, and image projection. It will not waste its resources finding out which package people prefer as an *objet d'art*.

A second tool that helps achieve understanding is the *Brand Positioning Statement*. This specifies precisely how the firm defines the product, how it differs from other products, how consumers can identify and evaluate the difference, who now uses the product, and those brands and user groups from whom the firm expects to derive additional business.

There are many ways to write a positioning statement, and the key is not so much style as it is general understanding and agreement on content. My favorite approach is shown in Table 2, p. 300.

A third tool one should have before attempting to evaluate a package is a precise, written set of *Design Objectives*. These should be consistent with the firm's overall packaging philosophy and brand positioning statement. But they also should specify exactly what is being attempted in a particular design effort.

Again, format is not important, but general agreement and understanding are. One format is illustrated in Table 3.

Table 3 (Brand) Design Objectives

1 Increase shelf visibility by _____ points.
2 Flag brand's new, better tasting formula.
3 Improve or maintain brand promise of:
   Good, long-lasting flavor.
   Coolness.
   Breath refreshment.
   High quality.
4 Specifically improve ratings among the _____ high volume user group.

These three tools, the company packaging philosophy, the brand positioning statement, and a set of project design objectives, understood and agreed to by all responsible persons in the organization, are the first three essentials to effective package evaluation.

EVALUATING PHYSICAL EFFICIENCY AND FORM

The best way I know to measure physical performance under nonlaboratory conditions is to place the product out for ordinary people to use at home.

Several years ago, I worked in the toothpaste business. The ammoniated and chlorophyll products had come and gone, and we were looking to be first with the next *new* item. Someone thought the answer might be in packag-

Table 2   Positioning Statement

(Brand)_____is the one brand of

(product classification)___that (differ-

entiating consumer benefit).

Consumers can know this is true because

(documentation/support).

(Brand)_____appeals most to people with
the following characteristics:

    Demographic: _____

    Behavioral: _____

    Values/Life Style: _____

It will get most of its new business from:
    Competitors: _____

    Characteristics: _____

---

ing and designed a sensational new package made of plastic. Its cap was on its side, and it had a screw closure. It looked something like Fig. 1, p. 301.

We put the product out for in-home testing and almost immediately began to collect consumer reaction. It was overwhelmingly favorable and, because we wanted to be first, we began releasing the product to market.

Then the results began to change. The cap didn't always close properly, and the paste began forming little piles on consumer sinks. The paste in the bottom of the container refused to flow *up* and out of the aperture. And the same consumers who were delighted with the product in Week One came to feel that they weren't getting either convenience or value in Week Six.

In the market, the product performed the same way: high initial acceptance was followed by disappointment and rejection. The greatest disappointment was ours, but out of it and others like it has come a list of suggestions that have saved our companies a lot of grief.

- *Take time.* Some packaging bugs aren't apparent or don't develop right away. Give it a real workout. Make sure testers have a chance to use up at least two packages.

- *Allow for a range of conditions.* No laboratory can duplicate the infinite variety of behavioral, sociological, psychological, climatic, and other physical variables that can emerge from three hundred representative families. You should provide for them.

- *Use packages assembled, filled, and wrapped on regular production*

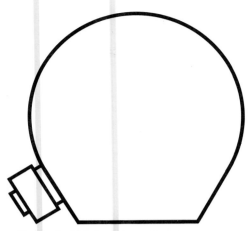

Figure 1   Toothpaste package prototype.

equipment. Models or pilot production product are all right for early experimentation, but may be too finely tuned to count on for valid responses.

- *Test against a meaningful control*. Don't supply only the new test package. Give them an old one to compare it to. Comparative responses are infinitely more reliable than those made "off the cuff."
- *Observe closely*. Don't rely solely on computer print-outs. Get into the field and see for yourself what is happening.
- *Concentrate on the practical*. You know your package is beautiful and well put together. But how faithful is it? And will it work?
- *Concentrate on the product*. If testers love your new package but think the product has depreciated or become less of a value, stick with the old package.
- *Watch for the exception*. You have to know what most people think because you can't get along without them. But keep your eyes open for that one person with a difference. She may be the one who can give you the most help.

EVALUATING ABILITY TO CONVEY INFORMATION

The rules for measuring a package's ability to convey information are not very different from those used to measure physical performance. Essentially you give the package to the consumer, let her read it, and then let her tell you what she thinks she has read. Here too are some well-learned suggestions:

- Measure communication both *before and after product use*. Use matched groups of respondents to determine what is being taken from the text alone—and how this is being reinforced or modified by personal experience.
- *Let the consumer do the talking*. Ask simple, direct questions and don't get in the way of the answers. She will tell you what the package is saying to her if you'll let her.
- *Really listen to what she is saying*. Watch for additions, modifications, distortions, personal connections, arguments, and deletions. Any or all should be included in your analysis.
- *Check your assumptions*. You may have done all kinds of research and you think you know all she should be getting from your message. If she isn't getting it, it may be that you're not saying it very well—or it may be that it's not as important as you thought it was.

EVALUATING VISIBILITY

A typical modern supermarket has over 30,000 square feet of selling space and stocks between 8000 and 10,000 items. A typical modern shopper spends about 40 minutes in the store and purchases between 40 and 50 items. Similar figures can be developed for other types of retail outlet.

There are several ways to look at these data, but one way is to divide the number of minutes the shopper is in the store by the number of items she has to look at. This tells us that the average supermarket shopper has about 1/4 second to look at the average item—and that that item needs all the help it can get just to attract and hold her attention.

One of the traditional tools for measuring attention-getting power or visibility is the Tachistoscope. This is a shutter-equipped apparatus that attaches to a slide projector and can be set to expose slides for brief, predetermined intervals. (I usually project pictures of test variants for 1/8, 1/4, 1/2 and 1 second intervals and ask the respondent to tell me what she has seen after each interval.)

Different researchers use T-scopes in different ways. Some show only one package at a time and check for element as well as logo visibility. Others show shelves full of merchandise and stress ability to spot the test item in a competitive environment. Some use small samples weighted to account for differences in speed of perception. Others count on the self-weighting effect of larger numbers. In any case, most do essentially the same thing: record what people say they've seen following short periods of package exposure.

For any who are new to T-scopes, I'll make the following suggestions:

First, *expose your package among others in its normal environment*. Competitive packages interact and can create visual impressions different from those they would create individually. Make sure your test package has a chance to interact with the same packages it will be contending with in the marketplace.

Vary the package's position on the slide. Your white package will be more visible between red and blue competitors than it will be between two white ones. Shoot at least three random competitive layouts to be sure you are testing under more than one market condition.

Use the whole slide. Don't minimize their impact or confuse the respondent by bunching the packages in the middle of the slide and surrounding them with white space.

Shoot your pictures against a white or off-white background. This will produce a slide that comes as close as possible to in-store conditions. Under no circumstances should most package goods be shot against dark blue or black velvet as though they were the crown jewels.

Eliminate distracting elements. Fancy racks and shelves can get in the way, and many plastics, cellophanes, and foils reflect photographer's light in ways that produce an unnatural, unattractive effect. A little discretion and a professional photographer should be able to produce a pure, nondistracting set of photographs.

There are also some things you should watch out for in *showing* the slides:

It seems obvious, but people have been known to send poor quality slides into the field. Be sure you actually *screen them for clarity and color correctness* before releasing them for testing.

Control the lighting in the test area. Unless you specify differently, you can count on the fact that the blinds in one test area will be left open while another test area tries to operate totally in the dark. And you'll get no clues from the data to help explain why the respondents in one area were able to see so much less than respondents in the other.

Be sure to *control projection distances*. Some researchers specify that the projector and screen be far enough apart to show the packages life-size. Others specify the same distance for all tests. Make sure you specify what you want, or again you will have a difficult time explaining differences in your data.

For the same reason, *control the viewing distances*. Always specify how far you want your respondent to be from the screen.

Don't forget to *provide for the respondent's comfort and privacy.* Let her relax and concentrate on the task at hand. Don't distract her with extraneous movement or with voices through the partition or over the transom.

Be sure to *provide adequate sample sizes.* Some experimental work has been done that shows trends beginning to stabilize at about the fiftieth respondent. I prefer to be a bit more conservative and specify at least 100.

If you do good tachistoscopic testing, you are likely to produce results that look something like Table 4. Test package A is significantly more visible than test packages B and C. At 1 full second, A provides a 22 to 27% increase in efficiency at being seen on the shelf.

Note too, that even test package A is not as visible as the well established old Competitor X. Everyone is familiar with X and expects to see it in this context. What is important is not that we didn't beat this competitor, but that test package A did such a good job of narrowing the gap.

Another popular way of measuring "visibility" is a Shelf Impact Test.* Here, qualified respondents are brought into a simulated store area and asked to walk past shelves of merchandise looking at the products in the way they would if they were shopping in a store (Fig. 2). They are then taken to an interviewing

*Developed by Opatow Associates, Inc.

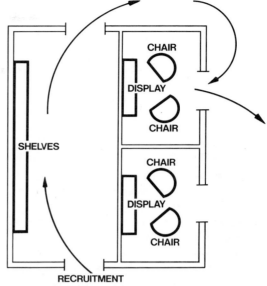

Figure 2   Shelf Impact Test.

area and asked unaided and aided questions about the brands they have seen.

Here too, several factors are important.

Be sure to *expose the product among other products in its normal environment.* You are trying to reproduce conditions that exist in the market. Don't mix things up and confuse the customers.

Again, *control the lighting and viewing conditions.* The test room should be lit to resemble the outlets you expect to sell in. Provide a clean aisle 5 to 6 feet wide for respondents to walk in. Eliminate anything that might distract from the shelves. Don't use spotlights, promotion materials, or special effects un-

Table 4   Percent "Seeing" Packages at Selected Time Intervals

	⅛ second	¼ second	½ second	1 second
Test package A	27%	38%	51%	66%
Test package B	18	30	46	54
Test package C	19	29	44	52
Competitor W	6	10	16	34
Competitor X	23	46	64	77
Competitor Y	3	8	15	29
Competitor Z	2	8	16	27

less you expect to match them in the marketplace.

Be sure to *control the time of exposure*. It's customary to allow up to, but no more than, 2 minutes to look at the shelves. Most respondents will take less than that, but more than 2 minutes permits "memorization" to overcome "impact."

Provide the respondent with privacy. Let her have her 2 minutes without being distracted by conversation or activity by other shoppers. When she gets to the recall section, make sure she gets neither help nor hindrance from voices heard over the partition.

If you do packaging-impact testing, you can expect to produce results like those in Table 5. Here too, one test package provides significantly greater efficiency than two others, and narrows the gap between our brand and familiar old Competitor X.

EVALUATING PACKAGE IMAGE

Let's move on now to tools that measure images or perceptions of end-benefits. Most researchers use open-ended questions plus some kind of word association or semantic scale device for this purpose. Table 6 is an example of *Controlled Word Association*. Respondents are shown a product/(package) and asked which of several words or phrases

Table 6 Controlled Word Association

(a) Which *one* of the following words best describes this product?
(b) Which two or three others also describe it?

	(a)	(b)
Fresh	( )	( )
Bland	( )	( )
Artificial	( )	( )
Strong	( )	( )
Cool	( )	( )
Natural	( )	( )
Weak	( )	( )
Tasty	( )	( )
Stale	( )	( )
Warm	( )	( )

best describe it. Some researchers specify the number of associations the respondent may select by asking "Which *one* of the following words best describes this product?" and, "Which two or three others also describe it?" Other researchers let the respondent select as many as she chooses. Either way, the heart of the approach lies in the number of respondents who select key goalwords to describe one test package and how this compares with the number who select the same words when asked to describe another one.

Table 7 contains examples of combination *Word and Number Scales*. Respondents are shown a product/(package) and asked to rate it from 1 to 10 on a series of characteristics. Respondents

Table 5 Percent "Noticing" Packages

	First Mention	Total Unaided	Unaided Plus Aided
Test package A	10%	45%	85%
Test package B	8	39	82
Test package C	7	34	80
Competitor W	3	12	75
Competitor X	25	58	99
Competitor Y	5	13	68
Competitor Z	5	14	72

Table 7 Word and Number Scales

Impressions of Product		
10 Excellent	10 Strong	10 Fresh
9	9	9
8	8	8
7	7	7
6	6	6
5	5	5
4	4	4
3	3	3
2	2	2
1 Poor	1 Weak	1 Stale

typically rate each product/(package) higher on some scales than they do on others. By comparing the ratings received by the test package to those received by another package, it is relatively simple to determine which design best meets its objectives.

Table 8 contains examples of a tool called the *Semantic Differential*. Here the circles represent points along a continuum between two descriptors. The respondent is told to put her mark on that point that best describes the product she is dealing with. These marks are then given numerical equivalents and tabulated as easy-to-compare averages.

It doesn't matter so much *which* of the association or scaling tools you use. What does matter is that you use the tool you select with sensitivity and skill. Here are several suggestions in that regard:

- *Use a standard format.* Select *one* that best meets your objectives and use it throughout your test. This will enable both interviewers and respondents to concentrate on the task at hand—and will produce a more clear-cut result. If you use the same format from test to test, you also will learn more about how it really works—and be able to make more reliable judgments based on your experience.

- *Describe consumer end-benefits.* You will produce a much more useful result if your scales concentrate on consumer or product-oriented issues like "tastes strong" rather than on package-oriented issues like "looks pretty."

- *Cover all the important end-benefits.* Go over your previous work. Talk with others on your team. Break apart general issues like "tastes good" into components like "tastes strong" and "tastes sweet." Do whatever you have to do to produce a list of benefits that covers the relevant consumer spectrum.

- Be sure to *randomize the sequence* of benefits. Don't bunch all your "taste" scales at the top of the page and all your "texture" scales at the bottom. Don't put all the "positive" words on the left and all the "negative" words on the right. Mix them up—and do what you can to make the respondent consider each one individually. This will reduce the "halo" effect and produce a more valid result.

Table 8 Semantic Differential

	This Product						
Fresh	O	O	o	o	O	O	Stale
Bland	O	O	o	o	O	O	Tasty
Artificial	O	O	o	o	O	O	Natural
Strong	O	O	o	o	O	O	Weak
Cool	O	O	o	o	O	O	Warm
For me	O	O	o	o	O	O	Not for me

- It seems obvious, but *don't crowd the benefits*. Paper is a good investment. Don't scrimp on it or you'll get mixed, blurred, or confused results.
- Don't miss out on chances to *measure your competition*. It is sometimes amazing how much you can learn about competitors merely by observing impressions respondents report taking from their packages. This information can help you in several ways, including helping you select the package that stakes out the best position to compete against them.

If you follow my suggestions, you can expect results that look like those in Table 9. Here, test package A beats B

Table 9   Mean Ratings

| | Test Package | | |
	A	B	C
For me/Not for me	7.2^{BC}	6.7	6.4
Fresh/Stale	7.5^{BC}	6.6	6.4
Cool/Warm	7.5^{C}	7.2	6.8
Tasty/Bland	7.2^{BC}	6.4	6.6
Strong/Weak	6.9	7.0	7.1
Natural/Artificial	7.0	6.8	7.2^{B}

and C on the "For me" scale. It is easy to see why. The product in Package A is perceived as offering the consumer significantly more freshness, coolness, and taste without cutting back on strength or naturalness.

EVALUATING SALES

It may be that your philosophy of packaging is, "We don't care about all that visibility and benefit stuff. Just put it in the stores and see if it sells." Well, if that is your approach, there are some suggestions even for you.

First, *make sure you use enough stores*. Your auditing firm can give you some valuable assistance regarding store-to-store and period-to-period variability. Make sure you set up panels large enough to produce a reliable result.

After determining how many stores you need, make sure you properly *match your test and control panels*. Don't test one package in Pathmark stores and another in A&P. Don't have all the large stores in your test panel and the small ones in the control panel. And don't confine the test packages to stores on the young, affluent, try-anything-new side of town. Balance your panels carefully and make sure they offer equal prospects for package success.

Take pains to *control your displays and inventories*. Don't build them beyond levels you can expect to achieve in the market place, and don't build them up in the test panel and simultaneously neglect them in the control panel. Be sure to service them frequently enough to prevent your test from being affected negatively by excessive out-of-stocks.

It only makes sense to *control other variables* as well. Don't pile special promotions, advertising, or selling efforts on top of your store test. They only make it next to impossible to read.

Allow enough time. Store tests take time and are rushed only at great peril. It may take a while for consumers to see that your new item promises more than it did before—and a while longer for them to confirm that it does or does not live up to these promises. Don't try to conclude anything from your data until the trends have been given time to stabilize.

Good store test results look like those on the top two lines of Table 10. In this case, test package A has produced a two point gain in share over a 4 month period and seems destined to move the Brand even higher in months to come.

From a testing standpoint, the results on the bottom two lines of Table 10 are also good. In this case, test package B has done nothing for the brand—and the manufacturer will probably be better off if he decides *not* to go with it.

Table 10 Test Market Share of Dollar Sales Trends

| | Four Weeks Ending: | | | | | |
	Jan. 28	Feb. 25	Mar. 25	Apr. 22	May 20	Jun. 17
Test panel A	10.4%	10.3%	11.3%	11.8%	12.0%	12.3%
Control panel	10.9	10.8	10.9	11.0	10.6	10.8
Test panel B	10.5	10.3	10.4	10.5	10.2	10.1
Control panel	10.9	10.8	10.9	11.0	10.6	10.8

IN CONCLUSION

So far in this chapter, I've made 38 specific suggestions as to things *I* think you should keep in mind while working with several standard research tools. My final eight suggestions are more general in that they apply across techniques and are stressed by most practitioners in the field:

First, *concentrate on your target consumers.* Don't do your research with your executives (who aren't typical), your employees (who are biased), or with a general audience (who don't give a darn). Do use people who are representative users of your product category and have a natural interest in your products.

Second, *research products, not packages.* Don't force the respondent into evaluating designs. She isn't an expert in either art or communications. She is, however, an expert at buying products for her family—and if you ask her about these products in *her* terms, she can be very expert indeed.

Third, *deal with one "product" at a time.* Don't ask the respondent to choose between alternatives in a way that could never be possible in the marketplace. If you do, she'll make her decisions about packages, not products—and move completely out of her area of expertise.

Fourth, *be as "normal" as possible.* Environment is important. Distractions are important. Keep things as simple, realistic, typical, and relaxed as you can.

Fifth, *research more than one dimension.* Visibility is fine, but you had better find out what it is the consumer sees in her mind when she's got your package in her hand. Image is also fine, but it doesn't do much good if you can't find it on the shelf. Neither visibility nor image will keep your product in distribution very long if it jams, breaks, or won't come out of the bottle.

Sixth, *use adequately sized samples.* Your researchers can help you with this, but make sure you have enough properly selected respondents or stores to give you statistically reliable results.

Seventh, *remain true to your philosophy.* I started this chapter saying it was important to have a well-thought-out position as to what a package is, what it is supposed to do for the brand, and what is supposed to be measured. Sometimes, after going through all that, someone will still ask you, "While you've got'em there, ask'em which one they like best." Well if that's not consistent with your philosophy, don't do it. Stick by your philosophy and your tools, and do a professional test.

Finally, *maintain good, effective working relationships.* The manager, the designer, and the researcher are all talented and dedicated to producing the right package. Talk to each other. Make sure you understand what the others are trying to do, and why. Find out what they are worried about, and how you can help them.

If you are able to do these things, you will be able to evaluate your packaging—and will provide some of the growth, efficiency, and artistic satisfaction we all are seeking.

CHAPTER TWENTY-EIGHT

The Package, Logo, and Corporate Image: A Multidimensional Research Approach

Davis L. Masten

The package, logo, and symbol should be coordinated to reach a goal: to build a favorable corporate image and make the company a success.

In many, if not most, corporations the roles of the package, logo, and brand or corporate symbol are not understood. Some executives tend to consider them mere identifying devices; others believe they are works of art, as if they are to be exhibited in an art gallery.

In fact, to the consumer, the package is the product. A consumer cannot taste a food product or test a new household item in the supermarket. The package is the manifestation of the product at the point of sale. An effective package, therefore, encourages or motivates the consumer to purchase.

The logo is a vital element, actually a major factor, on the package and has a highly significant role in building the brand or corporate image in all media. A logo is most effective if it is integrated or correlated with an effective symbol. The logo-symbol should be effective on the packages and independent of, or separate from, the package. It should characterize the brand or company because it represents the brand or corporation. A logo-symbol communicates instantly, whether favorably or unfavorably, the character of the corporation to consumers and stockholders.

The logo-symbol can express strength or weakness, progress or backwardness. It can promote trust or distrust, good will or lack of it. People are not aware of all this. The effect is on the subconscious mind.

Corporations differ on how they use their corporate logo-symbols.

Procter & Gamble uses its corporate logo on the back of the packages. Only rarely, for a new product introduction, does the Procter & Gamble name appear on the package face. Procter & Gamble has no desire to have consumers associate all of its products with the same company. Each Procter & Gamble product stands on its own, without the assistance of corporate endorsement.

General Foods has the same basic approach, except the corporate logo is used as a sign-off on all its advertising.

Borden Inc., on the other hand, has the Borden logo on the package face of all its consumer products. The main reason for this is to have consumers and the investment community recognize Borden as more than a dairy products company.

Hormel Company has some products with the Hormel endorsement on the package face and others that have the endorsement on the back of the package.

Most 3M products have 3M as the brand name or a 3M endorsement on the package face.

A package consists of several elements; each component contributes to its effectiveness. The elements are the brand name (in a logo), product name, selling copy, and color; on food packages, there is generally appetite appeal and descriptive copy. All the components have to be organized into a single or total image or gestalt. Each component has to contribute to the effectiveness of the package, and the organization of the elements, the total design, has to make its contribution.

For a new product, the first step is to get an effective name. A subjectively conceived or management oriented name does not have a great probability of success. An effective name can make a major contribution to marketing success.

When there is evidence from valid research that one of several proposed brand names will be effective, a designer is selected to produce several logo designs. This is both a creative and a research undertaking. Several proposed logo designs are put through objective, controlled tests. A symbol may or may not be integrated with the logo. Generally, it is advisable first to determine which of the proposed logos, independent of a symbol, is most effective in readability and in favorable associations. Then the effective logo is integrated with three or four different symbolic images. The logo-symbols are tested to determine which of the symbol designs upgrades the logo.

Now, the logo-symbol that comes out best in the tests is integrated into package designs. A designer generally creates from three to five distinctively different designs. Quantified qualitative consumer research is used to determine which of the proposed designs is most effective.

In many situations, the designer is provided information on colors that would be most effective—colors that are optically and psychologically appropriate for the specific product or brand.

Finally, the package is tested against competitive packages.

The preceding paragraphs outline various practices regarding the logo and package. Now I will analyze these practices.

How important is a name? The best

answer to this question is in well known cases.

The name Edsel was chosen for sentimental reason; it was a management oriented name. There were no marketing considerations in selecting the name. The name had deep meaning to the Ford family; it had no meaning to the public. It contributed much to the Edsel failure.

Dove is an example of how important a name is. Does anyone believe that the cleansing bar would be a great success if it were called Pigeon? Dove is a species of pigeon. But Dove has associations with peace and delicacy; a pigeon is a dirty bird.

Is a logo really important and how is a logo a psychological factor?

I have seen marketing research reports that showed logo designs for a luxurious cosmetic convey hardware or industrial products.

Is color really very important?

In a package test in recent years, a bright red package for antiheartburn medicine did well in the display tests (visibility, readability, eye flow or focal point sequence) but implied that it contained a product that would cause heartburn, not prevent it or cure it. This shows how important color is in packaging. There are hundreds of cases of wrong color use.

I have pinpointed problems related to new product introduction—name, logo, symbol, color, entire package, but what can be done about a package of a product that has been on the market many years and is losing ground? Reason for loss of sales is unknown, it could be the product, the package, or the advertising. Here we are concerned with the package.

First of all, we put the package through display tests—visibility, readability, focal point sequence, and we put the important competitive packages through the same kind of tests. These tests measure objectively the character-

istics of the package, the involuntary reactions of consumers; they are a means of determining how visible the package is to consumers at the point of sale, how readable the brand name and product information is, how the eyes move over the surface, and where attention is held and is not held.

The colors are rated also on the basis of accumulated research data; the ratings are of associations that consumers have with the dominant color or colors on the package and their retention in the memory.

Generally, some weaknesses and some strengths show up in the package. Recommendations are made to eliminate the weaknesses and retain the strong elements so that identity is not lost.

Complete revision is not advisable unless the product has lost almost all share of market.

For example, in the early 1950s research showed that Marlboro needed a radical change; share of market was insignificant. The completely changed Marlboro brand image and package is marketing success history. The new Standard Oil symbol—signs and packages—was a radical change. The growth of Standard Oil correlates with the introduction of the "oval with torch" red, white, and blue symbol.

To understand the role of a package, brand or corporation logo, it is imperative to understand the principle of sensation transference (see chapter 18 by Louis Cheskin). Sensation transference is the basic factor in the role of graphics in marketing. Its role in marketing is in the transference of sensation from the package to the product, or from the corporate logo to the company.

Consumers are not aware that they are affected by sensation transference (that they transfer the effect from the package to the product). Because they are not conscious of this, they are not defensive and they attribute the favorable associations with the package to the product the package contains.

Research conducted on an unconscious level shows that the taste sensation of a food changes if there is a significant change in the package.

All new designs—logos or packages—are not always more effective than the old ones. But new designs that are effective become potent marketing tools with consumers and may help security analysts recommend the stock.

When the logo was modified on some packages, sales increased dramatically because the new logo communicated to the unconscious mind of the consumer that the product was better than before.

We have also had research that showed a changed logo-symbol downgrading the brand. This research kept the client from making the change, which would have caused losses.

The incorporation of the corporate symbol on the package face has in some cases brought great increases in sales. All other factors remained the same, the increase could therefore be attributed only to the logo-symbol. In other cases the placement of the corporate symbol on the package downgraded the product and sales declined. Because there were no other changes, the introduction of the symbol was assumed to be the reason for the sales loss.

What reasons are there for placing the corporate logo-symbol on a package?

Often the chief executive wants to broaden the exposure of his corporation to consumers or to the financial community. Placing the corporation logo-symbol on a package is a valid way of achieving this objective, but it must be carried out with caution, on a product line by product line basis. If it has a negative effect on the sales of a product then the amount lost must be balanced with the corporate objective. Unfortunately, often a chief executive officer (C.E.O.) will not appreciate this prob-

lem, because he has little frame of reference for it.

Another reason for placing the corporate logo-symbol on a package is to upgrade the brand. This can be done by using the right kind of research to determine the effect of the logo-symbol. If it is effective, it should of course be used. If the corporation with the effective logo-symbol is well known, the endorsement helps the consumer identify the product. If the corporation has a good reputation, a favorable image, the recognition of the corporation inspires confidence in the product. Endorsement of products with a respected logo-symbol has been used successfully by many companies.

We have, however, often advised clients not to use such endorsement, although the corporate logo-symbol rated very high in total favorable associations. A company with associations of "high technology" and "chemicals" should not have its corporate endorsement on natural food or personal care products. A company that is known for its spicy, ethnic foods should not use its corporate logo-symbol to endorse a traditional, bland American food.

Also, a company can diffuse its corporate identity if it is spread over a great diversity of products. On the basis of research, we have prevented products from going to market with a corporate endorsement on the package face. One company that comes to mind is known for its expensive, very high quality products. Although in the short run the corporate endorsement would have boosted sales of this product substantially, in the long run it would have brought down the high quality image of the corporation; the other well-known products would have suffered, and there would have been corporate losses.

A few musts for developing a new package are the following. First, the package has to be functionally right for the product—a reliable container and easy to handle. It must be effective in display; it has to be visible in the context of competition; the logo and important points of copy must be legible; the design must have a primary point of focus, a place where the consumers eyes will fall first (ideally, this should be the brand name); other important elements should come into focus sequentially.

However, a package that is effective in display is not necessarily a good package from a marketing aspect. In fact, developing a package that will "pop off the shelf" is a comparatively simple task in comparison with developing a package that communicates and motivates consumers. We have tested many packages that marketers were excited about because they attracted attention. Many of these packages, however, never made it to the market because they were ineffective psychologically.

First, the logo on the package (and in communication media) must be readable. To our amazement, several times each year proposed brand and corporate logos that fail in legibility are sent to us for testing. If a logo cannot be read easily, the psychological effect, no matter how favorable, has no value.

It is the psychological effectiveness, however, that makes a package a potent or a weak marketing tool. The psychological effectiveness can be determined only by using the indirect method, by determining the kind of sensation transference from the package to the product. It must be remembered that the purpose of the package is to sell the product, not to hang in an art gallery.

There is a popular saying that packages that win design awards sometimes do not sell products. There is also much package research being conducted that makes design judges of consumers. Whether or not a package is aesthetically pleasing is of no importance. The package has to sell the product it contains, not itself as a work of art.

Much research is also conducted in groups. The focus group approach is widely used in market research. It is a worthwhile technique for learning consumers' language and for generating ideas about the product. But for packaging research it is worse than worthless; it is misleading. One of the main problems with focus groups is that respondents react differently in a group than they do as individuals. Also, in groups there are always leaders and followers.

For testing the relative effectiveness of several package designs even individual direct interviews are of no value. Rationally, the package is a container that protects and identifies the product. Consumers are not aware that they are psychologically affected by the package, that they judge a product by its package.

If you ask consumers whether the Morton Salt girl, Blue Bonnet Sue, a crown, or a crest has an effect on their purchasing, the answer would be "of course not." The people would consider the question ridiculous. If the Morton Salt girl, Blue Bonnet Sue, the crown, or crest were taken off their respective packages, however, sales of those products would drop substantially.

Consumers are not aware that they practice sensation transference, that the package motivates or does not motivate them to buy. You cannot ask them questions about the package. The consumers are only interested in the product. We must test the package by talking about the product. Package research should not only determine the effectiveness of a package but reveal why it is effective.

As discussed earlier, the effect of each component has to be known as well as the effect of the whole—the entire design. You cannot have an effective whole if every component does not make its contribution to the fullest possible extent.

There is much talk about qualitative and quantitative research. The fact is that qualitative and quantitative cannot be separated. Research must be qualitative, and it is not reliable unless it is quantified. Without a large sample size it is conjecture whether the qualitative difference is really different to a majority of consumers.

The corporate logo-symbol and package cannot be changed every few months as an ad campaign is changed. Consumers are reassured by the logo-symbol or package that they are getting the same product. To the consumers, the package is the product. A changed package design or logo-symbol may arouse suspicion that the product has been downgraded.

Marlboro retained the same package design for more than 20 years. Other packages, Tide, for example, have been modified but the identity has been retained; basically the Tide package is the same as the first one introduced after extensive testing in the late 1940s. It has been modified in response to competition about 20 times since its introduction but the basic "Tide character" remains.

Today's corporation executives know that for a product with a large share of market, a package change to increase effectiveness should be evolutionary. The change should attract and get new consumers without losing the old consumers of the brand.

On the other hand, a revolutionary change is indicated if a product has little share, or if the product has been losing share rapidly and research indicates that the problem is in the packaging. Some package designers do not like evolutionary change because they are unable to express themselves fully. They are restricted by the present graphics. Often designers must be reminded that a change in a package is not an opportunity for self-expression but an opportunity to increase sales.

Valid market research can give the designer direction about the strengths and weaknesses of the package (or corporate identity). With information about the weaknesses and strengths, the designer can produce with conviction and create to solve the problem at hand, which is selling.

"Corporate folklore" is a phrase I first heard used by the president of one of the major paint companies. He wanted objective research to rid his company of the "folklore" associated with his company's graphics. Some members of his staff thought the corporate graphics were fabulous; others hated it. Everyone on both sides of the issue had developed a rationale, each person "knowing" his or her position was right.

Many companies have folklore. Few executives know whether their graphics have effectiveness. This is not to say that there is a lack of design studios willing to create new graphics for corporate symbol and packages. There are many design programs carried out by some of the most prestigious design firms in the world. The problem is that many of the designs are not effective. In fact there are instances of expensive programs that have had a negative effect on sales. These design programs were implemented for innumerable reasons, most of them subjective. The point is that such mistakes were unnecessary.

Packaging folklore is disregarded by most new executives. I have seen many package changes made by new brand managers resulting in drastic decreases in sales. What generally happens is that the brand manager subjectively decides that it's time for a package change; the new package alienates the consumers and loses franchise.

The pressures on a brand manager are great. The brand manager inherits budgets for advertising, distribution, and so on for an established brand. There are few ways he can show management that he is doing something innovative. Package change becomes a means of demonstrating action. In many companies the brand manager can express his subjectivity, "knowledge," and "taste" in selecting the package. Often, top management does not interfere or shows little or no interest. "After all, the cost of a new package is less than that of an advertising campaign," said an executive of a food company.

Up-to-date companies are aware of the role of the package. Many top executives are establishing policies in which the package is given as much consideration as the advertising. They employ research programs to eliminate subjectivity. They have recognized that a valid research program is an investment, not an expense.

The role of the package has been the least understood. The package has in the past been the reason for many marketing failures. But this will not be true in the 1980s. Management demands complete information—display effectiveness tests and consumer attitude tests. The concept of "sensation transference" has been grasped by marketing executives; they know that consumers judge the product by the package and they appreciate the techniques used for determining the psychological effect of the package on consumers.

Most business executives are now aware that the logo-symbol is a vital factor. They know that the corporate logo-symbol characterizes the corporation; the logo-symbol represents and communicates about the corporation; its character is the corporation's character. The visual image is the corporate image.

The brand logo-symbol, which may or may not be the same as the corporate logo-symbol, is also an important element on the package; it is the most important element and must be a focal point, the first or second point of focus. The rest of the package design may sell the product; the logo-symbol sells the brand.

A multidimensional research approach is needed to determine the effectiveness of a corporate logo-symbol, a brand logo-symbol, a package—visibility, readability, eye flow, favorable associations, preference for the product or service.

Research guidelines are needed before designing begins, and research is needed to determine which of a number of proposed designs is most effective as a marketing tool. This means determining the weaknesses and strengths of the package or logo-symbol to make sure that strong elements are not eliminated in the redesigning. For a new product, it means providing information on images and colors that are favorably associated with the product.

The attitudes of consumers toward the product in the package, not toward the package design or logo-symbol design as a work of art, must be determined. This is where "sensation transference" enters. Only by measuring the sensation transference can the full effectiveness of the package be determined. Asking consumers which of a number of package designs or logo-symbols they like best can lead to marketing disaster. It is sensation transference that makes the package or logo-symbol such a powerful marketing tool. It is this multidimensional approach that has made packages and logo-symbols vital factors in successful marketing.

CHAPTER TWENTY-NINE

Package Design at Checkerboard Square*

Paul B. Schipke

THE COMPANY

Ralston Purina Company is a broadly based, nutritionally oriented food and feed company, with a primary emphasis in protein. The three business segments of the company are: consumer products, agricultural products and restaurants. To date each segment has operations in the United States and in foreign countries. The company is the world's largest producer of dry dog foods as well as of dry and soft-moist cat food and of commercial feeds for livestock and poultry.

Consumer Products

This business segment consists of pet foods, seafoods, cereals and crackers,

edible and industrial soy protein, mushrooms, green plants, institutional food-service and "Keystone Resort" operations. The company's principal pet food products are the largest-selling dry dog and cat foods in the world.

Agricultural Products

This business segment consists of animal and poultry feeds, soybean processing, grain trading, animal health products, breeder swine, and milk by-products operations. The company's domestic animal and poultry feeds consist of over 350 basic formulations; a variety of specialty feeds, animal health products, and

*©Ralston Purina Company, 1979

315

breeder swine products. The products are sold for consumption mostly by farm animals. Feeds are distributed in the United States primarily through a network of over 6000 independent dealers.

Restaurants

Restaurant operations consist of fast-service restaurants and specialty dinner houses. Through its commissary, the company's subsidiary, Foodmaker, Inc., processes, prepares, and packages a substantial portion of the total volume of food products sold by its fast-service restaurants.

Most of the company's products are distributed under the Ralston Purina trademark, and in addition under specific product trademarks such as Purina®, Chow®, Chicken of the Sea®, Ralston®, Chex®, Ry-Krisp®, and Country Stand®. These are identified by the Checkerboard symbol, which has been the company's principal trademark since 1900. Foodmaker's fast-service restaurants operate primarily under the name Jack in the Box®, and its specialty dinner houses operate under various names including Boar's Head®, Hungry Hunter®, and Stag & Hound®.

Throughout the years, the Checkerboard has been used in a variety of ways to identify Ralston Purina and its products and is one of its most valuable assets. It distinguishes Ralston Purina products from others and stands for excellence in food, animal feed, and health-aid products.

Today, as Ralston Purina places an even greater emphasis on the introduction of new food items, develops better agricultural methods, and invests in other expansion opportunities, it is important that the Checkerboard trademark be implemented through a consistent and unified communication system.

The Nine-square Checkerboard is Ralston Purina's corporate or "house" mark. As such, it is the key identifier. The traditional Checkerboard band may be used in conjunction with the corporate mark when desirable.

Currently the use of the Checkerboard trademarks in Ralston Purina product advertising and labeling is more flexible than uniform. Product concepts, the nature of certain print and television messages and so on, suggest that some flexibility in the use of Checkerboard identification is needed.

The original Checkerboard band played a very important role in identifying Ralston Purina products. It is quickly associated with the Ralston Purina Company and projects the imagery of animal feed and related products.

The development and usage of the Nine-square Checkerboard as Ralston Purina's corporate mark in the early 1960s has served to identify the human foods and other consumer products along with the animal feed and health care products. It readily identifies the product as a quality product from Purina.

There could be a negative aspect to a well-known corporate mark. It generally has a specific meaning in one category and it could possibly work against building unique images for individual brands in other categories. Here, the relative size relationship of the corporate mark and the product brand name becomes critical. The brand name must maintain its relative prominence in order to optimize the brand identity (personality) and recognition. For example, if the trademark is primarily associated with a specific consumer benefit, it might work against developing clear images outside of the specific benefit it is associated with.

In new product introductions, a brand might wish to capitalize on the company's strong quality trademark image or, conversely, deemphasize the corporate mark to minimize cannibalization of other brands.

A company must assess the value of its house mark and corporate signature for its individual brand needs. Each product must be allowed to stand on its own image and yet capitalize and benefit from corporate endorsement when appropriate.

PACKAGING

All packages directly identify the product as one from Ralston Purina. In certain situations where marketing objectives call for the product to be dissociated from Ralston Purina Company, the exception must have the approval of a corporate vice president and the chief executive officer of the corporation.

Unless approval for dissociation has been given, each package must include four identification elements:

1 *The Corporate Mark (Nine-square Checkerboard)*. It is company policy to apply the corporate mark to all packages, preferably on the front panel of the package.

2 *The Corporate Signature*. The corporate signature using the legal name (in addition to and separate from the corporate mark) must appear on a package at least once. The preferred position is to include it with the mandatory legal address line.

3 *The Mandatory Legal Address Line*. As required under the Packaging Law, identification of the manufacturer and the manufacturer's address must appear on the package. Ralston Purina Company's address line includes the corporate mark, the full legal corporate name, the division

designation, address copy, and the copyright line.

4 *The Mandatory Legal Copyright Line*. A copyright line must be included on all packaging first published after January 1, 1978. The copyright notice consists of the symbol ©, followed by the company name and year of publication. It is Ralston Purina's policy to incorporate the copyright line with the address line.

Optional communication elements which may be used on packaging include:

1 *The Checkerboard Band*. Because of the strong visual impact of the Checkerboard band and because it has been associated in many ways with Ralston Purina Company, the addition of the Checkerboard band as a design element on packaging is considered highly desirable.

2 *Endorsement Signatures*. The Ralston Purina name (with or without "company") may appear with the corporate mark as an appropriate enhancement for new products or established products other than Purina or Ralston brands.

Other considerations regarding packaging at Ralston Purina Company include:

1 *Communicative Signature*. When space does not permit the use of the full legal name, "Ralston Purina Company," the communicative signature, which includes the corporate mark, the communicative name, "Ralston Purina," and the proper division designation should be used on the side or back of the package. This does not replace the use of the legal name in the mandatory signature.

Approved Use of Corporate I.D./Copyright Notice

Acceptable positions of the copyright notice in conjunction with the Corporate Mark are illustrated below. A space equal to one interior square within the mark should be maintained between the mark and the copyright line. All examples are acceptable when the Corporate Mark stands alone. However, to maintain an uncluttered visual element, care should be taken to use Examples D or E when applying the Corporate Signature (Corporate Mark along with either the legal or communicative name).

The most commonly used and the preferred positioning of the signature elements has the name centered on one or two lines to the right of the mark. To accommodate different layout situations, however, the legal or communicative name may be positioned above the Corporate Mark as illustrated below. The space between the mark and name is equal to the width of one interior square within the mark.

This policy applies to print and broadcast.

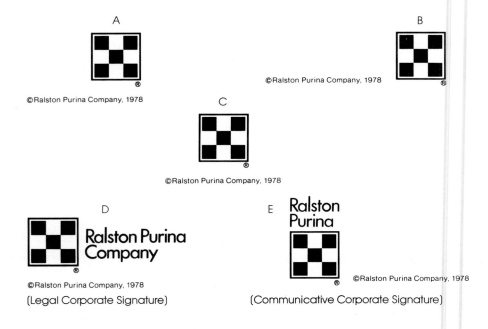

A

©Ralston Purina Company, 1978

B

©Ralston Purina Company, 1978

C

©Ralston Purina Company, 1978

D

Ralston Purina Company

©Ralston Purina Company, 1978

(Legal Corporate Signature)

E   Ralston Purina

©Ralston Purina Company, 1978

(Communicative Corporate Signature)

Minimum size for reproduction of the Corporate Mark varies with the reproduction technique employed. For most applications the acceptable minimum size is 3/8 inch square. For newspaper reproduction or rubber-plate printing, the acceptable minimum size is 1/2 inch square. Futura Medium is the approved typeface for signatures used in advertising, packaging, and promotional materials.

2   If the corporate mark and/or Checkerboard band are used on a package that is one in a series of flavors or sizes, a consistent design application of the corporate mark and/or the Checkerboard band should be used to maximize brand recognition.

3   *Approvals.* New packaging must be reviewed by the Packaging Committee consisting of representatives from Research & Development, Legal, Trademark, Purchasing, Marketing Services Control, and product management. The Chairman is the Director of Marketing Services Control. Accurate application of the corporate

identification system for all packaging changes is the responsibility of each product group. All new package designs must be approved by the identity coordinator of that operating group. This does not preclude other required approvals, such as Legal.

tion of new food items, development of better agricultural methods and other expansion opportunities, it is important that the Checkerboard trademarks be used with consistency to spearhead all of the new growth activities.

SUMMARY

Ralston Purina Company has facilities and personnel throughout the United States, in Canada, Mexico, Central and South America, Europe, Australia, and Asia. The company is eminently well positioned in people, products, and facilities. With emphasis on the introduc-

REFERENCES:

1 Ralston Purina Company Annual Report, 1977, Inside front cover and pp. 33–34.
2 Ralston Purina Company Annual Report, 1978, Inside front cover
3 Ralston Purina Company Graphic Standards Manual, ''Introduction'' and ''Packaging'', pp. 1–6

CHAPTER THIRTY

Corporate Image in Retail Store Packaging—
Testing Design
Validity in the
Marketing Environment*

Robert G. Smith FIDSA

For the well run business, whether manufacturing, retailing, or service, a professionally developed and integrated image objective will provide the rational structure around which all visual communications can be coordinated. It supplies the thread of positive continuity

*This article was developed from a broad series of marketing studies carried out by the Penney Company during which time Jay Doblin, based in Chicago, served as a consultant.

throughout every media of visual communication, from product to packaging, from company architecture to its advertising so each and every item builds on every other to improve communications effectiveness and at the same time reduce costs.

The image of a retail store, like any corporate entity, is formed by the complex of communications and interactions between the store and the public, including suppliers, government regulators, investors, and particularly the consumer, and everything she or he experiences about that store and its operations.

Retail store packaging is just one element in the chain of image/generating events. Like a link in a chain, effective packaging is neither the most important element nor the least important. It is simply an indispensable image factor within the chain of events in today's retailing world. It must be professionally executed, constantly reevaluated, and refined to meet changing market conditions. Each package must communicate those values that are both intrinsic and extrinsic to the product and satisfy the consumers' expectations within their perceptions of the particular shopping mode in which the package will be seen.

With their massive buying power, the national chain retailers have a unique opportunity to strengthen directly their corporate image objectives through their ability to control much of their basic packaging. In the past, this opportunity was largely overlooked, and packaging was often made to look similar to the design of the strongest national brand or, more likely, made so that the private label poduct would look less expensive. This was a business approach reacting to whatever leadership the national brand manufacturer produced, rather than a plan to capitalize on the use of private label merchandise to project a carefully conceived business plan and correlated image.

The use of post packaging (what the retailers apply at the point of purchase, such as wrapping paper and boxes), however, has long been used to reinforce the store image even though the merchandise itself might be of a national brand. The post packaging can add a substantial dimension of credibility and enhancement. For instance, an LCD watch boxed in a Tiffany gift package, a designer belt in a handsome Joseph Magnin box, or imported glassware in a

Crate & Barrel shopping bag are typical examples of excellent use of post packaging to build the image of the store and enhance the satisfaction of the customer at the point of purchase.

The retailer's concern about the imagery of the national brand packaging, over which it has no control, usually would be how the sales appeal, quality, and fashion impact of the particular brand would strengthen the image being developed for the store, or chain of stores. It might be very appropriate to feature the Pierre Cardin label, but not an anonymous name from Hong Kong. Copco, Dansk, or Cuisinart are names in retailing whose packaging will add to the authenticity and quality image of a gourmet specialty store. On the other hand, the same shelves lined with national brand products like Mirro or Pyrex, although these products are of very good value, would be neutral in supporting a "gourmet" shop image. National brand packaging is not exclusive to one retailer; however, it is obvious that the particular mix of national brand merchandise and its presentation at the point of sale does indeed have an important effect on the image of the individual retail establishment.

The ability of the national chain retailer to purchase a product to its specifications, thereby controlling its packing to help communicate its marketing strategies, is an important strength. The effectiveness of designing packages toward an image objective and its effect on company profits is directly proportional to management understanding of the part image plays in marketing.

To understand the corporate image in retail packaging better, it is important to understand the basic retail shopping modes, because the attitude of the consumer is quite different in each, and therefore what may be effective packaging in one mode may be quite inappropriate for another.

Before the 1900s, practically all of what we call today "department store merchandise" was sold in shops. The merchandise was produced in the back room and sold at the front. These shops, whether the milliner, the tailor, or the shoe maker were scattered from one end of main street to the other. (See Fig. 1.)

A consumer oriented integration of mutually supportive shops or merchandise specialties under one roof came into being with the advent of the department store around 1900. Macy's, Gimbels, and Marshall Fields are some examples. With the introduction of the supermarket concept during the 1930s, the integration of a different variety of shops for greater customer convenience (in this case, specializing in consumable goods, combining the bakery shop, the dairy, the butcher, the grocery shop, etc.) came into being. In point of of view and reason for being, the supermarket related more to the discount store (which appeared on

the scene in the 1950s), where cost and convenience are paramount, than to the department store, where service and fashion are the major factors. When the discount store appeared in the 1950s, it offered the customer a quick, convenient, "price off," no service, approach to distributing a large selection of *national brand* merchandise. The discounter appeared all across the country, pulling much of the hard goods business away from department stores and shops. As consumers became more affluent and more demanding in the late 1960s, they moved to the suburbs, adopted new life styles, and new values; they caused major crossovers in the type of merchandise sold and services offered. A realignment occurred: shops, departments stores, supermarkets, and discounters were regrouped into malls, dealers, servicers, and marts. This happened because certain merchandise is sold best in one kind of retail environment and not in another.

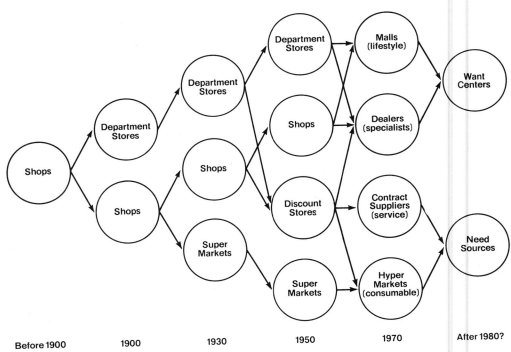

Figure 1   Chronology of retailing in the United States.

Discounters, hypermarkets, national chains, catalog houses, shopping centers are all part of the changing retailing scene, which will continue to metamorphose in the future as TV shopping and the mounting cost of energy and other technological and social changes affect our shopping patterns.

The kinds of merchandise sold in existing or future retailing concepts, however, all fall into two basic shopping modes, "want shopping" and "need shopping." These modes of shopping can be more clearly demonstrated when *wants* are subdivided into two groups that for convenience can be discussed as "life style" and "specialty" merchandise, and when *needs* are subdivided into "service" and "consumable" merchandise. It is obvious that the attitude of the customer when shopping for "wants" is quite different from his or her attitude when shopping for "needs". For example:

1 *Lifestyle Merchandise* Those items that provide social satisfaction and esteem such as fashion apparel, home furnishings, and leisure. They add to our quality of life. We can live without them, but life would be less enjoyable and fulfilling. Because of their symbolic overtones and wide price spread, shoppers usually compare before buying. The mall or regional shopping center is a natural evolution of the department store concept, which provides the ambience for life-style shopping. From one large building, they have developed into a strategically located complex of life-style merchandisers and services; some contain over 2,000,-000 square feet of air conditioned space and encompass as many as four major department stores and 200 other stores.

2 *Specialist Merchandise* Those vocational and avocational items that

bring added satisfaction to one's life and strengthen one's self identity. It includes products like sporting goods, cameras, and hi-fi equipment. These are highly performance oriented and require trained sales experts and specialized assortments to satisfy the new customer. Herman's for sporting goods, the Radio Shack for electronics, or Sherwin-Williams for home decorating are some of today's examples.

3 *Service Products* Those products that perform needed functions or services in the home—built-in kitchens, baths, heating systems, and so on. These products were sold first in department stores, then in appliance stores and discounters, and today there is a trend to builder or contract installation. Washers, air conditioners, kitchen appliances, and so on are becoming a basic need, another function of the house, like the furnace and hot water supply. Any brand will do, as long as it is reliable and good service is assured.

4 *Consumables* The sundry products necessary to keep us alive and functioning and purchased on a daily or weekly basis (e.g., food, beer, light bulbs, drugs, film, gasoline, and packaged soft goods). The consumer attitude when shopping for consumables is one of performing a no-nonsense chore. We buy them by the measure, rather than the style. Stores specializing in this category of merchandise are primarily the discounter, supermarket, variety store, and drug store. The newest and biggest we call marts, for example, K-mart, Target, and Pathmark. The new hypermarkets, like Jewel Grand Bazaar of Chicago, carry in addition to the usual consumables like food, toothpaste, or panty hose, many products associated with the discounter (e.g., small appliances, radios, bikes, and

toys), many of which have crossed over from being life-style items to commodities and find ready consumer acceptance in this new environment.

It is obvious that the packaging for each of these groups will require different emphasis. Each category of "want" merchandise and "need" merchandise includes nonbranded merchandise, national brands, and private labels. To research the effectiveness of the package properly, it is important to note how the packaging may be perceived by the consumer within the particular retailing environment. For instance, panty hose, which started out many years ago as a life-style product sold only in department stores, has shifted in the eyes of the consumer from a life-style product to a basic element of apparel. By the late 1960s it had become a commodity product that, when properly packaged and presented, as Hanes did so expertly with L'Eggs, changed the nature of shopping for this product from a full service life-style event to one of convenience. The professionally executed program pulled much of the business away from the department store with its costs of doing business to the discounter and supermarket with their efficient self-service concepts of retailing.

There are many ways in which packaging can function; however, its role must be evaluated for each distribution method. Another useful way to view retailing as it relates to packaging is a scale (See Fig. 2) starting with purveyors of need products or commodities on the left hand side and moving to sellers of pure want or appearance merchandise on the right-hand side. Your local AA Electrical Supply is an example of the retailer of pure need or performance products. No four color packaging, dramatic lighting, or luxurious carpeting, just specified products located by a salesman in work clothes. As we cross the scale, the customer is seeking more and more "wants" and less "need." A $200,00 diamond ring from Cartier would be an example of 100% appearance, or "want," merchandise; in this case the product is no more important than the ambience in which it is seen by the customer. An appropriate package would bear little resemblance to one for a set of electric fuses.

Just as the retail shopping mode can be segmented to help understand how the roll of the package functions, it is equally important to examine different consumer viewpoints. Consumer stereotypes have been studied and charted in many ways. One of the more familiar charts is made by listing from top to bottom, upper class, middle class, and lower class and on the horizontal axis old-fashioned to new fashion. In this relationship the mass consumer is central, with the elite, the superstars, kooks, and old fogies in the extreme corners.

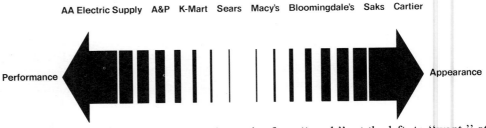

AA Electric Supply   A&P   K-Mart   Sears   Macy's   Bloomingdale's   Saks   Cartier

Performance   Appearance

Figure 2 Location of retailers on a scale ranging from "need," at the left, to "want," at the right.

Overlays can correlate the products they buy, the cars they drive, the periodicals they read, and the kind of clothing styles to which each will have an affinity and so on.

When shopping for "life-style" merchandise, the customer strikes a compromise between fashion and price within the context of individual self-image.

When shopping for "specialist" merchandise, however, performance is the most important product characteristic and must be balanced against the price. There is little market for low price, low performance merchandise. The ideal would be high performance merchandise at a low price.

Today's "service" equipment, including such products as refrigerators, washing machines, and air conditioners, is generally of an acceptable level of design and performance. High reliability in both the retailer and the product is therefore a critical index to value in the consumer's mind and must be evaluated against the price.

When shopping for "consumable" merchandise, the ideal tends to balance convenience with low price.

To illustrate these ideas, I will use some of the examples of the packaging and labeling within the framework of the JCPenney Design System. This is a concept developed to provide continuity over a wide spectrum of packaging and other visual communications media based on the corporate objective. The national chain merchant primarily develops its merchandise mix to the tastes of middle America with adjustments to meet local needs in both taste and affluence.

There are crossover areas that are in a constant flux as customer preferences change, as status products become consumables, and so on. The three major national chain retailers (to varying degrees) offer categories of merchandise in each shopping mode. All the merchandise is aimed at middle America, the standard of merchandise is raised as consumer affluence and sophistication demand. None of the merchandise would be at the ultra extremes, neither ultra fashionable nor ultra stodgy.

In the mode of "life-style" shopping the merchandise generally must speak for itself. As mentioned previously, because of the symbolic overtones and wide range in price, the shopper usually wants to compare carefully different styles, quality, and price points before making a purchase. Packaging for life-style merchandise is primarily labeling which should be designed and executed to enhance the quality and status of the item. The labels should provide quickly clear information including special status symbols such as famous designer identity, fiber content, and so on. This type of merchandise packaging, per se, is primarily of importance as a postpurchase enhancement, in the form of shopping bags, gift wrappings, and boxes (Fig. 3). In the most interesting and prestigious stores the well lighted and displayed merchandise dominates the presentation. Illumination can be compared to stage lighting, which makes

Figure 3   Postpurchase packaging (JCPenney)

colors sparkle and creates an ambience of excitement and enjoyment.

For "specialist" type merchandise the packaging becomes an important sales tool. Well designed packages are essential to project the image of technical skills and planning, which are intrinsic to specialist merchandise. The package not only visually projects the intrinsic and extrinsic values of the merchandise, it is an important link in building the image of the manufacturer (or in the case of the chain merchant, the seller). This binocular packaging (Fig. 4) and electronics packaging (Fig. 5) are suggestive of these qualities as they apply to one chain retailer. The images projected by this packaging are quite different from those of life style, service, or consumables. They illustrate the concept of the thread of continuity across a wide variety of products consistent with the corporate thrust, yet each has an integrity within the mode of shopping.

For service merchandise such as washers and dryers, packaging serves no other purpose to the customer than to ensure safe delivery. For the supplier and retailer, it can have the added functions of easy identification and stock control within the warehouse, and the package acts as a billboard for the retailer's name as the product is being delivered to the customer's house.

The packaging of commodity type products such as slow cookers, bag sealers, and toasters (Fig. 6) can serve to

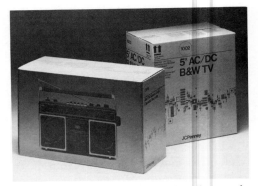

Figure 5 "Specialist" merchandise packaging (JCPenney)

enhance the product itself and, as part of a product line, have a considerable impact on other items of the line and the store itself. The decision to purchase this kind of merchandise depends on seeing and handling the product and feature identification when backup stock is stacked or on adjacent store shelves.

In consumable products the importance of the package is critical: for many products from milk to automotive supplies, from paint to sweat sox, the package sells the product (Figs. 7 and 8). It is the final interface between the customer and the producer.

The package must first attract the customers' attention. Its capacity to flag customers and interest them enough to examine the package further is obviously of critical importance. In this category

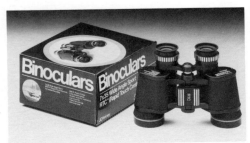

Figure 4 "Specialist" merchandise packaging (JCPenney)

Figure 6 "Commodity" merchandise packaging (JCPenney)

Figure 7   "Consumable" products packaging (JCPenney)

of merchandise, the synergistic effects of convincing copy and design reach their most rigorous test. The store presentation is usually neutral, with great quantities of shadowless fluorescent lighting, which is very effective for the no-nonsense chore of "need" shopping.

These ideas are not intended to solve individual packaging problems. They may be useful, however, to understand better the many roles packaging must perform and that what may test superbly in one situation may be inappropriate for another. For the retailer, this overview may suggest the importance of focusing beyond the simplistic battle cry of "maximizing profits" toward achieving this by a carefully conceived marketing strategy that can be communicated to the consumer through a corrolary image objective.

Figure 8   "Consumable" products packaging (JCPenney)

The image of the retailer that the consumer perceives is a homogenized potpourri of everything that is seen or experienced about the store and its operations. Psychologists tell us that about 80% of these perceptions are communicated visually. The various qualities identified in this chapter are only some of the many impressions blended in the consumers' minds to form their image of the retailer and the merchandise offered. The importance of this image, especially in determining in which store the consumer will expect to find the merchandise they are looking for, is suggested by market research findings indicating that in some product categories over 70% of the purchases are made in the first store shopped. For instance, the customer who is in need of replenishing their household supplies will form a definite image of the kind of retailer they are looking for, and then the particular store whose image suggests they will find this kind of merchandise quickest and best. In this shopping mode even the most skillful packaging design is not going to move this customer to buy a specialty product—a Nikon camera, for instance, or a life-style item like a silk blouse. On the other hand, the entrepreneurial retailer and manufacturer must be constantly alert to the fact that how an item of merchandise is perceived is constantly shifting in the eyes of the consumer. What is a luxury or status item in one time frame will very likely become a necessity in another as the consumer becomes more affluent, better educated, and more discriminating. Certainly the marketing of L'Eggs is a prime example. Calculators, LED watches, gourmet foods, and wines are others and the list expands by the day.

Design appropriateness is the sum total of intrinsic qualities plus extrinsic qualities as perceived by the consumer (Fig. 9). Performance, durability, safety, ease of use and maintenance, status versus price, age, sex, and formality are all intrinsic values developed in the production process but perceived by the customers in the look of the product, its packaging, and peripheral merchandising support. The consumers' perception is influenced equally by the extrinsic factors such as the pricing and return policies of the Company, the friendly smile of a salesperson, the product presentation, store location, that is, by all parts of the distribution process.

The product sold by a discounter will have a different extrinsic value than the same product sold in a store like Bloomingdales even though the intrinsic qualities might be identical. How the product is presented, its location within the store and the store's location within the community, its operation and activities all have an external but very real influence on the consumer's perception and acceptance of a product.

Carefully developed company marketing strategies that are communicated objectively to the consumer through a coordinated program of innovation, corporate design, advertising design, packaging design, product design, and all other visual communication media are vitally important factors in successfully selling merchandise today. They will be increasingly important in the future. Like the professional football teams that make the Super Bowl, there is no mystery in winning. It depends on the validity of overall strategy and the excellence of the individual team members' performance in each and every play within this strategy. The professional makes

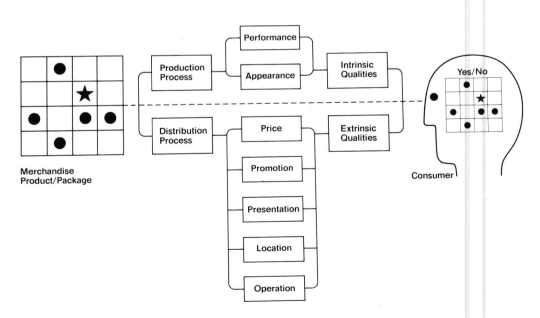

Figure 9   Consumer's perception and decision-making process

sure that his skills and actions coordinate with (and add to) the performance of his team as a whole.

For the retailer, manufacturer, or service business the professionalism of the design, coordination, and implementation of its packaging within the company image strategy is an important link in the marketing process that is pivotal to long term success. This is the context within which retail packaging must meet its test of validity.

CHAPTER THIRTY-ONE

Structural Package Design Research as a Corporate Service

Richard S. Ernsberger, Jr.

Corporate Packaging Research and Development has the responsibility for the conception, design, and development of innovative packaging systems and products possessing proprietary technology. The ultimate goal of a development effort is to support the realization of business opportunities. Following a brief comment on the delineation of the packaging research function, suppliers versus buyers, an overview of the organizational commitment and its impact will be provided. Next, specific functional roles satisfying various corporate needs will be highlighted. The ensuing remarks will

illuminate the packaging design, research and development function from a supplier's point of view, and articulate its role as a corporate service.

PACKAGING RESEARCH: SUPPLIERS VERSUS BUYERS

The character of the packaging design function can vary widely depending on the diversity and depth of the corporation's product lines and marketing strategies. Packaging organizations that supply products to the marketplace as well as those that buy these products both support packaging development activities. A packaging research group contained within an organization whose product lines represent packaging supplies as the primary product can expect a relatively high level of corporate support and controlling authority to pursue its development endeavors. In contrast, a group situated within an organization whose product lines employ packaging not as the primary product but as a product component can anticipate prejudicial differences influencing the functional control and administration of its project activities.

MANAGEMENT COMMITMENT

With respect to packaging suppliers, organizational commitment by higher management is a primary motivating element associated with a highly productive and successful packaging research function. Management's commitment will be reflected by:

- How the organization is structured.
- Project management skills.
- Direct management skills (specific tasks and activities).

With minor exceptions, varying the degree of management commitment will result in a similar variation of performance. In this sense performance is generally defined by:

- The time required to accomplish the project objectives.
- The amount and effective utilization of resources required for project execution and follow-through.
- The quality of the resultant product

as evaluated against the project objectives.

PERSONNEL RESOURCES

A second element responsible for the impact of the packaging research function lies with those personnel resources internal to the group. The level of job related skills, talent, and experience that can be brought to bear on research activities and the management of these resources is central to the success or failure of the group's performance. As with management commitment, a range of successful productivity dictated by the project selection process and timing will exist.

It can be expected that a packaging company with a high level of management support and possessing the human resources requisite to satisfying its objectives will continually bring to the marketplace profitable new products. Such an organization will be recognized as a clearly aggressive and innovative company. On the other hand, the lower end of the scale will be inhabited by a poorly organized group that is corporately sanctioned to function in response to internal problems and short term competitive developments. This latter scenario profiles a purely defensive and inhibited development function that is completely reactive, as opposed to being proactive. It can be expected that with this attitude new product developments will be at best short lived, of small impact, and infrequently experienced. It should be noted as well that the root of those exceptional successes will usually be found in an intelligent, aggressive, and creative individual whose perseverance, self-initiative, and vision will guarantee ultimately his loss from the company. This loss, of course, will bring back into balance a continuance of new product nonproductivity.

Within this range exist many combinations, each differing by level of corporate commitment as well as by the level of personnel resources contained within the packaging group.

MEASURING GROUP PROFICIENCY

How is the proficiency of a group determined and compared with contemporary and competitive groups? Perhaps an independent industrial survey or corporate sponsored investigation should be conducted to establish a yardstick for a competitive ranking of comparable packaging R&D groups. By identifying and enumerating R&D groups by their depth of experience, breadth of skill areas, degree of creativity, and industry recognition, new product development performance within the industry could be quantified. This system would be based on subjective evaluations, but if applied without prejudice, a fairly realistic and valuable planning aid would be produced. This evaluation technique would illuminate strengths and weaknesses.

These two components—management commitment and personnel resources—differ in many ways yet are nevertheless mutually dependent and both necessary in formulating the health, well-being, and success of the packaging R&D function. Developing and nurturing the improvement of the corporate commitment is a long range, slowly maturing, highly political process as opposed to that of enhancing the personnel resource level of the group. Although much can be discussed about corporate commitment, philosophy, and policies, our primary focus here is on the packaging function and the role it plays as a corporate service. Therefore, the following comments will be concentrated in that area.

THE IMPORTANCE OF A DATA BASE

Packaging R&D with its responsibility for the conception, design, and development of innovative products must have and must maintain a foundation of current and reliable knowledge relevant to the organization, the industry it serves, and the disciplines it employs in pursuit of its development objectives. This data base may be separated into two categories that include information on the state-of-the-art for technology as well as information pertaining to the organization. The state-of-the-art category will be generally self-maintaining because it is a natural function of the professional working environment with daily exposure to industrial information to stay up to date. Of course the level of competence here will vary according to an individual's interests and activities.

The second category of information, which encompasses that of the organization, is felt to be of more critical importance. The knowledge, understanding, and appreciation of organizational goals, strategic planning, product lines, and marketing approaches are inherent to and most important for a successful design function. Although state-of-the-art is considered self-maintaining, especially with an active group, adherence to informational currency pertaining to the organization requires the application of an above average level of effort. This is due to the dynamics of an organization and the proprietary restriction from, or accessibility to, accurate and complete information. Significant and meaningful information must be constantly probed for and verified. Proper maintenance and effective utilization of these data will help optimize and ensure the timely execution of project planning and implementation. Further illumination of this area and its effect on product development should become apparent and be more appreciated as development approaches are discussed.

PACKAGING RESEARCH IN THE ORGANIZATIONAL STRUCTURE

Multiple tiers of packaging research frequently exist within the corporate structure. Within larger corporations there exist business groups usually divided into divisions. These business groups are formed along similar product lines usually identified by:

- Product function.
- Manufacturing process.
- Customer orientation.

The individual divisions provide the front line profit and loss responsibilities. Their responsibilities focus on specific product and product line segments. At each level of this typical organizational structure (corporate—business group—division) there are associated packaging R&D entities. That is, a research group can be found at the division level as well as at the higher business group level. Also an R&D department or division may be located at the corporate level. Typically, the definition of each group's role will be determined by the amount of resources placed at the disposal of each group and the length of the average developmental activity. The divisional research group is usually chartered for *short term development* work (0–1 year dedicated to their specific product lines) as well as *technical service* activities. At the business group, or next higher level, the research function will involve itself in more of the *intermediate term development* activities (1–3 years) with a wider scope of interests. At this level more resources are brought to bear against increasingly complex, innovative systems development oriented around the business group's numerous product

lines. Research excursions at this level will concern themselves with primary processing as well as primary manufacturing and product systems development.

Corporate R&D generally assumes the responsibility for:

- *Exploratory Research* Investigations that will foster the successful discoveries of materials, processes, systems, and products not yet envisioned, for tomorrow's product development applications.

- *Long Term Research and Development* Projects (3–10 years) where the proper mix of personnel and resources are formally combined and programmed to produce a technologically superior process or product systems design, hopefully one that will maximize the market potentials of tomorrow.

- *Technical Development Support* Support to the business groups where the corporate level research resources (expertise, equipment and capacity) usually will exceed those of the lower levels.

Whether the organization contains one research and development group or many, the general avenues of approach taken to fulfill project planning, implementation, and the ultimate satisfaction of the objectives closely parallel one another. Approaches will vary depending on the latitude of the development activity. One end of the range will incorporate a scope of involvement encompassing all the research and development stages. Recognized phases of the product development cycle would begin with scientific research and continue on through exploratory research, advanced development, engineering development, and commercialization of the product system. Table 1 provides a definition of each of these research phases. The other

Table 1 Definition of Research and Development Stages

R&D Stage	
Technical support	Technical effort using existing knowledge and procedures to solve problems in existing products and/or processes and analyze complaints.
First commercialization	Perform detailed effort (design, procurement, installation, tests, etc.) required to bring an approved proposal into commercial operation.
Engineering development	Technical effort directed toward the engineering and economic evaluation of products, processes, and systems for commercial use. This effort involves construction and operation of demonstration units in commercial environments and/or trials on production equipment.
Advanced development	Experimentation and engineering effort directed toward proof of a process, product, or system concept, rather than development for commercial use. The work usually involves use of working models (i.e., pilot plants, product samples) and provides information on which to make an assessment and develop a judgment.
Exploratory research	Study and experimentation directed toward identification of business opportunities with important technical dimensions with the objectives of identifying issues and technical feasibility.
Scientific research	Study and experimentation directed toward increasing knowledge and understanding in those fields of physical and engineering sciences related to long term company needs. It provides a basis for innovation opportunities and development as well as fundamental knowledge for solution of problems arising from commercial operations.

end of the range, of course, would be responding to short term problems and performing technical support—fire fighting type activities.

APPROACHES TO PACKAGING SYSTEMS DEVELOPMENT

Approaches to project endeavors vary by their intended purpose and generally are profiled by their sponsorship, scope of development activities, period required for accomplishment, and the amount of resources needed to bring about project completion. Such approaches may be categorized by the following:

- Exploratory long-term product-systems development.

  1 Identifying a market need.
  2 Identifying a corporate need.
- Cooperative product-systems development.
  1 Identifying a customer need.
  2 Responding to an internal support development request.
  3 Responding to a customer support development request.
  4 Responding to a business support development need.

The exploratory long-term development approaches are usually generated by the research function, as opposed to the corporate development activities that come from within and are initiated by either the business groups or by their customers. Tables 2 through 7 provide a general check list of activities generic to each of the above approaches.

Table 2 Exploratory Product/Systems Development—Identifying A Market Need

- Maximize current state-of-the-art of technology, detect economic and market trends.
- Identify potential new product markets.
- Survey potential markets.
- Define market needs/product requisites.
- Qualify new product/market potential.
- Qualify level of commitment/resources to support new product development.
- Organize, plan, prepare, and present program development proposal for concurrence and approval.
- Establish reporting relationships and communications network.

Table 3 Exploratory Product/Systems Development—Identifying A Corporate/ Business Need

- Maximize knowledge of corporate/business product lines, strategic plans, and marketing approaches.
- Identify potential new products/market opportunities.
- Survey potential markets.
- Define market needs/product requisites.
- Qualify new product/market potentials.
- Qualify level of commitment/resources to support product development.
- Organize, plan, prepare and present program development proposal for concurrence and approval.
- Establish reporting relationships and communications network.

Table 4 Cooperative Product/Systems Development—Identifying A Customer Need

- Identify the need/problem/issue.
- Define the problem.
- Qualify the level of commitments/resources to support product development.
- Organize, plan, prepare, and present program development proposal for concurrence and approval.
- Establish reporting relationships and communications network.

Table 5 Cooperative Product/Systems Development—Responding To An Internal Support Request

- Identify, qualify, and screen support request.
  Does this request really need support?
  Level and intent of resource needs.
  Long or short term need.
  Can the requesting source do it themselves?
  Management support.
  Anticipated long term, significant business impact.
  Are the resources available to conduct the project activity?
- Define product/program development needs.
- Establish the level of commitment/resources required to support product development.
- Organize, plan, prepare, and present new product/program development proposal for concurrence and approval.
Establish reporting responsibilities and communications network.

Table 6   Cooperative Product/Systems Development—Responding To A Customer Request

- Screen via business relationships.
- Identify and qualify support requests.
    Does this request really need support?
    Level and intent of resource needs.
    Long or short term need.
    Does the business group have the capacity to perform the development request?
    Management support.
    Long term, significant business impact anticipated.
    Are the resources available to conduct project activities?
- Define product/program development needs.
- Establish the level of commitment/resources required to support product development.
- Organize, plan, prepare and present product/program development proposal for concurrence and approval.
- Establish reporting relationships and communications network.

PROJECT DEVELOPMENT ACTIVITIES

Packaging systems development projects are usually tightly focused on market targets. With any given package research project, activities generally start with an investigation of the various packaging concepts (beginning literally with the raw materials that are supplied to form the packaging materials), proceed through printing and other converting operations, and follow through to the design, development, and construction or modification of the mechanical systems that will support the total packaging systems concept. More specifically, once viable design candidates have been determined, handmade samples will be produced, evaluated, refined, and sent to marketing personnel or customers for their approval. Subsequent to these ap-

Table 7   Cooperative Product/Systems Development—Responding To A Business Need

- Identify, qualify and screen support request.
    Does this request really need support?
    Level and intent of resource needs.
    Long or short term need.
    Can the requesting source do it themselves?
    Management support.
    Long term, significant business impact anticipated.
    Are the resources available to conduct the project activities?
- Qualify business product/market objectives and plans.
- Define new product characteristics and prerequisites.
- Identify alternative new product concepts.
- Evaluate and recommend alternative new product concepts.
- Establish the level of commitment/resources required to support product development.
- Organize, plan, prepare and present new product/program development proposal for concurrence and approval.
- Establish reporting responsibilities and communications network.

provals, further refinements will usually be made after which sample quantities will again be produced for more involved package testing. When the packaging concept has been fairly well confirmed, design and development activities will begin to produce the mechanical system that will perform the form, fill, and sealing processes of this package concept. In an effort to provide a better understanding of this flow of events, examples follow that are associated with each of the two major avenues of approach—exploratory long-term, and cooperative, product-system development.

As an example of an exploratory long-term product-system development effort, a few years ago as packaging material prices began to increase significantly, the aspect of developing an alternative packaging system to metal cans, blow molded plastic jars, and glass bottles was preliminarily surveyed, resulting in the justification and initiation of research activities toward that end. Primary target markets included perishable processed food products with applications for both institutional and consumer packaging.

The preliminary survey of potential markets indicated applications for this concept in the following product areas:

- *Condiments* Salad dressings, mayonnaise, mustard, vinegar, acid-packaged pickles and relishes, sauces, and catsup.
- *Food Spreads and Syrups* Peanut butter, honey, seasoned butter, margarine, drink concentrates, and fountain syrups.
- *Powdered Foods* Soluble coffee and tea, drink mixes, powdered soups, and dehydrated potatoes.
- *Cosmetics and Toiletries* Shampoos, rinses, bath oils, mouthwashes, creams, and lotions.
- *Household and Industrial Chemi-*

cals Waxes, polishes, cleaners, insecticides, disinfectants, dry cleaning agents, solvents, soaps, and detergents.
- *Drugs and Pharmaceuticals.*

As can be seen, a fairly large potential for a successful alternative packaging concept would prove to be most profitable. Through further investigations it was determined that the key development objective would be a system design that would automatically form a lightweight, thermoformable plastic barrier liner inside a folding carton as an in-line step using production machinery. By satisfying the development objective, this packaging concept would overcome some of the disadvantages of competitive container systems such as rust, bent chimes, poor graphic treatment, breakage, excessive storage and shipping space requirements, limited access through the top opening, disposability, and the relatively large amounts of energy required to produce glass and metal containers. Furthermore, it was felt that this packaging concept would maintain product integrity and offer equivalent shelf life. Additionally, it would be a universal package utilizing a high quality paperboard material for strength and graphics, along with a formed plastic liner to meet the barrier requirements.

Using a laboratory thermoformer, square wooden forming mandrels were designed by means of which a plastic sheet was thermoformed into a folding carton for testing. The folding carton provided structural support to the thermoplastic barrier material during the thermoforming process and during subsequent package handling and distribution. The combination of carton and thermoformed liner produced significant advantages over existing thermoformed packaging. Prototype containers were evaluated for physical and barrier properties as well as rough handling characteristics. The mechanical system that performs the form, fill, and sealing processes was also developed during this period, resulting in a prototype piece of equipment. This equipment was designed and has produced test sample quantities of these containers in support of the systems evaluations. All in all, this packaging concept appeared to be considerably less expensive than the glass container and approximately equal to the blow molded bottle. (See Fig. 1.)

In summary, this packaging concept (which took 3 years to develop) provided the following advantages:

- A relatively small amount of high-barrier plastic is used in this design compared to an all-plastic package made of the same material.

- The folding carton of almost any design offers the structural strength of the combined package, while the plastic is required for moisture barrier protection, odor barrier, wicking protection, and top heat-seal closure properties.

- The plastic materials offer barriers that can be tailor-made to the customer's product requirements. Depending on the shelf-life requirements, the plastic can provide excellent MVTR and/or oxygen barrier properties.

- Full, overall six-color graphics on the carton board offer the customer a marketing image not possible with can printing, bottle labels, or plastic litho printing.

- Round glass jars and metal cans consume 33% more space and weigh more than a rectilinear container of this concept of equal volume; incoming empty glass and tin containers require expensive handling and warehousing space prior to filling, which this concept eliminates.

- Glass breakage problems are eliminated.

- This design offers superior disposability when compared to tin or all-plastic.

- Depending on the product and storage conditions, a year's shelf life may be expected.

By maintaining a currency in the state-of-the-art of technology, a market need was identified and defined and the systems concept qualified against which successful execution of development activities brought the concept to the point of commercialization (see Table 2). This packaging concept is currently under evaluation by major food processing companies.

The second avenue of approach for a package design activity is that of co-operative product-systems development. In the early 1970s, the packaging research group was requested by a leading fast food corporation to develop a completely disposable sauce dispensing package for use in preparation of one of their entrees. The sauce was produced and packaged by an intermediary company and shipped to the fast food service outlets in No. 10 cans. The sauce was removed from these cans and placed in a reusable polyethylene dispensing bottle. After inserting a plastic plug into the bottle base, it was then placed into a

Figure 1   Uni-Form Container

stainless steel caulking gun. Each time the gun trigger was pulled, ⅓ ounce of sauce was dispensed onto the sandwiches through a diffuser in the bottle neck. Approximately 3 man-hours per day per store were spent filling, removing waste from the hand filling operation, and washing these plastic bottles. The cost of the new package was determined by factoring in the cost of the No. 10 can and the dollar value placed on the labor savings.

It was decided to design a package that would fit the caulking gun, rather than to replace the whole system entirely. During the first 6 months of development, materials were selected and test designs were finalized, actual shipping tests conducted, and in-store-use tests evaluated. It should be noted that one of the significant contributions to this container design was the development and incorporation of a paperboard/PE/Foil composite. This material successfully satisfied the strict barrier requirements of the package.

The initial concepts were assembled and presented to the corporation's management group where tentative approval for the basic design was received. (See Fig. 2.) Upon approval, 5000 handmade packages were assembled and filled with sauce at the intermediary's plant location. These filled dispensers were packaged 24 per handmade shipping case and shipped to selected stores in the East, South, Midwest, and West Coast. In addition, a 90-day product shelf life study was conducted, the results of which were positive.

The next phase of the project was to investigate machinery suppliers to custom-build the equipment needed to assemble the package. One company was selected to build the tube body forming equipment while a second organization was selected to design and have the assembly machines built. A total of six tube formers and six assembly machines

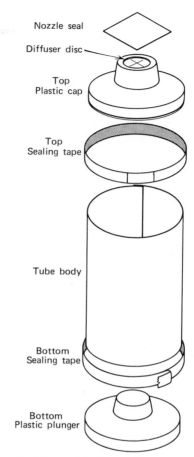

Figure 2 Disposable Sauce Dispenser Package

were required to satisfy the initial production quantities criteria.

Today quite a few more systems are in operation supplying economical and efficient disposable sauce dispensers that have significantly improved what once was a very costly processing stage of the food preparation cycle. In addition to corporate packaging research involvement in this project, several other research groups within supporting businesses and divisions were involved in the systems development and the production start-up activities.

Additionally, several of our divisions were also involved in the manufacturing phase. One division produced the foil laminated blanks for the tube body; another was responsible for the production of the plastic injection molded caps, plugs and diffusers; a third produced the corrugated shipping cases.

The preceding exemplifies a project that was initiated by a customer request which required cooperative (packaging R&D—business group—customer) interfacing and coordination that successfully defined the problem, and planned and executed development activities for its resolution (see Table 6).

The preceding examples, one of exploratory long-term product-system development and the latter, a cooperative research project, are generally illustrative of major development activities. Both took approximately 3–4 years to bring to fruition and cost 200,000–500,000 dollars.

On the other end of the scale is the short term technical development support type activity that may take as little as 3 weeks of part time involvement and cost 3000–4000 dollars. Such was the case when a request was submitted by a divisional research group to help support the development of a system designed to package hermetically a corrosive metallic product. The customer was interested in replacing the current metal container primarily because of the increasing costs of that system. As a result of an investigation of the product, its process, and the markets it serves, a combination polylaminated corrugated packaging concept was designed, developed, and tested successfully. This new concept required elementary sealing equipment which was easily incorporated in the production process. Cases such as this quite often present a situation where simple coordination of different research bases, affording a pooling of expertise, optimizes the packaging concept. In this latter technical development support example, once the resources were identified and organized, the pack-

aging design concept was quickly determined, executed, and realized. The resulting outcome of that effort has brought new and profitable business to the company, not to mention subsequent spin-offs of this concept in similar market areas.

SUMMARY

Distinct differences in packaging research and development philosophies do exist between organizations whose product lines are a form of packaging as opposed to those whose packaging is employed as an adjunct or component of the product. Whatever the differences, however, the health, well-being, and success of the packaging design group—its aggressiveness and innovativeness—will be a function of its management commitment and the personnel resources that are brought to bear against the project objectives.

The well-being and success can be further improved by actively maintaining a currency of information basic to the packaging design group's organization—goals, strategic planning, product lines, marketing approaches and plans—as well as to the state-of-the-art for technology. As the performance and success of the packaging design group heightens, so too an increasing positive impact on the corporation's well-being can be expected.

The scope and intensity of project endeavors will vary according to the level of reporting authority within the organization and the available resources that can be employed during project execution. Generally, avenues of approach taken in pursuit of these objectives will range from exploratory long-term systems development to the short term cooperative technical services support.

Corporate packaging research and development has the responsibility for the conception, design, and development of innovative packaging systems and products possessing proprietary technology. The ultimate goal of a development effort is to support the realization of business opportunities by delivering technology that allows the corporation to fill successfully customer needs. In this sense, packaging design as a corporate service is multifaceted in the functions it performs as well as the roles it plays. Yesterday, packaging's primary function was to simply contain the product. Today, packaging must provide the following functions:

- Containment.
- Protection.
- Unitization.
- Sanitation.
- Communications.
- Apportionment and Dispensation.
- Antipilferage.

As a design function, packaging research and development must take into consideration not only the application and satisfaction of the above functions but the environment affecting the packaging system when designing the package such as:

- Processes.
- Materials.
- Package design.

- Production.
- Distribution.
- Marketing.
- Consumer acceptance.
- Ecology.
- Energy.
- Economics.
- Regulatory affairs.

Currently we are seeing the penetration of new packaging systems into old markets, for example, the retortable pouch is soon expected to compete favorably with the metal can; the ovenable paperboard food tray can be used to reconstitute food in both conventional as well as microwave ovens; modified atmosphere packaging will, in most cases, more than double the current shelf life of bulk packaged fresh meat, poultry, and fish; and, aseptic packaging can extend the shelf-life of unrefrigerated liquid products from 6 to 8 months. As one writer has previously stated, the innovativeness that permitted the discovery of these concepts, as well as the ones before them, will lead to many more.

The packaging design, research and development function must be dedicated to the proposition that technology is a business tool that is backed up by the inventiveness and adaptability of its personnel and their collective capability to develop new packaging materials, containers, and systems.

CHAPTER THIRTY-TWO

Corporate Identity Research
Searching for the Hidden Meanings
Inside Your Name

Norman B. Leferman

More and more companies have turned to corporate advertising as a means of raising "the consciousness of their publics." Some companies seek greater attention and, ultimately, acceptability among the financial and trade communities. It is no doubt important for securities analysts and potential investors (pension fund managers or ordinary consumers), not to mention lenders, to have favorable impressions of a corporation. For these publics, corporate advertising might seek to improve awareness and familiarity of the broad range of products and services offered. Or it might seek to communicate specific attributes or benefits that are considered corporately important, such as technological leader-

ship, consistent profitability, social responsibility, and so on.

Other companies seek to improve their overall product image to consumers by creating and maintaining a corporate posture. To consumers, a company like General Mills might build an identity like Betty Crocker to provide an implicit endorsement of quality to every product and package she adorns. For consumer publics, corporate advertising might seek to communicate consumer-oriented attributes and benefits, that is, natural ingredients, convenience, good value for the money, and so on.

Whatever the objectives, corporate advertising is no small business. The Public Relations Journal, for example,

has reported six-media totals for corporate ad spending to be up dramatically during the 1970s from $149,500,000 at the beginning of that decade to $292,-700,000 for 1976.[1] With an investment of that magnitude, research must be dictating that corporate communications are necessary and that these ad efforts have had some measurable success.
What and how one measures are the subject of this chapter.

CORPORATE COMMUNICATIONS OBJECTIVES AND SYSTEMS

To build research methodologies, let's first define what we mean by a corporate identity *system* and consider some likely objectives. A corporate identity system, simply stated, is the company name and its method of presentation. In detail, the company trademark, as it is sometimes called, is a set of visual elements that collectively impart an image. Beyond the actual letters that spell out the company name, there is an opportunity to foster communications in the choice of type style and color and the use of symbols and/or nomenclature to modify the "basic" name.

The words "Burger King" are the symbolic meat set inside the hamburger roll that, together, are that company's identification system; the Ŗ symbol embedded in the name ECKEŖD on store fronts is a constant communicator of that company's business even without the additional Drug nomenclature that adorns each of their retail drug stores; the flour barrel head symbol utilized by Pillsbury is a constant reminder of that company's heritage. The motion and takeoff implicit in the Federal Express logo is entirely appropriate for an air-freighter. Even if the Hallmark crown has no real meaning, it comunicates a "good enough for royalty" message. The stepdown type configuration on the

name Pathmark conveys lower, marked down prices consistent with their discount operations. (See Fig. 1.)

Thus a corporate identity system often consists of three elements:

1   *A Logotype* A stylized lettering form.
2   *A Symbol* Incorporated with the logotype to communicate a positive element of the company.
3   *A Color Scheme.*

Research can study the totality or any element.

At any point in time a company can find itself in need of a new/improved corporate identity. Stephen Bowen, in writing about TRW's successful decision to use corporate advertising, describes a fairly typical set of objectives:

1   To significantly increase the awareness and understanding of TRW among our target audience in the key markets.
2   To increase TRW's appeal as an investment.
3   To emphasize TRW's growing and important involvement in electronics.[2]

It is important to be mindful, though, that corporate identities are not static. Just as a company needs to expand and/or diversify into new markets, so too can a company find itself in need of a new name. Corporate name changes—one form of a new corporate identity system—sometimes are the best way to communicate to various target markets that corporate changes are taking place. In fact, as Frank Delano wrote, consideration of a name change is totally unnecessary when:

●   Your company is not growing rapidly.
●   Your company is not expanding into new markets.

Figure 1 Expressive Corporate Identity Symbols

- Your company is not introducing new products.
- Your company is not merging with or acquiring other companies.
- Your product mix is the same today as it was 10 years ago.
- You're satisfied with your current market share.[3]

An examination of the use of animals as corporate symbols by Walter Margulies, for example, quite revealingly demonstrated the need for continual assessment of corporate identification. After tracing Elsie the Cow's rise at Borden from use in a minor ad campaign in the 1930s to major celebrity status in the 1940s, he explains her return to pasture: "But a cow, no matter how famous, cannot easily represent a company that makes potato chips and wallpaper . . ."[4]

Margulies similarly wrote to the need for change at RCA: "RCA's Nipper, with his head cocked to the sound horn of an antique 'Victrola' was RCA to everybody. But Nipper could not adapt to the company's more sophisticated electronic and entertainment products."[4]

In concluding his article on animal symbolism, Margulies reminds readers that symbols are not timeless, while he calls for research. "All corporate symbols have essentially the same goal: more effective communication. All are vulnerable to time and change. Thus it is critical to monitor all communications for their impact and distinctive strengths."[5]

As researchers, we can and should achieve far more than just monitoring the communication values in a corporate identity system. Not only should we be able to note that the swirling water symbol in the Whirlpool Corporation's identity system (Fig. 1) may have outgrown its value as the company grew from home laundry equipment to microwave ovens, but we should also provide creative direction for the development of a new corporate identity system.

To research corporate identity adequately, one must understand the design criteria by which a company's corporate mark was created.

Paralleling a "Communications Checklist" suggested by Frank Delano, researchers should address:

- The effects of replacing brand or service names with generic, descriptive, or geographic terms.
- The effects of utilizing the parent company's name in all or most operations.
- The breadth of applications' appropriateness of identity and nomenclature system from stationery, forms, and brochures to packaging and rolling stock.
- The impact, visibility, and image overtones of a new identifier or trademark.
- The extent to which a new nomenclature system could help individual product advertising to promote other, unadvertised company products.[6]

The following sections outline how we can study these dimensions and some typical results.

The Totality of Corporate Identity

When a member of the target market is asked to think about your company, what products or services come to mind? What general attitudes do they have? What, if any, product/service attributes do they expect? What possible new products and services would they consider appropriate expansion areas? The answers to these questions are the current identity. Where these thoughts and images came from is history, for the moment.

One large multiproduct corporation periodically takes the pulse of its consumer targets in a continuing telephone tracking study. While each product manager is interested in having the brand improve its impact in a specific product marketplace (i.e., increase levels of brand awareness within specific product categories), corporate objectives are less "selfish." The company depends on selling many products. To the extent that satisfied customers of one product may be the best prospects for another product, the corporate objective is to:

1 Increase general awareness of the brand in a broadly defined category of competition.

2 Increase the number of correct products that are spontaneously associated with the brand.

Standard unaided and aided awareness questions are asked. Instead of asking consumers to name brands in a product category, however, they are asked to name products made by various brands.

Table 1 displays some data from one specific point in time and suggests a danger in identity research. Identity, like awareness, is in part a function of advertising and distribution. Don't make the mistake of doing any less thorough a job on identity research—different results can be found in different markets.

Similarly, don't be satisfied with surrogate measures. The data show that consumers in two markets have the same depth of company familiarity but at different awareness levels. Further inspection would show new products which have been advertised in Market A, while Market B has traditionally had better distribution and sales of established products.

Another company has learned the need to validate measurement scales in its identity studies. Broadly stated, correlating overall attitudes with "agreement" on specific corporate attributes can reveal a set of priorities—"being open to the public," "being energy conscious," and "caring about the public interest"—may be of PR value to a vocal minority, but obviously are not the key to a favorable overall attitude for the company described in Table 2. Since the level of favorable attitudes far exceeds the levels of association on those attributes, we must conclude that you can favor a company even if they are not "energy conscious" and so on per se.

Table 2 shows another set of interesting data. The white collar professionals were asked in personal central location interviews, via a five point scale, how familiar they were with the company. This level of *claimed* familiarity did not correlate well with the ability of this target market to name products the

Table 1 Brand Identities Can Have Different Meanings in Different Markets

	Total Sample (1198)	Market		
Base: All Respondents		A (402)	B (397)	C (399)
Awareness				
% Naming brand first	5%	8%	4%	3%
% Naming brand on unaided basis	18	25	18	12
Familiarity				
Mean number of products associated with brand	1.8	2.0	2.0	1.5

*Table 2   Claimed Versus Proven Familiarity Can Lead to Different Image Conclusions

Base: Aware Of Company	Total Sample (400)	Claimed Familiarity		Number of Correct Product Mentions	
		High (126)	Lower (274)	Two or More (141)	Fewer (259)
Overall attitude:					
Favorable	73%	85%	67%	76%	71%
Unfavorable	1	2	—	1	1
Uncommitted	26	13	33	23	29
% Agreeing that company . . .					
Improves product quality	80	87	77	83	77
Is a leader in R & D	77	88	72	87	73
Makes increasing important products	77	83	75	85	74
Stands behind its products	75	82	71	79	73
Provides fairly priced products	71	81	67	76	69
Has good financial record	69	85	62	70	69
A good investment	67	76	63	71	65
Is well managed	65	75	60	70	63
Cares about public interest	65	74	60	65	66
Is energy conscious	59	62	58	61	59
Is open to the public	50	58	46	50	49
Claimed familiarity:					
High	32	100	—	35	29
Lower	68	—	100	65	71

company manufactures. The greater importance of this finding is to note the degrees of difference in attitudes. That is, people *claiming* a high degree of familiarity tend to exaggerate positive perceptions of the company. This will be most important when we consider how to test new corporate identity systems. Independent samples should be matched on the most "honest" bases.

As. mentioned earlier, the corporate identity *system* is often essentially a name form with style, symbol, and color. However, the totality of an identity derives from a much greater pool of resources. Corporate advertising, independent of product specificity, certainly has an impact on a company's identity. However, we have also seen cases where strong TV advertising for a single

product brought so much attention to its parent company that favorable corporate overtones (Table 3) were stimulated. The mindful researcher will recognize the launch of every new product, package, or major ad campaign as an opportunity for studying the impact on corporate identity. Perhaps the situation could have been so reversed that consumers developed negative feelings about a company in spite of its good product advertising intentions.

One clear way to avoid this possibility with new products is to study consumer perceptions of appropriate/inappropriate expansion areas in advance. If the company studied in Table 4 had just finished developing a new line of salty snacks, the time would appear right for a changed corporate identity. In this ex-

Table 3 Product Advertising Can Also Affect Overall Corporate Identity

	Percentage Point Change In:		
Average Base	Ad Markets (804)	Control Markets (402)	Net Gain
Aided awareness of company	—	+ 1	− 1
Have positive feeling about company	+ 3	− 4	+ 7
Have favorable feeling about company's products	+ 3	− 3	+ 6
Recall advertising	—	− 2	+ 2
Products associated with company			
Advertised product	+ 7	+ 2	+ 5
Nonadvertised product 1	+ 2	+ 4	− 2
Nonadvertised product 2	+ 7	+ 2	+ 5
Brand names associated with company			
Advertised brand	+21	− 1	+22
Nonadvertised brand	+ 2	—	+ 2

ample, however, the company was trying to plan ahead.

This type of study can be helpful to R&D people as it identified a broad charter of products that can be offered at highest acceptance levels within the current corporate identity and a list of study areas to neglect for the moment.

Note also the discriminating specificity that can be obtained:

- A *candy bar* would be moderately appropriate, other types of candy would be largely inappropriate.
- Cold beverage mix 1 would be highly

Table 4 Extent to Which Current Identity Precludes Expansion to Other Areas

	% Rating Company		Net Positive Rating
Base: All Respondents	Appropriate For (202)	Inappropriate For (202)	
Hot beverage mix	91	4	+87
Instant meal drink	90	3	+87
Meal bar	82	1	+81
Energy foods	82	4	+78
Cold beverage mix 1	79	6	+73
Dessert	55	18	+37
Candy bar	53	22	+31
Cookies	44	20	+24
Cold beverage mix 2	41	28	+13
Cold beverage mix 3	39	31	+ 8
Other type of candy	21	57	−36
Salty snack	9	68	−59

appropriate, while mix 2 or 3 would be far less appropriate.

Regardless of who is interviewed or how the data are obtained, it is important to know your *total* identity today.

The next is to learn the extent to which those attributes are wedded to any particular graphic elements in the current corporate identity system.

Testing the Corporate Identity Elements—Equity Testing

Once the total corporate identity has been established, management must decide whether that basic identity is currently accurate with room for growth, "close enough" so that incremental additions to the identity will suffice, or whether an entirely new identity direction is required. The solution to minor identity changes might be corporate advertising that builds on the current identity. When more elaborate change is required, management may elect to have a new corporate identity system developed.

Perhaps the most important pieces of research that can be conducted at this point are equity tests—a series of studies in which the key graphic elements of the current corporate identity system are evaluated. The spirit of this research is to identify the current elements that are, or can be, most closely allied with the identity that management wants to project. After years of being identified with a specific symbol—together, typically, with millions of communications dollars—a successful company doesn't want to throw away its investment. A new or refined system will seek to preserve the equities (i.e., contemporizing the White Rock Girl) of the current system.

Some of these equity decisions will necessarily be drawn from the amount of corporate identity change that is sought. A corporate identity change for Federal Express, for example, could include a change in type style, a change in nomenclature, and/or change in color. Equity testing might show, however, that all of Federal Express' *unique* competitive capabilities are embodied in its unique corporate color scheme, but that the stylized "s's" in "Express" do not carry any particularly positive message. Therefore, if research indicated the need for change, design criteria might be built around preserving the current color keys, but not necessarily the logotype.

In one study, the element of type style was isolated tachistoscopically. Matched samples of consumers were shown a familiar brand name on one of that company's typical packages. All respondents saw this package in a competitive display environment. Half the respondents saw the brand name on the package set in a Roman typeface; half saw it set in its current script. After repeated high speed timed exposures, the cell who had been exposed to the Roman typeface attributed spontaneously more variety, overall quality, and skin benefits to the brand than had the other respondents. Similarly, when respondents were probed, with the shutter held open, to rate the brand relative to the competitors shown in the display, consumers in the Roman typeface sample gave the brand superior ratings on reasonable pricing, value for the money, and overall quality (Table 5).

It is important to keep in mind that the brand name being tested was so familiar and so well established that almost three quarters of the consumers that participated in either cell had had prior trial experience. Thus the fact that a different, unfamiliar, type style could even tie (and this Roman type style actually outscored the current) suggested that the equities in the current corporate identity system did not revolve around the typeface.

Table 5 Lettering Style plays a Subtle Role in Brandmark Imagery

	Attitude Toward Brand In:	
Base: All Respondents	Roman Typeface (157)	Current Script (150)
Spontaneously associate brand with:		
Variety	19%	11%
Overall quality	23	17
Skin benefits	20	14
Rate brand vis-à-vis competitors:		
Much more reasonably priced	38	27
Superior value for money	62	57
Superior in overall quality	26	22

On the other hand, when similar studies were conducted to assess the importance of brandmark location, we noted a high degree of equity in the current horizontal usage of the name rather than in a proposed vertical presentation (Table 6).

Equity testing can be done in a variety of ways. Tests can be elaborate to employ controlled store tests, magazine portfolio tests or simpler, forced exposure techniques. Even focus groups can be useful in identifying the key equities in a corporate identity that must not be destroyed in the creation of a new system.

A good way to judge the relative equity of different parts of an identity system is to ask consumers (on an unaided or aided basis) what they remember about the name—what it looks like, any symbols they can recall, color schemes, and typefaces. For example, how many consumers know whether a current corporate name is:

- All capital letters or in upper and lower case.
- In blue type on a white background, red type on a white background, or white type on a red or blue background.
- In script or block letters or italics.
- Preceded by, followed by, or not adorned with a symbol.

Table 6 Brandmark Location Also Plays a Role

	Attitude Toward Brand When Name	
Base: All Respondents	Applied Horizontally (151)	Applied Vertically (150)
Spontaneously associate brand with:		
Variety	19%	11%
Overall quality	21	17
Skin benefits	15	14
Rate brand vis-à-vis competitors:		
Much more reasonably priced	30	27
Superior value for money	61	57
Superior in overall quality	27	22

The extent to which consumers know these answers is, in some manner, a measure of the liability inherent in changing a corporate identity system. These must be carefully weighed, as RCA must have pondered their lovable Nipper before deciding on a totally new system. Sometimes equity testing provides a set of transition rules which will guide management through a two or three step process during which a drastic change is softened by intermediate, less drastic steps.

Proposed Changes

This last section will provide some examples of element tests that have been conducted to document target market sensitivity to changes in a corporate identity.

One company found a great variance in the personality they projected in different typefaces. After timed exposures to the company name printed in black ink on a white background, consumers greatly varied in the extent to which they associated the company with products for men, products for women, or products for both sexes (Table 7).

Another well established company, a manufacturer of furniture products that used an animal symbol in its identity system, found that its symbol was precluding company consideration for certain rooms of the house. Its identity was so strongly allied to certain attribute dimensions that a proposed "broader" symbol actually led to even more narrowly defined perceptions (Table 8).

The preceding example notwithstanding, a new symbol *can* improve brand acceptability. In a forced exposure study, matched samples of consumers were exposed to one of three corporate identity systems, each as the only label on a hand drawn package of one company product. In all three cases the brand name was executed in the same color and in the same typeface. The name's environment and modifier symbol differed. As the data in Table 9 show, one of the proposed symbol systems led to significantly higher net positive purchase interest. This is particularly impressive not only because of the brand's level of establishment, but also because half the respondents in each cell were current users of the brand.

It is presumed that anyone undertaking an evaluation of corporate identity elements knows which attributes and benefits are most important to the target market. Not at all unlikely are instances where one symbol will be a better communicator of certain attributes, while an alternative symbol will more strongly bring other attributes to mind. In a recent study for a food company (Table 10), consumers seemed to express that the symbol that portrayed the most healthful foods was not the best symbol to use for communicating family taste fulfillment.

Table 7 Extent to Which Corporate Typeface Can Impart Personality

Base: All Respondents	Extent to Which New Company's Products Are Appropriate for:		
	Men (141)	Women (141)	Both (141)
Typeface 1	4%	74%	22%
Typeface 2	20	11	69
Typeface 3	64	4	32
Typeface 4	34	17	49

Table 8 Symbols Can Dramatically Change Perceptions of a Well Known Identity

Base: All Respondents	Current Symbol (301)	Proposed Symbol (301)
Net positive* appropriateness of products for:		
Child's bedroom	+ 9	(-19)
Adult bedroom	- 60	-67
Office/study	+55	+75

*Percentage point difference of consumers rating products appropriate minus those rating products inappropriate.

SUMMARY

From a marketing standpoint, the study of one's current corporate identity is at least four-dimensional:

1 Exploring current total perceptions.
2 Assessing expansion opportunities.
3 Identifying current elemental equities.
4 Diagnosing elemental strengths and weaknesses.

The development of a new or improved corporate identity system should take into account not only who the company is today, but what it wants to be known for tomorrow. Of course, the relative importance of the category attributes should be considered also. Researchers and managers should keep several points in mind:

1 A new corporate identity system represents both risk and opportunity. While it may broaden or focus perceptions among some publics, change in and of itself may confuse the current user. Therefore, current users should be a readable subsample in any test.

2 Just because it is unfamiliar, a proposed corporate identity system is likely to lose any paired comparison test with the current system. Corporate identity systems should be tested monadically among matched samples of consumers. A new, never-before-seen system should be declared a significant winner if it adds or refo-

Table 9 Symbols Can Improve Purchase Interest

	Exposure to Current Name With		
		Proposed Symbol	
Base: All Respondents	Current Symbol (50)	1 (50)	2 (50)
Purchase interest			
Positive	68%	74%	62%
Neutral	20	22	28
Negative	14	4	10
Net positive	54	70	52

Table 10 Different Symbols Can Communicate Different Features in an Established Brand

	Evaluated Brand with Proposed	
Base: All Respondents	Symbol A (101)	Symbol B (99)
Feel products would be:		
Wholesome and nutritious	52%	32%
Made with all natural ingredients	47	32
Free of artificial preservatives	48	34
High in quality	52	47
Tasty to entire family	31	39

cuses communications areas without giving up anything important.

3 Because a corporate identity system comprises many graphic elements, each area of variation can be isolated and tested. If an equity test suggests that the current color scheme has great impact value, then new typefaces should be tested against the current typeface, all in the current color scheme. Symbols can be tested "in outer space" to determine the kinds of attribute perceptions they evoke in a category. But they must also be tested with the company name set in a comparable form so that each alternative symbol has an equal opportunity to enrich and/or draw directly from the company name.

Underlying all of these suggestions, of course, is the need to understand your objectives. The mechanics of testing corporate identity systems are relatively easy, be they exploratory or diagnostic. The tricky part is to know what you are seeking and to recognize the answer when it has been found.

REFERENCES

1 Peg Dardenne, "The Cost of Corporate Advertising," *Public Relations Journal,* November 1977, pp. 22–24.

2 Stephen N. Bowen, "Solving an Identity Crisis at TRW," *Marketing Communications,* July/August 1977, pp. 34–37.

3 Frank Delano, "Corporate Name Changes: Don't Overlook the Obvious," *Advertising Age,* June 18, 1979, p. 67.

4 Walter P. Margulies, "Animals Going Out as Corporate Symbols," *Ad Age,* November 7, 1977, p. 60.

5 Ibid., p. 60.

6 Delano, pp. 67–68.

BIBLIOGRAPHY

Stephen N. Bowen, "Solving an Identity Crisis at TRW," *Marketing Communications,* July/August 1977, pp. 34–37.

Barbara Boer Capitman, *American Trademark Designs,* Dover, New York, 1976.

David E. Carter, *Corporate Identity Manuals,* Art Direction Book Company, New York, 1978.

David E. Carter, *Designing Corporate Symbols,* Art Direction Book Company, New York, 1975.

Peg Dardenne, "The Cost of Corporate Advertising in 1976," *Public Relations Journal,* November 1977, pp. 22–24.

Frank Delano, "Unified ID System Can Aid Perceptions," *Ad Age,* July 17, 1978, pp. 44–45.

Frank Delano, "Corporate Identity: An Investment Worth Protecting," *Ad Age,* January 1, 1979, pp. 16–17.

Frank Delano, "Corporate Name Changes: Don't Overlook the Obvious," *Ad Age,* June 18, 1979, pp. 67–68.

Solomon Dutka, "How to Measure the Effects of Image Communication," Audits & Surveys, Inc., New York, 1973.

Les Luchter, "Packaging Places, People, Products," *Marketing Communications*, **3,** No. 3 (May 1978), 23–27.

Walter P. Margulies, *Packaging Power*. World Publishing Company, Cleveland, 1970.

Walter P. Margulies, "Total Communications: Exhaustive Revamp Concept for Smart Retailers," *Ad Age*, January 3, 1977, p. 16,

Walter P. Margulies, "Animals Going Out as Corporate Symbols," *Ad Age*, November 7, 1977, p. 60.

Harold H. Marquis, *The Changing Corporate Image*, American Management Association, New York, 1970.

James Pilditch, *Communication by Design*, McGraw-Hill, New York, 1970.

Ben Rosen, *The Corporate Search for Visual Identity*, Van Nostrand Reinhold, New York, 1970.

James T. Rothe, "Corporate Image and Corporate Name . . . An Inquiry," *Atlanta Economic Review*, February 1970, p. 28.

Elinor Selame, *Developing a Corporate Identity*. Chain Store Publishing Corporation, New York, 1975.

Sally Urang, "Corporate Names: A Tendency Toward Alphabet Soup," *New York Times*, April 15, 1979, p.F3.

PART FIVE

Design Research in Special Markets

CHAPTER THIRTY-THREE

Techniques of Conducting Package Design Research with Children

Donald A. Cesario

In writing this chapter consideration was given to illustrating the principles of conducting package design research with children by using the case history approach, whereby a particular research project is discussed from its inception to its successful completion.

It was felt, however, that this approach would be limiting. There is likely to be too much variation from one project to another, or from one product category to another, for any one case history to do justice to the complexities of research with children. For example, the methodological design one would employ and the questions one would ask

testing one package design would be quite different from those used to test three alternative package designs. Similarly, package design research in the candy bar field would have a different set of considerations from package design research for toys. Consequently, it is believed that the astute researcher would benefit most by understanding certain basic principles of conducting research with children and that, armed with this understanding, he or she would be able to apply these principles to the particular needs of any specific research project.

This chapter, therefore, discusses

general techniques of conducting research with children aged 4 to 12, and wherever pertinent makes special reference to conducting package design research. It is divided into six sections:

- Understanding the limitations of children.
- Techniques of conducting qualitative research with children.
- Techniques of conducting quantitative research with children.
- Conducting research with children within the broader context of the family unit.
- Suggestions for establishing a plan of package design research among children.
- Concluding remarks.

UNDERSTANDING THE LIMITATIONS OF CHILDREN

The researcher who wishes to conduct package design research among children is often confronted with specific problems that require specific solutions and are frequently different from those encountered in conducting similar research with teenagers and adults. These problems are mostly due to certain developmental limitations of children. If these limitations are understood and respected when designing and implementing research among children, the solutions to the problems will become obvious.

Let us therefore review the kinds of developmental limitations that confront children and then offer possible ways of accommodating and overcoming these limitations in research.

The following discussion is based on the author's own experience as well as on the theories of cognitive development in children propounded by Jean Piaget.

Verbal Limitations

One obvious verbal limitation is that of vocabulary. This affects the words children use to express themselves as well as those they can comprehend. It also affects their use and understanding of words to express nuances of meaning, their ability to comprehend certain kinds of humor, like puns and satire, and to connect language with complex thoughts or ideas.

These verbal limitations are more prevalent at the ages of four and five than they are at the ages of 11 and 12, but in any sample of children at any age level, some children will be less advanced in verbal skills than others, and the researcher must always keep in mind this variation of verbal ability.

Cognitive Limitations

Children are subject to limitations in various areas of cognition. Let us examine some of these areas.

Quantity

In its most simplistic form, children generally understand unit quantity—one, two, three, and so on—but only up to a realistic point. Quantities of hundreds are less well understood than tens, and thousands are less well understood than hundreds. Another type of quantity measurement that can cause children problems in comprehension is quantitative equivalency, of size, volume, or a combination of both.

For example, suppose children are confronted with one package containing a certain volume or quantity and two packages each containing half the volume or quantity of the one larger package. Some children will perceive the size/volume equivalency in one way, other children in the opposite way. Some children may perceive that the two smaller packages are better than the one larger package, because they think they get twice as much as in the one package; others may perceive that one package is better, because it holds twice as much as either smaller package. In both cases, children have not comprehended the notion of equivalency, that two can equal one or, vice versa, that one can equal two. The child who is quantity oriented will choose the two smaller packages.

Similar to the concept of equivalency of size and volume is the concept of componentcy, or the whole versus the parts. Children may not always understand that the parts are distinct from the whole, or that the parts form the whole or that the whole is composed of the parts.

This is not only true when applied to physical objects, but also when applied to attitudes: children cannot always differentiate or divorce their attitude toward the whole from their attitude toward the parts, which often leads to halo effects in certain types of research questions.

Time/Frequency/Periodicity

Children have difficulty understanding various aspects of time, frequency, and periodicity. To the child, today is distinct from yesterday, and tomorrow is distinct from today. But when time frames are expanded, distinctions can become blurred. The distinction between today and last week, between last week and last month, and between last month and last year are increasingly difficult to comprehend and to define accurately. Similarly, future time frames pose problems—tomorrow versus next week versus next month versus next year.

Serialization/Progression

A difficult concept for children is that of serialization or progression, that one

thing is an outgrowth of another, which itself is an outgrowth of something else. Figure 1 is a schematic representation of how a toy manufacturer has fractionalized his product array in pyramid fashion and has thus set up a difficult structure for children to comprehend. If one then wishes to measure children's awareness of the product line, one must be aware that problems will exist in measuring this awareness. At what level in the diagrammed structure is brand awareness to be measured? The manufacturer may consider the series name level as being the "brand", but the child may perceive the toy name as the brand name.

Cause/Effect Relationships

Children are not always cognizant of cause/effect relationships, especially in hypothetical situations outside the realm of their experience. This can also occur in situations reflective of their experience. It is not surprising, therefore, that many children may answer "I don't know" when asked why they would act in a certain way or have made a certain choice between package design alternatives. This is because the child has not been confronted with this situation in real life.

Abstraction Versus Concreteness

Children are very literal and tend to accept things at face value for what they are or for the way they are presented. As a result, what is concrete is far more relevant and is better understood than what is abstract. It is difficult for children to abstract from the concrete. Thus to ask children what the ideal product or package would be presents a difficult concept. They will answer only in terms of what they know and with what they have had experience. Similarly, it is easier for children to deal with comparatives than with absolutes.

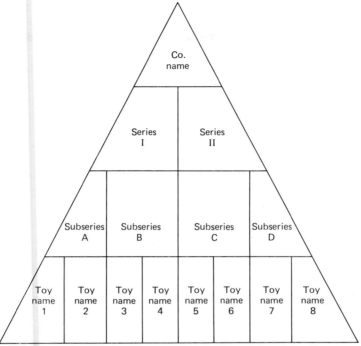

Figure 1   Fractionalized structure of a toy products line.

Experiential Limitations

Children start out living in a very ego-centric universe and as they mature the sphere of their universe increases; the prenatal environment gives way to the crib, which expands to their home, then to the street in which they live, then to a neighborhood, a city, a state, a country, a universe. At each stage of maturation not only does the children's sense of their universe broaden, but the people they encounter increase. At the beginning they are aware of mother, then father and siblings, then relations outside the household. When they go to school they interact with peers and authority figures other than parents. They start to have dealings with shopkeepers, etc. They become increasingly independent, both socially, politically and economically.

Researchers must be aware of the experiential stage a child is at so that they do not attempt to measure an experience still outside the realm of the child.

Psychomotor (Physical) Limitations

As children mature, so do their psychomotor or physical capabilities: awkwardness gives way to dexterity; children can manipulate objects more easily, like opening and closing packages or constructing architectural forms.

As physical ability increases, so does physical endurance. And as mental ability increases, so does children's attention span.

These, then, are some of the limitations in children's developmental stages of which the researcher must be aware if valid, reliable, and representative research with children is to be conducted.

Let us see how these limitations affect how the researcher should conduct package design research and what should and should not be expected of children in a research situation.

TECHNIQUES OF CONDUCTING QUALITATIVE RESEARCH WITH CHILDREN

Conducting focused group sessions or one-on-one, individual depth interviews with children can be an effective means of information gathering, provided the researcher respects children's limitations and capabilities and uses them to advantage. Let's see how the researcher would use this knowledge of children to design the study, choose the subjects to be discussed, conduct the interviewing or moderating, and interpret the results.

Sampling Design

Choosing a Sample

If the researcher's needs call for interviewing both boys and girls, then up to the age of 12, the sexes should be interviewed separately. In mixed groups, boys tend to show off in front of girls, whereas girls tend to take a passive role vis-à-vis boys. Among 9–12 year olds both sexes seem to be "anti" the opposite sex sociologically. This is the age span where each sex thinks its opposite members are "dumb," "silly," or "babyish."

The most productive minimum age is 6 years old. Below the age of six, both verbal and nonverbal means of communication are difficult for children, unless the 4 or 5 year old has had some nursery school or kindergarten experience and has learned how to interact with other children. At this young age the researcher must rely heavily on observational techniques and nonverbal means of communication. In package design research this may mean observing which of three

new package designs a child reaches out to first, which would be an indication of which design is most attractive and appealing to the eye.

In group sessions, the age span of the children should be no more than 2 years: 6 and 7 year olds in one group, 8 and 9 year olds in another group, so that the cognitive development and verbalization skills will be homogeneous. Older children in a group with younger children are embarrassed, whereas the younger ones are intimidated by the older ones.

It is advantageous to interview one age span lower and one age span higher than the proposed or "known" target group age. Children may be at one age chronologically, but at another age in terms of their behavior and/or attitudes. Children at any age aspire to emulate older children, so if the target group upper limit is 12 years old, one should investigate what the 13 and 14 year olds think, because that's where the 12 year old's head is.

Similarly, if the lower limit is 8 year olds, interview some 6 and 7 year olds; they want to be like the 8 and 9 year olds and will be in a year or two.

Length of Session

Ideally, group sessions or depth interviews should be a maximum of three-quarters to one hour for four and five year olds, one to one and one-quarter hours for 6–8 year olds and one and one-quarter to one and one-half hours for 9–12 year olds.

Longer sessions at each age group result in fatigue, boredom, silliness, and lack of attention; consequently, the quality of information toward the end of the interview drops sharply.

Topics To Be Covered

As age increases, the focus of the sessions or interviews can be broadened. The researcher, however, must strike a balance between introducing a variety of topics to maintain interest and taking into consideration short attention spans, limited experience, and the possibility of introducing confusion if too many topics are covered.

The Interview Situation

Environmental Considerations

The *mise en scène* is very important. Unlike adults sitting on chairs around a table with a microphone in the center, children's groups should be conducted in a very informal arrangement, with space for physical motion. A functional arrangement would consist of sitting on the floor in a circle, with provision for a surface on which children can write or draw.

In personal interviews, it is imperative that the parent or sibling not be present so the child will not be intimidated.

Moderator Qualifications

Of utmost importance is that the moderator/interviewer be able to communicate well with children. This entails being able to come down to their level without being condescending or patronizing; being able to control the group in order to maintain cohesiveness (some children are hyperactive), yet at the same time not stifle their exuberance but rather channel it toward the positive goal of communicating information; being one of the group, which relaxes the children and establishes rapport, yet maintaining the role of leader.

Questioning Techniques

General Techniques

It is important to establish at the start what the interview/session is about, what is going to transpire, and what is to be accomplished; this allays fears of the unknown. Children should be told that they can leave the session if they are no longer interested or willing to participate in the proceedings. This makes the children commit themselves to cooperate.

It should be established that the session/interview is not school, that the children are not being tested, and that there are no right or wrong answers. Everyone has an opinion and his or her opinion is important. A proper group session or depth interview will combine both verbal and nonverbal techniques of eliciting information.

Verbal Techniques

Shouting Out Answers: To encourage independence of thought and action and to avoid "me-too-ism," children can be asked on the count of three to shout their answer to a question, such as "Which of these three packages do you like best?" The moderator can proceed from child to child asking "What did you shout out" and then follow up with additional questions about the package mentioned.

Secret Votes: To elicit information from shy children, they can be asked to whisper to the moderator or interviewer their "secret vote." This information can be utilized later to draw out each child.

Creating a Story: The moderator/interviewer can start a story and ask each child to add to it in turn. The composite story that results is often very revealing. For example:

MODERATOR:	Mary saw this box of toys that . . .
CHILD 1:	. . . had lots of pictures on it.
MODERATOR:	And . . .
CHILD 2:	. . . she wanted the box
MODERATOR:	And . . .
CHILD 3:	. . . and she asked her mother to buy it.

MODERATOR: What happened next?

CHILD 4: Her mother said "no."

MODERATOR: Why did her mother say "no"?

CHILD 2: Because the toy looked too hard for her to play with.

Sentence Completions and Word Associations: Children can be asked to complete a thought or to tell the first word that they can think of when given a key word. Their answers can then be explored for meanings.

Second Guessing: Children can be asked to explain in their own words what they think another child is trying to say or is acting out.

Word Fights: Two children who have different opinions or choices can be asked to shout out at each other the reasons for their opinion or choice in an attempt to convince the other child.

Nonverbal Techniques

Role-playing/Acting: Children can be asked to assume roles and to act out situations. Some role pairs and situations might be: mother and child having a discussion at the dinner table; child and salesman acting out a purchase situation; child and sibling acting out a situation where a conflicting choice of TV program or product must be reconciled.

Single role-playing situations might be: a child playing with her doll where the child assumes the role of mother and the doll that of the child; a child as TV announcer who is asked to make up a commercial for the product or package being investigated.

Drawing Pictures: Children frequently express in drawings what they cannot express in words. They can draw objects they know, putting into pictures their

understanding of product concepts, product line extensions, images of products as conveyed by package design.

Picture Projectives: These can be used to provide a stimulus to which the child can react. Pictures of events, such as a picture of people around a dinner table, could be used to elicit behavior and attitudes about food packaging. Nondescript pictures of children could be used to help a child give a description of the personality profile of a certain consumer segment.

Analysis and Interpretation

An analyst should observe the group sessions, because often the key to understanding and interpreting what is said lies in attendant facial expressions and body movements, which the analyst can record while observing.

The analyst should be acquainted with the meanings of current language, such as "bad," which to children means good, or "heavy," which means important or meaningful.

It is beneficial to have a transcript of the session wherein each comment is identified by the name of the child who made that comment. This can be accomplished by having a court stenographer present at the session, who records verbatim what each child says. This way, the analyst can follow the thought of each child individually from start to finish and can identify at what point in the interview and because of what stimulus a child might have changed their mind and whether that was a "me-too" response or their real feeling. The court stenographer can also be trained to record nonverbal reactions, like grimaces.

Equal stress should be put on evaluating both nonverbal responses and verbal responses.

Consideration must be given to what is not said or what is left out of the child's responses. Often this indicates a negative attitude or a lack of comprehension, which the child expresses by avoidance rather than by confrontation. For example, in playing back a commercial they saw, children might leave out a character or deemphasize the character's roles in the commercial, which would indicate negative feelings.

The internal bench mark of the analyst is most important. How do these groups and the information obtained compare to previous groups and previous findings? Do they substantiate each other in part or do they conflict totally and if so, why?

Finally, it is helpful to keep in mind the distinctions between children at various age levels in terms of their cognitive development and judgemental ability. This often explains why a child responded in a certain way.

QUANTITATIVE SURVEYS

Methodology

Sample Size, Geographic Distribution, and Interview Location

Generally, the researcher can use smaller and more localized samples with children than with adults. The age of the children limits the range or repertoire of responses that are likely to be uncovered. Also, in today's world of immediate mass communication, geographic differences tend to be minor; children today share a much more common culture at a much earlier age than in the past. For this and other reasons centralized location interviewing is likely to have fewer sampling biases than when used in adult interviewing.

Sample sizes should be large enough to allow comparisons between boys and girls and between younger children (6–8) and older children (9–12), since findings often differ by sex and age.

Personal Interviewing Preferred

Personal interviewing is the preferred method of data collection with children and is likely to be the only viable technique for package design studies. It establishes rapport between the interviewer and child and allows visual aids, which are often necessary, to be used to communicate with children.

Monadic Versus Paired Comparison

The monadic test design is often used when surveying adults. However, it may have some weaknesses when used to survey children.

If the task is to measure which of several package designs has the greatest potential appeal or consumer acceptance, a monadic design is recommendable only if the stimuli have sufficient dissimilarity. Children respond most readily to the broad, general aspects of a stimulus and cannot always perceive or verbalize the subtleties, as can an adult. If the stimuli are too similar, a monadic design is likely not to differentiate the stronger from the weaker alternatives. This difficulty is further compounded by the need to use scales that have fewer points than are used with adults.

If a monadic design is unavoidable, it is advantageous to include, along with absolute measures, some comparisons to a "control." For example, in a monadic test of various packages, the child should be asked to evaluate the test packages relative to current, familiar packages.

The researcher can now evaluate the performance of the test alternatives not only on the basis of absolute ratings, but also in terms of how strong each one is relative to the "control."

The Importance of Controlling Test and Nontest Variables

Because we live in an imperfect world, the methodological designs of research studies are sometimes compromised. The researcher may wish to test three alternative package designs, but because of budget or time constraints may have only one design available.

Therefore, the one design is used to represent the three alternatives. First the respondent is asked to evaluate the available design; then the respondent is told to imagine that the design has been changed in a certain way and is asked to evaluate the "new" design.

Although the researcher might need to rationalize using this technique with adults, it should never be attempted with children. Children's limited ability to abstract, and their literalness, would make it almost impossible for them to disregard a part of what they see and to substitute mentally something else.

Similarly, if showing children two alternative package designs in a paired-comparison test, the researcher should make sure that the two designs differ only on those variables that are being tested. For example, if size of package is not a test variable, both package designs must be equal in size, as children will tend to choose the larger size package.

Privacy of the Interview

When interviewing children it is imperative that the parents, siblings, or friends not be within earshot of the interview, as this will intimidate children and bias their responses. The interviewer should tell the parents what is going to be asked of the child to allay their anxiety and request the parents leave the area. It is recommended that an interview not be conducted if the condition of privacy cannot be met.

Interviewing Personnel

Experience has shown that interviewers who are good with children can be male or female, young or old, white or black. What characterizes a good interviewer are the same qualities that identify the good focused group moderator, as well as the ability to speak slowly and clearly, to modulate the voice, to enunciate or emphasize key words in the question, to refer to the child periodically by name, and to encourage the child in a nondirective way.

Questioning Techniques

Length of the Interview

The researcher should attempt to limit the length of quantitative questionnaires to a maximum of half an hour. Although it is possible to keep a child for one hour if the interview is interesting and designed to include activity, such as card sorting, the responses to later questions are suspect, as are the ethics of the researcher.

Moreover, children's attention span being short, the researcher should avoid long questions with multiple parts, such as taking the child through a battery of five package designs and asking overall and attribute ratings for each one. If this is necessary, it is better to split-sample and use a round-robin design.

Nonverbal Techniques

It is important to be concrete with children. They interpret what they are asked very literally and their definitions of words are very precise. Moreover, they may have the answer to a question, but not have the words to express that answer. Therefore, whenever possible, nonverbal questioning techniques, which are especially important for the under-9-year-olds, should be relied on.

Pictures can be used in conjunction with words. The younger child who is shy or has difficulty with vocabulary can point to a picture to express an answer, whereas the older child can respond with words.

An example of the use of pictures is shown in Fig. 2.

Figure 2 Use of pictures in nonverbal questioning.

Scale Questions

Scale questions can be adjectival, adverbial, or numerical. Ideally, adverbial and adjectival scales should be no more than three or four points as children cannot verbalize subtle differences in their attitude, even though differences may exist.

Whereas an adult scale may be

Excellent
Extremely good
Very good
Good
Fair
Poor

reaslistic scales for children would be

Great

Okay

Lousy

or

Like it a lot

Like it a little

Don't like it

Numerical scales can measure a wider range of an attitude and therefore can have five or six points, but should employ symbols familiar to children, such as stars or check marks.

Figure 3

One can combine numerical and adverbial scales, such as

Figure 4

and, pictures and scales can be combined, as follows:

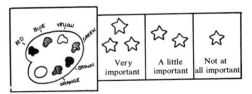

Figure 5

Time and Frequency Questions

As noted earlier, these types of questions pose special problems because younger children under 9 years old have little conception of time and frequency. It is necessary to give the child a point of reference. For example, in a recent survey on books done in October, children were asked how many times they had gone to the public library since school started that term. The resulting answers were interpreted as "number of times within the past month," since the school term started about the beginning of September.

Figure 6 illustrates another method of asking a frequency question. This method uses a calendar as a point of reference, an item with which children are familiar.

Quantity Questions

A question often used in market research is the constant sum allocation. The respondents are told they have 10 points or 10 dollars and are asked to allocate this quantity between two or more variables. This type of question is unsuccessful with children because of their limited perceptions of the aspects of componency.

Variables to be Evaluated

The researcher will no doubt have a fairly comprehensive idea of the variables important to the product/package being evaluated. Unless prior exploratory research, such as focused group sessions, has been conducted, however, or unless the researcher is very experienced with conducting research with children on the particular product category in question, it is possible that some variables important to children may be inadvertently omitted from the survey instrument.

Some variables that experience has shown to be important to children in packaging design (and of which the researcher may not be aware) follow. The list is by no means exhaustive.

1   The degree to which the package design is commensurate with product image and usage behavior.

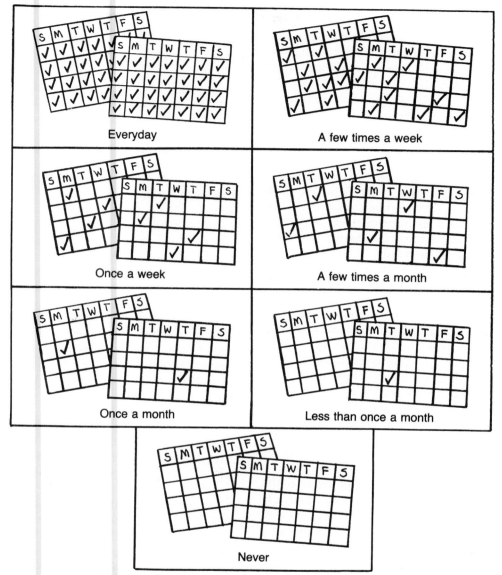

Figure 6   Frequency questioning through nonverbal means.

2 The ease/difficulty of opening and closing the package.

3 The amount and kinds of information and instructions for use on the package.

4 The size and color of the package.

5 The amount, kinds, and position of photographic or artistic elements.

6 The transparency or opaqueness of the package.

7 The degree to which the package evokes perceptions of the product's appropriateness for the child's age and sex.

8 The fun image the package evokes.

Sequence of Questions

If possible, the questionnaire should alternate between questions requiring an oral response and those requiring some activity to answer, such as card sorting or crossing out picture answers. This will help avoid boredom and keep the child's concentration focused on the interview.

In surveying adults, the common procedure is to ask first for an overall rating or preference and then for a rating or preference for individual attributes. This technique does not work well with children because of their difficulty in understanding componentcy.

If the adult sequence is used with children, a very strong halo effect results. If the overall rating chosen is top box, the top box is chosen for the attributes. If package A is preferred overall, package A will be preferred for individual attributes. This is because children have difficulty in differentiating their overall opinion from their opinion of the variables that go into shaping that overall opinion. Children also feel the need to uphold and reenforce their overall opinion, so they give the same rating or preference down the line.

The researcher should reverse the sequence for children's studies, first asking about individual attributes, then asking for the overall rating or preference.

Each question should be a logical outgrowth of the preceding one. If the subject matter changes, one should have a brief introduction for the new area of inquiry. Questions should proceed from easy to difficult and back to easy near the end of the interview.

Analysis and Interpretation

The same modes of analyses as applied to adult surveys can be applied to children surveys. Children can be classified into segments based on continuums, although they often segment into fewer groups than do adults.

The criteria applied to evaluating the performance of products, packages, or commercials in children's research is usually more stringent than with adults, as children tend to be more enthusiastic than adults. Thus whereas among adults the benchmark for potential market success might be 6 out of 10 choosing the top two boxes on a scale, it would be more realistic and prudent to use 7 out of 10 among children.

Occasionally the researcher is confronted and confounded by a seemingly contradictory finding: for example, a product being perceived as too sweet and at the same time too sour. In such a situation the researcher must discard the adult's rational and logical view and try to understand the world through the eyes of the child, where black and white can exist simultaneously. What the child is saying is that the taste of the product is not distinctive in one direction or the other.

CONDUCTING RESEARCH WITH CHILDREN WITHIN THE BROADER CONTEXT OF THE FAMILY UNIT

The researcher is often confronted with the need to do package design research with children as well as with teenagers and adults. In doing so, there may be certain family interaction patterns and methodological considerations that should be taken into account. Often, the latter bear on the former.

Family Interaction Patterns

The researcher should consider whether the purchaser and the primary user are likely to be the same individual or different individuals. For many product categories, the child or teenager is the primary requester of a product, the mother then exercises the role of primary purchaser, and the child then assumes the role of primary user. At each stage, the package design must have the appropriate appeals. Therefore it is often necessary to research whether the elements of package design needed to excite a child to request a product and those needed to persuade the adult to make the purchase are similar or are different and, if different, how they can be combined in one package design.

The physical nature of the packaging itself may have variable appeals to different ages in the family. Children may want transparent packaging so that they can see what they are requesting whereas mothers may prefer an opaque packaging that allows reclosure. Portability may be a concern to children, but not to adults. Children, teen-agers, and adults may have different orientations to, and perceptions of, the color spectrum. A child might want to see a simplified picture of what can be made with a construction toy in order to feel capable of accomplishment, whereas a photograph of such a simple construction might lead the adult to feel the price of the set is not a good value.

It's possible that the reverse situation may prevail: mother may desire a photograph of a simple construction so she feels that if she buys the set her child would be able to accomplish it, whereas the child might think a simple form to construct makes the toy "babyish."

In food products, frequency of usage is often increased when packaging allows children to prepare the food without adult help. Thus the directions for usage that a child needs for single servings would be different from what mother needs for bulk preparation.

Many other similar types of questions and how they affect research methodology when conducting family research studies will become obvious to the researcher as the nature of the product category is considered.

Methodological Considerations

The two major methodological considerations are whether different members of the same family unit must be interviewed or whether separate samples of unrelated children and adults would suffice, and the degree to which the wording of the questionnaires and response categories must be identical for the different age groups.

In regard to the first consideration, family samples versus unrelated samples, the critical determinant for using a family sample is the presence or likelihood of interaction effects within individual family units that are likely to result in compromises between adult and child, or in one family member overruling another. If, however, the product category is one in which the parent almost always accedes to the child's wishes, such as in the cold cereal market, then the researcher could interview separate samples of children and mothers who are unrelated to each other. In this case the adult sample is being used to check out that there is nothing seriously confusing or objectionable about the package design among adults.

Wherever possible, however, it is recommended that parent-child dyads be used as this allows for a much more sophisticated analysis of the data. If analysis of the aggregate samples indicates a large discrepancy between parents and children, then the researcher has the option of reanalyzing the data on a family unit basis.

In regard to identical questionnaire and response category wording, it is almost impossible for all questions to be worded the same for both children and adults.

As discussed in the preceding, fewer scale points are used with children than are used with adults; children's language skills are less developed, and so on. The researcher should not sacrifice greater sensitivity of measurement among adults just for the sake of having identical wording between the two samples. Furthermore, the analysis and interpretation of adult data must perforce be different from that applied to the children's data.

SUGGESTIONS FOR ESTABLISHING A PACKAGE DESIGN RESEARCH PLAN AMONG CHILDREN

A quality plan of package design and research among children should consist of the following stages:

1 Predesign research (creative input phase).
2 Creative design and fabrication of the package (or packages) in prototype form.
3 Qualitative diagnostic evaluation of prototype package design(s).
4 Modification of the design(s) based on research findings.
5 Quantitative evaluation of modified design(s).
6 Final modifications of design(s) before marketing.

Predesign Research (Creative Input Phase)

A quality package design research plan should start off with some predesign research. Professional designers often create fairly complete package designs, which are then subjected to consumer evaluation and which are found to be unacceptable in totality or in large measure. Consequently, the designer goes back to the drawing board and starts all over again at square one. Much time and money would be saved if the marketer and package designers would start out with a small scale research study that has as its objectives creative input for designing the package(s).

This is especially important if the child is to be the major purchase stimulator and/or user of the product and its package, and the marketer or the package designer has limited experience in the child market in general or in the specific product category in question.

One effective approach is to conduct some focused group discussions with children using existing packaging for competitive products as a point of reference. Gathering insight as to how children view and use current packaging in the product category can help the designer avoid repeating mistakes already made by others and might even lead to uncovering possibilities for innovative package designs that can yield competitive advantages.

The use of existing package designs to stimulate discussion is necessary when we recall that children have difficulty thinking in abstract ways. To ask children what kind of a package would be good for a particular product will yield little usable information, but asking them what is good and bad about an existing, concrete package, will yield much information. Depending on the age of the child and his familiarity with the product category, it may even be possible to ask the child for suggested improvements. However, even if this is not possible, the analyst should be able to infer improvements based on the child's reported attitudinal and behavioral patterns.

Armed with this information, the package or packages would be designed and fabricated in prototype form and then be subjected to qualitative diagnostic evaluation.

Qualitative Evaluation of Prototype Packages

One can cite many examples in society of the failure to communicate clearly and to concretize ideas into actualities. This is especially true when adults attempt to transform into concrete forms ideas gathered from children.

It is therefore advantageous to see if what the child said has been interpreted and executed properly by the adults. This is much less costly to do in a small qualitative study than in a larger scale quantitative study.

As in the predesign phase, qualitative research is a viable choice for preliminary diagnostic evaluation of the prototype designs and for input into constructing a quantitative questionnaire to be used in evaluating the modified design(s).

Quantitative Evaluation of Modified Designs

At this stage in the design and research plan, the researcher and designer should have a fairly accurate notion of the broad range of attitudes and behavior patterns that might exist in the marketplace concerning these package designs.

It is now time to quantify these behavior and attitude patterns and, if more than one alternative design has been executed, to choose the strongest among these alternatives for eventual marketing.

The guidelines for conducting quantitative research among children discussed in this chapter should provide the researcher and marketer with the information necessary to maximize the effectiveness of the packaging.

CONCLUSION

This chapter has covered only the surface of the myriad conditions that pertain to survey research with children. The key to surveying children is to respect them as individuals, to respect their limitations, and to try to see the world as they do.

Children are highly reliable reporters of their own behavior and attitudes, even more reliable than an adult reporting on behalf of a child. The researcher who is willing to experiment and to build up a knowledge of children via research will have a competitive edge over his colleagues and will have fun in the bargain.

CHAPTER THIRTY-FOUR

Problems in Communications Testing with Unusual Samples

Anthony Armer

The problems caused by unusual samples in packaging and commmunications research are essentially the same as those in other forms of research, except that testing is generally restricted by the need to use visual stimuli; thus telephone interviewing is normally not an available technique and interviewing by mail may not be possible. This chapter attempts to indicate the various types of unusual samples encountered, the problems each presents, and some of the possible approaches to solving these problems.

DESCRIPTION

Samples may be considered unusual in two ways: They may come from a population that is difficult to locate, or they may have special characteristics that make it difficult to obtain correct information and/or attitudes from them. Some groups fall into both categories (for example, specialists). The first category includes:

1 Users or owners of low incidence

brands, products, or categories, such as owners of home videotape recorders, people who have eaten licorice in the past week, former subscribers to *Popular Mechanics*.

2 People with unusual demographics, such as men earning over 100,000 dollars annually, women in the last trimester of pregnancy, men age 30 to 35 with a master's degree and at least two children.

3 People with unusual psychological or personal characteristics, such as people who play strategy games at least once a week, supermarket shoppers who do not consider any one store their "regular" store, people who use newspaper advertising for information and believe television advertising to be unhelpful.

4 People in specific occupations, positions, or avocations, such as people in large companies responsible for the production of printed material, people in hospitals making the decision as to which brand of disposable electrode is purchased, chess players with at least eight points toward Master, urologists.

The second category includes children, ethnic groups, specialists, and family combinations (for example, mother-child, husband-wife). Specialists, such as physicians in specialties or chemical engineers, were listed in the first category also, but present problems in both areas.

PROBLEMS

The relative rarity of a desired population to be tested creates the problem of difficulty in location. Specifically, potential respondents may be spread geographically and/or may be difficult to

identify. Either problem increases the cost of research and generally extends the time needed to complete a study. Geographic spread may also limit methodology, since assembling respondents at a central location, either for group interviewing or to show them something difficult to transport, such as a television commercial or a supermarket aisle, may be out of the question.

Special characteristics that can interfere with obtaining correct data from various groups must be considered in methodology, questionnaire design, personnel selection, tabulation and interpretation of results.

Children

Not only is vocabulary a consideration in designing questions for children, but attention must be paid to the difficulty children have in conceptualizing and abstracting. They sometimes will focus on one detail of a picture, for example, and ignore the overall communication.

A child may understand the words, but not the significance of a question. Further, he may understand the question, but lack the vocabulary to answer it. Children also are often less likely to give a true unbiased response because of a general desire to please adults, a tendency to exaggerate, or fear or discomfort concerning the interview and its surroundings. Children typically have a limited attention span, so that interview length must be watched.

Ethnic Groups

In populations such as the Mexican-Americans, Italian communities, kosher households, language barriers must be overcome. Even if it is acceptable to exclude non-English-speaking people, the interviewer may be faced with respondents of limited vocabulary, who may misunderstand questions and who may have trouble articulating responses. Further, answers may be subject to misinterpretation due to a respondent's incorrect choice of words. Language problems are also present because of cultural idioms in subgroups who presumably speak "English," such as blacks, naturalized Americans, or Appalachian rurals. For example, in current vernacular "bad" can be a strong favorable response from an urban black.

A second problem is one of rapport. It may be difficult for a middle class white interviewer to get honest responses from a low income black, either because the black does not trust the interviewer or because the cultural difference encourages the respondent to give answers that will "please" the interviewer.

There is another aspect of cultural bias to be considered. In many communication research studies such bias is at least part of what is being measured (for example, Do white models arouse negative reactions among blacks? Are any foods in an illustration violating cultural habits or regulations?), but in some cases the bias may interfere. For example, in most Oriental cultures it is considered polite to nod agreement and smile, even if one does not understand the question.

Specialists

There also may be language barriers in interviewing specialists, who have a tendency to speak in jargon. Further, they may in some situations give the answers that they feel are appropriate to their position, rather than expressing true opinions to a "lay" person. Physicians, for example, may avoid disagreeing with an official AMA position, regardless of their own feelings; businessmen may only be willing to express the company line.

Another difficulty may be in conducting an interview at the respondent's place of business, although that may be the only practical location.

Family Combinations

The primary problem with family combinations is to get them to interact in the interview situation as they would in real life. The most frequent manifestation of interference is one person, normally an adult, and usually male, speaking for the others, effectively cutting off the opportunity of the others to express different views.

SOLUTIONS

Rarity

If the problem is simply collecting enough people meeting given criteria, several approaches can be considered. One is the use of lists sold by professional list services. Lists of special groups such as architects, presidents of large corporations, and donors to conservative causes can be purchased for so much per thousand names and addresses. In some cases phone numbers can be included for an extra charge, and in some cases titles, rather than names, are what is provided.

Other similar sources are the subscription lists of specialty publications, lists of prospects compiled by such publications, and membership lists of organizations. For example, golf publications can provide lists of subscribers, presumably golfers, either on a random basis (every "nth" name) or for a limited geographical area. It should be noted that all of these list sources vary widely in quality, that is, in the percentage of names still at the designated address or telephone number who still qualify on the desired criteria. It is advisable to check on the degree to which purchased lists are updated.

Some lists can be collected through secondary research. For example, births, marriages, and deaths are published in newspapers; purchases of residences are on public record; some hospital records are available; government sources are often gold mines of information.

Especially when looking for users of a particular low incidence product category or brand, the use of an omnibus study may be warranted. Many research companies conduct such studies quarterly, or even monthly, in which they ask a large number of respondents questions on assorted unrelated topics. Single questions can be bought that will produce a list of users of aerosol mothproofing, families in which the husband does the grocery shopping, wearers of soft contact lenses. Research is done by mail, by telephone, or in person, depending on the research company. It should be noted that some research companies use large national panels of 30,000 or more households for which they may have the characteristic desired already filed in their data bank. The limitation of this source are the time involved in obtaining the list and the necessity of using that particular research company for the actual research.

Another source for manufacturers looking for users of their own products is warranty cards or other inserts. Such cards can ask for additional simple screening information such as age or type of store in which the purchase was made. An incentive should generally be offered for return of the card except for warranty cards: the implied requirement of return to obtain warranty protection should be sufficient. Product insert or warranty cards have a high degree of bias, however, as evidenced by the low return rates.

The preceding methods of obtaining respondents may solve the problem of locating appropriate respondents, but often they do not help when it is necessary to bring respondents to one or a few central locations to show them something. In metropolitan areas it is often possible to hire buses to bring respondents to a central location, especially if they can be interviewed in groups. Another approach is to advertise for respondents, describing the qualifications and specifying a fee. The fee should take into account driving and parking costs as well as the length of time the respondent is needed.

Data Collection Difficulties

Children, as suggested in the section on problems, require care in designing questions, taking into account vocabulary levels, difficulty of abstracting, generalizing, projecting into the future, and attention span. Often questions such as those using scales need thorough explanation with examples and sample questions for the child that indicate to the interviewer whether or not the child understands how to answer.

Turning the interview into a game, or at least a happy experience, helps overcome some of the obstacles, such as fear of unfamiliar surroundings and a testing environment, desires to please or astonish adults, and the short attention span of children. For example, there has been widespread use of "smiley-face" scales for attitudinal measurements, and projective techniques are often used.

It is important to use interviewers who work well with children; generally such interviewers are nonthreatening, sensitive to children's moods, and observant.

Another recent development is the use of autonomic devices to measure the "real" reactions of children, as well as other groups, devices such as galvanic skin response, pupillary dilation, and voice pitch analysis. There is not widespread agreement on the validity of these measurements.

Ethnic groups, whenever possible, should be interviewed by interviewers of the same ethnic background. When this is impractical, as it often is, care should be taken to use interviewers with experience in talking with people of that ethnic background, preferably ones who do not have negative or strong superiority feelings toward that group. Such feelings can lead the interviewer to misinterpret responses and generally limit the rapport between interviewer and respondent. It may also be advisable to have someone connected with coding responses be familiar with the idioms of the ethnic group to avoid misinterpretation.

The jargon problem in interviewing *specialists* can be approached in two ways—either by using interviewers who are previously trained in the special vocabulary and knowledge, or by having the interviewers insist that answers be rephrased in language they can understand. An interviewer who knows how to "play dumb" can often elicit a great deal of information. Getting a specialist to give the interviewer enough time at his or her place of business is generally solved by advance appointments. Doctors and similar professionals may insist on charging the price of an office visit.

Dealing with *family combinations* requires the same skills required for moderating focus groups: the ability to draw out the shy without frightening him and to prevent dominance by the extrovert without totally silencing him, keeping people interacting without fighting or just agreeing with one another, and so on.

SUMMARY

The basic problems with research among unusual samples are difficulty of location and difficulty of obtaining correct data once located.

Relatively rare populations may be spread too thin geographically for efficient interviewing. They can often be located with purchased lists, data on public record, or use of omnibus interviewing. Advertising for respondents and bussing to central locations is sometimes advisable.

Children pose special problems in collecting and interpreting data due to their limited language skills and possible atypical reactions in an interviewing framework. Special questionnaires and interviewers are generally required.

Ethnic groups pose problems in collecting and interpreting data due to language barriers, idiom unfamiliarity, and lack of rapport between interviewer and respondent. It is advisable to have personnel who are either members of the ethnic group or who have worked closely with it involved in questionnaire design, interviewing, and interpretation of responses.

Specialists can be difficult to locate and may speak in jargon difficult to decipher. They are usually located through appropriate lists. Jargon is handled either by people familiar with it or by insistence on speaking in "lay" terms.

The problem of controlling the interview situation when family groups are interviewed is handled by interviewers skilled in working with group dynamics.

Thus problems in communications research among unusual samples can usually be overcome by allowing additional time and money and by anticipating the inherent inaccuracies and misinterpretations that can come from not recognizing the special nature of the sample.

CHAPTER THIRTY-FIVE

Testing the Teens

Lester Rand

At our last count (1978), American teen-agers (13–19 years of age) spent a total of $32.2 billion annually, an increase of $3.5 billion over the year before.

This is the largest single annual increase in this organization's 26 years of monitoring teen-age expenditures—almost a billion dollars over 1977s record $2.6 billion dollar jump over the prior year.

Teens are taking these giant spending strides although their total numbers are receding, with individual youngsters tightening the slack. Quite obviously the spending habit has become so ingrained over the years that it takes fewer young men and women to mount higher figures than the previous year. There were almost half a million fewer teen-agers in 1978 than in 1977 yet the total teen-age population was able to disburse $3.5

billion more despite this handicap. Inflation helped of course, but it was primarily a major effort on the part of young people that really did the trick.

HIGH DAY-TO-DAY SPENDING

To bring these numbers more sharply into focus, we see that 13 through 15 year olds spend an average of about $12 per week on day-to-day items while putting aside a little over $3 per week in order to accumulate enough money to pick up more expensive items later on such as cameras, radios, athletic equipment, stereos, hand calculators, and bicycles.

Meanwhile, 16 through 19 year olds are more than doubling this pace, disbursing a little bit under $28 per week on the average for day-to-day living expenses. They put aside an average of about $8.50 per week to acquire more expensive items.

Quite obviously, teen-agers are prompting cash registers to ring merrily throughout the United States.

Who are these consumers and how have they become so economically powerful at such a young age?

What's behind their rise?

Before discussing the various techniques which we use for testing their receptivity to packaging styles, it would be well to glance backward over our shoulders and see how the teen-age market developed. This tends to cast a light on teen spending behavior.

Since we work closely with teen-agers and are constantly analyzing their purchasing styles we are inclined to feel that people in general know all about them. After all, there are millions of parents who view teen-agers close at hand and just about everyone comes into close contact with them in one way or the other.

In reality, however, adults are mystified by youthful extravagance since most grownups were teen-agers during years of economic scarcity, and thrift was heavily accented.

So let's see what caused this giant to emerge.

THE TEEN-AGE MARKET— BACKGROUND

The affluent teen-age market is a combination of three forces that came together almost simultaneously:

1 A greatly increased birth rate following World War II.
2 General national prosperity.
3 An overall youth-oriented society that placed extraordinary emphasis

on looking and acting young, with a consequent premium on being young.

Then, of course, there was the advent of television and a variety of technological developments as well as mobility and more educated parents, which enabled young people to become more knowledgeable and sophisticated at an earlier age.

Also, a more permissive society allowed teens to be extraordinarily demanding and to secure virtually whatever they desired.

Relative to the rising birth rate, in the late 1930s and mid 1940s the yearly number of babies born was between two and one-half and three million. In the early 1950s, however, the annual number of births began its spectacular ascendancy over the four million mark—a definite bellwether of things to come.

For eleven straight years (1954–1964) the annual number of births rose over the four million mark.

Stimulated by this numerical explosion, the teen population swelled to a high of just under 30 million in 1975 and then receded to 29.7 million the following year, dropping to 29.4 million in 1977 and off to 29 million in 1978.

Spending Increases

In keeping with the greatly augmenting gross national product in the 1950s, teen spending, which heretofore had been comparatively lackluster, bolted sharply ahead. Families were earning more and consequently acquiring a much greater inventory of possessions than ever before. Young people fell right into this groove and did likewise. The booming birth rate was joining forces with a skyrocketing prosperity to create a teen spender of unprecedented dimensions.

Total annual teen-age disbursements rose gradually from 5 billion dollars in 1950 to 7 billion dollars in 1953 and then up to 12 billion dollars in 1959, 16 billion dollars in 1964, 23 billion dollars in 1970, 25 billion dollars in 1974, hit close to 29 billion dollars in 1977 and zoomed further to the 1978 mark of over 32 billion dollars.

Trend Setters

However, the most impressive monument to youths' economic ascendancy is the intensity with which industry has geared its styling, advertising, and marketing to teen-age tastes and preferences.

One of the chief motivations behind this new trend is the unprecedented 'youth kick' evident throughout the United States.

With about half the population aged 27 and under and the other half striving through diet, dress, cosmetics, sports and automobile ownership to achieve a youthful image, it is no wonder that the teen-ager finds himself at stage center. After all, he has what everyone seems to want.

The largest manufacturing industry in the United States, automobiles, has swung over to a policy of styling its product to conform to the adolescent flair. The entire clothing industry is heavily youth oriented. Moreover, records, movies with a string of disco products on tap, and most of the rest of the entertainment world mirror the teen-age syndrome.

Then, too, the cosmeticians, those weavers of glamour and fantasy, have zeroed in on teen girls, making them a leading target of their wares.

Teen-age Persuasion

However, teen-age economic strength extends beyond their own, individualized purchasing.

Within the family, we find, teen-agers have a strong say about almost everything that is bought for family use and pleasure. From what we have seen, their influence stems chiefly from persuasion and a surprisingly keen knowledge of products being offered in the marketplace.

Many young males, for example, are so thoroughly grounded in automobile mechanics that fathers often ask a son's advice before purchasing a car. Boys in general appear to have a far greater knowledge than adults about the engineering qualities and performance of various makes of automobiles, and fathers throughout the United States seem aware of this. On the selection of colors and interiors of cars, however, daughters' preferences may be persuasive.

In a survey we conducted a few years ago, the influence of adolescents on car purchases alone approximated 15 billion dollars per year. The overall effect of youths' persuasiveness on parental purchases may very well hover between 50 and 60 billion dollars a year—speaking conservatively. Moreover, the rapidly increasing capitulation of adults to teen-age tastes indicates that the end is nowhere in sight.

Young people have become so accustomed to an affluent life that they were undeterred by the business slowdown in 1974–1975 and still busily urged more cautious and conservative parents to purchase household and family products. What adults may consider postponable luxuries are looked on as necessities by youths raised in a highly modernistic environment.

Stimulating the Economy

In a nationwide survey conducted by us in the spring of 1975 ("Stimulating Consumer Demand: The Role Young People Play"), we found that almost two thirds of all teen-agers (63%) urged their parents to buy products for the home and for the family despite the economic downturn at that time. This contrasts with the early 1950s when only about one fourth of American teens maintained that they exerted any sort of buying pressures at home. As the postwar boom intensified and teen-agers became more economically significant, the role of teen-agers within the family became far more prominent. Not only were teen-agers increasing their own personal spending, but they were actively engaged in seeking to get their parents to buy the newest and latest products being offered.

Teens became quite adept at persuading parents to buy all sorts of home and family products during the booms following World War II. Adults tended to be cautious and careful, remembering the more difficult economic period of the recent past and wondering about the future. But kids had no such hang-ups and proceeded to whet parental appetites and inject enthusiasm for a spectacular array of new products. This lobbying proved highly effective then, and it is helpful today as young people keep up their drumbeating for all sorts of merchandise.

With this type of background, teen-agers can be relied on to 'do their thing' whenever a recession threatens.

Continued Importance

Now that teens have been discovered by industry and have become a prime market, the young consumer will in all likelihood remain an important sales target despite declining numbers.

This is because of sharply increasing per capita teen income and spending as well as the overwhelming emphasis we continue to place on youth and on being young. Also, the total number of teens, even at its lowest point, will probably never dip below 21 million. A spending force of this dimension is well worth

cultivating, and once a market has been highly developed and recognized, companies will continue to strive to provide it with a variety of products.

The startling increase in population—spurting ahead at the incredible rate of one million a year between 1965 and 1972—caused industry to ooh and aah over the sales potentials such numbers suggested. But teen-agers have always been around, and it only took this population explosion to bring them to the fore. Now that they are so highly visible, there's little question but that they will remain so.

So this is the teen-age market. It has in the 1960s and 1970s grown numerous and financially strong. Now let us look at the varying methods for determing the differing preferences of this market.

RESEARCH TECHNIQUES

We employ a variety of tried and accepted techniques for conducting effective marketing research among this nation's teen-age population.

The technique used depends largely on the type and sophistication of information being sought.

We maintain a large cadre of student interviewers in high schools, youth clubs, and Junior Achievement units who have been carefully selected and trained to conduct interviews among their contemporaries. Since the 1950s we have found that under certain circumstances young people are inclined to speak more easily to people their own age. They are less suspicious of the questioner and consequently are more candid in their replies. We also utilize this method on college campuses.

One great advantage of this approach is that it is a lot easier to find large numbers of young respondents in their normal habitats. It is also generally less expensive than searching for teens in their homes.

Interviewers are supervised by teachers or youth club workers who see that questionnaire instructions are understood and that individual personal interviews are satisfactorily completed with forms properly filled out and returned. Supervisory personnel are also involved in the validation process.

Although this technique has been time-tested and is popular, it is generally limited to a survey calling for a comparatively short interview with fairly simple objectives.

Brand penetration is a good example.

1500 or so teens in varying parts of the United States may be asked to identify their favorite deodorants, how often they use them, and how long they have had such preferences. A list of reasons for desiring deodorants may also be shown the respondent for check-off purposes, this list may be open-ended in order to round up unusual and perhaps significant reasons for brand preferences. A few similarly simple questions can very well be asked in addition to demographic data.

We have found that the youthful interviewer, when closely supervised, can accomplish this type of assignment satisfactorily. He's not bogged down by a long and unwieldy questionnaire which is all too frequently difficult to understand. The purpose of a survey is to secure marketing data. But all too often companies seek too much information. This makes the questionnaire too long and the consequent interview rambling and unstimulating.

Better and higher quality information can be elicited by a concise, interesting approach. We continually urge companies to adopt this more relaxed form of questioning on the theory that the end product will have less quantity but undoubtedly more usable information.

Professional Interviewers

It is naturally necessary at times to probe more deeply into the youthful psyche in order to secure a broader profile of the consumer under investigation. This calls for a longer, more complicated questionnaire with open-end questions. Under these circumstances we work with professional adult interviewers. They are more patient and better equipped to interview in greater depth.

We may be probing attitudes towards life insurance, savings banks, or perhaps big business.

Although the questionnaire is longer with the interview running 25 minutes to half an hour, it is still most important to vary the questions in such a manner as to make the whole interview interesting and, if possible, entertaining.

These days we invariably use professional interviewers at the shopping malls where they frequently have easy access to large numbers of teen-agers. By a quick screening process they can usually find the type of respondent they are looking for.

On some occasions, however, interviewing is conducted in areas where young people congregate. This may be at sporting events, around schools, near fast-food establishments, at play yards, or at clubs.

A more expensive approach is house-to-house interviewing where the residence of teen-agers is determined beforehand.

Focus Group Interviews

Our favorite technique is focus group interviewing. This enables us to go out into the field and unearth a great deal of marketing information rather easily and not too expensively.

The focus group should have the relaxed informality of simply speaking with your neighbor and finding out what he is thinking. We vary our number of participants from four to eight depending on the situation.

We may be trying to discover teen-agers' interests in certain snacks—and facts about snacks in general—but we attempt to pace the session in such a manner that the teens find themselves discovering a great deal about themselves and, most importantly, how young people in different sections of the United States live and think. As they are informing us, we are informing them. Since we conduct focus groups as well as quantitative surveys throughout the nation, and are up-to-date with what young men and women are up to, we are in an excellent position to liven up such sessions with a great variety of data teens find absorbing.

More than most age groups, teen-agers are interested in other teen-agers. They want to know the styles, preferences, and thinking of teens in different cities, states, and regions throughout this country. They wish to know how similar or different they are.

Various Tests

When testing the acceptability and possible success of a package which is under consideration, we invariably commence with a focus group interview. We highly recommend this procedure.

For example, if we are measuring the effectiveness of a soft drink bottle or can, we invite approximately eight teenagers (four boys and four girls), aged 13–15, to sit in. The setting is usually in a large living or recreational room atmosphere where the respondents can lounge comfortably on sofas and easy chairs. Other sessions are held composed of the same number of teens, aged 16–19, equally split between the sexes. Sometimes we find all male and all female panels preferable, as mixed groups

can prove inhibiting to some young people. Two or three focus groups usually indicate which composition is working best.

The moderator might start off by discussing snack situations or those times when soft drinks are consumed. This is an easy subject for young people of this age to handle since they are deeply involved in the intake of all sorts of sodas. It's something they know a lot about.

Prior to getting deeply involved in the subject at hand, we might delve into new movies teens prefer, disco dancing, or roller skating. This is just a brief skirmish to put everyone at ease, and we get back to business rather quickly.

We then introduce several different soft drink bottles and cans with the packages under investigation included. The teens are requested to give their top-of-the-mind opinions of the various candidates. They are asked what they think of the different items and why they might select one over the other. This is not a quantitative assessment. We are merely cultivating different thoughts and ideas that can be of great value in advertising or presenting the product at the point of sale.

The thought flow is invariably quite heavy. The respondents may be tentative and hesitant at first as they, like most people, never really bother to sit down to analyze why they use the products they do, and in many instances may not even recall the toothpaste or soap they've been using for years.

But someone makes a telling comment and another follows it up, and pretty soon the conversation starts to build. Gradually people begin remembering things they had never really thought about seriously.

As the various reasons for usage unfold, we find that the packaging is not considered particularly critical. Teenagers do not consume one Coke after another or reach for a Pepsi because they are intrigued by the bottle or container involved. More important, these drinks have been around a long time and are highly regarded, friends like them, they are consistent in taste, and of course they prefer the taste.

So why bother to test for packaging in this instance?

Because desirable and eye-catching packaging adds an important dimension to the drink's preferability. Standing alone, packaging is not of the utmost importance. The number of teen-agers who buy a leading soft drink simply because of the way it is packaged is extremely small.

A somewhat unknown soda, on the other hand, could very well be purchased because of the eye-catching shape of its bottle or an overall color motif that is beguiling. At any rate, we consistently find that colorful and unusually appealing packaging will prove a magnet and draw teens to it whether they buy or not.

Stimulating the Discussion

The panel moderator will stimulate the discussion around the way the various exhibited drinks are bottled and canned as well as the appeal of the packaging. The panelists aren't aware that they are being pointedly tested for their thinking on packaging. This aspect is, to them, merely one small part of the overall discussion about snacks, snacking situations, and the displayed products.

It really has to be done this way to sustain personnel interest and thus hopefully secure what may be considered the true thinking of the respondents concerning soft drink packaging. If the sessions were merely a lengthy inquiry into packaging per se, apathy would soon set in. There really isn't that much of an unforced nature to talk about. Pretty soon the kids would start to make up

things and grope for ideas in order to appear productive. But they wouldn't be giving their honest feelings on the subject.

On completion of the first focus group interview, the tape is replayed and analyzed in order to secure a 'feel' of the thinking being developed as well as of the direction in which future discussions may go. The most important points that are highlighted are investigated further in the next group session, a building process that is repeated from group to group.

After about five or six such sessions, the principal outlooks and underlying beliefs of the young people we are interviewing are pretty well solidified. Further discussions may be held in various parts of the nation, if this is deemed necessary and desirable in order to test the validity of the predominant hypotheses.

Although these concepts may seem increasingly valid as they gain support in various panels, we recognize the fact that we have still only spoken with a comparatively small number of subjects.

In order to present valid findings, it becomes necessary to quantify the obtained data.

Quantification

The hypotheses obtained in the focus group sessions are incorporated into a questionnaire—preferably a concise one—in order to nail down what we believe to be the attitudes of teen-agers towards the packaging we are testing. The questionnaire is administered in malls and around schools by adult interviewers. In addition to the formal questionnaire, pictures of the various bottles and cans used in the focus groups are shown to the respondents. Five hundred interviews are conducted at various interviewing points, either regionally or throughout the United States, depending on the client's needs and feelings concerning geographical differences, product distribution, and advertising.

These procedures may of course be varied. Marketing research is not an exact science despite the fact that final reports presented to clients are impressively packaged and the statistics contained within seem exact and precise. The questions used, the skill of the interviewers, the way questions are asked, how they are interpreted and analyzed, and many other factors enter into the picture and influence the final results.

Consequently we find a great deal of experimentation.

Different Procedures

We invite a small group of teens into a supermarket and ask them what they think of the packaging and displays in the soft drink section. Or we may go into a restaurant and bring up the subject as we order drinks.

Why not have a simulated party with the usual snacks and soft drinks?

We have found that while young people—and this undoubtedly goes for adults as well—will express definite choices and preferences when shown a range of products in an artificial interviewing situation, their wishes can be quite different when inside a store and exposed to virtually thousands of other products. The items that seem so important when isolated in an interview become almost lost inside a market, which is where the real selections are made.

Several years ago we were testing one of the many types of new pens that are constantly being introduced. Since the teen-age market would be a prime target for this product, the way teens regarded it was very important. We showed the pen to various groups of teens and queried them closely about it. It was new and unusual and receptivity was high. Most of the respondents said

they would like to own one. But products aren't sold in this manner.

When people go into a store they are confronted with an enormous number of competing brands. They don't have time to find out about everything and test one item against the other. As a result, the information that we found out about the pen under ideal laboratory conditions had a certain amount of value in the marketing process, but the manufacturer had to restrain his euphoria over the test results. The marketplace is the true test. As it turned out the pen did quite well, but not as well as the survey enthusiasm lead an unsophisticated marketer to believe.

New product success as well as new packaging success cannot be easily predetermined by what samples of the population—no matter how accurately researched—say beforehand. People are very likely to say they will buy something when they know they don't actually have to take out the money and do so. After all, they'd like to have most things.

It is the function of the marketing researcher to discover what people want through a variety of questions and discussional leads. If he is successful, this can be highly advantageous to his client's marketing procedure, but it does not guarantee success. We bring this up to show that in research experimentation is important.

How many focus group interviews should be conducted? Where? How many panelists should be included in each group? How many respondents in a quantitative study? Should it be regional or national, and where should the interviews be conducted? Then there's phone interviewing and mail. These, as well as others, are the various decisions that have to be made when testing teens for packaging or anything else.

In the case of the soft drink study that we discussed before, quantification might not be considered necessary if we felt that the data being developed were sufficient. The client may have just about made a decision and needed a little more backing from the public. Perhaps it was a question of seeking various hypotheses and ideas to be used in advertising and marketing. On the other hand, focus groups could have been eliminated or minimized and a quantitative study conducted at the onset. It all depends on the situation and the information being sought.

Much the same procedure as we outlined concerning soft drinks is utilized in measuring teen attitudes and preferences regarding just about any type of packaged goods being evaluated.

OTHER PACKAGING

Cosmetics and grooming products are sold in large numbers to the teen-age market and we are consequently called on to check the packaging preferences, among other aspects, of fragrances, lipsticks, deodorants, powders, shampoos, hair dressings, and so on.

This is an area where packaging is extremely competitive since a highly acceptable container gives the impression that the product within has extraordinary capabilities. In fact, if the package is really outstanding, the young purchaser is inclined to retain it long after the product has been used. This is particularly true of exotic bottles and boxes.

It is hardly any great surprise to learn, as we did in one survey, that 86% of all teen-agers interviewed throughout the United States felt that packaging is very important. It might, however, be startling to discover that 79% of the teens believe they frequently purchase products because they are influenced by the way they are wrapped or packaged. Quite often, teens buy package over product.

Packaging in this instance also includes the way the product is colored and shaped. We consider these figures exceptionally high because when queried about purchasing behavior people generally like to give the impression that they are careful shoppers looking for value. They're not inclined to admit that they can be taken in by flash and glitter and appeals of a nonfunctional nature.

Teens, however, tend to consider packaging quite functional. It gives them the feeling that the product seems better and is therefore better. Something packaged in silver and gold colors conveys the image of wealth and quality. Teenagers find bright colors attractive and cheerful. There is also the matter of anticipation. A beautiful package strongly suggests that there is an item of great desirability within. Teen-agers are extremely susceptible to such lures.

GULLIBLE BUYERS

The magnetism of packaging to teens is explained by their general purchasing gullibility. They are continually buying things that they have little or no use for. Their impulse shopping is a big weakness they wish to overcome. Over and over again they save up a few dollars to acquire something for which they have absolutely no need. After a few days they don't even look at it. Their rooms and closets are loaded with junk.

In one survey it was found that 69% of the teens interviewed said they bought expensive items they seldom used. Some three fourths of this contingent said they wished they could overcome this problem.

We don't mention this in order to suggest that manufacturers take advantage of such teen buying practices. It is merely a description of the way young people who are just entering the purchasing arena tend to act. It is part of the maturation process. The mere fact that they are fed up with such shopping indicates that they are gaining sophistication. A large number of people, however, never do get over these habits and go through their lives acquiring merchandise that they have little use for.

As we noted at the beginning of this chapter, however, teen-agers have acquired an enormous amount of purchasing power. Boys and girls have shed their traditional, heretofore insignificant role as primarily candy, ice cream, and soda purchasers and have assumed a loftier and catered-to status as shoppers for such costly items as cameras, stereos, hand calculators, radios, motorbikes, athletic equipment, typewriters, TV sets, and cars.

Moreover, teen income is largely disposable. In most instances they are not required to support themselves, to pay rent or mortgage interest, buy insurance, or pay the types of bills parents are confronted with. Because they have little access to credit, it is difficult for them to get into debt.

So teen-agers are out there buying. The marketer who is well versed in catching their eye via the right color combination, logo, design, and shape will in all likelihood find young people reaching into their pockets. They want things.

CHAPTER THIRTY-SIX

Evaluating
Package Effectiveness
in Hospitals

Julie H. Williams

This section of the *Handbook* is devoted to packaging research in segmented markets. The hospital does indeed present special challenges to designers, marketers, and researchers alike. Products used in hospitals have two customers whose needs must be met—the user (doctor, nurse or other medical professional) and the end-user (the patient). They must be designed and marketed within the bounds of newly amended industry regulations for medical devices. Marketing research of any kind is new to the field and requires modification of the traditional approaches to data collection. Moreover, the health care system as a whole in the country is undergoing

an identity crisis whose implications the astute marketer cannot ignore.

The chapter begins with a brief review of the state of the art of medical marketing research and proceeds to a discussion of the function of medical packaging and how it interacts with its environment. There are also sections on the special requirements for conducting packaging and other types of research in hospitals as well as a summary of the marketing and research objectives of the packaging research with which we are familiar. Although involved on a daily basis with health care research over the past 8 years, the author found this assignment challenging simply by virtue of the infrequency of packaging studies. The "detective" work proved informative and we are left with the overall impression that medical marketers increasingly recognize opportunities to adapt consumer marketing strategies to their own needs and those of their special customers.

MEDICAL MARKETING RESEARCH: THE STATE OF THE ART

The health care market has historically done little marketing research. The exception is the pharmaceutical industry whose major representatives have had research departments for 20 or 25 years. Manufacturers of hospital supplies and equipment, however, have lagged behind both consumer and industrial companies in applying research to the marketing decision process. This includes even diversified corporations such as Johnson & Johnson, whose consumer and pharmaceutical companies have used research successfully to develop and market some of the country's best known brands (Band-aid,* Johnson's Baby Shampoo,* Modess,* and Tylenol* to name a few) but whose hospital divisions and companies have staffed marketing research departments only recently. A survey by the American Medical Surgical Market Research Group of its membership in January 1979, fixes the marketing research department of the average $75 million and smaller medical supply company as having existed for only 5 years; those of larger companies have existed for about 10 years.[1] Not only are medical supply company research departments quite new; they have fewer staff members, smaller budgets, and more limited activities than their pharmaceutical counterparts.

The interest in health care research

*Trademarks of Johnson & Johnson.

has been spurred in part by the entry of several major consumer products marketing firms into the industry including Colgate-Palmolive, Bristol-Myers and Procter & Gamble.[2] Also, more marketing jobs in health care firms are being filled by people with consumer product management backgrounds—people who are accustomed to the information research can provide to help them manage their brands. The days of a sales representative "testing" a new product with a couple of friendly customers definitely seem to be numbered.

If research providing even basic market information has been absent from the health care field, packaging research has been almost totally absent. The manager of a major pharmaceutical company's research department described packaging research in his 10 years with the firm as happening "almost by accident." The American Marketing Association's 1978 Survey of Marketing Research shows usage of packaging research among consumer and industrial companies, financial services, advertising agencies and publishers and broadcasters as follows. (See table 1)

While we know of no comparable data for health care firms, the files of a well known manufacturer of a wide variety of hospital supplies yielded fifteen references to packaging research in nearly 500 projects over a 10 year period (April 1969 to September 1978). Our own company has been involved in approximately a dozen packaging studies for four different companies since 1972.

In what kinds of marketing research does the health care industry engage then? Pharmaceutical and medical supply companies use research quite differently.

For medical supply companies most marketing research is product related—evaluation by the user of the acceptability of a new product or design modifications in an existing product. Form

Table 1   Doing Packaging Research* (Design or Physical Characteristics)

	Total	Consumer Companies	Industrial Companies	Financial Services	Advertising Agencies	Publishers Broadcasters
	60%	83%	65%	39%	79%	55%
(Sample size)	(798)	(186)	(200)	(106)	(57)	(51)

*Reference 3.

definitely follows function, and the key question usually is Does it work? Information over and above the establishing of basic functionality/acceptability (e.g., usage and benefit data to assist in the positioning and communication processes) are often slighted or totally ignored to concentrate more completely on function. One of the most common reasons for product testing has been the evaluation of cost improvements as a way to help offset price increases and improve slipping margins. The specific area receiving the most attention is material substitution (e.g., synthetic nonwovens for gauze and muslin).

The manufacturers of medical supplies also do a considerable amount of concept screening for new product ideas. Rarely is this done quantitatively as it would be in a consumer package goods company, however, Focus groups are the method of choice. The reasons for shortcutting a more comprehensive method are probably the high cost of personal interviews with medical respondents (especially physicians) and the difficulty of sorting out multiple buying influence in the market.

Finally, the collection of usage and share data through primary research methods consumes a substantial portion of a medical supply company's marketing research budget. Secondary data sources are limited in both number and scope. There is only one firm providing a hospital audit, and there is no syndicated product information at all on distributors of medical supplies (a service

comparable to SAMI, for instance) or for market segments such as the physician's office and nursing home. Although sizable in terms of dollars, the medical and diagnostic equipment markets are not audited either. Even in the hospital area there are products such as pacemakers that fall into the gap between medical devices and drugs and thus escape tracking. In our experience, this absence of reliable and complete market monitoring constitutes the single greatest frustration for the medical marketer.

Compared to the one or two person department in a medical supply company, pharmaceutical research departments resemble those of consumer package goods companies in size and scope of activities. Most of their research budgets are spent on various types of attitudinal and promotional research, the latter category encompassing copy research and studies of ad effectiveness. Much of the research is done on the market (versus the product) including measurement of market potentials, market share, and sales analysis. Product testing is, of course, out of the question, and secondary sources are well developed and reliable.

THE FUNCTION OF MEDICAL PACKAGING

Let us set aside for the moment the idea that the medical marketer currently uses the research department to satisfy his or her most basic information needs and consider packaging, the subject of this book.

In analyzing the absence of packaging research in the health care industry two factors must be considered—the package itself and the environment that encompasses both the situation in which the product is used and the marketplace.

To begin at the beginning we must examine the function of a medical package. Is it the same as that of a consumer package goods package? Does its design and the execution of that design merit the careful attention a new product to be sold in a grocery store or other type of retail outlet will receive from marketing?

Boyd and Westfall mention that the growth of self-service has dramatically affected the function of packaging.[4] The package must not only protect its contents and deliver the product to the consumer in a viable state, it is also a key factor in point of purchase impact. Packaging carries the product's message to a potential buyer as well as creating an image for the manufacturer. It must do so in a busy, cluttered environment where many messages assault the buyer's senses simultaneously and compete for his attention.

In contrast, the health care industry seems to regard medical packaging as strictly functional. The functions include:

- Maintaining the product's integrity in the shipping, distribution, and storage processes.
- Providing the doctor or nurse (that is, the user) with the necessary information to use the product properly.
- Protecting the patient (the product's end-user) from the results of package abuse.

Of secondary importance, if considered at all, is using the package to provide the doctor or nurse with an added benefit such as convenience, thereby distinguishing it from competition or the broader objective of building overall company image.

The engineering department of a major manufacturer of medical supplies describes the criteria around which his company's packages are designed. Not surprisingly, the process is more systematic than creative. The four major considerations are:

- Is the product sterile?
- If sterile, how will it be sterilized? The method of sterilization (steam, etylene oxide, or gamma radiation) will dictate which packaging materials can be used.
- What is required to keep it sterile until presentation? Carefully controlled studies simulating the rigors of shipping and handling are conducted on materials not used in other packages the company makes.
- Finally, what is the medical community accustomed to in this product category? Is there a reason to deviate from the packaging norm?

In view of this last point it's not difficult to understand why packaging in a given product category is so similar across brands. There are basically three types of sterile presentations for disposable medical supplies: the peel-back package/bag, the tear string package/bag, and the tray.

The aesthetics of a package (its graphic design) can be supplied in whole or part by engineering, marketing, and/or an outside design firm/advertising agency. Graphics are not considered to be as critical a design element as they are in consumer packaging because medical packages are rarely displayed in a competitive environment, and they usually arrive in the user's hands *unpack-*

aged. The interviewed firm had several guidelines for its package graphics. The firm wanted the package to be "professional" looking, maintain a "family" image, and make no claims (except those required by law) on the package exterior. Other companies, no doubt, have similar guidelines, which could explain the predominance of two color combinations (often red and white or blue and white) and uncluttered package exteriors.

Government regulations, of course, also play a role in medical packaging. Rules for the labeling of drug packages and package inserts leave little room for creativity. Medical device packaging is currently subject only to regulation regarding the sterility claim (adulteration and/or misbranding) although this may change for certain classes of devices as specified by amendments to the Federal Food, Drug, and Cosmetic Act of 1976.

Finally, the user can influence medical packaging in ways other than the one alluded to earlier (that is, familiarity with design). Several of the dozen or so packaging studies conducted by our company were the direct result of customer complaints in the areas of sterile presentation and ease of opening. The customer's other main area of concern is disposal. With the advent of single-use items in almost every product category, the hospital has become a garbage can for litterally tons of unrecyclable paper, plastic, and metal. Assistance with disposal is an area of potential opportunity for manufacturers and one that appears to be largely unexplored.

MEDICAL PACKAGING IN ITS ENVIRONMENT

Beyond the emphasis placed on function by both the manufacturer and the user, there are a number of other factors that diminish the importance of the packaging of medical products. Considered as a whole, the following factors indicate why medical marketers place their priorities elsewhere.

- The health care industry is one of the, if not *the,* most highly regulated businesses in the country.

- It is still a "hands-on" business; personal contact with the customer by a sales representative is the single most important part of the selling and image building process.

- There are multiple buying influences for more and more medical supplies. The user who would benefit directly from a product or packaging improvement may have no say in its purchase.

- Many of the products used in hospitals are never seen in their packages by their user. Even when they are, however, medical packages almost never compete side by side for the buyer's attention as do packages in a retail situation.

- The scrutiny under which the health care system as a whole has been placed is forcing the medical marketer to deal with issues much more fundamental and broader in impact than packaging.

Let's examine these points one by one.

Industry Regulation: The pharmaceutical industry especially has little leeway with what and how a claim can be made about a drug. Once a patent has been granted and the 17 year clock starts winding down, a drug is as likely to be held off the market over the contents of the package insert as the company's ability to establish the product's safety/efficacy. New legislation will soon force manufacturers of the formerly unregulated medical device category as well to establish the safety of products ranging from pacemakers and orthopedic implants to tape and tongue depressors. Thus the same regulations many industry experts hold responsible for the so-called "drug lag" will soon apply to marketers of all types of products in the health care field. Rules and regulations must certainly put a damper on the creative spirit not to mention their drain on financial resources.

The Sales Representative: Health care sales is still a very personal business. The physician sees representatives of the companies whose drugs he buys on a regular basis, and hospital purchasing agents' offices are crowded with their suppliers. *Advertising Age* estimates the 1979 expenditures for advertising in the health care field will be close to $1 billion with the greatest portion of these dollars being allocated to the sales force.[5] It is not uncommon for a large pharmaceutical house to have a field force in excess of 1000 men and women, and the industry as a whole has more than 20,000 full-time sales representatives.[5] Deployment of resources in this manner obviously influences how advertising dollars are spent. Sudler & Hennessey, the largest health care ad agency in terms of gross income in 1978, breaks down its billings with 30% devoted to detailing and sales material; journal advertisements are second.[6] Image in the health care field is largely the result of hands-on sales and service. A company is its people, not its packages or displays.

Multiple Buying Influences: The user of a medical product (the doctor or the nurse) doesn't select those products the same way he does products for use in his nonprofessional life. As a matter of fact, users have no influence on the purchase of more and more of the products they use daily. Hospital purchasing departments, frequently banded together in buying groups for added clout, exercise substantial control over the selection of many medical supplies. Several strong for-profit hospital corporations even warehouse and distribute their own goods. Committees within the hospital fight to standardize usage of the same products department by department with an eye to keeping the budget under control. They also conduct evaluations prior to purchase in which several competitors' products might be put into use in order to choose the one that best meets the hospital's needs. This can be a very effective barrier to a company not included in the evaluation that tries to get its product considered at a later date. While there must still be examples of the prima donna surgeon for whom special items are stocked, this situation is rapidly becoming a thing of the past.

The "Naked" Product: The physician especially does not see many of the products he uses in their packages. In order to accommodate technique, a sterile product is almost always presented to him by a nurse out of the package, ready for its application. Brand awareness of all but the tools he considers absolutely essential to his trade, therefore, is alarmingly low, especially to marketers accustomed to consumer levels of awareness. A related issue is the hospital that supplies its own packaging. High volume, easy to process items such as sponges are often bought in bulk from the manufacturer, sterilized in the hospital, and "put up" for the various departments that use them. Under these circumstances, the hospital assumes responsiblity for the packaging and follows guidelines such as the Association of Operating Room Nurses' *Standards for Inhospital Packaging Materials*[7] and the standards for Central Service as outlined in the *Accreditation Manual for Hospitals* published yearly by the Joint Commission. Products handled this way have no commercial packaging as far as the user is concerned.

No Competition: When the doctor or nurse does make a buying decision on his own (for the office, for example) it is usually directly from a sales representative or perhaps from a distributor catalog or price list. Rarely do medical packages compete side by side for the purchaser's attention as they do in a retail situation. Even if the product is not "naked," then, it is either all by itself or not physically in front of the purchaser.

The Health Care System in Flux: The scrutiny under which the health care system has been placed in our country today strongly influences a marketer's priorities. Packaging seems to come out very near the bottom of the list and probably rightly so. State and local governments in ever-growing numbers are effectively forcing the substitution of generic drugs for their proprietary cousins. In low risk, high volume categories such as sponges or underpads, hospitals are beginning to respond to cost pressures by openly buying on price instead of quality. In the race to control the cost of health care, third party payers are picking and choosing which diagnostic procedures they will pay for. Professional Standards Review Organizations rule on whether or not an institution will be allowed to provide a given service or add new beds. Hospitals are starting to look at themselves as marketers for the purpose of adequately and efficiently serving their consumer and generating revenue in ways other than filling beds. In the midst of these not always clear or compatible forces, it's not all that hard to justify the "wait-and-see" attitude of some health care marketers.

USE OF PACKAGING RESEARCH IN HEALTH CARE MARKETING

Despite the fact then that little packaging research is done compared to other types of medical marketing research, the projects with which we have been associated directly and/or those unearthed from cooperative companies' files illustrate a broad range of objectives. Table 2 summarizes the projects by marketing objectives, research objectives, and methodology, and gives examples of specific studies where possible. Our survey of the medical marketplace was by no means complete, and we cannot be sure it objectively represents the universe. It is impossible to say with any high degree of certainty, therefore, what the most frequent application of medical packaging research has been. Experience tells us, however, that it has probably been the need for an immediate solution to a specific packaging problem that jeopardizes an existing brand's franchise, or the desire to effect a cost improvement through packaging. What is done the least frequently is quite easy to identify, however; it is an evaluation of a package's graphic effectiveness using any one of the many techniques that might be applied to a consumer package. We found only one example of such a project and it was undertaken when a division of a large corporation became a freestanding company and Marketing wished to identify the best of several alternative designs for the packaging of its major product line. The company had to establish awareness of a new name as well and chose packaging as an avenue of communication. Not listed on the Table are any examples of attitudinal research on packaging since we have chosen to classify such projects as attitudinal (or motivation) research. A successful product whose package "becomes" the product (e.g., plastic IV containers) inevita-

Table 2 Types of Medical Packaging Research

Marketing Objectives	Research Objectives	Methodology
Revise to Correct Problem Change a package that has been generating customer complaints. Revision could involve a specific component or entire package. For example—bandage whose sterile delivery to the wound site is compromised by its package; packaging for surgical drapes that doesn't survive shipping and causes storage problems in the hospital.	Determine user *preference* for test versus current. Decision rule: test *better than* current.	Actually put test package into use among current customers.
Cost Improvement Change a package to effect a cost savings. Examples— automate a packaging process for hand-rolled plaster of Paris bandages; eliminate expensive cellophane shrinkwrap from prefilled needle and syringe packages.	Determine user *preference* for test versus current. Decision rule: *equal to or better than* current.	Put test package into use among current customers (and possibly among competition, depending on additional information sought).
New Package Design Design a suitable package for a new product or product line. Examples—partial-fill containers in a company's new line of all plastic IV containers; choose exterior design for an electronic thermometer.	Determine user *acceptability* of the new package. Criteria for acceptance usually based on functionality. If a control is available (e.g., competition), a decision rule may be written embracing the control.	Put test package into use among target audience.
Communication Effectiveness Determine the communication effectiveness of the graphics of a package or packaging line.	Establish visibility, eye movement, and other applicable ocular measurements for various packaging elements. Criteria for acceptance based on analysis of individual and cumulative scores of the specific technique being used.	Expose test package alternatives to target audience under strict controls the specific techniques demand. No usage necessary.
Make the Package the Product Provide the customer with one or more benefits not offered by competition through packaging. The marketing objective can be to establish, protect, or increase the franchise. Loss of patent (and the corresponding influx of generics), competing in a market that has very few manufacturers (e.g., sutures), or the challenge of the marketing of a commodity product are all motivations for this marketing strategy. Examples include Baxter Travenol's original plastic IV container, Viaflex, and Tubex, Wyeth's injection system for its IM drugs. container, Viaflex, and Tubex, Wyeth's injection system for its IM drugs.	Determine user *acceptability* of the new package. Criteria for acceptance usually based on functionality and comparison to a control (which could be the current or competitive packages). Decision rules variable depending on specific marketing objectives.	Put test package into use among both current customers and competition.

bly generates a flurry of this type of research from competition.

DOING RESEARCH IN HOSPITALS

Due to the emphasis on functionality, most medical packaging studies will require putting the product into actual use. Research among medical professionals, especially in the hospital environment, has some important differences from consumer research. These differences should be taken into consideration whatever the type of study or data collection method to be used. The key differences are:

Terminology: Medicine has a language of its own that must be incorporated into questionnaires, concept statements, and other study materials. The researcher has to use the technical terminology of medicine to be taken seriously by those being interviewed. This doesn't change any of the rules for building a good questionnaire, however. An analogy might be translation into a "foreign" language—"consumer-ese" to "medical-ese."

Intimidating Environment: The hospital is an intimidating environment for even the most assertive marketing research professional. Interviewers describe having the feeling of being constantly on the verge of making a mistake while not knowing exactly what it will be. Coping with the strange and unfriendly surroundings can produce feelings of anxiety. Consequently, specially trained "executive" type interviewers are necessary. They need to have highly developed interviewing skills, a comprehensive briefing for each project, and it's a good idea to expose them to the environment in which they will be working to desensitize them somewhat. Another way to approach this is to hire medical

professionals to do the fieldwork. Our company has had good success with former nurses.

Structured Environment: As a research company operating in the hospital environment, it's hard to know who invented bureaucracy—the government or the medical profession. Hospitals are organized along strict lines that can seem as inflexible to an outsider as a caste system. Ignoring channels of operation (and especially the power structure) when setting up a project will almost always lead to trouble. The name of this part of the game is definitely play by *their* rules. Unfortunately these rules vary from institution to institution and an interviewer will need to know the lay of the land before she can operate effectively.

Accessibility and Tolerance: Doctors and nurses, hospital administrators, and purchasing agents are hard to reach, and they generally don't have a lot of time to talk when you finally do reach them. It is necessary for an interviewer to work her schedule around theirs and, because she's interviewing them on the job, questionnaires must be briefer than those used for consumer research. The issue of tolerance has implications for test design as well as for questionnaires. We have found it impossible to do more than a paired comparison type evaluation in a hospital, although consumers regularly taste and/or use three or more products. For some categories even a paired comparison is impossible. A proper evaluation of plaster of Paris, for example, will take 12 to 16 weeks, and an orthopedic surgeon's patience has been tried sufficiently by then to make him an uncooperative prospect for a second product placement.

Multiple Influences: As referred to earlier, the purchaser and user of a medical product are often not the same person.

It is increasingly difficult to sort out purchase influences with the advent of Product Evaluation Committees and the trend towards group purchasing. We know of no current techniques that successfully model the multiple buying influences in hospitals for medical supplies. The picture certainly differs by product category, with pharmaceuticals and equipment requiring capitalization being the easiest to deal with. In order to cover all the influences, multiple interviews are often necessary. Even this doesn't completely solve the problem since one is still left with the question of how to weight the various responses.

Smaller Sample Sizes: For usage studies in particular (both product and package), sample sizes slightly smaller than those typically used in consumer research are possible—75 per variable instead of 120, for example. Seventy-five respondents should be enough to measure differences (if they exist) using a five-point scale at the 95% level of confidence. The reason is the homogeneity of the profession due primarily to sameness of training. Standard deviations are low for medical scale data—often below 1.0 on a five-point scale.

For the user of medical research the implications of these differences are:

- Research that is somewhat higher in cost than consumer research (probably comparable to high quality industrial research).

- More time required to do the field work.

- The potential for compromise on amount of information being sought, although not necessarily the quality of this information.

DOING PACKAGING RESEARCH IN HOSPITALS

Keeping in mind the guidelines for hospital research in general, there are a number of additional issues to consider when fielding package research.

The Usage Situation: A glance at the methodology column in Table 2 shows how frequently a medical package must actually *be used* in order to be evaluated properly. It is important to make the usage situation as "real" as possible. This is best accomplished by prior observation in hospitals and pretesting in order to provide the field workers with the best study protocol. Hospital procedures can differ, and it may be necessary to collect data in several usage cells to accommodate the most common procedures. Some examples follow. If the package is normally opened with wet hands (as a roll of plaster of Paris would be) make sure it is evaluated that way. Or, if the package is operated with one hand because the other hand is occupied (a tube of sterile lubricant, for instance), take this into consideration in the study protocol. If it is unreasonable to expect a nurse to refer to lengthy setup and usage instructions in the real world, don't bias the test by providing them. A client of ours wanted to solve his package's problems by assuring the nursing staff of a lengthy in-service ("medical-ese" for training and education). Nurses told us the category is not normally in-serviced and doubted the hospital would want to take the extra time to do this, especially since there were products available that required none. On the other hand, if in-servicing is required, be prepared to do it for the study. If the interviewer cannot be trained in the procedures, a representative of the sponsoring company should be tapped. For complicated packages for which a learning period is advisable, the methodology should provide for a trial period of a week or two, depending on frequency of category usage.

The Use of Prototypes: Packaging research with almost any objective will involve the use of a prototype, possibly one that is handmade for the project. It is important that the prototype be as representative of the finished package as possible and very durable since it is likely to be handled and examined at close range. If the package must actually be used over a time in a clinical setting to obtain a fair evaluation, the contents obviously must be safe for use. Consideration must also be given to whether the package is reproducible on a production line. We have seen many product managers frustrated over Manufacturing's inability to reproduce the results of a pilot line.

Use of a Control: While it may be impossible to avoid using a commercial package as a control, you should be aware that respondent familiarity can have an effect on the study results that is even more marked than in consumer research. In a consumer package test one would expect a well-known brand to receive quicker recognition by more people in a shelf display than a brand new test package. In the world of medicine the analogy is comfort. If ever there was a group which has developed rituals and is reluctant to make unnecessary changes, the medical profession is it. A new package could suffer unjustly simply because it's different.

Sampling: Although sampling is vitally important to any marketing research study, the power structure in hospitals presents the medical researcher with a dilemma not usually found in consumer research. When clearing a medical project through channels, the interviewer will encounter lots of "experts." The tendency will be to try to push her as high up the ladder as she will concede to go to get respondents. However, in packaging in particular the best respondents will usually be users of the product, and these people may well have a comparatively lowly status in the overall

hospital structure. LPN's, technicians of various types, and even nurses aides and orderlies may be the best evaluators of a medical package. A physician who sees many products only after they've been removed from their packages is no judge of functionality.

Just as choosing the best qualified respondent is important, so is choosing the correct hospital area. Many products are used in more than one area and a package may be called upon to meet different criteria for different areas. This is particularly true of specialized areas such as the Emergency Department and Operating Room where speed and technique are emphasized.

Observation: Observation, a rather infrequent method of data collection in consumer research, can play an important role in medical packaging research, especially if the package will be used only once or twice and not over a period of weeks. Observation may be necessary to determine, for example, if a respondent is opening or handling the package correctly when she cannot be depended on to know this herself. Or the researcher may want to determine how easily a particular task can be performed. Boyd and Westfall[8] point out that one of the drawbacks of this method of data collection is that the observer is human and, therefore, subject to error. Moreover, an observation of the ease or difficulty of a task is subjective by nature. In medical research there is the added difficulty of training the interviewer well enough in the particular procedure to be an accurate and informed observer. Even a nurse interviewer will not be familiar with all types of techniques and procedures. Observation, by the way, usually would be used not alone but in conjunction with more conventional direct questioning methods.

Application of Decision Criteria: Like product evaluations, much packaging research will lend itself to the establishment of a decision rule or rules. The formulation of these criteria should take place before beginning the field work, since they could influence the questionnaire design, sample size, and/or research method. They are based, of course, on the marketing (as opposed to the research) objectives of the project, and it is vital that the researcher understand the issues well enough to assist in their formulation.

The Questionnaire: Although questionnaire design will depend on the specific objectives of the study, when the project involves usage of the product and its package we favor the following format as a minimum:

- Importance ratings (or forced ranking) of the various elements of the category's package. This must be done *prior to* a respondent's exposure to the test variable(s).

- Intent to use or intent to recommend for use scale ("top box" analysis is useful as in consumer research.)

- Open-ended questions on likes/advantages, dislikes/disadvantages/improvements. Medical respondents, like most people, enjoy talking and can feel hampered by an entirely structured interview. Furthermore, open ends are an effective way to check that something important hasn't been overlooked in the attribute lists.

- Rating of the packaging elements on the hedonic scale of choice (five-point, nine-point, etc.).

- If using a control, preference for each packaging element and preference overall.

If the package will be used over time and the questionnaire is self-administered, provisions must be made for the form to be filled out after sufficient time has passed and learning has occurred. We've found self-administered questionnaires useful as a way to accommodate a hospital's schedule, demonstrate our flexibility, and reach particularly inaccessible respondents. They must be carefully edited by an interviewer, however, and recontact could be necessary.

REFERENCES

1 *Survey of Market Research Operations,* conducted for the American Medical Surgical Market Research Group, January 1979, by Custom Research, Inc., Minneapolis, Minnesota.

2 L. Edwards, "P & G Tries Lucrative Health Care Field" *Advertising Age,* **48,** No. 52 (1977), 1.

3 D. W. Twedt, Ed. *1978 Survey of Marketing Research,* American Marketing Association, Chicago, 1979, pp. 40–44.

4 H. W. Boyd and R. Westfall, *Marketing Research Text and Cases,* 3rd ed., Irwin, Homewood, IL, 1972, p. 10.

5 J. Levere, "Media on Their Marks in Race for Ad Dollars". *Advertising Age,* **50,** No. 15 (1979), S-10.

6 N. Giges, *Advertising Age,* **50,** No. 15, (1979), S-10.

7 *AORN Journal,* **23,** No. 6, "Standards for Inhospital Packaging Materials". (1976), 978.

8 Boyd and Westfall, p. 160.

CHAPTER THIRTY-SEVEN

Design Research for Bilingual Packaging

Arline M. Lowenthal

When manufacturers make the decision to market their products in countries other than their native country or in bilingual countries, heed must be paid to the pitfalls of bilingual packaging. Packages and packaging communicate and promote the product as well as provide instructions on usage while serving as containers and outer wrappings. The following areas must be addressed when considering communications in bilingual research.

When designing bilingual packaging and its associated research, the input of a native resident of the country or areas concerned is of utmost importance. A common pitfall is overreliance on bilingual personnel who happen to be in house and conveniently available. For example, a company makes a decision to produce a bilingual, Spanish/English package for a product to be marketed in Mexico. It so happens that one of the executives in the marketing department is a Mexican-American. He was born in Mexico, is completely bilingual, and although he has lived in the United States since he was 16, he still visits his family in Mexico. He appears the ideal executive to use on the product. In some ways he is. In other ways, overconfidence in his bicultural abilities may endanger the success of the entire project. Potential problems may include all or some of the following:

- Any person living for many years in another culture inevitably acquires some of that culture. Not only is the individual often unaware of this cultural merging, but it is frequently extremely difficult for him to determine where one culture ends and another begins.

- A person educated in another country will tend to think on a different plane from that of compatriots in his native country. A Mexican attending college in the United States will not only tend to be more highly educated than the majority of the Mexican consumers we are trying to reach, but may think in different terms from people educated to a similar level in Mexican institutions.

- Because a person visits or maintains frequent contact with residents of his native country, he does not necessarily maintain a familiarity with all the trends, fads, current events, and minor changes in custom and idiom that are continually taking place within a culture at any given time. Our nonresident native might easily use a word, phrase, or image that would have strong topical connotations leading to a negative, misleading, or even ridiculous interpretation.

- When a person is away from a culture for a period of time, there is a tendency for recollections of that culture to be imperfect. As a result, our Mexican-American resident of the United States may perceive Mexico in an overly romantic light, exaggerate some aspects of life, or be unaware of certain aspects of life just because they played no part in his boyhood recollections, and were never called to his attention during his later visits.

- Groups of expatriots in a foreign place will tend to develop a subculture with mores and values of their own. This can lead, for example, to a package ideally suited to the Mex-

ican-American population of New York City, but inappropriate to Mexico itself.

The very nature of these problems makes them extremely difficult to detect without the assistance of someone extremely close to the culture in question. In the example cited, our bilingual executive may save himself and his company time, trouble, and cost by taking a copy of his first draft down to the basement of the building and trying out his ideas on the Mexican cleaning woman who only arrived in the United States three weeks ago.

In the same way, while a good bilingual research team can be invaluable in developing bilingual copy and packaging, it must be remembered that the team's expertise is in research, not in developing and editing bilingual copy. The fact that a knowledgeable member of the research team may be able to point out problem areas in a piece of copy even before commencing research does not replace the research itself. Let us suppose, for example, that a research team conducts research on a set of Spanish instructions. They find that certain phrases are inappropriate for some reasons, but come up with other phrases that would be more suitable.

This will be valuable information, but further research must be conducted with the revised copy in case the *combination* of these new phrases creates an undesirable innuendo.

The culture of the countries in which the package will be offered is of the utmost importance. Symbols on packages, colors, logos, and artwork being considered must be studied to be sure they do not offend or ridicule the culture of the country in question. Even when the language is the same, the cultures may differ. For example, Spain, Puerto Rico, Cuba, Mexico, and most of the Central and South American countries have vast cultural differences although they are all Spanish-speaking countries. The research must be structured to determine any necessary culture-oriented changes that must be made. The colors used in bilingual packaging must be carefully researched, since they assume a cultural value. The values and image orientation of various colors are different from country to country, and considerable thought should be put into research regarding color in bilingual packaging.

Just as there are cultural differences, there are also idiomatic differences within the same languages; thus the same bilingual package may need the input of natives of three or four countries or areas to avoid error-laden copy or instructions. In French, for example, the meanings of various words differ in France, in French Canada, and in Haiti, all of which use French as a main language. Even within France there are provincial differences and regional dialects and idioms that can give words connotations they do not have elsewhere.

The same problem exists in Spanish, since the same Spanish idiom can mean something entirely different in Mexico, in the Carribean Spanish-speaking islands, and in parts of Spain. An example of this is the idiom "Ahorita." In Mexico it has the meaning of immediacy, "right this *very* minute." In Puerto Rico, however, it means "when you get around to it, no hurry, later will do." If you want to convey "right this minute" in Puerto Rico, you would use "Ahora mismo," which would bring prompt attention. Thus, one would have to take care when writing Spanish package copy urging consumers to "buy this new package right away."

As for the name of the product (brand name or logo), means of communication can be responsible for the success or failure of a given product. Acronyms sometimes spell out words of lewd or offensive nature and thus the product may either be too well remembered or forgotten intentionally.

A consultation or secondary workover of the copy by several "native" experts of a given country, area, or region would be a worthwhile investment to avoid embarassing errors in copy that could turn out to be incorrect idiomatically or, even worse, off-color or double in meaning. It is true that since we in the United States are aware of most of our slang and other words with dual meanings (e.g. grass and pot) we are careful about using or not using them in a name or trademark. Unless, however, considerable secondary and possibly primary research is conducted, this could prove quite problematic in a foreign language.

The area of legality must be considered especially when dealing with the packaging of patented medicines, pharmaceuticals, food, and other products where a risk factor is present. It is important to insure against failure by complying with existing local packaging laws. The retention of an expert in labeling for a given country and region is a worthwhile idea, as is the use of secondary research to obtain competitively similar products for a comparison of the labeling language used in the same type of products by other manufacturers.

The point of "habit" and "practicability" is one that must be addressed in performing bilingual package design and the research for it. To research simply the desirability of the package design is not enough. Present packages being used for similar types of products should be carefully researched to determine whether "habit" and/or "practicability" are reasons for selection of the package. Because of weather, customs, ease of transportation, and so on many countries use different packaging materials from those used in the United States. Plastic bags are used in place of paperboard cartons in many Latin countries. Plastic is easier to get than is paperboard in Mexico and other Latin countries. Thus the habit is formed based on practicability, and a package that can answer the needs created by these two concerns should be designed and subsequently tested for the discussed points. Also if packaging material is to be different from that normally expected by the consumer, it should be researched for its effective functionality as well as its habit-breaking appeal. This is an example of one product type, but it happens in many different items; thus some advance research is necessary before starting the original package design.

Included in the package design is usually the task of writing instructions for use of the contents. It is a difficult enough task to write this necessary information in English, but when one considers the varying educational and comprehension levels of most of the countries for whom these instructions may be written, one must stop and research ways of creating more meaningful and usable copy. We were faced only recently with a task of instruction simplification in English. The key problem for the market segment, as shown through research, was the word "insert." When altered to read "put in" (along with other modifications of the instructions), greater understanding of the instructions was achieved, and the repurchase level for the product rose to previously expected levels.

The same problem exists to an even greater degree in bilingual package design, and thus research should be conducted to assure total comprehension of instructions by the consumer. We have found many creative ways of offering these descriptive or instructive sections through qualitative and quantitative research. It must be conducted, however, by a trained researcher familiar with psychographics and idioms of the researched group.

Thus if all normal steps for package design research are carried out and these additional cautions are taken under advisement, a successful package design should result and contribute favorably to the overall success of the product.

PART SIX

International Package
Design Assessment

CHAPTER THIRTY-EIGHT

Coordinating and Analyzing Multinational Package Research

George M. Gaither

GENERAL CONSIDERATIONS

Package research abroad and package research in the United States are conceptually the same. The specific objectives of the research may vary from one country to another because market and marketing conditions vary, but the underlying *purpose* does not. What *does* vary, often sharply, is the means by which that underlying purpose is achieved.

The problems of conducting research abroad are administrative as much as they are technical. The existence of the technical problems is self-evident, and

therefore in a sense less critical: technical problems are highly visible and by extension very likely to be thought through carefully and resolved before a project is undertaken. Administrative problems are less in evidence and therefore less likely to be given the careful consideration they deserve. Moreover, in many ways administrative problems are more subtle and complex, more *difficult* to resolve than the technical ones.

It is clear that there must be full understanding of the technical implications of modifying a survey technique so that it can function in a different cultural environment. It is sometimes less clear that there must also be full understanding of what the facilities and the state of the art in general make it feasible to *do* in a given country; what the educational levels and general cultural background make it feasible for respondents to *understand* in a given country; indeed, what the package design objectives should *be* in that country, given the marketing/retailing/competitive realities that exist there. These are administrative as much as technical concerns.

If the market is reasonably well understood and the research facilities there are reasonably competent, technical questions are no more of a problem internationally than they are domestically. The need for special concern shifts to the administrative sphere. But if there is *inadequate* knowledge about a country and/or the facilities there are *not* reasonably competent, then both technical and administrative concerns share the same high priority. In every case, the administration of a project is of critical importance.

Because of the apparent difficulties, it is easy to be discouraged at the outset of an international project. However, these difficulties are far from insurmountable. Some of the most sophisticated packaging research in the world has been and is being conducted outside of the United States. There is a growing body of expertise in this country that is focused specifically on the international field. Virtually anything can be done that sponsors of research may need or call for.

Recent packaging research has covered problems and used techniques that range from simple measurements of attitudes toward disposable containers to the effect on product image, on a worldwide basis, of differing packaging materials, to the development and design presentation of a brand name in the Roman alphabet that will convey the desired image even where the Roman alphabet is unintelligible. The body of data and of experience is growing every year, as the following chapters indicate. Some of the pitfalls, and some of the ways to avoid or overcome them, are touched on in this chapter.

THE NEED FOR COMPARABILITY

Multinational research usually requires comparability of data from one country to another. However, what is technically feasible in one country may not be feasible in another.

The techniques to be used in a given project must always be chosen with a view to the countries involved—the cultural context in which they are to be applied, and the facilities that are available to apply them. One of the enduring dilemmas of multinational research is the need to reconcile the requirement of comparability with the almost infinite variety of situations in which the research must be done. Again, the problem is administrative as well as technical.

Even a matter as apparently simple as scaling is subject to cultural variations. Semantic scales encounter the problems of translation; when subtle wording is involved, it is virtually impossible to render all concepts accurately in many languages. Different cultures give different values to numerical scales, and many do not use them effectively at all; the education that makes their use simple to sophisticated populations is not present in others. (In one African country it was found that the only scale respondents could understand and use was a drawing of a hill, with the scale represented by the distance of a man's figure from the top.) The recognition of the existence of differences such as these, and the potential existence of others, is fundamental to ensuring that the right survey design is used and that comparability is in fact achieved.

Not all international research is or should be multinational, of course, and not all multinational research covers the full spectrum of international markets. A distinction must be made between the requirements of a single study in a single country and of a study limited to a single region, or of one that covers the full gamut of possibilities. The basic study design may be radically different in each case because of the range of potential variation in the capability to apply it. If a single country is involved, the techniques may be as sophisticated as that country permits. If several countries are involved, the capabilities of all of them must be taken into account.

The objective of each project must be examined carefully, with this basic concept in mind. Some studies are clearly limited in their geographic scope and can never be used outside the country in which the research has been conceived. Others may seem to have purposes that are geographically limited, but in fact they are not: they will be found to have broader applications at a later point in time. (Many companies, for example, develop normative data from repeated applications of a single research design, and this concept can apply to international markets as well as to domestic ones.) Yet if the technique selected for one country is one that cannot be transferred to another, the benefits deriving from that study will always be limited to its initial application—simply because no other country can replicate it. This is a mistake that should never be made.

If the study is multinational from the outset, or if it may become a benchmark for other measurements in other places and at other times, then the technique selected for that study must be one that is broadly applicable. In effect, the lowest common denominator must be chosen.

To a great extent, these considerations relate to the structure of the questionnaire and the analytical plan that is envisioned. However, sampling is a major consideration as well. It is of course essential that the universe of study be comparable from country to country, and that the sampling method be free of bias—it must provide an accurate cross-section of the market or market segment it purports to represent. But as distinct from the questionnaire or from the analytical design, sampling method does *not* have to be identical in every country. If a probability sample is used in one country, probability samples do not have to be used in *every* country. The method most likely to yield an accurate representation at reasonable cost is the method that should be chosen.

Different countries tend to emphasize different kinds of sampling techniques. In Latin America, for example, probability or modified probability sampling tends to be the most effective, because the data that make quota samples accurate simply do not exist in reliable form in most of those countries. Conversely, the French, the British, and many other European countries tend to use quota samples much more than any other method—because they do have the necessary inputs to make this type of sampling accurate; because it is more cost efficient; and because it is the technique with which they are most familiar.

In sampling terms, therefore, comparability of data can be achieved by comparability in the definition of the universe of study. It is not necessary that the specific sampling method be identical from country to country.

TECHNIQUES THAT ARE LESS THAN UNIVERSAL

"The lowest common denominator" in research design implies simplicity, and in fact, the most universally usable techniques are the simplest ones. Studies that require elaborate facilities or highly specialized personnel can rarely be extended beyond those places that already have such capabilities—and there are relatively few of these. Outside of the industrialized nations, there are almost none.

Simplicity in design need not imply inadequacy or inaccuracy. Nevertheless, some of the techniques that are not advisable for a study that plans to cover the full spectrum of international markets, now or at some later point, include the following:

1 *Methods Requiring Special or Unusual Equipment, Especially If That Equipment Is Not Compact and Easily Portable* It is highly unlikely that sophisticated equipment will be available in all of the countries in which the study is to be conducted. It is especially unlikely to be found in the developing nations. It must be brought in, and the transportation of such equipment is difficult and costly. In addition, customs regulations in the developing countries make it exceptionally difficult (if not impossible) to gain entrance, even for a brief period of time. Other factors (e.g., differing electrical systems from country to country) compound the problems of importing electrical devices of any kind.

2 *Methods Requiring Exceptionally Capable or Sophisticated Interviewers* The caliber of field personnel varies greatly from one country to another. Some of the finest interviewers in the world can be found outside of the United States, but these are usually limited to the industrialized countries and certainly are not to be found in all research agencies even there. In developing countries, some may be barely literate. A truly multinational design must leave as little as possible in the hands of the interviewers themselves.

3 *Methods Based Entirely on Scaling* Language differences are such that any elaborate form of semantic scaling is exceptionally difficult to render in identical fashion in every country under study. Obviously, certain basic words are common to all languages (good/bad, hot/cold, male/female, etc.), but the nuances implicit, for example, in multidimensional mapping must be handled on a country-by-country basis or at least on a language-by-language basis.

Numerical scales also must be used with caution; not all cultures react the same to all scales. Although they are not usually the subject of commercial research, there are major segments of the world's population for whom numbers of any kind have little meaning. More commonly, in a practical sense, there are major segments of the world's commercially active population that are unaccustomed to working with or thinking in terms of numerical differentiation. Finally, certain cultures are more predisposed to respond positively, or to respond negatively, than others, with the consequence that different average scores on the same scale in two different cultures may really have the same meaning–or the same average scores may have different meanings.

4 *Methods Requiring Expert but Subjective Evaluation* These methods can be useful, but must be used with caution. Group discussions, unstructured depth interviews, and so on, even when they are done well, must be analyzed by a person highly familiar with both the technique and the culture at hand. It is exceptionally difficult to apply these methods multinationally, even though they

may be of great value in a given country or region. It *can* be done, but the director or user of the research must be aware of the pitfalls and of the danger of comparing conclusions arrived at subjectively by different analysts in different cultures.

The most dramatic problems, and the most dramatic differences between cultures, are found in the developing nations. The more remote the country, the more "unusual" that country is likely to be. In a truly multinational study, the problems these countries generate must be taken into consideration if the overall objectives are to be achieved. At the same time, however, it would be a mistake to place so much stress on these developing countries that some broader or more complex objective fails to be fulfilled. The developing countries should not overbalance the industrialized countries, when there is no special need for it; the tail should not wag the dog. When this danger threatens, it may well be more advisable to eliminate the developing countries from the project, thus making it possible to use more sophisticated techniques without loss of comparability—or to cover the developing countries as a supplementary project, recognizing their lack of comparability and utilizing different techniques that have been developed especially with them in mind.

The number of countries in which truly sophisticated research can be done is surprisingly limited. Outside of the United States and Canada, only Japan and Australia in the Far East, and only the more advanced countries of Europe, fall into this category. (Spain and Portugal, for example, do not. The Eastern European countries do not; even Italy can be a question mark.) In Latin America, the Middle East, and Africa (with the exception of South Africa), facilities are generally very limited indeed.

GUIDELINES FOR TECHNIQUES THAT CAN BE APPLIED

The list of techniques and devices that are difficult to employ seems almost all encompassing. It is not, however; straightforward, structured, simple questionnaires are universally applicable.

Sophisticated devices and techniques can be applied wherever it seems feasible to do so on a country-to-country basis. However, for cross-cultural comparability, any multinational study must contain at least a nucleus of straightforward questions—questions that can be asked anywhere. The additional information that may be garnered from more elaborate methods can represent an extra dividend in a given study; it can even be a valid subject for experimentation, so that the extent and direction of biases can be detected. However, it should always be viewed as secondary to the hard-core data based on questions that even illiterates can understand, when a truly multinational project is involved.

The need for simplicity in the phrasing of questions is just as important as the need for simplicity in the project's overall design. In this case as in so many others, language differences usually do not permit accurate rendering of complex phrasing. Even where these language differences can be overcome, other factors often intervene. In many countries, for example, several languages or dialects are spoken, some of the languages are not written ones, and either interviewers or interpreters make verbal translations from the base language as the interview progresses. (For example, in Singapore—a relatively sophisticated market—interviewers must be prepared to interview in four languages: English, Malay, Tamil, and Chinese. The Chinese—usually but not always the Hokkien dialect—is not a written language for most of the interviewers or the respondents, so that interviewers transliterate the responses. Other countries have even more extreme problems of this type. Nuances of phrasing are clearly lost in these situations.)

Monadic testing in these circumstances, for example, is especially suspect. Whenever possible, testing should be comparative: a preference declared between two alternatives is clear in its meaning, but a statement that a package or a product is "very good," or that it achieves an average rating of 4.6 on a six-point scale, may have entirely different meanings in different places.

When more than two alternatives are being considered, it is always safe to have them placed in order of preference or in order of their association with some concept. Again, monadic ratings are difficult to interpret cross-nationally, and when multiple alternatives are involved these individual ratings are usually not sensitive enough to permit clear differentiation among any but those that fall at the extremes.

Obviously, not all package research can be based on questions such as Which do you like best? even when the lowest common denominator is being sought. It does not have to be *that* low. However, whatever is done must be *clear*—not subject to misunderstandings; wherever possible it should offer choices rather than separate evaluations. If the imagery of a given design is being evaluated, for example, certain fundamentals can always be assessed no matter how primitive the society involved and no matter whatever *else* may be assessed on a secondary or an experimental basis.

Such questions as Would this design be used more by men or women? or Which of these products would be used more by well-to-do people and which would be used more by the poor? are universally understandable, present clear alternative choices, and make reference to variables that are pertinent to any society.

Nevertheless, just as it may be a mistake to rely solely on monadic devices, it would also be a mistake to discard this valuable technique completely. There is too much evidence of its utility in the United States and in other sophisticated markets to cast it aside lightly. In addition to the fact that it is a perfectly valid technique in and of itself when both the research facilities and the cultural level of the population are adequate, it can under certain circumstances be an invaluable *secondary* analytical tool in developing countries as well. Monadic ratings of specific dimensions of a product or package, coupled with forced-choice data, can give added insights into the reasons for choice. One logical consequence of this is the rather widespread use of protomonadic formats—designs in which each entity is evaluated monadically and then rated comparatively in the later stages of the interview.

There are, of course, many reasons for the limited reliability of purely monadic testing among less sophisticated populations. The lack of familiarity with numerical sequences has been mentioned, as has the desire of some populations to please—to the extent that even their "negative" responses are clustered at the upper end of the scale. However, one general concept helps in reducing this problem and stimulating a more flexible range of response: in developing countries, it has been found that the more "real" the item to which respondents react, the more likely they are to respond realistically to it. Thus an actual cigarette package is more likely to be reacted to realistically than a cutout package face panel; a cutout package face panel is more likely to be reacted to realistically than a drawing of that package face panel, and so on. Descriptive terms and concepts are the most difficult of all to stimulate meaningful response. Coming to grips with relatively abstract concepts, and evaluating them by relatively abstract yardsticks, is exceptionally difficult for the uneducated populations of the world, but the difficulties are reduced if the research design can provide a greater measure of tangibility.

FIXING OBJECTIVES

Just as facilities and applicable techniques may vary from one country to another in a multinational project, the objectives of the project may vary from country to country as well. It is axiomatic in package design that the first requirement is to identify the objectives of that design—the marketing function it is being asked to perform. The same is true in the international applications of that design; yet often an objective that is appropriate for one country may not be appropriate for another.

This may seem to be self-evident. Often, however, the goals for a package design are fixed only with the world's more sophisticated markets in mind, and those goals bear little relevance to goals that are appropriate for a developing country. The research design either should be modified to encompass these countries with differing needs in differing ways or should be modified structurally. A project that at first glance is perceived as multinational may not be; there may be two or even three projects involved, because the goals of the package design itself, on inspection, may be different for different parts of the world.

These kinds of distinctions relate primarily to the conditions of the market in which a package design will perform. It is essential to have some grasp of the kinds of consumers that exist in a given country—their literacy or lack of it, their general level of sophistication. It is equally important to know the nature and intensity of the competitive environment, as well as the retailing context in which purchases can occur. In sum, packaging research must begin in the framework of at least some knowledge of the package's environment. The more

different the market seems to be from the American market—or the home market of the designer, whichever that might be—the more fundamental the questions that need to be asked.

If the package design is seen primarily as an additional stimulus to purchase, in what context do purchases occur? Is this an item that will move from self-service shelves or will it be marketed from small stores where the product will be marginally visible, if at all?

If the package design is seen primarily as a conveyor of image, what image is desirable in a given market? What criteria do consumers apply to product selection in this case?

In many instances, especially in developing countries, one of the key functions of a package is to instruct the consumer in the product's use. What information, if any, is needed? How does this differ, if at all, between the information needs of the sophisticated and ignorant consumer?

The container is in some cases an item that is perceived as having value in itself. To what extent is this a factor? How is the product stored by the consumer? Is the container used again—or should it be?

These are self-evident questions. What is not always self-evident is that they should be asked in each case for every country. The answers may alter the nature of the study to be done—or the country in which it will be done.

It is unlikely that *definitive* answers can be obtained before a study begins; usually not that much information is available about many of the world's markets. However, simply raising the subject for questioning ensures that the possibility will at least be logically considered.

In cross-cultural research, nothing is a given. The implications of form or color that are "well-known" in the designer's home market may be quite different in another. Prestige may come

more from self-effacement than from os-
tentation. The man may be the supermarket shopper rather than the woman.
Questions such as these need not be
asked in every survey; clearly this would
be an impossible task. However, nothing
should be accepted simply because it is
"known," unless it is known for that
particular market in which the research
is to be done. An open mind and sensitivity to the *possibility* of difference
make for a successful multinational project.

STEPS TO TAKE

In a practical world, the conduct of an
overseas research project—whether in a
single country, a region, or multinational—should follow a certain logical progression. Although some steps may be
omitted or the sequence may occasionally vary, in essence the key steps are
the following:

1 *Determine the Scope of the Research*   Is the geographic scope the
   right one? If a study is for a single
   country, should it be made compatible with prior or future research in
   other countries? If it is multinational,
   should it be? Are the marketing needs
   for the several countries involved
   sufficiently compatible for a common
   design to be used?

2 *Determine the Countries in Which
   the Research Is To Be Done*   If the
   research is designed to solve a specific problem in a single country, this
   question does not arise. However, if
   it originates from a strategic need—
   if, for example, it is designed to establish a design, or design criteria,
   for a world-wide product—a choice
   must be made among the many countries that might possibly be the location of the research. Not every country can ever be covered, nor should
   it be. Regions, however, must be

represented. Costs and, of great importance, the quality of local facilities
available for the research must be
taken into consideration in this
choice. Facilities make for feasibility
and are often the determining factor
in selecting one country for the research instead of another.

3 *Seek Out the Facilities Through
   Which the Work Will Be Done*   Two
   basic approaches are possible: to deal
   with a research firm that specializes
   in multinational research or to go
   directly to overseas firms in the individual countries under study. The
   middle road—using a central firm,
   but one that does not have international experience—is also possible,
   but in essence this is simply a variant
   on the concept of "going direct." In
   this situation, an inexperienced central firm simply acts as agent for the
   study's sponsor, and performs the
   same functions that the sponsor himself would otherwise perform: it
   "goes direct" to local firms in overseas locations, with no international
   input of its own.
   No matter how contact is made, there
   are two key points that must be
   learned about the overseas research
   firms that may do the work in a given
   country—aside from the universal
   essentials of cost considerations and
   assurance of professional integrity.
   One is to know the types of physical
   facilities that are available—special
   equipment that may be needed or
   desirable, for example. The other is
   to determine the general level of
   technical sophistication that can be
   found among both the professional
   and field staffs of the firm.

4 *Establish the Initial Design of the
   Project*   If time permits, the initial
   design is best presented in outline or
   draft form to the overseas research
   agencies that will be involved. Comments and suggestions should be in-

vited, so that obvious anomalies can be avoided. An exchange of views will enhance the central office's knowledge of local market realities and will avoid any major pitfalls in the overall approach to the study.

5 *Send Out the Final Design* When the design is finalized, it should be sent to the overseas firms with comprehensive instructions. These are not field instructions as such; they are written at and for a higher level of understanding. It is appropriate to explain the objective of the questions as well as to give specific instructions for their application. These instructions are for conceptual guidance as well as for clarification of meaning for purposes of translation; they are not directions for the field personnel. The specific field instructions can be better prepared by the local agency than by the central sponsor.

6 *Prepare Coding Instructions* One of the acid tests of multinational control is the ability to achieve compatible coding. Open-ended questions, if there are any, are of course the hardest. The central office should prepare a "core" code book with essential codes for multinational purposes, but should still leave flexibility for local variations. One way to do this is to establish the codes for one column, but to leave a second column open for every open-end question. Even the "core" codes should be based on the sum of local inputs. The individual overseas agencies should be asked to remit hand tabulations of the first few questionnaires to the central office, so that the core codes will be meaningful.

7 *Prepare a Tabulation Plan* In most cases, unless the central office of the sponsor has its own facilities, final tabulations are carried out overseas. Again, comprehensive instructions

are called for. As little as possible should be left to chance; it is almost axiomatic that anything that is not spelled out in advance will contain a surprise—invariably an unpleasant one—when cross-national comparisons are made.

8 *Complete the Analysis* The final analysis may be done at the central office or by each overseas agency individually. In a multinational project, central analysis is essential. Nevertheless, the comments of the overseas agencies should be elicited: their input can add greatly to the perceptiveness of the central analysis, as well as avoid misunderstandings as to the implications of the data.

CONCLUSIONS

Good packaging research can be conducted anywhere in the world, but not all *techniques* can be used anywhere in the world. The best and the most universally applicable techniques are the simplest; language and cultural problems are best overcome by the design of carefully formulated, structured questionnaires; respondents—who may react differently to similar questions in different cultures—can understand and respond to questions that offer them choices better than to questions that ask them to make an independent evaluation of a single concept or product or package design.

Administrative considerations are as important as, if not more important than, technical questions in the conduct of multinational research. Often more errors are made because of improper coordination from a central level, or inadequate understanding of cultural differences, than because of technical errors per se.

CHAPTER THIRTY-NINE

Package Testing: The State of the Art in Canada

Irving Gilman

Prior to the 1950s, the "art" of package testing had yet to earn such a dignified designation. Almost totally ignored as a research technique, package testing in Canada, Europe, and the United States remained a catch-as-catch-can operation until mid-century.

At that time the marketplace assumed new shape and substance. The era of self-service had begun, and with it accelerating new product development. Consumers, now given a warrant to

Illustrations by Florence Sacks
The author gratefully acknowledges the assistance of Patricia Merker, Director of Research, ARI, in the preparation of this article.

"search and seize" any brand or product of their own choosing, were afforded ample opportunity to take or discard at will. The prosaic package, once wrapped (and hidden) by the neighborhood grocer, suddenly emerged front and center in customer milieus.

Literally miles of products vied for ascendancy on supermarket shelves, influencing brand distributors to exert every effort to place their products in preferred locations. Adding to their burden was the daily arrival of new, improved, and "super-improved" brands from their competitors. The level of activity and anxiety increased accordingly, as did the astuteness of bemused consumers who missed nothing of the "contest" then in progress.

The pressure on competing manufacturers was intense. Product research expanded, new marketing programs were initiated, media advertising proliferated on the tube, in print, and particularly at the point of purchase, where the once ignored package eventually earned recognition as a significant communicator in its own right.

Today, U. S. marketers are investing, annually, over one billion dollars in packaging research, a figure matched proportionately in Canada (approaching 100 million dollars). It is generally conceded, however, that Canada took the initiative in packaging research, and is still considered "ahead" in package design.

The influx of "new Canadians" in the 1950s and 1960s generated increasing demands for multiple-use products expected to accommodate a wide variety of practical, cultural, and ethnic requirements.

Canadian manufacturers were quick to meet these needs and to enlist the aid of top designers whose task it was to encourage product acceptance through innovative development of the "wrapping." Recognizing that the relationship between package and product is extremely complex, Canadian marketers set out to ensure consumer acceptance for both.

When it became evident that the package was a *communication* as important in its own way as a TV commercial or double-truck ad in the local newspaper, a new marketing concept was born. Today, no contemporary manufacturer would think of launching a marketing program without considering all the advertising and promotional ramifications inherent in the product/package relationship. Because of this, the package test, an art that has evolved quickly to add dimension and depth to Canadian research, has assumed new significance and status.

A DUAL BURDEN

In one sense, however, the term "package test" is oversimplified, failing to suggest that the product *inside* the package also is being evaluated. Regardless of what may be perceived during the actual test, the "box" is not empty. Psychologically, it still encloses the product.

Canadian marketers were among the first to recognize the psychological implications of packaging research. They knew that visible or not, the product never is far from view—either in a supermarket or in a research environment. Understanding who uses the product, what consumers hope to find when they buy it, and how they expect its integrity to be reflected honestly in communications (media or package), will have a distinct bearing on how respondents and/ or consumers react to the "wrapping."

In corporate echelons the "wrapping" is the image projected by the company. A package also must project an image, preferably one consistent with that of the corporation. Thus the package assumes a dual burden—not only must it "speak" for the product, it must reflect the integrity of the manufacturer. In this sense, the package performs as both mirror and messenger for producer and product, respectively.

As appreciation of this phenomenon began to emerge on the Canadian scene, the consumer was observed as the only one who could tell the manufacturer what he wanted to see reflected in this "mirror." Like the Queen in "Snow White," the consumer was perceived to need reassurance of that of which he is still uncertain: Is the product *really* the "fairest of them all?" The packaging was seen to play a vital role in allaying his uncertainty.

"COMING TO YOU, LIVE"

The package, as Canadian marketers soon perceived, is not an inert object. Although unappreciated for years because of a utilitarian image that masked its true potential as a viable communicator, the package eventually was revealed as a rather complex organism. Once the mask was lifted, this merchandising "frog" was found to be a "prince" of some distinction. It needed only to be brought to life.

Imbued with specific sensory and tactile appeals, the package invites curiosity and involvement. Unlike other communications, it is a palpable force that can be touched, handled, opened, closed, reused, or "reconstituted." Because of its many visual, psychological, and mechanical characteristics, the package excites interest, elicits discussion, and on occasion incites wrath. In short, it testifies to the truism that art often imitates life (Fig. 1).

With this enlightened perception of the package firmly in view, Canadian manufacturers and marketers began to coordinate programs that would put the package to work as the brand's good will

Figure 1

ambassador. Before the package's credentials could be assumed to be correct, however, it was necessary to seek approval from those who would be buying, using, or seeking the enclosed product.

Taking a cue from the premise that "art often imitates life," designers were careful to vest the package with characteristics expected to emulate those of its users, while ascertaining also that the package provided easy access to the product. Thus a beer bottle might have a rugged, "no-nonsense" design, a perfume a slim, sophisticated package, a box of toys an unobstructed view of the treasures inside.

Alternative designs, considered to be "on target" for potential customers, were finally readied to make their debut in a marketing microcosm called the "package test."

THE EYE OF THE BEHOLDER

At first, packaging research in Canada followed prescribed patterns practiced by most in the marketing community. The idea that much might be gained by researching a design *concept* before submitting the finished *design* emerged later rather than earlier in the marketing environment.

Nevertheless, the package finally was receiving the attention it deserved, although some years would elapse before packaging research achieved the level of sophistication it enjoys today. Early testing depended primarily on eliciting opinions of packages already committed to production—in the sense that the package perceived as "best" would be utilized essentially unchanged. It was not long, however, before Canadian expertise in the business environment responded to the need to evaluate the design's emotional as well as practical appeals.

The "eye" of the beholder, it was discovered, could be trusted to react to the tangible characteristics of a package, but more important, to respond subliminally to stimuli that could substantially influence acceptance or rejection of the package and, by extension, of the product within it. Respondent panels, initially composed of women, but now including men, teen-agers, and children were discovered to be effective platforms for studying attitudinal and behavioral responses to a wide variety of design concepts.

In this way, aesthetic, functional, and psychological appeals could be measured with a higher degree of accuracy, and the package design could be reformulated to meet consumer expectations more closely. While paired comparison tests still functioned to answer questions about preferred designs and package attributes, they, too, were enhanced to evoke considerations of new concepts and to provide valuable direction for Canadian designers.

Quantitatively tabulated and compared, these tests essentially were qualitative assessments that provided important psychological and sociological insights for package design.

MORE ABOUT THE "EYE"

The use of panel sessions and paired comparison tests have proved their value over the years, and are now supported by technological methods that have gained increasing acceptance in the 1970s. The "eye," more than the "beholder," has become the focus of many researchers.

With the advent of sophisticated instrumentation that can record eye movement within millionths of a second, a new frontier in package testing was revealed. The way was open to expedite the testing process dramatically. Whole

series of designs could be evaluated in a very short time merely by exposing each concept for a prescribed number of seconds and allowing the machine to track the direction of a respondent's eyes as they were drawn (or not drawn) to specific design attributes.

Important information concerning the relative impact of design elements — graphics, color, and so on—could be elicited in this manner. However, Canadian packagers did not earn their illustrious reputation for design merely by studying research results. Their creativity was triggered by change and an understanding that social, cultural, and economic movements give rise to new expectations in consumer environments. They understood that it is necessary to know both *what* the consumer likes and *why* he or she likes it.

SEPARATE BUT EQUAL

Canadian strength in package design developed from a belief that the package is an integral part of the entire product/marketing/communications syndrome. The health of the organism is dependent on the health of any part of it. If one element in the organism falters, the body as a whole is threatened.

With this in mind, Canadian marketers reinforced the concept that packaging must assume a separate but equal role in ensuring that the complex task of bringing a product to market is not inhibited by merchandising traumas that interfere with manufacturer capability. The package, along with other elements in the marketing organism, is expected to perform an increasing variety of functions to insure the health of all.

This is as true in industrial environments as in consumer areas. The industrial package carries an additional responsibility in that it is exposed to an "expert" whose experience and expectations regarding the product and its relationship to the package are highly specialized. The package will be expected to be more than a reliable communicator—it also will be studied for efficiency (storage, convenience), durability, and maneuverability. In addition it will be expected to facilitate handling, shipping, and order selection.

As to consumer packaging, designers are aware that its message must be personal as well as practical. Regardless of how varied the product mix (size, weight, grades, etc.), the package must be able to tell the consumer something he wants to hear, or he will look for another brand. One of the package's primary jobs is to encourage purchase and then *repurchase*. Supported by promotional and advertising communications, in store or in the media, the package and its design are integral parts of a marketing network committed to sales, profit, and growth.

Canadian designers know, then, that a test of a package design is not a simple exercise of choosing between packages A and B; a whole range of choices is involved in response to a package, influenced ultimately by the consumer's ever-changing reaction to the psychological and environmental forces pervading his environment.

THE CONSUMER AT LARGE

Responding to the rather kaleidoscopic pattern that characterizes Canadian society, manufacturers and marketers addressed their efforts to creating a marketing mix that would meet the diversified requirements of a broad spectrum of cultural and ethnic Canadian types.

The problem was never simple, and it was exacerbated during the 1950s and 1960s when a new wave of emigrants found its way to Canadian shores. In Toronto alone, more than one third of the population comprises first and second generation citizens, and 22% of the total population are considered "new Canadians."

Reactions to the Canadian "melting pot," particularly by those who are not quite assimilated, vary in emotional and philosophic intensity. The diversities between the incumbent and the arrival sometimes are found difficult to reconcile, creating new shades of differences among a citizenry already diffused with colorful language, racial, and religious hues. This makes the Canadian consumer difficult to categorize, and increases the burden on the manufacturer to standardize his marketing approaches. The package designer, together with the marketer and advertiser, choose instead to innovate.

Observing that most Canadians share a common European heritage, the marketing community opted to concentrate on the positive appeals of this heritage, rather than to wrestle with its differences. Package design, in some instances, took its direction from European models, which enabled its creators to innovate on a wider and more imaginative plane. At the same time, Canadian designers stayed abreast of what was happening in contemporary Canada to make sure their concepts reflected the changes emerging in every area of the consumer's environment.

In addressing the vast "melting pot," the designer was aware that his offerings must communicate both traditional and contemporary meanings with which his many audiences could identify. That he achieved high levels of success in this area is now a matter of record (Fig. 2).

The Canadian consumer performed as beneficiary and benefactor in elevating the stature of Canadian package design. Seldom reticent in his attitudes toward the marketplace, the perceptive Canadian demonstrated approval and

Figure 2

disapproval with equal aplomb—at the point of purchase, as a participant in a research program, sometimes in direct contact with the manufacturer or representative.

Salvaged from a "direct mail" communication to a cereal product manufacturer, the following "complaint" is framed in an original setting.

Choosing to vent her spleen in verse, an anonymous customer points out what she considers a serious flaw in package design:

> *I like the 'zip' atop the box which*
> *opens it with ease,*
> *But in the name of common sense,*
> *Would you inform me please*
> *What good's a 'zip,' despite its*
> *speed,*
> *That only works one way—*
> *You cannot close the carton, sir,*
> *Now, what have you to say?*

The answer, unfortunately, is lost to posterity, but there is little doubt that the poem helped to inspire a significant modification in package design.

THE PACKAGE AS SURROGATE

Early indifference to package testing was not entirely because of a failure to appreciate the package's potential as communicator. The manufacturer already had a "communicator," several in fact, and had no reason to view the package in this guise.

The neighborhood grocer, local ap-

pliance dealer, "lady" in the dress shop, drugstore, five and ten, were all important agents for the manufacturer in persuading customers toward specific brand purchase.

As the trend toward self-service continued to expand, however, a major shift occurred. The salesman disappeared. Not only in supermarkets, but in clothing stores, discount houses, and variety stores, self-service overtook the salesman's role. With the consumer in charge of his marketing destiny, the manufacturer was left without his "communicators."

This was not necessarily fatal, since manufacturers had little control over retail salesmen. They accepted the fact that a sale could hinge on the salesman's mood of the moment. Nevertheless, an important line of communication was broken, and it became necessary to find a replacement.

Enter, the package. With increasing numbers of salesmen vanishing over the horizon, it became evident that the package might have to be enlisted as surrogate. Designers and packagers began to investigate how the package could be imbued with a personal identity that would encourage a more vital relationship between the customer and the "box." They were no longer content to view the package merely as an aesthetic accompaniment to the product, but as a marketing messenger that could deliver the product story effectively.

In the late 1950s and early 1960s the package began to come into its own and to prove its value as a dependable surrogate. In many ways, it offered manufacturers a measure of control that they had not previously enjoyed with their "live" salesmen. Canadian marketers and designers quickly recognized they could "build in" qualities that provided the package with superior "sales abilities," not always found in its live counterpart.

Primarily, designers wanted the

package to communicate a message of concern and to perform as an accurate reflection of manufacturer regard for his customers. A poorly designed package, they noted, tells the customer that nobody cares, while a well-designed package is proof that the manufacturer cares about both product and customer and is willing to make an effort to please.

To make sure that the message would be understood, manufacturers and marketers began to expand their package testing programs until today, packaging research in Canada has achieved an enviable record of accomplishment.

"TESTING, TESTING . . ."

Before unisex appeared on the contemporary scene, package testing was entrusted mainly to women, on the assumption that women were the shoppers and men, the bill payers. While the cultural climate in Canada tended to support this view, dramatic sociological changes in the 1960s altered the marketplace ambience.

Men began to infiltrate women's domain. In the kitchen and at the market, males increasingly could be seen preparing meals, choosing foods, comparing prices. Their assimilation into other "female" environments was noticed also. Products previously thought to belong to the "woman of the house," began to make inroads into male environments. Hair sprays, colognes, "baby shampoos," neck chains, and similar products met the changing needs of men who had grown their hair, discarded neckties, and in general had adopted a more flexible attitude toward the world and their roles.

As a result, package testing programs also assumed a more heterogeneous quality. Men were invited to scrutinize other than industrial or "male-oriented" products and packages, joining consumer panels to add selective insights to the study in progress.

Children also joined the ranks of respondents. Toy, doll, game, and food manufacturers understood the power exerted by young customers in the marketplace, and respected their need for safe, recreational equipment and wholesome nourishment.

As the Canadian consumer's knowledge and sophistication grew, so did packaging research. By 1970, the product/package relationship was viewed as a significant alliance, inspiring comments like the following:

Effective utilization of packaging as a marketing tool requires recognition from the outset that "packaging" and "marketing" are both variables of extraordinary scope, complexity, and dynamism. . . . The demands upon successful packaging are complex.

Information about the product must be understood by different publics (distributors, retailers, and a variety of consumers). . . . The body of knowledge and information from which the designer must draw . . . is vast. He must be aware of the production line and its capabilities; he must be completely familiar with the product life cycle . . . he must speak in multi-lingual imagery to a variety of publics. . . . Consumer market research and testing can add supplemental and valuable material to the analytic and evaluative design stage.[1]

In eastern Canada, "multilingual imagery" necessitated multilingual graphics so that the requirements of both Ontarian and Quebecois could be accommodated. French/English instructions, titles, and so on called for incorporation into designs without sacrifice of the package's vital communication function or interference with its important aesthetic and visual appeals. Thus the package test evolved from a simple "like and dislike" exercise to an evaluation of some complexity.

While testing the package's mechanical attributes continues to be important, it is evident that package testing cannot confine itself to mechanics. The psychological appeals of the package (positive or negative) introduce a wide range of variables that require strict evaluations, especially now when Canada is expanding its cultural and political horizons at an ever-increasing rate.

For example, growth of international trade in Canada presents a variety of new and special packaging problems. In addition to questions involving the package's functional aspects, issues of legality, language, and cost become more crucial. Further, packages designed for overseas shipment must contain appeals capable of reaching audiences from border to border, whose emotional and practical bases can vary. Because of the intricate nature of international marketing and the problems it presents to the packager, marketing personnel also are put to the "test."

Canadian manufacturers realize that in domestic as well as international areas the growing needs of package design and research require a more specialized approach to each. Consequently, there has been a rising interest in packaging technology, accompanied by a general improvement in training methods for involved personnel. Universities have been enlisted to offer special courses in packaging or packaging research and on-the-job training programs have expanded. As a result, more sophisticated design factors, requiring complex and often extensive evaluations, have been introduced into package testing areas.

Package designs, like manufacturer communications, are expected to motivate purchase. Therefore, any evaluation of the package and its components must determine in what ways the design is seen to encourage:

- *Orientation* Spontaneous consumer response to the package and its message. Does the design invite further investigation of the product, or does it confuse or offend the viewer?

- *Identification with Design and Informational Elements* In what ways does the design enable consumers to relate to the product, reassured that their emotional and practical needs are being met?

- *Involvement* The intrinsic appeals of a good design will invite further examination and convince the consumer that the time he invests for perusal will pay dividends in enlightening him about the product. The extent of satisfaction derived from such an "investment" often determines the eventual acceptance or rejection of the product.

- *Mood matching* How do the package and its design reflect the mood and nature of the product as conceived in the consumer's mind? Does the design run counter to consumer expectations on this level, or does it support his feelings about the product and help him to relate to it positively?

- *Sensory Stimulation* The impact of sight, sound, and touch can be powerful inducements for buying or rejecting a product. The tactile and sensory appeals of a package are paramount in encouraging product trial.

- *Credibility* A package's design and graphics must be complementary to the extent that each reinforces the product message in a unified and consistent manner. It is important to know in what ways package design convinces consumers that the message is logical and valid.

Ultimately, the package and its design must be tested to discover whether marketing, communications and merchandising objectives are being met. Is the message of quality or function coming through? What status appeals are intrinsic to the design, and are they triggering the desired response among selected audiences? Finally, is the design doing the job it was elected to do as an integral part of a strategy for improving the product's marketing and sales potential?

FINDING THE ANSWERS

As packaging research began to mature in the Canadian marketing community, it became evident that one test would not suffice to study the full impact of a package design. It is common practice today to subject a design to several tests before the decision-making process is finalized. In this way, the element of risk is substantially reduced, and the marketing strategy reinforced. Different elements of the package and its design often require utilization of several research techniques. Some of these techniques are:

- *Word association* tests and a variety of projective inquiries eliciting various degrees of emotional response to design and graphics.
- *Quantitative* tests conducted among large samples to determine differing tastes and buying patterns regionally or nationally.
- *Readability* tests, which help to provide data regarding consumer comprehension of informational elements incorporated in the package design.
- *Paired comparison* tests, eliciting responses to alternative package designs and determining whether design concepts are triggering the consumer reactions anticipated by designers and marketers.
- *Panel* tests, where attitudinal/behavioral responses to a variety of designs and packages can provide insights for design modifications to meet the changing needs of Canadian consumers.

- *Electronic optical* testing, including film and videotape, valuable in determining a package's visual or color appeals and in discovering the relative impact of design elements through electronic measurements of eye movements.

All these tests enable designers to identify problems or to discover new ideas for design development that might have gone unnoticed otherwise. For this reason, many in the marketing community feel it is extremely important that the designer play an important part in planning test programs. Since many variables, including packaging attributes, affect a product's marketing success, manufacturers increasingly are consulting designers at all stages of a product's development. The designer's contribution also is considered vital when an old product is being evaluated for modification or change.

Cost, production, choice of the right packaging materials, and advertising impact are all related to design. The package test paves the way for a more successful marketing result, but the final "test" is in the hands of the Canadian consumer who will determine the product's fate in the marketplace.

THE CANADIAN CONSUMER

The package's debut in consumer or industrial environments follows an exhaustive study of how and where the product will be used, stored, transported, or placed to best advantage in retail or other outlets. The trade helps to provide answers to these questions, but in recent years it is the consumer who has proved to be the manufacturer's strongest ally. While consumer resistance has stiffened in the marketplace and Canadians are less easily wooed to purchase, their changing demands, when heeded, provide important directions for product or packaging development.

The contemporary Canadian has become vocal to an increasing degree — especially during the 1970s. Regardless of background or ethnic heritage, many consumers have joined in common cause to make sure their demands are heard.

Unwilling to feel alienated in an environment that strikes many as far too depersonalized, large segments of the population choose to make a bid for the individual. The economic and environmental pressures of the late 1970s have bred a "new Canadian" whose practical, social, and psychological orientations to himself and his society have little to do with immigrant status (the other "new Canadian").

Among the young, opportunism and hedonism are common characteristics. As to the mature, intentions to preserve the status quo despite economic, energy, or political "crises" are manifested in efforts to reap hard-earned rewards "now" before "conditions" deteriorate further. Research conducted among both segments confirms these responses, but also indicates that they will be looking to the marketplace to provide the products that will make their lives more meaningful. Young and mature Canadians need reassurance that authorities in the marketing community are making every effort to meet their practical and psychological requirements.

The package and its design can play a significant role in providing this reassurance. Along with the manufacturer and the marketer, the designer is compelled to present the product in a manner that invites perusal and also cuts through growing negative expectations that the consumer is "being taken."

A customer, already irritated by Canadian "problems," does not want his frustrations compounded on any level. A poorly designed package, difficult to open or impossible to read, will only confirm his worst suspicions that "nobody gives a damn." His response to the product will be influenced accordingly.

While the package test in its various forms is expected to eliminate these problems and increase the product's acceptability, accelerating cultural, social, and technological changes introduce another set of variables with which the package designer must contend.

THE TV TEST

The influence that television exerts on the lives and habits of Canadian consumers is extensive, but never more so than when persuading consumers to purchase. In its own way, TV provides the ultimate panel for evaluating a package or product. Although conventional testing is conducted in a structured environment, with a carefully selected sample, the TV "package test" is not so equipped. There is no "national probability sample" among television viewers — at least none that can be counted. Yet for hundreds or thousands, television is affording an initial opportunity to see and evaluate the product and its package design.

While marketers and distributors jockey with their competitors to place their products in preferred retail locations, unrecorded responses to these brands are ticking off in every town and hamlet. As the commercial or "package test" draws to its conclusion, the consumer, without moving away from his television set, already is making judgments about a brand. By the time he arrives at the store, he cares little whether the product is in a preferred or "unpreferred" location. If he wants it, he will find it. If he doesn't, he won't bother to look (Fig. 3).

It becomes increasingly important, then, that package design be able to pass

Figure 3

this most crucial TV test. The package must incorporate elements that communicate the product message in two-dimensional, as well as three-dimensional environments. Color and graphics viewed in the hand or on a supermarket shelf can take on a new and less favorable image in "black and white" or on a color set of dubious quality.

For this reason, more creative attention to line is critical. Distorted color will not necessarily diminish the appeal of a televised package or product, but a poor line can negate the merits of a package design as surely as it can spoil the appearance of an otherwise fashionable garment. The role of color is significant in package design, but through the lens of the television camera, line and style project a more dependable image.

THE COLOR TEST

Color is a powerful communication element, especially today when more liberalized attitudes toward color are manifested in every part of the Canadian environment. Colors never found in the spectrum have emerged at the hands of creative designers who have taken full advantage of new developments in material and dye technology. In clothing, puritanical grays and browns have been replaced by more frivolous hues, and a once conservative Canadian can now be found sporting an outfit that 10 years

earlier he might have considered flamboyant.

In the provinces and in the United States, the color explosion has taken off in several directions at once. Ignited by the spark of change, freer and more relaxed attitudes toward color reflect the consumer's growing need for self-expression. During the late 1950s and into the 1960s and 1970s, consumers of every class and station acknowledged their acceptance of color as a desirable asset in clothes, homes, and in the marketplace.

The effect on package design has been significant. Designers, once confined to create within more conservative norms, have been able to rid themselves of browns, blacks, and bilious greens, substituting more vibrant shades in package design. Packages, along with customers, have become more multi-hued in "dress," responding to a culture that advocates innovation and individuality.

The use of color always entails risk, however, regardless of social or cultural mores currently in fashion. Like taste, color is a totally subjective experience, conjuring up a host of positive or negative associations for the viewer. Also, like taste, color invites a fickle response—what is appealing or attractive today, may be boring or passé tomorrow.

At present, change is very much a part of the Canadian character, and it has imposed its own demands, exacerbating those already invoked by economic and environmental pressures. The unwanted but not unexpected result is that the Canadian consumer has become increasingly uncertain about what he wants or doesn't want and whether it will do him good to adopt a permanent posture on anything.

Inflation has made him a more belligerent customer, and reinforced his intentions to buy warily. At the same time, he is eager for reminders that the manufacturer is providing some positive opportunities to "enjoy" at the dinner table, at home, or in selected leisure situations. The "color" he seeks for his life is both practical and psychological.

The package designer is in a unique position to deal with this duality. He will want to stay apprised, however, of what is happening within the consumer's emotional color spectrum. What shades of difference can be detected that must be matched in a skillful rendering of color on the package? Which colors carry a positive message for today's consumer, and which communicate negative or inappropriate information? The package test provides answers to some of these questions, but in Canada's fast-moving society more comprehensive explorations of color are required.

A "color test," conducted before commitment to design is made, can prove to be an exciting and enlightening experience. Much more than the consumer's color preference is elicited. Color delivers the same impact as an ink blot, evoking a multitude of associations that provide important insights into the consumer's psychological and practical orientations. Since the package is the instrument that "orients" the consumer to the product, color can be the catalyst that triggers closer identification with the brand.

For example, a major beer manufacturer utilized color in just such a manner. Designing a label for a bottled beer product, the manufacturer and the designer planned to communicate a historical as well as quality message. The label had no fewer than six colors, including black and white, all incorporated to tell the story of Canadian growth in a 3 by 4 inch frame. A yellow stagecoach, a red car, a brown locomotive, were some of the elements that demonstrated Canadians in motion, providing purchasers with an intriguing view of themselves as well as of the product.

The label encouraged discussion at home, in bars, in sports stadiums and parks, where consumers enjoyed "making up" their own versions of what was happening on the label. A creative blend of color and graphics also helped to make the bottle an aesthetic enhancement in leisure settings, and a "stand out" in merchandising centers. The brand prospered.

Today, package design is involved with more abstract appeals, but even sophisticated renderings of a product message must include color and design elements that embrace, rather than alienate, the consumer. Canadians admit they already feel alienated. A package design that abstracts its message to the extent that it becomes meaningless, in effect is saying that the consumer doesn't count and that "art for art's sake" is paramount. Contemporary Canadians do not appreciate the implication.

By current standards, the previously mentioned beer label might appear old-fashioned or naïve, but it serves as a significant reminder that a package design can offer more than a product message. It also can tell the consumer something positive about himself. As the 1970s drew to a close, Canadians indicated this as an urgent need.

THE FOUR "S'S"

Years ago a philosopher arrived at the hypothesis that "the more things change, the more they remain the same." The marketing community has long been aware of this paradox. While consumer behavior may change and attitudes shift with the breeze, change is deemed acceptable only when it supports the constant and the familiar. A consumer, intent on being *au courant,* still seeks what he has always sought—comfort and approval. He will reject a "change" that affords neither.

Whether wandering through the aisles of a supermarket or participating in a package test, the consumer seeks the answer to one basic question: What will the product do for me? The same question could be posed another way. How will the product improve my physical, social or sensual existence? Today, a bikini may have replaced the fig leaf and a penthouse the cave, but the consumer is still governed by four basic needs:

- Subsistence.
- Status.
- Sex.
- Survival.

How change will be manifested within this frame will determine the extent of consumer individuality, but also will provide the "same" reassurance man has required since he took his first faltering steps as *Homo erectus*.

In today's Canadian environment the consumer's step is more confident. The questions he raises about the product's value to himself he expects to be answered in large measure by the package. Aware of this, Canadian designers keep a keen eye on the past, while they consider consumer needs for the future. In this way they know they will be afforded a more definitive view of the present and of the permutations that control consumer behavior in the market.

"MODERN TIMES"

Change has imposed severe demands on Canadian consumers. In a period of cultural and economic flux, the individual is apt to rearrange his practical and psychological prerogatives, in effect, to "march to a different drummer." Basic needs may remain the same, but the ability or willingness to meet these needs

may differ. So it is with today's Canadian.

Recent research conducted in the Canadian environment indicates that subsistence and survival have taken on new meanings in the 1970s, modifying attitudes toward sex and status. Many feel that the citizen's "life support system" is in jeopardy, and for this reason both the young and the mature have adopted a more opportunistic approach to "survival." "Living for today" has become the new maxim for contemporary Canadians, who look to the marketplace for products that will allow them to "subsist" on a more comfortable and enjoyable level.

Consumer priorities also have shifted. With the threat of economic and international crises ever present, the significance of status is minimized. Anxiety and uncertainty have become the "great equilizers" among rich and poor. No one feels immune to the threats in his environment that, since the 1960s, have persuaded Canadians to a deeper appreciation of individual needs and a refutation of value systems which discriminate between "white," "blue," or any other "shade" of collar. Today, "status" is described as how an individual perceives himself, rather than how he is perceived.

The paradox of "modern times" is that the consumer becomes more demanding as he becomes more accepting of social, sexual, and cultural behavior. His new self-awareness permits him to respond fully to sensual as well as practical pleasures, strengthening a resolve to gratify his needs by any means available. When this attitude is carried to the marketplace, the demand is reinforced at the product level. The consumer expects this new sense of freedom to be reflected in the product message. In his lexicon, "freedom" equals "integrity," an ability to communicate honestly, unintimidated by what "others" in the establishment might think.

Today, Canadians tend to view manufacturer integrity with growing skepticism. Thus the marketing community is continually challenged to convey a message of concern that carries conviction in the marketplace. The package designer also shares the responsibility to meet this challenge, and to heed more closely contemporary demands for accommodating the four "S's."

Canadians imply they are not willing merely to "subsist"; they want to enjoy. Food, clothing, and home products are expected to communicate a concept of renewal and reward, rearticulated in the package design. Similarly, all elements of the marketing mix will evoke considerations of how each conforms to new consumer meanings for status. Condescending or patronizing messages will not be seen to acknowledge the consumer's growing respect for himself as a significant and enlightened individual. He will be encouraged, however, by communications that invite realistic consideration and involvement.

Although it would be an exaggeration as well as a denigration to refer to the package as a "sex object," its role as a reliable "seducer" has long been recognized. The tactile and sensory appeals of a well-designed package always have been excellent communicators, and they are especially significant today when Canadians are more in touch with their feelings, finding such appeals reliable as facts in forming personal or practical evaluations. Heightened sensual awareness also has made Canadians more receptive to the aesthetic (Fig. 4).

In these areas, perhaps more than others, the package will play an increasingly important role in persuading consumers to purchase. Its sensual (and sometimes sexual) tangibility is an asset that consumers respond to positively. Even before the "age of enlightenment," customers bought a product solely because they liked the "shape" of the

Figure 4

bottle or box. Today, positive orientations toward the package are likely to be reinforced on many levels, inviting new considerations for design and distribution testing strategies.

THE FOURTH "S"

Subsistence, status, sensuality, all are components of individual efforts to survive—preferably in a comfortable and enjoyable manner. When survival becomes "uncomfortable," the individual retaliates. Canadians along with many other citizens of the world are finding survival "uncomfortable," and they are highly critical of the authorities that they feel are responsible for their dissatisfactions. Authorities in the marketplace always have been logical targets during difficult times, and today's "times" are indeed difficult.

Although it would be unfair to suggest that the Canadian consumer comes to the marketplace "loaded for bear," he has an arsenal of emotional equipment that he will activate if he feels his practical needs are being ignored. Consciously or unconsciously, he expects the marketing community to accommodate the elements that enable him to survive. Survival is a complex process for today's consumer, incorporating a wide range of physical and sensual gratifications from which *he,* not *others,* will determine his status as a responsible individual.

This is the man (or woman) who will be asked to participate in a package test. As the French would say, *formidable.* The demanding consumer also has become more perceptive, however, with insights for marketing and package design that can help to improve the "positioning" of manufacturer and customer in the highly charged Canadian environment.

ADDENDUM

Although marketing environment demands impose severe challenges on all who participate, each challenge carries an element of irony that is not lost on the professional observer. For the marketing community, an old axiom remains particularly apt. "The improbable we do right away, the impossible takes a little longer."

Some humor has been added over the years, scribbled on the backs of conference programs, meeting agendas, and the ubiquitous but indispensable "legal yellow." A few originals have survived; for example, the following "inspirations of the trade":

A literary "doodle," composed during a meeting of manufacturers and packagers:

> *A package is a sometime thing,*
> *A "throw-away" like cord and string,*
> *But if we don't design it well,*
> *Let's face it boys, our stuff won't sell.*

Found on a table following a seminar on package testing:

> *Tell me quick and tell me true,*
> *(Or else my love, the hell with you)*
> *Not how this product came to be,*
> *But, what it will do for me.*

Left on a dais at the conclusion of a conference on package design was the poem "Le Package":

> *It should be good, it could be bad,*
> *It must be happy, never sad,*
> *If red's the color, that's all right,*
> *But bilious beige? A dismal sight.*

> *Shape's the thing, but what's surefire?*
> *Phallic, oblong, round or spire?*
> *And how do graphics pass the test?*
> *A plain brown wrap, I think, is best.*

Rescued from the chair of a package designer, participating in a panel discussion on the product/package relationship:

> *When shoppers heed the siren call*
> *Of "Sale," or "Mark-down," "New" et al.,*
> *How do we know what makes them buy?*
> *Their cash on hand or practiced eye?*

> *What is the shopper's great design,*
> *And how does it compare with mine?*
> *The answer is not clear to see,*
> *Unless the label's tag says "Free!"*

REFERENCES

1   Vernon L. Fladager, *Packaging as a Marketing Tool,* McGraw-Hill, New York, 1970.

CHAPTER FORTY

Package Design:
The Importance of its
Role in the EEC
Marketing Environment Today

Michael Sarasin

How important is package design in the widely differing and increasingly complex marketing environment of the European Economic Community? Is the role of package design within the EEC in any way different from that of the United States and if so, why?

I believe these are key questions. They will be considered in this chapter, but due to the varied nature of the background of the EEC itself and the policies of the different companies involved, clearly there is no universal answer. Nevertheless, after working on a number of specific national and international package design projects with Cato Johnson, design consultant to Ralston Purina in Europe, I believe we have learned a number of lessons and evolved a certain approach. Most of the experience gained has been in the Benelux, French, and German markets, and comments are limited to fast-moving consumer products. I feel that the following remarks could be useful background information to a consumer products company or package design agency about to embark on a similar exercise in the EEC.

A first proposition would be that package design in the E.E.C. today is no different from that in the U.S.A. in one important aspect: it is rightfully regarded as a critical part of the total marketing mix and as such conveys intended or perhaps unintended visual symbols, brand images, and so on. In the special circumstances operating within the EEC, however, the importance of effective package design takes on an added dimension. I believe these special circumstances arise as a result of a number of external rather than internal constraints and mainly concern factors such as language, legal barriers, and media availability.

Before discussing these external factors, I should point out that in marketing and package designing within the EEC one is concerned with a total market of around 257 million people living in nine different countries and speaking six main languages. It follows, therefore, that in marketing fast-moving consumer products such as convenience foods, soups, or petfoods, it is usual to find that consumer attitudes, behavior, and usage habits in relation to identical products often vary from country to country and sometimes within different regions of the same country. The critical marketer or package designer needs to understand these differences in consumer perception and usages and modify (or not modify) his packages and designs accordingly.

Suffice it to say that a detailed market research program to determine consumer attitudes and usage habits is a mandatory first step before embarking on any packaging project, and this ordinarily forms an essential part of the briefing. The overall subject of the differing consumer attitudes and usage of different products is obviously outside the scope of this particular discussion, yet its importance cannot be overemphasized.

The first important external factor is *language*. As mentioned, the nine countries of the EEC have a combined population of 257 million. Theoretically, this means that any truly EEC-directed package design project should take into account six national languages.

A practical solution to this problem could be to find a short, relevant brand name that at least identifies who or what you are for the countries you are aiming at. Ideally this brand name can be a primary trademark that can be registered and at the same time adds to, or at least is consistent with, the brand image or use of the product. With this common primary mark you could then add some common visual elements to your package design so that it becomes a truly international brand, differing from country to country only in the national language section that spells out the key positioning/benefit to the consumer.

If all this sounds too good to be true, you are right: it totally ignores some of the practical problems that legislate against such a simple solution. One simple fact is that the six main languages usually required to communicate just don't travel too well across frontiers if you try to compromise or rationalize them into one. It is a real creative and legal challenge to find a multinational brand name, especially when it must have positive product connotations.

A somewhat earthy but direct illustration of the problem of transferring one country's brand name across borders is shown in Fig. 1. This example quickly illustrates the potential problem that could arise for this French brand of soft drinks should it decide to enter the United Kingdom with its French primary mark. One could give other examples of this type of language problem. There must be other reasons, however, to explain why so many international companies operating in the EEC have identical products bearing different primary and secondary descriptive marks and totally different package designs.

Figure 1 This French soft drink's brand name could create difficulties if it were marketed in the United Kingdom.

Some of these decisions may have been based on a corporate objective, but we can be sure that *legal* factors are a great deterrent to uniformity in marketing within the EEC—whether it is desired or not. One thing is clear in this complex area: there is no uniform EEC legal framework existing at present that can simplify matters for marketers (although debates and draft legislation continue to work toward uniformity in legislation in such areas as advertising, trademarks, and packaging sizes).

The net result is that in planning for a new product introduction in a number of EEC countries, you still need to check the respective legal aspects in each country. What are some of the things you may discover? First, for your hypothetical product name, the trademark registration and protection laws vary from country to country, so it's almost impossible to be 100% secure with any international mark.

Perhaps one of the best known examples of this was the great soap powder war fought between Unilever and Henkel over the trademark, Persil, which was first registered in Paris in 1906. (See

Fig. 2.) According to Common Market competition law, the holder of a trademark in a member state is not entitled to prevent goods from being imported and marketed in that state when they lawfully bear a trademark of the same origin placed on them in another state. Henkel used the trademark, Persil, in Germany, Belgium, Luxembourg, The Netherlands, Italy, and Denmark. Unilever used it in the United Kingdom and France. Both companies fought for exclusive use of the name in those countries where they owned the trademark. Henkel fought to prevent imports of cheaper Unilever Persil from the United Kingdom into Germany, while Unilever France tried to prevent imports of Henkel's Persil from Belgium into France. After lengthy negotiations both companies have agreed with the EEC Commission to end the dispute; one company will use the trademark in red and the other in green.

Even if you have no problems with a brand name, however, varying laws governing packaging and product description may prevent you from positioning or describing your product similarly. For example, if you market margarine in Belgium, you are obliged to package it in a specific container.

La Paz, a Dutch manufacturer of

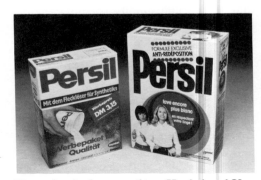

Figure 2 Arch competitors Henkel and Unilever find themselves competing in the same product category with a common brand name.

cigars, finds it necessary to describe somewhat differently the identical cigar that it sells in Germany and neighboring Belgium. The Wilde "Havana" of Belgium becomes the somewhat less exotic Wilde "Cigarros" in Germany. (See Fig. 3.) One could give many other examples to illustrate legal constraints, but we must consider another major external factor affecting the EEC marketing environment—the media.

The nine countries in the EEC spent around $8 billion on measured media in 1975 compared with $24 billion in the United States. The United Kingdom, Germany, and France accounted for around 80% of the total. Once again it is the fragmentation of the EEC into nine countries, and the national language problem, that create a media nightmare and make it almost a necessity to use a country-by-country approach for even the most "international" advertising campaign.

Someone unfamiliar with the EEC media situation might well accept this for poster, newspapers, radio, and magazines, but expect television to be unaffected. To start with, however, commercial television is not allowed at all in Belgium and Denmark, and in practically every other country commercial air time is limited severely in total and is often placed together in blocks. For example, in France commercial air time on the main TF1 channel averages 12 minutes per day. In the Netherlands, no commercials can be screened until 8:00 p.m. on one channel, and just before the 9:30 p.m. news on the other. Advertising is limited to three blocks of 335, 340, and 225 seconds. A spokesman for the Dutch Association of Advertisers commented thus, "It is quite obvious that the longer the blocks become, the greater the possibility that audience interest cannot be captured for the whole series of advertising messages that are sent out." Finally, there is a total ban on commercials

Figure 3 Different laws cause La Paz to describe an identical cigar differently in Belgium and Germany.

on Sundays, Good Friday, Ascension Day, Christmas Day, and days of national mourning.

Other countries also impose similar constraints in terms of limited time blocks given to commercials, and restrictions on certain categories of products. For example, in France they ban any advertising for foreign airlines and tourism, alcoholic beverages, tobacco, housing, and margarine. In the Netherlands, a stylized toothbrush must be shown in all confectionary commercials. In Italy luxury goods may not be advertised because there might be too great a contrast with the lives of some of the viewers. Therefore the following products are either forbidden or restricted: records and tapes, cars and motorcycles over 125 cc., jewelry and furs, pet foods, matrimonial agencies, and last but not least, funeral parlors.

In spite of these restrictions, most TV commercial time is heavily oversubscribed because of its very scarcity (with the sole exception being the United Kingdom). Therefore, in Germany, France and the Netherlands you must book your TV time in the early autumn for the entire next calendar year and it is rare for the serious advertiser to get the amount of time he would have liked anyway. Most large advertisers will be perfectly happy to get one spot per week.

Finally, in North America few people are concerned about the spillover effect of advertising except the Canadians. In Europe, however, people get very excited about it for all kinds of reasons. You may recollect that in Belgium commercial television is banned; nevertheless the Belgian viewer can switch to seven channels on his set where he can get commercials put out by two French, two German, and two Dutch stations and one Luxembourg station. Similarly, in the small area between Switzerland and Northern Italy advertisers are using German TV to reach the Swiss, Swiss TV to reach Italians, and Italian TV to reach the Swiss. To cap it all, there are TV stations in Monaco and Yugoslavia aiming at the Italians. The net result of all this is that TV allocation, buying, and coverage is a very hectic business that can thwart the best laid international intentions.

I hope the preceeding has highlighted some of the *external* factors that can affect the marketing and packaging environment. These are important and could well influence decisions, but what of the *internal* factors that influence marketing in the EEC today? How important and decisive can they be?

I have selected only one internal factor—company objectives. In a nutshell, one can say it is still the most decisive and important single consideration influencing marketing strategy and therefore packaging design. In this respect, companies operating in the United States and in the EEC are no different even though the external marketing environment may influence the final outcome differently.

My proposition is that within the EEC there are many companies operating similar international businesses but whose company policies have determined that they have essentially either national, multinational, or international brand images—at least as expressed in

their packaging. One can look at a number of examples, bearing in mind that those selected have been chosen to illustrate a point and that no value judgment is intended or should be construed.

First, the case presented is that of an international company that has elected to go the "national brand image" way. (See Fig. 4.) Three packet soup brands are shown—Batchelors in the United Kingdom, Royco in Belgium, and Blå Bånd in Denmark. Each of these package designs is entirely different, without any family resemblance, and yet they are all brands from the Unilever company. The presumed strategy here: to build up an international business via strong national brands without any recourse to international packaging/company identification or common brand name.

Second, a case that illustrates a multinational approach is presented. This can be defined as a marketing/packaging strategy that recognizes that individual country requirements are different and yet sees advantages in incorporating certain common elements such as brand name, logo, and corporate symbol. (See Fig. 5.) The example shown is Nestlé's Maggi brand of packet soups marketed in Belgium, France, and Germany. Attention is paid to local consumer/legal differences, but an obvious corporate look is present in the common logo, symbol, and design, which provide an interesting contrast to Nestlé's competitor, Unilever.

Figure 4 An example of Unilever's national branding approach in packet soups.

Figure 5   An example of Nestlé's multinational branding approach in packet soups.

Figure 6   An example of Campbells' international branding approach.

Finally, an example of international packaging is given. Here a key objective of the marketing/packaging strategy is to present a truly international image that does not vary according to country and moves easily across borders. Drink products, whether they be whiskies or Coca-Cola, are classical cases, but since we have used consumer packaged goods, I have used an example of canned soups in Belgium, the United Kingdom, and Germany. (See Fig. 6.) From this one can immediately identify a simple, strong international package design and image that needs no further comment as far as Campbell's intentions are concerned.

To conclude, marketing within the EEC is different from marketing within the United States. Whereas in the United States one may have regional differences in market and consumer segments, a company can reconcile these differences without resorting to different products, packages, brand names, and languages. Such a standardized one-package design approach can rarely be a realistic option within the EEC for reasons that are clear by now. In fact one of the key management problems of marketing within the EEC is the constant struggle to balance different market/consumer demands against the economies of standardization. To err on one side encourages duplication of effort and time and may lead to short production runs, large inventories, and maybe a fragmented image. To err on the other side means you risk overlooking significant national consumer differences in your quest for low unit costs and long production runs.

Under these conditions, allied with the media restrictions referred to previously, well researched packaging design is a vital element that can play a critical role in visualizing a company or brand strategy to the consumer. What role packaging design will take on will normally be a decision for top management based on how it sees its past, present, and future corporate strategy. This is one universal proposition that does not change whether it be the United States or the EEC.

CHAPTER FORTY-ONE

Planning a World-Wide Package Design System

David K. Osler

Imagine that you are a global shopper of groceries—a jet age consumer. You walk into a grocery store in Germany, Japan, Latin America, or the United States, and without too much difficulty you can find your favorite brand of biscuits. You find what you want because of consistent recognition—a uniformity of appearance that is made possible by a package design system.

Universal product recognition comes from a carefully planned design system of brand identification. It enables international marketers to promote and market their product line, while reinforcing the leadership of any given brand anywhere in the world.

This strategy is mandated by the need to maintain product profitability in an economic climate of world-wide infla-

tion. The labor intensive food industry, for example, experiences high costs in production, distribution, and materials. It thus makes good business sense to reduce the variables in package design when selling in different geographical areas where the shoppers come from different cultural backgrounds.

Because graphic design speaks a universal language, a unified design format is essential to product recognition. This form of communication assumes verbal and nonverbal qualities. On the verbal side are brandmarks and copy that comprise the informational part of the design. Thus with the artful use of color, line, and pictures, nonverbal communication is achieved by graphically organizing the verbal elements into a coherent and attractive presentation.

The objective of the design system is a format that is sufficiently flexible when applied to various package sizes, shapes, and materials. A design system benefits a line of products of many varieties — especially those with differing amounts of advertising support.

Manufacturers do not advertise every product equally. So customers, transferring recognition from well-known, highly advertised products to lesser known varieties with less promotional support, become a prime consideration in planning a design system.

The essential objective in planning is the effect of the total presentation when seen as a display by the customer. Everyone is familiar with the so-called visual pollution rampant in most supermarket displays. Aisle upon aisle of gondolas burst with packages of all colors, shapes, and sizes. Consider the customer's visual problems in trying to cope with such a jumble of images. Within this confusion is the opportunity to develop "eye pause" areas where the brand can be seen with clarity and attention. This is a crucial function of a package design system.

In many countries the practice is to display items by product type, not by brand. In spite of this, the comprehensive effect of design repetition allows a visual carry-over for products scattered in a shelf display.

Grocery merchandising techniques long ago established the benefits that accrue to corporate identity at point of sale. Television has enhanced this factor in marketing to the extent that the package and its design become the focal point in a television commercial. The visual impression that the viewer carries from the television screen to the point of sale is created through penetration of the consumer's consciousness and recall of the message.

The purpose of a television commercial is to inform the viewer of the product's quality and values. However, the impact of a strong, memorable display of the container when the consumer is in the store will help make the sale.

Corporate identity at point of sale has a marketing responsibility for product acceptance and consumer good will. An easily recognizable corporate identity makes it possible to bring new products to market at lower promotional and advertising expense.

Corporate identity at point of sale also encourages impulse purchases. Consistent identification enhances the display control of the grocery shelf area. The value of a carefully designed brand identity system is twofold. First, it can "push aside" brands in adjacent displays. Second, it permits the customer to shop within the system display by fixing her attention on a planned brand design presentation to which she is loyal. A successful experience using one product encourages the shopper to try others within the brand display.

Finally the food manufacturer's greatest protection and security rests in his reputation as a producer of quality products. This factor is relevant to all aspects of package design. The quality projection of packaged goods is most important. A maker's reputation is built on the consumer acceptance of his products. Quality of presentation is basic to the brand's competitive position. The attention to many details comprises packaging. The best affordable materials, the best printing reproduction of the design, the convenience of opening, reclosing, and storing the package, the copy that informs the customer of the product's uses, and the protection by which the package maintains the product's integrity bring a satisfying experience that confirms the value of the purchase. Most important of all, these factors confirm the shopper's good judgment. Magnifying that projection by repeating recognizable package facings is where a design system performs its basic function.

Dynamic changes have occurred in food merchandising techniques since World War II. Two major factors contributed to the dynamic growth and expansion of the food industry. One factor was the population explosion with concurrent shifts in population to the suburbs. The other was the rapid growth of the food industry.

After a slow beginning before World War II, the supermarket-style sales outlet came of age. Mass merchandising offered a higher standard of living at lower prices than the independent "mom and pop" store. As supermarkets continued to gain public acceptance, new food products proliferated; later, nonfood merchandise entered the picture as an important factor in supermarket marketing techniques. Grocery retailers saw the opportunity to improve their markups by offering nonfood items such as housewares, toiletries, health and beauty aids that complemented the regular food lines. Convenience was a strong motivation. The shopper could purchase items at the food store for use within the home, while saving time and effort.

Suburban living and the wider ownership of automobiles created a new mobility. Larger stores offered more products, and the era of self-service marketing had arrived. The shopper was offered an abundance of products from which to choose. The need for a distinctive design system became even more imperative.

The concept of mass marketing through supermarkets began in the midwestern United States during the 1930s. After its acceptance nationally, the idea moved overseas around the 1950s. Because of well-entrenched buying habits, a relative lack of land for construction, and limited automobile use, the supermarket concept caught on slowly overseas.

Population growth forced food marketers to rethink market potential. In the larger European cities, increased living costs forced families to move to the suburbs. In most countries, supermarket growth followed the pattern of the United States. Supermarkets reached Latin America, the Middle East, and Australasia. The idea of one-stop shopping tended to concentrate where burgeoning economies provided the population with increased discretionary income.

The layout and appearance of store interiors generally followed concepts already established in the United States. In the late 1960s in northern France, for instance, planners seeking population dispersion built satellite towns or villages around the outskirts of large urban areas, such as Paris.

The French food industry saw a major opportunity to use the open space between these new towns to set up *hypermarchés*. The concept was to house a comprehensive assortment of life support products that included not only food, but also would offer every material product necessary to sustain existance "from the cradle to the grave."

This kind of outlet devotes about one third of the floor space to food products, primarily in-house brands. The "hypermarket" concept depends on vast areas of low cost real estate for the structure and accompanying parking areas. Low retail prices attract customers from surrounding towns and villages.

The merchandise is presented in large mass displays. Many products are sold out of metal bins, packed by the manufacturer. Most of the packages are stored above the display fixtures on pallets. When the products in shelf displays are sold out, fork lift trucks lower shipping cases of additional products.

Besides the store's usual product shelf presentation, it becomes a self-contained warehouse. The air space above the conventional shelf display gondolas is used for temporary storage of shipping cases. Efficient materials handling keeps down labor costs. Computers monitor all aspects of inventory control.

In this environment, a package design system is a standardized design plan that unifies disparate elements of the package into a cohesive unit. It applies most to several different varieties seen in the mass shelf display. A design system adds unity to the entire display.

The totality of impressions is greater than the individual packages displayed. Therefore the total impact of collective masses seen in repetitious pattern is greater than the quantity of items comprising the presentation. To the viewer who sees the display, the overall range of items is magnified above the actually available number of items.

Massing a brand in a display creates a unified appearance. It simplifies brand identification. Within this array of merchandise, a shopper becomes more aware of the variety of individual products offered. She shops accordingly. This is especially important where the shopper can visually relate to strong, heavily promoted items and to lesser known varieties. Having experienced good value for the former, the shopper can go ahead and purchase the latter with confidence. This value transfer demonstrates the essential worth of a design system.

The first consideration in planning the package system is ranking various design nomenclatures according to how the design will be emphasized. Since we are dealing with a package design system that will be implemented world-wide, a prime consideration is how to accommodate brand names and copy in different languages. Sometimes there are obscure connotations or meanings of words that may not be favorably received in the local lexicon and should be investigated for suitability.

Assuming that research has already

weighed the relative merits of the corporate trademark or brand name, it becomes the focal point of the design plan. The brandmark becomes the common denominator for all packages. Empirical knowledge of the time/use of the brand name or trademark offers the best assurance for the efficacy of that strategy.

It is extremely important that consumer familiarity with the local brand not be interrupted during planning and implementing a new design strategy. If an improvement in display impact can be incorporated when the brandmark is integrated into the system, this will enhance the total design.

After seeing to the principal design emphasis, the next step is to develop a graphic design format that can be adapted to the shapes, sizes, or styles of container to which it must be applied. Here the theory of ''less is more'' is crucial. Simplicity of the graphic arrangement is fundamental to a comprehensive system. Unforeseen design elements, therefore, can be anticipated and accommodated without disrupting the basic graphic structure of the design format. Flexibility of design adaptation is a key issue in the planning stage. A good design allows the addition of different elements, such as slogans, premium offers, promotion deals, and such.

In any package design, designers work with various graphic elements such as color, line, texture, pictures, size and form, and copy. In planning a global design system, color is important. It is the first package design element that a shopper readily identifies when asked to recall the features of a container seen on the shelf. Using one color can involve the entire surface of all packages of a brand line. In the case of a line having strong selling individual packages, each of different colors with entrenched consumer identification, familarity with these colors should carry over to the new system.

The key to brand penetration via package communication is using a high visibility color for the brandmark. Certain colors have higher visibility than others—warm colors project toward the viewer, cool colors recede.

If a line of products being planned as a design system is less than 10, consider using a single color for all packages. Remember that the products may be scattered in a food department or throughout a supermarket. It is therefore necessary to maximize the visual effect of the product range offered the consumer.

This technique is most appropriate when the product range is grouped together in a mass display. The name of the game in the supermarket jungle is "presence." Many data have been supplied on consumer time exposure to a package display. Any planned design strategy must carefully take color into account.

In the average supermarket selling between 8500 and 10,000 items the best way to achieve "presence" is to use color for shelf area control. A single color makes a powerful design statement in presenting a line of merchandise. Color is the emotional communication with the consumer. Therefore it must relate properly to the nature of the products being sold, so that the shopper's response will be favorable.

"Line" refers to the graphic organization of the other elements of design into a clear and distinctive whole. Packages are usually displayed on horizontal shelves. They are stacked vertically. A linear design treatment takes these fundamental factors into account. It is especially beneficial if a graphic design can be endowed with a "bridge" effect. A bridge effect is carry-over of the design motif or format from one package to the next in a continuous pattern as they stand side by side on the shelf.

"Texture" refers to package design areas that have a detailed pattern suggesting tactile qualities appropriate to the product image. On an opaque package where the product cannot be seen, pictures are essential. In a self-service shopping environment, the package serves as a "silent salesman" and the design itself becomes both the sales promoter and the advertisement at the point of sale.

The picture says to the customer: "Buy me, I am the best value for your money." Therefore the picture must be convincing. Distortions or misrepresentations of the product being portrayed must be avoided. While a glamorous photograph of the product may close the sale, the customer may feel deceived if the contents do not live up to the perceived promise of the package. The one-time buyer is lost for good.

Containment is the package's traditional function. Size is determined by marketing considerations, such as product price, usage, and production, distribution, and home storage requirements. Form follows the function of the container as a protector of the product.

"Shelf life" is a basic requirement of the self-service supermarket. The product must be delivered to the customer in a satisfactory condition. A can of peas has a different shelf life from a carton of frozen peas. The respective packages are designed to meet the marketing requirement of products with different shelf lives. Again, simplicity of size and form is necessary to avoid costly production, distribution, retailing, handling and storage complications.

Because we are dealing with a product that is marketed internationally, copy is an important consideration at the local level. Planning should anticipate any national language requirements. This is especially important where a local company has been acquired by the parent organization. Projecting the local brandname within the framework of the design is highly desirable so that customer identity with its franchise of familiarity is not lost.

Depending on how much advertising promotion supports the product line, local language communications must deal with product identity, slogans, and mandatory or regulatory copy imposed by government requirements.

A major contribution to success in implementing the design plan is made by designating individuals at each local subsidiary who will be responsible for liaison with corporate headquarters. These individuals also coordinate supplier services within the company. This requires a thorough understanding of the corporate objectives involved in the plan, as well as the ability to preserve local marketing prerogatives. These individuals must have the authority to back up directives. Generally, a company that has had a successful business on the local level is structured with sufficient personnel to assist in solving marketing, production, sales, purchasing, or legal problems. In today's business climate, the greatest responsibility is marketing—planning sales of the product. This function is the focal point in competitive selling. It must bear the burden of any successful enterprise.

Because implementing the package system involves all these activities, coordination is essential. For example, the responsibility for insuring that all mandatory copy conforms with government regulations is handled by the legal department. Where legal problems concerning copy occur, the adjudication must be left in that area of expertise so that product copy is not vulnerable to censorship.

The judgment factor in controlling and coordinating these functions for the best performance of the system requires careful diplomacy. This is especially necessary where the will of the distant parent corporation is imposed on local personnel. The corporation wants the plan carried out, but it must also respect local customs that haven't been identified in the planning stage. Otherwise resentment may ensue and additional problems may be created.

In a marketing oriented consumer products company, especially where a wide range of products are manufactured and distributed, a brand or product manager is responsible for sales of a product or group of products. This person is responsible for carrying out his part of the design plan. Because his is part of a total plan, however, it is necessary to be sure that he fully endorses the entire plan. He must be convinced of the plan's merits while being allowed to offer criticisms or suggestions. In this way, both the local company and the parent corporation will have enthusiasm for its objective.

In implementing a package design system, a design control manual serves as a useful device for ensuring its control and the preservation of its integrity. Such a manual sets down guidelines of design usage for those who deal with the details of design adaptation and reproduction.

CHAPTER FORTY-TWO

Developing New Products in the United Kingdom

George I. Suter

MAJOR TRENDS IN NEW PRODUCTS DEVELOPMENT (NPD)

In common with most developed countries, NPD has been through a classic boom, a recession, and a moderate boom. Up to 1973 most companies were falling over each other to launch capital intensive new products, regardless of competition. A total reversal was signaled by the miners' strike, the 3 day week, and the recession commencing in January 1974, when all manufacturers virtually halted NPD. It was not until mid 1976 when NPD came to the front and then it was characterized by low risk projects. Now (1979) the scene is business as usual but in a much more accountable, realistic style than the heady years up to 1973.

METHODOLOGY

Methodology has notably returned to the simple and excludes ostentatiously sophisticated (and expensive) techniques.

Qualitative

Most companies are now standardizing on six to eight standard length group discussions, (1½ to 2 hours, eight respondents) often with a pilot group or two to improve concept boards and product. There are few requests now for the 40 group interviews or 200 depth interviews for any single job that were commonplace before 1973.

Quantitative

It is now more usual to standardize on 4 hall days for a typical hall test (somewhat similar to the U.S.A. Mall Intercept). Often these are carried out simultaneously with 1 or 2 day semidepth interviews on the same material. The large-scale multivariate and other analyses have largely disappeared in favor of a much simplified Item × Use technique. Norms, as in the five point Try, Serve, Pay scales have fallen into disrepute.

Placement

Placement tests are becoming more standard, more extended, and often replace test markets.

Concept Versus Creative Execution

In the days when it did not matter, concept and creative execution became hopelessly combined. Advertising agencies classically produced initial concept boards looking very much like press advertisements. Respondents' reactions were to creative execution not to concept. Wherever possible, the rule has been to have a clear separation.

Test Marketing

Test markets have been reduced in favor of extended placement and other tests. The Research Bureau Ltd. Minivan is one of the best established techniques of measuring repeat purchase, especially on novel products. Our group has pioneered an effective shop testing system, primarily for testing off take of different packaging types.

OWN LABEL DEVELOPMENT

Retailers own labels (private labels) are classically me-too products with different graphics. Own labels grew rapidly to 27% of grocery turnover up to 1976. However, their quality was often poor, not least because they were consciously cheaper than their branded equivalents. Design was often inconsistent and positively amateurish. The main supermarket chains clearly gave their own labels a bad name.

In the last few years a positive change has occurred. Marks and Spencer, Sainsbury's and Boots were models of good quality and good design. Own label ranges were rationalized to major sellers of reasonable quality, with better margins and a positive image.

International Stores placed all their own labels under an umbrella, Plain 'n Simple, concept representing consciously good value but treading dangerously near an appearance of cheapness/poor quality. Marks and Spencer developed a three-pronged approach: (1) standard St. Michael own label, (2) semibranding (e.g., the Care range: Bath Care, Hair Care, Salon Care, etc.), (3) brand mimics (e.g., the Ivory Range). (Fig. 1 and 2)

Naturally the brand mimics ape the brand's quality feel and substitute in sales with them while the semibrands

Figure 1   Own label semi brand: Hair Care.

Figure 2 Own label brand mimic: Ivory Collection.

middle ground of cosmetic acceptance above its cheap and cheerful Evette own label and competing with the main brands. This very major exercise involving brand mimic development, physical pack design, color choice, display, dispensers, and advertising was all developed and researched as an entity by NCK for Woolworths in the United Kingdom and has now extended to Canada. (Fig. 3, 4, and 5)

are obviously superior quality own labels. In this way, the shop can give consumers the full range from cheap and cheerful to expensive and classy but do so more with their own brands, while giving the consumer the range she wants.

Another classic own label development was the Tu range developed by Norman, Craig & Kummel (London) for Woolworth's. This range occupies the

Figure 3 Tu range of cosmetics.

Figure 4 Display unit for Tu.

Figure 5 Tu blister packs for pegboard.

BRAND EXTENSIONS OR WHAT'S IN A NAME?

A classic research and marketing problem is to what extent a brand name can be extended across new products. In some cases, however cunning the presentation and however subtle the research, one could not overcome the product identity of the brand. Thus Oxo could not be used on a range of wines for cooking since it implied beef flavored wine in a cube. However, Ajax means cleaning power and extends very happily to pan shiners, general purpose liquid and cream cleaners, and even window cleaner (although some respondents thought it might scratch the windows). The result is an obvious range of six products where the new ones were launched without advertising on the back of Ajax and where respondents honestly thought Ajax pan shiners, for example, were an old established brand like Brillo. (Fig. 6)

A rather different use was faced by Brooke Bond with coffee. Brooke Bond's excellent image with tea was readily transferred to coffee. However, the market was so large at £200 million (at retail sale price in 1977) and fiercely competitive that the company had to supplement its good name with a product distinction. One was the use of refill packs that were well received in research but made little difference in the market-refills are a classic item to show up better in research than the market). The other was All Brazilian content—this was found to be of low importance in initial research but proved much more important in the marketplace. The reason is that people still do not believe that instant coffee is pure and instinctively consider it chemical and a mishmash of the cheapest ingredients. The Brazilian Blend name and the coffee plant graphics, when reinforced by advertising, added up to an image of purity and naturalness above that of the well established Nescafe and Maxwell House. Brooke Bond quickly succeeded in obtaining over a 5% share (£10 million, equivalent to a U.S. brand of $100 million), which exceeds the minimum viable share. Unfortunately, the fierce price competitiveness of this market forced Brooke Bond to undercut the brand leaders price, and profitability after advertising was not good.

NEW (PHYSICAL) PACKAGING

The 1970s have seen the introduction of a large number of completely novel physical pack types (as opposed to shape or graphic design) especially for liquids. Many of the more novel packs were so evidently doomed to failure as to be

Figure 6 Ajax range extension.

obvious nonstarters. Whatever the consumerist or overly socially conscious respondents might say in qualitative or quantitative research, consumers in the market place were very satisfied with cans and bottles.

The more novel the pack, the more obviously it was doomed. Thus the ICI Merolite (a plastic bag in a composite can) hardly got off the ground. The stand up foil laminate pouch packs scored better on interest but had no basic rationale and have largely faded away.

The retortable pouch was on confusing middle ground. Research was partly favorable—the product seemed fresher, smarter, easier to store and cook, and so on. At the same time, it was seen as expensive, perishable, and novel. Indeed, well structured research showed that the disadvantages clearly outweighed the advantages, with the possible exception of gourmet meals. The Japanese success and strong support from Metal Box, however, caused many companies to try vigorous entries into this sector where the pouch was promised to replace the can. The results in the marketplace were disastrous— Brooke Farm fruit and vegetables were followed by sauces almost in desperation. Heinz made a more cautious, primarily catering oriented entry. Marks and Spencer own label did well for a while before gradually withdrawing. The research was right all along, and no real opportunity existed.

The Swedish aseptically filled Tetra-Pak was worthwhile. The product was significantly better in taste especially in the crucial areas where retorting damaged flavor most, in milk and fruit juice. The initial tetrahedral pack was unpopular with consumers and trade and failed. The Tetra-Brik (rectangular carton) has been highly successful, however, both for ultra high temperature flash sterilized milk and fruit juice. Again, the research was right; it was

also clear. Unlike with retortable pouches, manufacturers were much more cautious—some clearly felt "once bitten, twice shy." Eventually it was the dairies and Adams Just Juice who entered the area seriously (Just Juice alone have sales of £3 million) although these companies had no history of being innovative.

On a much less innovative front European Marketing Surveys, Limited was involved with Coca-Cola's introduction of the 1½ liter plastic bottle. The bottle had no notable disadvantages but little advantage compared to glass in consumer research, and so the question had much more to do with sales substitution within the brand and against Pepsi and own label, than with consumer acceptance. The research was carried out using a shop test with a close control over facings, pricing, and stocks. After a 4 week bench mark the test pack was introduced with 10 supermarkets for 12 weeks, while a matched sample of 10 supermarkets acted as control. The results were convincingly successful and the new pack was launched nationally, with equal success.

CONSUMER SEGMENTATION

In many cases a product can seem new because it is an old product specifically aimed at a distinctive consumer group. Very often such products are almost the same formulation/physical product as general purpose products and simply use the new label to communicate to a specific consumer segment.

Slimmers are always a classic consumer segment. The secret of a successful slimming product is good taste. Since most low calorie products taste very inferior, the answer is to give an impression of slimming without saying so. Findus have been successful with a disarmingly honest but effective positioning with "Calorie Counters," which not only makes no low calorie claim but goes out of its way to state that it is merely "counting" calories, not saving them. This excellent positioning appeals to nonserious slimmers (who are numerically superior to serious slimmers) and it does not discourage nonslimmers.

Equally successful was the launch by Bass of low carbohydrate beer. Bass first briefed this concept and product development of a low calorie beer in September 1972; it took 5 years to launch Hemeling in 1977. In the intervening period low calorie (50% or less) beers were developed but were totally unacceptable in taste. Fortunately, low carbohydrate (and "Lite") were perceived as having the same meaning as low calorie. The low carbohydrate formulation was technically easy and of good taste. It was simply fermented to the limit but had the same strength and calories. Much then depended on the product claim and creative presentation. Examples of early concept boards and the finished pack are shown and Hemeling has not only performed very well but been followed by many other Lite beers in the United Kingdom. The same concept has performed well in the United States. As can be seen, the final creative execution is so "laid back" as to be obscure—this being another case where

the product claim/distinction was secondary to good taste. (Fig. 7)

Unlike the United States, Britain has had little success in children's products because (there is some evidence) the British are less indulgent with children, unprepared to pay extra or to duplicate adult and children packs. Indeed, National Food Survey data clearly prove that children are fed cheaper foods (e.g., potatoes and bread) than adults. Children do need to be coaxed to keep clean, however, so expensive, fun toothpaste or soap are justified as better than grubby kids. To this end, Super Matey was developed by NCK with Aspro-Nicholas. The original pack was almost as subdued as Hemeling. This proved wrong, however, since kids want ostentatious fun while adults prefer their concepts to be less garish. Hence Matey was redesigned with even clearer children orientation than shown in Fig. 8.

No chapter on consumer segmentation would be complete without more detailed coverage of the mechanics of that most special of all segmentation exercises—life style (also known as psychographics) as applied to toiletries and cosmetics (note it has also been used in drinks and tobacco). A nice life-style example preceded Shulton International's launch of Mandate where a target category of male life style was isolated as the target market.

Figure 7 Concept and creative development of Hemeling.

Figure 8 Matey bubble bath before redesign.

Table 1 Typical Life-Style Questions

	% agree
A house should be dusted and polished at least three times a week	35
I always use the same brands when I can get them	48
I like spending most of the time at home with the family	45
I dislike the idea of a housewife and mother going out to work	22
I make an effort to worship regularly	4
In a job, security is more important than money	47
Living together can be useful preparation for marriage	64
If I was offered soft drugs or pot I might try it	24
As a country, we give in too easily to foreigners	58
If I were a magistrate, I would give stiffer penalties than most	33

These and other statements were factor and cluster analyzed to give six viable groups of like thinking persons. One such group, code-named the Tony group, was the selected target for Mandate. (Tables 1 and 2)

Table 2 The "Tony Cluster"

DEMOGRAPHICS

Tony is the youngest of the men with an average age of 34. The unmarried members are a third of the cluster and have an average age of 24 while the average married age is 38. Mostly upmarket but some unskilled workers and unemployed. Well educated. Half of students over 16 fall into this cluster. The cluster is the second largest.

CHARACTER

Tony is serious and ambitious in his job, not satisfied with what he is earning yet. He can see an easier, more status-filled life ahead of him, which he looks forward to. He's stylish but not very particular about points of hygiene. He's a girl watcher, even more so when he is married.

HOME

When he's young, he'd very much prefer a flat or bedsitter to living with his parents, and when older he would like his own house. He doesn't like fussing over housework, and when at home prefers his wife as a companion rather than stuck in the kitchen.

Table 3 The Tony cluster and aftershave

ATTITUDE TO AFTERSHAVE

He likes good smelling aftershaves, which he believes make him attractive to women and superior to other men. He likes expensive products for special occasions and keeps at least three brands in his bathroom.

Table 3   Continued

IDEAL PRODUCT

Sexy, long lasting scent. Romantic and high quality. Adventurous, modern, and good for special occasions.

PRODUCT USAGE

95% of the Tony group use aftershave and they are the heaviest users of the six groups. The group has a strong liking for Brut, but is the weakest group for Old Spice usage.

Clearly Shulton needed to target Mandate to this segment where their Old Spice brand was weakest and where product positioning was most distinct. The subsequent positioning and selection of Sacha Distel as international presenter followed what had become a clear, tight creative brief. The launch copy was a clear parallel to the description in Table 3. Sacha Distel states, "It's sophisticated, long lasting and very sexy! Whether I'm working or simply relaxing, Mandate says what I want to say. It speaks my language—even sings my songs."

NOVEL PRODUCTS

Most new products are not new but are simply similar products in slightly different clothing. Even some products that are completely novel to the retail market may have been around for some time in other areas, for example, filled milks were in catering and sun screen cream

Figure 9   Mandate aftershave.

was as ethical pharmaceutical. In retail consumer terms, however, these products seem new even if technically they are old, and the job of marketing is to repackage and change the distribution.

A nice example is filled milk, an inexpensive milk substitute, which had been available as a cheap alternative to fresh milk for many years, including St. Ivel Miracle, which was available in 2½ kilograms or larger bags for caterers. However, the real miracle turned out to be packing in 11 ounces (making 5 pints) plastic bottles looking like milk and called 5 Pints. Even the name came out of the group discussions, while each research stage perfected the milklike positioning. The results were excellent with this one brand selling £10 million at retail sales price (equivalent to a $100 million brand in the United States) at a good margin. The positioning was so good that Cadbury's, the natural competitor with a major share (with their Marvel brand) of the £15 million skimmed milk powder market, found it difficult to improve, and their subsequent carton of four sachets called Pint Size could be only a limited success because St. Ivel had their product just right—packaging, name, brand, appearance, and product.

A parallel story in a different area was the launch of Uvistat Sun Screen. The product was previously available on prescription from doctors in plain blue tubes. The product was in no sense dangerous, however, the required doctor's prescription and the consumer problem of sunburn in spite of sun tan cream during the couple of weeks when the average Briton actually sees the sun, meant there was a real consumer need for it. This proved the most straightforward of NPD exercises—the product was repackaged to look over the counter rather than ethical, displayed on view in chemists, and supported by a lip protection product. The difference between screening and tanning could be confused—but then this was the product's Unique Selling Proposition. (Fig. 10)

As always, genuinely novel products understandably grab the marketing man's attention. The Brooke Bond flavoring development was a classic attempt to match the product to consumer needs as shown in textbook but rarely seen in practice. A formal problem tracking exercise was carried out in two stages, first qualitative and then quanti-

tative, to identify and then rank the nature of the problems in cooking meat. High on the list was lack of flavor in roast chicken plus indications that it was dry and took too long to thaw. Of course these are opportunities, not answers.

Further down the problem tracking list came wine for cooking. This was not an important problem but it did prove to be technically feasible, not least because of the development of successful wine flavors and alcohol enhancer—not good enough to drink, but certainly good enough to enhance a chicken casserole and make it taste like coq au vin. The product formulation required little research after achieving an optimal flavor on judgment. The first research was group discussions on concept and two pack forms as shown in Fig. 11, a bottle (actually a crown capped Babycham bottle) and a foil sachet. This showed good proposition acceptance, made it quite clear that a screw top (use some, keep some) bottle was vital, and was extremely cautious about use of the Oxo branding, which gave an impression of beef flavored wine. The two bottles on the right of Figure 11, one without Oxo, one with a subdued Oxo, were then used in red and white flavored wines. Subsequent research established other flavors and trade research was used to define the unfortunately "licensed only" distribution. The product would have sold better near spices or salt than wine, but

Figure 10   Uvistat Sun Screen.

Figure 11   Cooking wine research packs. Bottle on left used in group discussions.

Figure 12   Final cooking wines and follow-up product.

that was the law. The final packs in Figure 12 performed well at a good margin although distribution, especially visible display for impulse purchase, proved extremely problematical.

Finally, the closest to a really new product that grocers have seen in the last few years was the launch of TVP products. First in were Crosse and Blackwell with Mince Savour—intended to be mixed 50 : 50 with minced beef—and bound to be minor because of the low use of minced beef in Britain, the product's inconvenience, and very minor price advantage. Cadburys and Spill-ers had moderate success with a canned meat substitute called Soya Choice and Soya Mince. This reflected the research, which clearly indicated that consumers found soya perfectly acceptable as a meat substitute, with no need to pretend that it was meat.

Brooke Bond followed the dry route and first tried a mince additive as a cheap substitute. It had the same disadvantages as Mince Savour, compounded by the coy camouflage of the name; we were not introducing a meaningless Country Meadow—this was at best obscure and at worst misleading. Thus the first priority was to reformulate as a 100% mince substitute rather than a 50% meat extender. However, the irrelevant Country Meadow was still the lead-in and the pack was excessively bleak and unfoody. The final design learned these lessons—Country Meadow was dropped, as was protein from the title since Soya Protein sounded too much like a health product (cranks this way!) and the package design execution was much more atmospheric and foody. The national launch went smoothly and with considerable success.

CHAPTER FORTY-THREE

Packaging Design Research in Sweden:

SIX CASE HISTORIES

Ingrid Flory

Business conditions are about the same everywhere. You have to know your market, and you have to find out how to satisfy your customers. These very basic conditions are valid in Sweden also.

If the product is directed to retail consumers, it is most likely that it is packed and that the package is a vital part of the product. It is the package that conveys all information and projects the associations that may create product acceptance and sales success. Packaging design research is a helpful tool in finding out how well the package communicates the benefits of the product.

I have been asked to give a survey of packaging design research practices in Sweden. While thinking it over, I decid-ed not to present a broad survey in general terms but rather to relate a few case histories that have a common core of interest and uniqueness. They are not all success stories. Two of them describe products that are not yet on the market. Two other cases do not deal with products at all but with practices. Four cases out of six deal with genuine Swedish products and at least two of them have been given new packaging to please and attract the American market. Let me review them all briefly:

1 The CLASSIC Coffee story represents a well known and exclusive brand in need of "revival."

2 SALVEKVICK represents a wide

range of adhesive bandages, an institution and a "must" in many Swedish families. Wanted to "clean up in the design jungle" and to go international.

These two stories date back to 1975 or before, so the results are known. All packaging design work was done in Sweden.

3 ABSOLUT Vodka, good old Swedish aquavit made by the state liquor monopoly, is making a controversial but professional move directed at the American market, backed up by substantial design and test work. A thorough preparation for a tough market. The full result is not yet available.

4 RAMLÖSA mineral water looking for new markets for its product—the classical "business meeting" refreshment—from one single source. Created a new bottle in a new blue color, a first from the PLM Group. The Americans did all the research.

5 Packaging design research has also been used to solve, or at least interpret, *legal matters* by stating the degree of recognition—identification—that could grant a trademark an exclusive in the marketplace.

6 Finally, packaging design research has also been used in combination with purely technical research in *consumer ergonomics* in order to find out how to satisfy basic strength requirements in opening and reclosing of packages.

The last two cases may also result in considerable economic savings if their methods are further developed.

Market research techniques are almost the same all over the world. In addition to the product and its market conditions it is the use and interpretation of them that differs. It is this use and application of Swedish design research that makes it unique.

Exhibit 1   Classic Coffee

Product: Classic brand ground coffee in 1 pound metal can.
Producer: Arvid Nordqvist AB, Stockholm.
Characteristics: New package design for established product in standard coffee can.
Research Institute: Skandinaviska Marknadsinstitutet AB, SMI, Stockholm.
Test Method: 1. Distribution analysis 2. Package design research 3. In-store market tests with
                  interviews.
Advertising Agency: Ted Bates, Stockholm.

Background

Classic brand coffee has been on the Swedish market for many years and is a well established brand from a company with an excellent reputation. Sales had developed well but the management wanted to take measures to maintain its position but also to improve sales, primarily by appealing to new consumer groups.

A series of tests were carried out to help the management in its decision making process:

1. A distribution analysis to help identify the structure of actual buyers.
2. A design test in order to select two out of five alternative new designs.

3. An in-store market test with interviews to check the effect of the cans in the market place—on the shelves in the food stores.

Distribution Analysis

The distribution analysis clearly indicated that buyers of Classic coffee were a very homogeneous group of people, loyal to their brand. Buyers were primarily from the Stockholm metropolitan area, in rather high income brackets, and slightly older than the average coffee consumer. The image of the coffee was high quality, good taste, value for money, and expensive.

The management now considered changing the package design. A new design should offer a natural platform for an advertising campaign and should emphasize the image of Classic coffee as being exclusive and expensive. A very important part of the brief was that the new design should clearly indicate, by the color of the can, different roasts (dark versus medium).

The next step was therefore to ask the designers at the advertising agency to work out alternative package designs within the limits of size and material of the standard 1 pound coffee can. They were free to elaborate with visual factors—color, surface design, pictorial elements, typography—without abandoning the overall concept of a high quality, exclusive brand. The purpose of the research was also to assess the "family image," the conformity of appearance, of the alternatives.

Design Research

The design research was done by the Scandinavian Market Institute (SMI) in Stockholm. The respondents were contacted by telephone and invited to the SMI office. They were exposed to five different can designs, and trained interviewers questioned them according to a questionnaire. 59 interviews were carried out, half of these with regular Classic coffee buyers.

The test gave some very clear results and helped the management to make necessary decisions. A dark package, it appeared, reflects the image of a strong, full-flavored, luxury coffee, appreciated by people who like good food and extra rich coffee. Significant differences in the attitudes to the five samples could be measured by semantic scales: strong/weak, expensive/cheap and so on and the cans that had a particular "family-line" appearance could be singled out. Two of the designs were finally chosen for in-store tests.

In-store tests

The tests were run in six food stores in the Stockholm area. The regular cans were removed and replaced by the new ones in two different designs, one for dark and one for medium roast. No particular sales promotion activities were allowed during the test period. There were interviewers present in each store during 5 days. Their job was to contact consumers who had just bought the new can as soon as they had passed the cashier and ask for a telephone interview on the next day. Nothing was revealed about the purpose of the contact and the planned interview. Both the buyers in the store and other customers who had noticed the cans were addressed. A short interview was carried out with them also, following a special form. The questions covered were, in short:

Frequency of buying Classic coffee.
Why did you buy this time? (open question)
Impressions of the new can (positive/negative).
What did you think (while still in the store) about the new design?
Did you think the coffee had changed too?
Positioning of the can.
Probability of buying and reasons.

Result

The new cans did not attract new buyers during the five test days. Only 3 out of 113 respondents were first time buyers of Classic coffee. 80% of the nonbuyers who had noticed the new cans were also regular Classic users. However, many positive comments were given about the new design and the new colors. The respondents said the cans were beautiful, eye-catching, and appealing. Thus the in-store test confirmed the result of the design test and indicated that the image of the product was exclusive and expensive. Many respondents said spontaneously that the different colors made it much easier to choose the right roast. This factor was considered very important.

The new design was well received by the consumers and achieved the goals set by management. The company therefore decided to change to the new package design. This has enabled them to reinforce, and slightly improve, its position in this highly competitive market segment (Fig. 1).

Figure 1   Classic coffee can. The old design to the left, the new to the right, in two different colors, dark brown versus golden brown to indicate type of roasting.

Exhibit 2   Salvekvick Adhesive Bandage

Product: Salvekvick Plåster
Producer: Cederroth AB, Stockholm.
Characteristics: New packages for established products in old and new markets.
Market Research Institute: AB Marketing Bo Jönsson, Stockholm.
Research Method: 1. Semistructured qualitative study 2. Package design research using tachistoscope and interviews.

Background

The Cederroth Company has been producing and marketing adhesive bandages on the Swedish market for many years and keeps a dominant position. During the 1970s the company has also expanded into new regional areas, particularly in Europe, directed by its corporate headquarters in Geneva, Switzerland. All their marketing activities, including change of products, packaging, and conceptional platforms have been subject to thorough market research.

The management decided to make radical changes in the design of all Salvekvick packaging. The purpose was twofold:

1   To differentiate the package design sufficiently to make differences in types of adhesives obvious to the consumer.
2   To modernize the design.

The most acute problem was felt to be that of identification—how to use design devices to signal differences in product properties—and that of finding the right mix of such design elements as illustration, copy, colors, and size. Some bandages are made from textiles, others from plastics; some are precut, others are not. Quantity, size, and mix also vary. The problem was to determine whether the color coding would help or confuse the consumer.

Methods

Three different tests were carried out. The first one was a qualitative study with 20 semistructured interviews. The respondents were housewives with children under the age of 15. The purpose was primarily to identify how people perceived adhesive bandages, their habits and attitudes, and what functions they considered important. The purpose was also to find out if there were any associations of product types with current color codes. This study provided certain bench marks for the following work.

The next study was a packaging design test on a number of alternative new designs for Salvekvick. The purpose of this test was to get individual views on three alternative designs as to identification of product type, brand name, bandage types, and the associations they revealed. The purpose was also to single out the best design of the proposed three and compare it with the package already in use.

The research method used was a qualitative study, quantifying some of the results. Personal semistructured interviews were carried out. Part of the interviews used a tachistoscope in order to determine the appeal of certain elements of the package design and the time required by the respondents to identify these elements. 80 interviews were conducted. The result was a very definite endorsement of one of the new package designs.

At this time the product and packaging strategy of the company was given a different, broader approach, and management decided to expand the use of the present design to

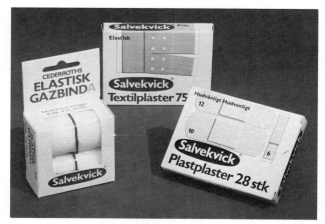

Figure 2 Salvekvick Adhesive Bandage comes in several varieties. The new pack design gives guidance to the consumer.

the international market. It was therefore necessary to conduct a third package design test to determine the best design and the best concept for international consumers. The management was undecided about which design to choose for the Scandinavian market: the international one or the new design that had been singled out in the previous test. It was therefore important to collect additional information about the reaction of Swedish consumers to the two competing package designs.

Again personal interviews were carried out with housewives having children aged 2 to 15 years. They were asked questions about the importance of various characteristics of bandage styles. They were also asked for their spontaneous reactions to a number of package designs, for their overall preference among two distinct alternatives: the international or the new Scandinavian design.

From the result it was obvious that one of the packages was preferred and that two thirds of the respondents preferred it while one third preferred the other package design (Fig. 2). The company acted accordingly.

Comments

These studies show the importance of testing management opinions and ideas on the consumer. Such testing is particularly important if the product appears in different geographic markets where attitudes and preferences could be different from the home market. Other studies have clearly shown that, generally speaking, a favorable association in one country could well be unfavorable in another. Historical and cultural differences call for great care in international marketing.

Exhibit 3 Absolut Vodka

Product: Absolut Vodka intended for the American market.
Producer: Vin & Spritcentralen (the Swedish Liquor Monopoly), V & S.
Characteristics: Established product, new package, new market.
Agency: see below
Market research group: see below

Background

V & S, a state monopoly for production of liquor in Sweden, had decided to expand by exporting.

The company had no tradition whatsoever in marketing its products abroad. But they had good products, so here is how they started and how they finally ended up on the U.S. market.

Phase Zero: Exploration

V & S started by taking a look at products and packaging in the liquor field around the world. An international market consultant (Ilmar Roostal, Switzerland) was commissioned to conduct unstructured field and desk research that resulted in the following conclusions:

Most interesting market: United States
Most interesting product: Vodka (Akvavit)
Potential consumers: Young people, high income, urban areas.
Prerequisite: The products must be at least as good as the best on the market, and have a distinctive profile. The packaging must be of at least the same technical standard as that of the competitors.
Basic principle for the brand design: high quality, associations to Sweden, high status, strong character, natural tie-in to advertising.
V & S confirmed the decision to go international with the Swedish akvavit and to launch it as "Vodka." The United States was to be the first country to try the Swedish vodka.

Phase One: Concept Work and Basic Design

The basic work on concept and design was done in Sweden. The assignment was given to the advertising agency Carlsson & Broman AB in Stockholm, which is known for successful work in the beverage area. The agency's first job was to create packaging concepts and a platform of ideas for the launch. After a presentation by the agency to the management of V & S it was decided to evaluate the proposals on the American market.

As a first step in exploring the reactions of vodka drinkers and purchasers to the idea of vodka imported from Sweden, the market research department of N.W. Ayer ABH International was asked to conduct an initial focus group exploration. Three focus group interviews among qualified consumers were conducted. The purpose of the group interviews was to study the following factors bearing on the marketing of Swedish vodka:

1 How vodka is used and when and why it is used in preference to other "white" liquors, specifically gin, white rum, and tequila.

2 The perception of differences between vodkas and how these perceptions affect brand selection.
3 Experience with and the image of imported vodkas, all of which require the purchaser to pay a premium price.
4 What qualities consumers would want and expect in a Swedish vodka to warrant trial and regular purchase.

The method used in the group interviews, which were taped in their entirety, consisted of having a moderator review each subject of the discussion outline and guide the general flow of the discussion. The research company says in its report:

"The purpose of focus groups is to stimulate an exchange of views by the participating respondents in such a way as to bring out the range and intensity of the attitudes they bring with them and to see how these attitudes might change as different points of view are expressed in the course of a casual, free-wheeling conversation. . . . As a source of new ideas and the manner in which existing ideas are accepted or rejected, focus group sessions often present findings which are useful in themselves, given a sufficient number of such sessions. Another function of focus groups is to get the raw material from which to construct a formal questionnaire to be administered to a statistically valid cross-section."

Absolut Vodka Focus Groups

The group interviews were carried out in the Philadelphia area with 35 participants, 23 men and 12 women. The respondents had to meet the following qualifications:

1 Used or purchased vodka for home use within the last month.
2 Regular purchasers of premium brand of liquor, although not necessarily vodka.

In addition, at least two members of each group were to have some acquaintance with an imported vodka, either as purchasers or triers.
All of the interviewed persons had total household incomes of $ 15,000 or more, the median being in the $ 20,000 to $ 25,000 range. The average age was 42, and the mens'

job titles identified them primarily as professionals or members of middle management.

The result of the group discussion revealed certain facts about how American consumers regard Sweden, vodka, and the concept of a vodka product from Sweden.

The next step in this explorative research was to conduct field studies in American bars and to observe competitive bottles of vodka. Most bottles seemed to be copies of the well known Smirnoff brand and to have an east European character. It was considered unrealistic for the Swedes to make a more fanciful or more elegant bottle; it was decided to create a totally unique bottle and label.

The Swedish firm of Carlsson & Broman developed five alternate concepts that related to Sweden and which would be eye-catching on liquor shelves. Samples of the bottles in wooden prototype form were shown to people who professionally market liquor in different countries. The test clearly showed that compared to the others one of the bottles was outstanding.

Phase Two: Consultations

Again the Swedes went over to the United States, bringing with them the sample bottles, to talk to creative experts and to make arrangements to have research on American consumers performed to test the bottle, the name, and the concept. The bottles were also shown to American liquor distributors. At this stage the original five alternate concepts had been reduced to just one in a couple of variations. Modifications were made as a result of reactions by those who had been consulted.

It was emphasized that the bottle must have a prestige connotation as American consumers often look for symbols and buy according to their life style. It was also thought that the vodka by itself was rather uninteresting. The bottle had to be in line with the liquor traditions of the country but also convey associations with the simplicity and beauty of Swedish glassware.

Phase Three: Testing

In this phase a series of concept, package, and product tests were carried out on the American market by American market re-search companies. Modifications of package, products, and assortment were made.

One such project with consumer developmental groups was conducted by N.W. Ayer ABH International. Total number of respondents was 57. They were all frequent drinkers of vodka. Two campaigns were evaluated, both based on the same bottle. Two "distracter" campaigns were also shown. Proposed label copy was also exposed to each group along with the first campaign. For each campaign, after initial exposure and before group interaction, respondents were asked to fill out a questionnaire detailing their purchase interest, likes, dislikes, and the likelihood of purchase as a gift. They were also asked to rank the campaigns.

The next step in the process was to undertake a qualitative research study to test initial concept and packaging approaches in order to form the basis for developing the advertising and packaging for the brand.

Martin Landey, Arlow Associates, Inc. had been commissioned by V & S and by Carrillon Importers to develop the advertising and the packaging for Absolut Vodka. A quality study by MPi Qualitative was initiated to test the final concept and packaging approaches. The group sessions were carried out in New York and Los Angeles. Further changes in concept and bottle prototype were made as the knowledge increased.

Phase Four: Test Marketing

Time had now come to produce "real" bottles in glass and to fill them with the beverage. Considerable technical problems occurred in this phase when producing the bottles in accordance with the original concepts. The bottle was to look like traditional crystal ware, but produced on modern equipment. The clear white color (flint) was a must, but very difficult to obtain in Swedish bottle manufacturing. (Swedish sand results in a greenish glass because of its iron content.) Finally the sand necessary for the production had to be imported from France. It had also been decided that all facts about the product should be conveyed in a decoration applied directly to the outside of the bottle (See Fig. 3). This technique is very difficult and requires special equipment of very high quality.

Figure 3 Absolut Vodka bottle designed specifically for the U.S. market, with product identification directly on the bottle surface.

Further modification had to be made in the bottom and on the neck in order to adapt the bottle to the filling lines. There are still problems to be solved.

Comments

Some people feel that packaging design research is performed as an insurance for those who do not dare make decisions. It is also said that new ideas are often met by a negative response from most people and that good ideas will thus be defeated. This case shows that an original concept, created by intuition at a very early stage, could survive through all decision and evaluation processes.

However, packaging design research should be a valuable instrument in assessing message communication to the consumer. Testing can also help eliminate negatives. Some of the research conducted on Absolut Vodka also had the purpose of creating acceptance of the product and the concept by those people who were to work with it in the American market.

The product was finally introduced on the American market in the fall of 1979. The initial response has been positive.

Exhibit 4   Ramlösa Mineral Water

Product: Ramlösa, a natural mineral water from one particular spa in southern Sweden.
Producer: Ramlösa Bottling Company in the Beijer Group.
Characteristics: Established product, new pack, new market.

Background

Ramlösa and the packaging manufacturer PLM had been doing business for years. Independent of each other, they had both investigated the market potentials for mineral waters and found that it was very large in Europe and definitely underdeveloped in the U.S. On both continents consumption was going up —although from different levels; in Europe this was partly caused by the poor quality of tap water.

Health and calorie consciousness also seemed to contribute to the favorable outlook particularly in the United States, so Ramlösa decided to develop the export market potentials and to make the unique Ramlösa mineral water available outside of Sweden. The United States was chosen as prime target.

Program

The owners of Ramlösa, the Beijer Group, could furnish financial support and general export market know-how. The group was also established in the United States and could help open doors and give practical advice and assistance.

The PLM company had its glass products division develop a number of package concepts for the bottles, all in flint glass.

Ramlösa went to the United States to obtain reactions to the packaging concepts. They returned home with two directives that were completely different from the original ideas:
1   The shape and size of the bottle should be in line with similar products on the international market.
2   The bottle should be made in blue glass. (A major imported mineral water brand in the United States is Perrier of French origin, bottled in emerald glass.)

Package Development

It was decided that the Ramlösa bottle should be marketed by the Schlitz Brewing Company. All contacts with all parties involved in the United States were handled directly by Ramlösa.

The package design process concentrated on two aspects: shape and color. Especially the color was very important for the image of the product but also difficult to handle.

Market tests were conducted in the United States in order to analyze the use of (and attitudes to) mineral water and of existing products. Package design research was undertaken concurrently with the various phases of the creative process. These tests will not be reviewed here as they do not reflect Scandinavian practices.

The basic idea of the shape was outlined by Americans who also had plastic models developed in order to define the concept for the Swedish glass manufacturer. The characteristic shape of the bottle was determined early and did not create any problems.

The blue color, on the other hand, was difficult to achieve for various reasons. One reason was that blue glass had never before been used in commercial bottle production in Sweden or anywhere else in Europe.

Glass production is a 24 hour continuous process. The number of colors are kept to a minimum, traditionally three, sometimes four. Change of color normally creates very large production losses and is generally avoided.

PLM turned to the manual crystal ware industry for help. They could make the appropriate shape in sample sizes, and they were also willing to make the samples in blue glass of handmade crystal ware. The color—or rather the exact shade selected from 700 variations—was produced by the Glass Research Institute, which could produce small sample runs of glass in a range of shades.

Pieces of blue glass were shown to decide on the exact shade, identical with that of a certain ashtray—the model for the color from the very beginning!

Finally, after many Atlantic flights, discussions and technical headaches, the final decision was made, and the PLM company decided to "set aside" one of its smaller ovens in order to run a production test. The result was successful and the producer, Hammar Glassworks, is now the only glass plant in Europe to make blue glass bottles commercially (Fig. 4).

Figure 4   Ramlösa mineral water from Sweden in new blue glass bottles for the U.S. market.

Comments

This project started in September 1978 and was due for introduction on the American market a year later. In the meantime market tests and technical development had been run simultaneously, supporting each other on both sides of the Atlantic. However, one aspect was never tested—the blue color.

People responsible for the project believed so strongly in the color that they felt they could do without testing.

The market launch has been delayed due to unexpected complications created by American multipack requirements, but the original reaction from test marketing has been very positive.

Exhibit 5   Use of Packaging Design Research in Determining Identification of Trademarks

Product: Consumer product
Client: Trade Mark Advisory Board, Stockholm Chamber of Commerce
Characteristics: Research technique to assess degree of identification of trade marks, distinctive designs, and so on.
Research group: AB Marketing Bo Jönsson, Stockholm
Test method: Quantitative method with structured interviews following questionnaire.

Background

A trademark can be protected by registration. Exclusive rights can also be granted if the brand has become "well known," that is, has established a certain level of recognition. Distinctive designs or features other than trademarks, for example, package design or package shapes, can be granted the same protection according to Swedish trade mark law.

The criterion of identification has also been used in a case of identical copying (plagiarism) which was brought to the Swedish Market Court, claiming that consumers could be misled as to the commercial origin of the product with reference to article 2 of the Marketing Practices Act. The crucial point in cases of this kind is to assess the degree of identification. Packaging design research has proved to be a helpful instrument.

Methods Used

The traditional way to assess the degree of identification of a consumer product has been by indirect methods, that is, by questioning producers and distributors. The same method had been used to identify the target group who bought and used the product. Today quantitative methods are available and accepted, and data are collected directly from the consumers. During the past 5 years this technique has been used in approximately 75 cases by the Trade Mark Advisory Board in Sweden. The function of the Board, and the research technique which it has developed and used, have been described by Dr. Sten Tengelin, in Nordiskt Immateriellt Rättsskydd, NIR, 1975: 2-3. The case history presented here refers to this source.

The Trade Mark Advisory Board is organized by the Stockholm Chamber of Commerce. Its function is to make investigations and to give advice in trade mark and identification matters to member companies, lawyers, and patent agents.

If the problem is related to the consumer market, the procedure will be to conduct desk research and to work out a questionnaire. The desk research covers matters such as:

1   Groups to which the product is sold.
2   Quantity sold to each group.
3   Identification of buy decision maker (man, woman, child, age).

4 Channels for sales promotion.
5 Time the product has been on the market.
6 Sales.
7 Marketing costs.
8 Regions where product is sold.
9 Samples.
10 Competition.

The most important issue is determining who the consumer is. This is necessary not only for identification purposes but also for practical reasons in order to define the target group, decide on sample size and sampling procedure, and so on.

The design of the questionnaire is given considerable attention and often test interviews are conducted. Typical questions are:

Have you heard of the products Alpha, Beta, and Gamma?

If you have heard of product Alpha, what sort of a product is it?

How is it used? Can you briefly describe the package?

The word Alpha—do you feel it is a trademark, that is, the brand name of a product from a specific company or just a product category designation used by several producers?

Do you know the name of the producer?

Result

The Trade Mark Advisory Board reports the result of the survey in a written statement. It also draws conclusions as to the degree of identification of the mark or other distinctive product or package designs or features. The Board has set up a scale to facilitate and standardize this assessment:

Rarely known	less than 15–20%
Fairly well known	20–50%
Well known	50–80%
Very well known	80–95%
Generally known	more than 95%

The extent to which the product is known is the central part of the statement as it reflects its position in the marketplace. This is also the factor for deciding whether the product has such a degree of identity that it satisfies the objectives of the Trade Mark Law. The Board facilitates—some say anticipates—the court procedure.

Application under the Marketing Practices Act

Packaging design research has also been used in a case submitted to the Market Court for violation of the Marketing Law. Company A claimed that Company B was misleading the consumers as to the true commercial origin of the product—a plant nutrient—the reason being that product B was packed and sold in a bottle very similar to that of product A in size, color, shape, and labeling.

A structured interview test was ordered by company A and conducted by an independent market research group of good reputation, (IMU, The Institute for Market Research in Stockholm) in order to ascertain consumer identification of the two products, brand names, and producers.

The survey revealed that people mixed up the two competing products presumably because they were very similar as to form, size, color, design, and label. The Market Court found that pack A was unique and established and that pack B interfered by misleading the consumer. The Court decided that company B was prohibited under penalty of a fine to use pack B. Reference: Market Court Yearbook 1974:5.

Comments

It has proven useful to apply quantitative research techniques in legal interpretation of industrial property rights. An interesting application of packaging design research could be envisioned in the following area:

How does one identify the criteria that make a product reach such a level of recognition that it can be considered identified under the Trade Mark Law? How does one assess with a reasonable degree of accuracy the factors—if any—that would facilitate and speed up the recognition phase? Could it be proven that a particular package design is easier to recognize and remember than others, and how do you measure this in advance? If the product could be considered well known and recognized at an early stage, then marketing efforts and costs could be limited in time and value. This offers a field for important basic research work in the familiarization process in the packaging field.

Exhibit 6   Consumer Ergonomics

Product: Ergonomic research focusing on the package opening function.
Producer: The Swedish Packaging Research Institute.
Characteristics: Packaging design research on functional aspects resulting in normative data
     for opening strength and forces.
Test method: Mechanical measuring of the forces required by consumers to open packages.

Background

Very little work has been published by ergonomists on the subject of package design. The principles and methods of ergonomic research, however, can be readily applied to this field.

Even at the early creative stage the designer must be aware of the functional limitations within which he has to operate so that he is working on a realistic basis. Two sets of factors are important: ergonomic and psychological.

The Swedish Packaging Research Institute, which is partly sponsored by the government, has conducted research work in consumer ergonomics since 1973. Several reports have been published, for example, on the use of hand movements of the typical consumer and of the strength of forces required to lift or pull in opening. Reports have also been published on consumer attitudes toward different activities in the life cycle of a package based on interviews about gripping, transporting, storing, opening, emptying, reclosing, and disposing. This supplies information about the relative importance of various functional aspects of package closures.

Research Work

A pilot study clearly showed that the opening movements were especially problematic, particularly the movements of *turning* a vacuum lid, *pushing* in a detergent box, and *pulling* a strip on folding boxes. The main study therefore concentrated on these aspects.

Special instruments were developed to register the torque of the closure and the applied forces. The measuring instruments were applied on:
1   Two glass jars of different sizes.
2   Lids (large and small, grooved and smooth, high and shallow).
3   1 liter soft drink glass bottles.
4   Washing powder capsules.

A sample of 200 men and women were drawn from the normal population, aged 20 and up. Another sample of 90 individuals represented people with different handicaps. Responsible for the project has been Gunilla Jönson, Ph. D. Mechanical Engineering, former assistant professor, Michigan State University, U.S.A.

Result

The study gives detailed data for the turning, pulling, and pushing forces that people from a normal population are able to apply in opening consumer packages. The data can be used to estimate how many individuals—of what characteristics—can manage to open a package if the required force is known. The study suggests such norms for the packs in the sample and states that 95% of a normal population should be able to *comfortably* open a package by turning, pushing, or pulling.

The study also identifies the forces required to open today's packages. 60% of the men and 95% of the women cannot comfortably open today's packages. The opening force required to open very frequently used packages is therefore too high for the normal population and must be lowered. The new and lower norms that have been suggested for normal consumers are, however, still too high for handicapped persons and will prevent most of them from opening. The report states, however, that it seems totally unrealistic to develop norms and packages to satisfy not only the normal population but also the handicapped. The conclusion is therefore that special opening aids should be recommended as a more realistic alternative for the handicapped rather than adapting the packages.

Comments

The most important practical result of this study seems to be the fact that norms for opening in terms of consumer ergonomics can be established within relevant parameters and that these parameters are particularly age and sex. The sample size can therefore be reduced and the number of participants in a consumer panel for testing opening aspects can also be reduced. The testing procedure will therefore cost less than it does today.

CHAPTER FORTY-FOUR

Intercountry
Comparability in
Multinational
Research Studies

Thomas T. Semon

Managers in multinational enterprises often find themselves in one of two situations: making decisions on the basis of research studies already completed in several countries or having to plan a multinational research project.

It is dangerous to assume intercountry comparability, and many factors that affect comparability are mentioned and discussed in this chapter. There is no simple, foolproof system for comparing survey data across countries. It is possible, through careful planning and design, to eliminate many sources of potential noncomparability when setting up a multicountry study. It is also possible, in drawing conclusions from multicountry study results, to set up a systematic process for identifying and assessing the effect of factors that may impair comparability of the meaning of results.

The "research studies" discussed are those involving interviews. There are at least three other types of studies: observation studies, secondary-source research studies, and audits. They have their own problems of comparability across countries, but these problems are less severe and complex than those involving interviews because they are less dependent on verbal communication.

COMMUNICATION AND TRANSLATION

These are familiar terms, but they require comment. Communication problems exist not only between the interviewer (or self-administered questionnaire) and the respondent but also between the headquarters and local executives of the marketer, between both of these and the central research supervisor, and between the supervisor and his subcontractors or affiliates. They occur both in the design and the reporting phases of the project. The "supervisor" may be either an internal research manager or a general research contractor/coordinator.

The need for translation is merely an added complication in the communication process. Every language has its own specific areas of greater and lesser communication efficiency. The efficiency of language as a communication medium can be measured by its ability to convey meaning precisely and unambiguously in as few words as possible. Unfortunately, the areas of relative inefficiency most likely to cause problems do not coincide among different languages.

English, perhaps more than any other Western language, has a facility for concise neologisms and idiomatic expressions that are difficult to translate concisely into other languages without losing some important nuance of the meaning, a fact that has caused many problems and considerable grief especially to advertisers whose slogans could not survive translation. It is difficult for persons who speak only their own language to realize that a dictionary is a reference book of approximations, not of precise correspondence except for simple, physical objects and actions.

A simple example relevant to packaging research is the word "appealing" or "appeal." An American manufacturer who wants to evaluate the appeal of a new package in Germany or France will find that there is no German or French word that precisely corresponds to "appeal" or "appealing." It is easy to ask how well the package is "liked" or whether it "pleases" people, but not whether it "appeals" to them, since "appeal" denotes some element of attraction in addition to merely pleasing. A design can be "liked" without being "appealing"; liking is more passive. It is no wonder that the English phrase "sex appeal" has been taken over into so many other languages; they have no concise equivalent.

In addition, the Indo-European languages' common roots encourage errors by careless or less than wholly competent translators. It would be tempting to translate the German *eventuell* as "eventually," but it would be wrong; the word means "possibly." Similarly, *actuel* in French does not mean "actual," but "current." In both languages, the words that look like "concurrence" mean "competition." The meanings are obviously related, but their emphasis has evolved and shifted over time, and they are no longer equivalents.

Better known, but still worth noting, are confusions arising from differences in definition. A British gallon (like the Canadian one) is 20% larger than a U.S. gallon. A short ton in the United States and a "ton" on the European continent both equal 2000 "pounds," but the European "pound" (half a kilogram) is 10% heavier than ours. Most confusingly, what Europeans (and most others) call a billion is our trillion. Most Europeans use the 12 hour time system when talking, but the 24 hour system in writing; a verbal appointment for "4:30" almost certainly means 4:30 p.m., but a written confirmation would usually be for "16:30."

INTERCULTURAL DIFFERENCES

Psychology and Expression

Language itself is only one aspect of the problems of communication in multinational research. Independent of language, there are pronounced cultural differences even between neighboring countries; these cultural differences can result in response differences that are easily misinterpreted if taken literally; they affect even nonverbal responses. These are, naturally, broad generalizations, and they sound suspiciously like stereotypes, but they are real. Given a rating scale from 1 (worst) to 10 (best), the central tendency for the rating of the respondent's *favorite* toothpaste (department store, industrial supplier, etc.) will be lower in France than in the United States and probably higher in Brazil.

Verbal scales accentuate these differences. There are no reliable, comparative studies of the relative incidence of "yea sayers" versus "nay sayers" in different countries, but they do differ. Many Europeans regard the culturally desirable optimism of Americans as evidence of foolishness and immaturity: they were brought up to consider it smart to be suspicious, reserved, at times almost superstitiously reluctant to voice optimistic sentiments or praise. Again, these differences are generalizations; obviously, there is a wide range of positive/negative attitudes within the population of every country, but the proportions differ.

The use of verbal responses and verbal scales adds the translation problem on top of the attitudinal one. A position in an importance scale such as "somewhat important" can be readily translated into German *(etwas wichtig)* but not into French, where the closest approximation may be *d'importance modérée* (of moderate importance), which is not quite the same. If the scale is short, offering only four or five choices of response, the precise wording of the highest and lowest choices becomes crucial: more Europeans than Americans will tend to avoid what they may regard as very strongly positive statements and prefer the more moderate second position instead. A country norm must be used for reference.

More elementary than these differences in reaction style, but closely related to them, are differences in the attitude toward the interview itself. Again, these are broad, oversimplified generalizations, but they are based on realities. On the whole, Americans may well be the most relaxed and truthful survey respondents anywhere. A large majority will agree to tell a perfect stranger what their income is, and the degree to which the response is incorrect is more a function of superficiality than of dissimulation. Many Europeans would (if they agree to answer at all) severely understate their income; in less developed countries, all but the wealthiest respondents would be more likely to overstate it. Again, these are tendencies, not rules.

A parallel difference is observable in business research; industrial and commercial respondents outside the United States are, on the whole, less cooperative in surveys and far more concerned with confidentiality. They also are less ready than American businessmen to share industry information through industry associations.

Lack of familiarity with surveys can be a problem in some of the less developed countries. According to one South American research organization, it is difficult to have good interviews done not so much because the interviewers are not competent, but because the respondents do not know what is expected of them and thus they give distorted answers in an attempt to live up to the presumed expectation. Inexperience with the role of respondent became evident in the course of a cosmetics packaging study, where women were asked to evaluate the packages of several competing lines: the women were unable to separate their opinion of the product from their opinion of the package, in spite of carefully designed instructions. It is worth noting that when the study design was submitted to the local organization for cost estimating, they strongly recommended major changes in the design, which their client rejected in order to "maintain comparability."

Habitual Competitive Contexts

There is another class of intercountry differences, closely allied to cultural differences; these differences can be called contextual differences. Mustard in a tube would be a novelty in the United States, but it has been available and widely used in Switzerland and some other European countries for decades. Americans are used to portion-pack jellies and preserves in restaurants, but not portion-packs of high-priced premium brands, widely available in Europe. Many more Europeans than Americans still shop for groceries daily, walk to and from the stores, and take along their own collapsible shopping cart or string bag. Refrigerators tend to be smaller, and freezers fewer. All these differences from the U.S. patterns are diminishing, but they are still important.

As a general tendency, reusability of containers is a matter of greater concern and appeal outside the United States. South American women, in a group evaluation of powder packaging, coupled their admiration for the favorite package with a complaint that it was not reusable as a flower vase. Returnable bottles with the old-fashioned wire-clasp porcelain stopper are still used in Europe for some brands of beer and cider. Overpackaging is a subject for satirical cartoons and editorial comments on waste in many European countries. Such attitudinal differences are generally recognized by marketing personnel in these countries, but, as a matter of wishful thinking, are sometimes disregarded by headquarters in the United States.

Differences of this kind must be considered in the design of questionnaires and in processing the responses. It is often essential to give foreign respondents answer choices that would be unimportant in the U.S. In a study of office machines, the manufacturer wanted to exclude noise level as an in-use characteristic to be evaluated, on the basis that it was unimportant in the U.S. At the author's insistence, it was included, and turned out to be one of the four characteristics of greatest concern in most countries.

PLANNING FOR COMPARABILITY

The Analysis/Interpretation Problem

The executive confronted with a set of parallel studies conducted in different countries and analyzed as a group cannot be expected to resolve these problems without a great deal of firsthand experience with these countries. This experience is rare. To avoid possible serious misinterpretations (by applying an American perspective to foreign survey results) it is advisable to have the results interpreted by a competent local analyst in each country or by a specialist in multinational research.

In effect, each country's results must be interpreted in the light of norms and experience standards for that country.

This sort of country "customization" takes place as a matter of course when selecting competitors to be used as comparison bench marks or foils in each country; it is equally important in the analysis process.

Performance Standards

The problems discussed illustrate the desirability of performance standards. The term is familiar through the long-term argument regarding building codes: should codes specify actual materials and measurement, or should the codes be set in terms of the actual performance of the materials used? The latter provides flexibility in the use of new materials and technology.

In multinational studies, performance standards are useful as a supplement to detailed, standardized specifications. They provide an objective criterion for evaluating the merit of suggestions or recommendations for changes from individual countries and also permit individual country researchers to make more useful and appropriate recommendations than they could otherwise.

Planning: Functions and Agents

Perhaps even more than in most other activities, advance planning in multinational research is preferable to ex post facto solutions. Table 1 provides for identification of what *agent* performs what *function*. It sets up a planning framework for each country to be included in a multinational survey. The terminology is explained and discussed in the following sections.

Problem Definition

Researchers never cease to be amazed at the number of large, costly studies that are conducted without a clear management definition of the problems to be addressed or a clear statement of the hypotheses to be tested. Important as such clear definitions are in surveys conducted in one country, it is not hard to imagine what can happen in a multinational effort that is not well planned and coordinated by a central executive or team.

A multinational survey may have different purposes and different foci of application for its results. These must determine the organization of the research and the roles to be played by various "agents." Is the "client" for the information the headquarters team, or is the survey one that is conducted largely for the benefit of local operations in the various countries?

There is little question that the inter-country comparability requirement has both positive and negative effects:

Positive: □ Improves utility of information for centralized,

Table 1 Functions and Agents in Multinational Studies

	Headquarters (U.S.)			Foreign Country		
	Corporate Mktg/Product Management	Corporate Research Staff	Outside Research Agency	Corporate Mktg/Product Management	Corporate Research Staff	Outside Research Agency
Problem definition	□	□	□	□	□	□
Study design	□	□	□	□	□	□
Analytic plan	□	□	□	□	□	□
Coordination	□	□	□	□	□	□
Processing instructions	□	□	□	□	□	□
Processing	□	□	□	□	□	□
Report, conclusions:						
Multinational	□	□	□	□	□	□
Individual Country	□	□	□	□	□	□

headquarters-based strategic planning.

□ Establishes standards that might not otherwise be met.

□ Makes possible centralized processing.

Negative: □ Can reduce the specific utility of information for individual country operations of the company.

□ Can discourage or undermine initiative and creativity of individual-country staff.

There are so many variations in the way authority and responsibility can be divided between headquarters and foreign operations that a comprehensive discussion of the pros and cons of central control would be futile. Suffice it to say that these are largely determined by the usual criteria of intracorporate power: Who has budget authority? Who decides policy? Questions of national chauvinism also arise: the Barcelona office, not well versed in research, may welcome the technical expertise from headquarters, but resent its exercise of control. If the overseas staff has research competence, they may vigorously criticize the specifications set down by headquarters. A clear division of functions must be established between central (headquarters) managers and the company's staff in the various foreign countries. The latter can play a valuable role in multinational surveys, but that role must be carefully spelled out in complete detail of function and responsibility. The interests and priorities of XYZ's Brussels office are not necessarily identical with the interests and priorities of the headquarters executives who are in charge of the survey. It is important to set up such procedures that the survey benefits from the specialized local knowledge of the

Brussels office, without being affected by its specialized parochial interests.

Study Design

Two major components of study design are the questionnaire and the sampling specifications. The questionnaire, naturally, is subject to all the potential problems in language and communication that have been discussed in the preceding sections. The best results in standardizing across different countries are obtained by the double-translation method that most international research organizations use: a standard questionnaire is sent to each country, translated there, and the translations are returned to the central office. They are then retranslated into the original language, preferably by a native of the foreign country, who has not seen the original standard questionnaire. The retranslations are then compared with the original standard questionnaire, and all differences carefully evaluated to make sure they are not significant. Where significant differences are found, they are traced back to see whether they arose in the original translation or in the retranslation.

Sampling specifications—how to qualify the persons to be interviewed and how to select them—are perhaps best written on the basis of performance specifications (discussed in the preceding) rather than detailed instructions. A headquarters researcher unfamiliar with a foreign country can easily write instructions that would be just right for the United States, but either not feasible or inappropriate for the other country. To avoid this problem, what is needed is to set down in detail what the sampling *objectives* are: what kinds of people, in what proportions, are to be included; what possible biases in sampling it is important to avoid. The contractor for the actual work in the foreign country is then asked to submit a detailed plan of how these specifications will be achieved; this plan can then be discussed and perhaps modified.

Analytic Plan

The analytic plan, of course, is the basis for the questionnaire design and the sampling specifications. If intercountry comparability is an issue, that is because the analytic plan calls for either amalgamation of different country results into a whole, or for comparison of study results between countries, or both. In either case, the analytic plan, at least in its broad outline, is a centralized responsibility.

Coordination

The responsibility for coordinating the individual-country parts of a multinational study must be assigned specifically. It can be broken up into several parts (questionnaire, instructions, selection of individual country contractors, processing), with different parts to be handled by different agents, but centralized overall responsibility is likely to yield better results because it avoids possible misunderstandings as to who is responsible for what. Executives unfamiliar with the operating details of a coordinated multinational study almost invariably underestimate the importance of centralized coordination, as well as the amount of work involved.

Processing Instructions and Processing

The standardized questionnaire will of course have standardized lists and codes. Processing instructions must deal with questionnaire editing, open-end questions, and the detailed tabulation plan. It is preferable to have editing performed in the individual countries, unless all responses are in terms of numbers or checkoffs.

Editing is one of the steps that is often overlooked. If editing rules are not standardized, the result can be great inconsistency that in a multinational study may appear as a country difference but is actually just a result of different data treatment. Naturally, someone has to check that the rules are followed also.

Reports and Conclusions

Reports and conclusions should take as much advantage as possible of local experience in each country. It is often desirable to have a dual analytic and reporting scheme; in this case, each country's results are reported in terms of responses to a standardized set of management questions, for headquarters use, and each country prepares an individual report whose emphasis is determined by local interests and conditions.

"Agents"

The "agents" assigned to the performance of each of these functions can vary a good deal, and the company's own staff availability is an important factor. Only a few very large companies, for instance, have their own research staff in each of many countries, but where such research staffs are available, they will surely be involved.

The agents are divided first on the basis of their location: United States or a foreign country. It is assumed here that the U.S. location is the corporate headquarters. Three general types of agents are distinguished: corporate marketing or product management personnel, corporate research personnel, and outside research personnel or agencies. The last type, outside services, can perform various types of services: consultant, coordinator, or survey contractor for individual countries.

The utility of the function/agent matrix is in planning. First, for the project as a whole, who will do what, and what are the lines of authority? Second, using a separate matrix for each country, how will the general plan be adapted to the particular circumstances prevailing in that country (in terms of personnel, for instance)?

RESEARCH ALREADY COMPLETED

Suppose that the VWX company has a new marketing manager and one of his first problems is a decision on a new package design for product P. The product is sold in seven countries. Over the past year, package appeal has been researched through consumer studies in four of them. In three of the four, a new design N has been evaluated, but each country ran the test on its own, using different questionnaires and general procedures. There is no time to do more research. Now what? One solution, of course, is to make the decision on the basis of the manager's good judgment, which is what he was hired for in the first place. His good judgment, however, tells him (as does decision theory) that available information should be utilized.

The way to attack this problem is through orderly organization, which defines and narrows down the areas where judgment must be applied:

STEP 1   For each "Study Descriptor" in Table 2, write down, in detail, what the *best practical* procedure would be if one could start from scratch.

STEP 2   For each country separately, determine how each "Study Descriptor" was in fact handled, and write in this description.

STEP 3   For each difference between Step 1 and Step 2, make a judgment as to its effect on the results.

STEP 4   For each country separately, review all the judgments made in Step 3 to arrive at an overall assessment of what the results might have been if your preferred design had been used—or whether the data are useful at all.

This procedure, much of which is best carried out by research personnel, develops overall judgment through an orderly process with good heuristic potential for improving the quality and consistency of judgment.

SUMMARY

Although the discussion in the preceding pages may seem discouraging, that is not

Table 2   Form for Assessing Effect of Procedural Differences

Study Decriptor	Best Practical Choice or Method	Actual Used in Country _____	Assessment of Effect
Criterion (i.e., most important) question:			
Wording			
Response scale			
Questionnaire context			
Competitors evaluated			
Timing of study			
Interview locale/mode			
Respondent qualification			
Data treatment			

its intent. Intercountry comparability is a necessity for multinational marketing information. It cannot be achieved perfectly, but then, information is almost never perfect. Even the comparability between two supposedly parallel studies done a year apart in the United States by two different research organizations is not perfect.

The aim of this chapter has been to identify as many as possible of the special factors that may impair comparability, so that the planner of multinational research and the user of the results are alerted to specific problem areas and can take them into account in planning and interpretation. In addition, suggestions made as to research organization and individual country survey assessment should be of practical utility to planners and managers.

One bright spot in the picture is that the caliber of market research facilities in foreign countries has improved greatly in the past decades and will probably continue to improve. In most countries it is now possible to have surveys conducted competently by responsible contractors who have had prior experience in dealing with U.S. clients or research coordinating agencies. As a result the degree of intercountry comparability that can be achieved in 1980 is considerably better than it was in 1960, or even in 1970.

CHAPTER FORTY-FIVE

Testing Product Imagery in Latin America

John E. Pearl

In this chapter product imagery will also comprise the testing and measurement of both company and product image. An expanded interpretation could be extended very appropriately to include service and location imagery.

The first efforts to conduct any amount of product imagery evaluation in Latin America were, to my knowledge, made by Elmo Wilson, founder of International Public Opinion Research Associates, which later changed its name to International Research Associates, the oldest and largest international network of market research companies. Mr. Wilson was diligent in setting up the necessary regional infrastructure which enabled IPORES, now known as INRA to conduct image studies in Latin Amer-

ica, mainly in Argentina, Brazil, Colombia, Panama, and Central America, for various types of clients such as the United States government and the Standard Oil Company of New Jersey, now known as Exxon Corporation.

Areas covered under these earlier research projects, the first of which date back to the early 1940s, included items such as lube oils, gasoline, internationally broadcasted radio programs, and attitudes towards foreign governments and various forms of governments.

Without question the United States of America is still the most advanced country in the world, in all fields, including the field of survey research. No new techniques have been developed in Latin America in the field of survey research; rather the development of techniques and the improvement of such techniques has taken place in the United States and researchers in Latin America have borrowed heavily from the United States. Thus the state of the art in Latin America lags behind the United States anywhere from 3 to 10 years. Countries in Latin America where research techniques are more up to date are Brazil, Mexico, Argentina, and Colombia, in that order.

Currently the main research sponsors in Latin America are the regionally based companies that are subsidiaries of United States and European based corporations, the domestically owned companies, the United States based corporations, which often sponsor research assignments whose results are to be delivered directly to the United States (bypassing the management of their own local subsidiaries), and the United States government, also in that order. A number of European based corporations are beginning to do a limited amount of product image research in Latin America; these deal mainly with tobacco, pharmaceutical products, and consumer products. As an exception, the Chilean government, which was repeatedly accused of violating public rights, undertook a public opinion study early in 1975 that included several countries, among which were Colombia, Venezuela, and Mexico. The study revealed the existence of a very unfavorable image of the Chilean government. A second follow-up study was undertaken approximately half a year later, but the results were unsurprisingly similar since the military junta had taken no steps to warrant a favorable change in the attitude of the general public.

SAMPLING TECHNIQUES

The sampling process inherent in product imagery research differs in no signif-icant way from the normal sampling procedures applied when implementing a regular market research study. Therefore, the two classical approaches of quota sampling or random sampling are used. Sometimes a variant such as stratified disproportionate random sampling is used.

In product imagery research the limitations and principles of sampling in general are as applicable as they are in general market research. Therefore it is important to bear in mind such basic principles and to follow the same design procedures that would apply to any other survey research study of comparable specifications.

QUESTIONNAIRE DESIGN

In testing product imagery the accuracy of questionnaire design is crucial to the results of the study. Of prime importance to the success of the research effort is the adequacy of the research instrument to the study objectives. All efforts should be made to ensure that answers provided by the questionnaire will fully satisfy research objectives. A number of guidelines must be borne in mind at all times in order to be able to develop a top notch research instrument.

The Research Problem

1   The researcher should have total and fully clear knowledge of the problem he wishes to solve through research. This is important if a proper two-way communication base is to be achieved.
2   Make an objective evaluation of whether the problem on hand has any meaning to the interviewees, since it is hard to do proper research about problems not fully understood by the source of information.
3   If there are reasons to believe that

the subject under investigation is not adequately known by the source of information or to certain segments of the population, then appropriate methods, such as filter questions, should be devised to exclude these from the general sample frame.

4   The project director should ask himself constantly if any improper assumptions are being made.

Open Questions

1   As a rule open-ended questions are undesirable and are best placed in the pretesting of questionnaires rather than in the formal research instrument. Also as a rule, adequate pretesting should in most instances eliminate the need for open-ended questions.

2   Throughout the pretest, the researcher should be on the lookout for opportunities to convert open-ended questions into closed questions.

3   Questions should be worded to simplify the respondent's expected job of expressing his replies in terms of the expected categories or classifications.

4   Even if the question is open-ended, the most frequently anticipated replies may be coded; thus only some replies will be left open for future coding.

Dichotomous Questions

1   Avoid the implicit alternative. Whenever two alternatives are possible, these should be included fully and clearly within the context of the question.

2   State the negative option in detail. Whenever there are two alternatives these should both be stated clearly within the text of the question.

3   In the case of those questions that offer two alternative answers to a given argument or dichotomous proposition, both alternative answers should be stated clearly within the text of the question.

4   Provision should always be made for all "don't know" and "no opinion" answers.

5   Normally the choices made by the interviewee should be mutually exclusive. If this is not possible, provision should be made for "no opinion" or "no answer."

6   All possible answers must be considered.

7   When advisable, not only the respondent's opinion, but also the intensity of this opinion should be recorded. The use of scales is recommended in all of these cases.

Multiple-Choice Questions

1   Alternative replies must be mutually exclusive.

2   No alternative should be dismissed a priori. However, alternatives can be restricted as long as such restrictions are borne in mind at the time results are being analyzed.

3   It should be decided ahead of time whether a single answer or multiple answers are expected from the respondent, who in turn should be made aware of such expectation.

4   When four or more alternatives are contemplated, the respondent should receive them on a printed card or other visual aid. All alternative answers printed on a card should occupy randomly distributed positions to avoid positioning bias.

Dealing with Interviewees

1   Avoid offending the intelligence of interviewees with overexplanation, but avoid also being insufficiently clear.

2   Try to use good grammar, but avoid sounding unfamiliar.

3   Do everything to avoid confusing the interviewee and to help him with intelligent questioning, proper visual aids, and, when necessary, adequate supplementary explanations on the part of the interviewers.

4   All questions should be clear and specific enough to be understood by those respondents in the sample with a lower IQ.

Semantics

1   Use as few words as possible. Almost any question can be asked in less than 25 words.

2   Use the simplest word that will properly convey the desired meaning.

3   Remember that certain words of common use may differ in meaning from one place to another or from one segment of the population to another. For example, some commonly used words in one Spanish speaking country may have an entirely different and even embarrassingly offensive meaning in another Spanish speaking country.

Miscellaneous

1   When certain words need to be emphasized by the interviewer, this is best accomplished by underlining such words.

2   Punctuation should be held at the necessary minimum, and no word should be abbreviated.

3   For the sake of future repetitive studies, try to envision how a question will sound 10 or 20 years from now.

METHODS OF INTERVIEWING

It is generally accepted that there are six basic formal research methods: tele-
phone, mail, direct observation, experimental design, depth interviewing, and personal interviewing. In product imagery research, personal interviewing is the most widely accepted, usable, and used method.

In spite of being the most widely used, the personal interview method is subject to a number of drawbacks, mainly:

1   Time and cost factors are usually high.

2   The human element can cause all sorts of distortions or biases. Interviewer-induced bias, however, can be minimized by proper design of the questionnaire and adequate briefing and training of the interviewer force.

The former two are among the most commonly raised objections against the personal interview method. However partially true they may be, these drawbacks or limitations are offset by the advantages inherent in the system, among which are:

1   A very high response rate is achieved. Appointments can be made for interviewers with particularly hard to reach respondents. Overall refusal rates are very low, seldom above 5%.

2   A nearly theoretical random sample can be achieved with the personal interview system.

3   In certain types of studies the generated information is more trustworthy since the interviewer is always available to clarify or explain any doubts the interviewee may have with regard to question interpretation.

4   Of particular importance is the fact that the interviewer can register his own observations as to apparent age group, income level, and many other

factors that cannot be observed nor registered by mail or telephone.

5   Visual aids can be presented to the interviewee.

6   In personal interviewing the interviewee cannot know what questions lie ahead and thus that source of possible distortion is eliminated. The mail questionnaire does not have this advantage.

7   Spontaneous reactions are more easily obtainable from the respondent.

8   Rapport can be gradually built up by the interviewer; this will help to evade barriers that inhibit respondents from replying to certain sensitive questions.

A Case in Site Research

A particularly difficult case comes to mind when dealing with image research in Latin America. In 1978 the International (Colombia) Resources Corporation, better known as INTERCOR, was faced with the problem of building a community from grass roots for 4100 employees and their families at El Cerrejón in La Guajira, an arid region in northern Colombia devoid of any attraction and well known as a smuggler's paradise. The location is the site of enormous coal deposits that will be dug out in open pit mining fashion. The company needed to measure the image of the location, the degree of possible antagonism towards the region, and the likelihood of people going to work there for a number of years.

To solve the problem and give INTERCOR management valuable insight into the situation a study was designed to fulfill the following research objectives:

1   Describe image of La Guajira in measurable terms.

2   Assess negative factors related to working in La Guajira and living there.

3   Determine ambiance and accommodations desired by those willing to go.

The study, which was designed and carried out in a joint effort by Interamerican Research of Colombia and Rescon Corporation of Pittsburgh, Pennsylvania, comprised 2200 interviews with 1900 heads of households and 300 spouses. Of exceptional length, the questionnaire required well over 1 hour of the respondents' time. It measured many aspects related to the image of La Guajira and provided the answers and clues to develop countermeasures that would enhance the image of the region and to offset any negative aspects when they were found.

This is a case where product image research, in this case the product being a working location, proved to be a tool of extraordinary value in aiding an enterprise that will initially invest well over five billion dollars.

CHAPTER FORTY-SIX

Package Design Assessment in Latin America: the Venezuelan Experience

Andrew Templeton

Package design assessment in Venezuela is an area of marketing research that brings to mind the frequently quoted "curious incident of the dog in the night time."* An informal survey of marketing research companies and major manufacturers indicates that less than 1% of marketing research expenditure in the past 5 years has been dedicated to the evaluation of packaging performance in the marketing mix.

At this stage the technically inclined reader might well consider passing to the next chapter, however, it is felt that the constraints to the development of this area of research in Venezuela may well be typical of many emerging economies and consequently of interest to practitioners. What this chapter will attempt to do is to describe what kind of package assessment is carried out, why and how it is conducted, and why relatively little is done.

VENEZUELA AND ITS ECONOMY

As most readers will be aware, Venezuela is a petroleum producing country,

*" 'The dog did nothing in the night time. That was the curious incident.' remarked Sherlock Holmes." A. Conan Doyle, "Silver Blaze," in *The Complete Sherlock Holmes,* Doubleday, New York.

founder member of Opep and a leading exporter of petroleum products to North America for over 50 years. It is the seventh largest country in Latin America with an area of some 350,000 square miles of territory. It is bounded on the north by the Carribean Sea, to the west by Colombia and to the east and southeast by Guayana and Brazil. An important influence on the socioeconomic development of the country in the past 5 years has been the vast extent of its land frontiers with these neighbors.

Official estimates of population based on projections from previous censuses give a total population for mid 1979 of 13,515,000. It is almost certain that these figures underestimate by at least one million and by possibly as much as two million the real population of the country because of the large numbers of illegal immigrants whose arrival in the country has been caused by the economic boom of the years after 1973 and facilitated by the ease of entry through the land frontiers referred to previously. It is evident that if the United States finds it difficult to control immigration from poorer Latin American countries, it must be much more difficult for Venezuela, where the illegal immigrant with his cultural and physical similarities finds it much easier to merge into local society. This extra

population has been referred to at some length because, as will be seen in the following, its existence has affected the demand for all goods and services but in particular mass consumer goods, which in turn has influenced the marketing planning of manufacturers.

During the past 15 years, and more particularly in the years 1973–1978, the Venezuelan economy has seen rapid growth. Table 1 indicates changes in some of the key indicators. Before commenting on these indicators, it is necessary to discuss one further aspect of the Venezuelan economy that is germane to our subject. Until the early 1960s, packaged products consumed by Venezuelans were largely imported, principally from the United States. Government policy since that time has been directed to the substitution of locally manufactured goods in order to obtain a greater input of Venezuelan labor in products consumed by Venezuelans. An open policy (until the implementation of the Andean Pact agreement in 1973) coupled with the prosperous economy and hardness of the currency led many multinational companies that had previously enjoyed prosperous export markets in Venezuela to set up local manufacturing or assembly plants. In some cases these industries were wholly owned subsidi-

Table 1   Key Indicators*

	1961	1966	1970	1973	1979
National income (Bs. million)	19.6	30.3	44.1	63.6	143.4 (1977)
Population (millions)	7.5	9.0	10.4	11.3	13.5
National income per capita (Bolivares, thousand)	2.6	3.4	4.3	5.6	11.3 (1977)
Private automobile ownership (thousands)	242.8	390.2	614.6	784.0	1182.0 (1978)
TV homes (thousands)	265.1	514.4	817.3	1094.5	1776.8
% of grocery business self-service outlets	6	24	36	42	46

*Sources: National Income Figures: Banco Central de Venezuela; Population: Statistical Office; Automobile Ownership: National Transport Office; TV homes: Datos C.A. Audience Survey Dept.; Grocery Business: Datos C.A. Store Audit Dept.

aries of multinational corporations, in some cases, there were associations with local capital, and in other cases there were simple technological licensing agreements. Whatever solution was adopted, for the purpose of this analysis the end result has been the same. Thus in 1979 as in 1960, the majority of packaged products Venezuelans consume were foreign rather than national. In many cases brand names are international although multinationals will also develop local products and use local names for formulas sold under other names in other countries.

The typical Venezuelan consumer will usually buy a detergent made by Procter & Gamble or Colgate, use Lux or Camay toilet soap, smoke cigarettes made by British American or Philip Morris, drive to work in a Ford or Chevrolet and watch TV on a Sony or a National while drinking Pepsi-Cola. The major exceptions to this "dependency" are in products such as beer and basic household staples such as rice and corn flour.

Reviewing this brief and necessarily superficial description of the Venezuelan economy in a marketing context, we return to the incongruity of our opening sentence. In an economy where there is an active presence of sophisticated international marketing companies, where per capita income is the highest in Latin America, where growth rates are among the highest in the world and where population is highly urbanized, why is so little packaging research carried out? The following section will attempt to answer this question.

THE CONSTRAINTS

It is obvious that packaging research is relevant only in terms of marketing criteria.[1] We are not interested in package design assessment by the board of direc-

tors or the chairman's wife although this is by no means uncommon in Venezuela.

Schlackman & Dillon have commented, "The criterion is related not only to the marketing strategy of a given brand, which is always critical, but to the marketing environment in which the brand must live."[2] It is to the marketing environment that we must address ourselves when examining the constraints to the development of research on problems of packaging in Venezuela.

It is possible to divide our examination of the marketing environment into two broad periods, first the period 1960–1973 and then 1974 to the present. In each of these periods somewhat different constraints have affected the development of packaging research. Although there are obvious overlaps, the difference in emphasis in the marketing environment justifies the separation of the periods.

As has been mentioned, from the early 1960s there was a shift to import substitution, particularly in areas that would provide employment to the Venezuelan labor force and that were feasible within the limits of Venezuelan technological capacity. In 1960 the population of the country was approximately 7 million and per capita income about $700.* With the small proportion of the population that was economically active and the uneven distribution of wealth, this meant that the newly established manufacturing and assembly plants were faced with markets that were by any standards extremely small. At the same time productivity of the relatively unskilled and nonindustrialized labor force was low and distribution costs were high. Despite protection, the manufacturer was presented with obvious cost problems that encouraged him to use as far as possible already developed

---

*Nevertheless, even at that time this was the highest per capita income in Latin America.

resources from larger markets within his sphere of operation. These resources included packaging. Few, if any, attempts were made in the early years to determine whether a package design of proven success in the U.S. Midwest was really appropriate for Venezuela.

Cost contraints have always been of vital importance in the development of sophisticated marketing techniques in any small market, and nowhere is this more evident than in the area of research. From retail audits or large scale consumer surveys to focus group discussions there is no reason why costs in markets with populations of 10 million should be be similar to, if not higher than, costs in countries of 100 million, yet obviously the actual or potential sales revenues to justify these costs will in most cases be proportionate to population size.

Nevertheless, in Venezuela of the early 1960s it was not only cost considerations that limited the growth of packaging research. During this period both multinational and local companies were investing increasingly high proportions of their budgets in other areas of marketing research. Retail audit services that were started in 1958, limited to the major cities of Caracas and Maracaibo, had been extended by 1966 to a complete national audit. From early beginnings in 1955, television and radio audience research had developed by 1965 to coverage levels equaled only in the most sophisticated developed countries. General consumer surveys on habits and attitudes were commonplace by the mid 1960s, and a high proportion of new products had been subjected to consumer acceptance tests before being launched. Why did package testing lag behind?

Manufacturing companies that had previously exported to the Venezuelan market intuitively felt that consumers were primarily concerned that the local-

ly manufactured product would not be as good as that which they had previously purchased from abroad. This was by no means an unreasonable view. Awareness of the limitations of quality control and production facilities was one factor, but so was what is probably a third world syndrome: the conviction that local manufactures cannot be as good as their imported equivalents.* Thus the objective became to achieve, as far as possible, a sense of continuity between the product locally manufactured this year and that which was imported last year.

Clearly one of the best ways of obtaining this was to leave the symbols untouched, and thus brand names and packaging usually remained the same and the necessary additives such as "Made in Venezuela" were printed as discretely as possible. In the case of some products, for example high quality toilet waters and colognes, manufacturers were able to import their packaging materials from abroad while mixing and bottling the product in the country. Consequently (to the consumer) the product seemed to be imported as the outer package could still legitimately bear the slogan "Made in France."† The author was engaged in research studies in the early 1960s to determine how far labeling and instructions for use of mass consumer goods should be made worded in English as well as Spanish although the products were to be directed to an entirely Spanish speaking market. In the same period research indicated to a processed food manufacturer the need to return to a snap top closure from the more convenient twist and turn as supermarket cus-

*Research in Venezuela has shown that this applied to manufacturers but not to raw materials or produce of the soil.
†High priced cosmetics still use imported packaging (1979)

tomers were opening jars on the shelf to test the quality of the locally manufactured product, thus destroying the vacuum.‡

By the second half of the 1960s consumers were becoming more accustomed to locally produced international brands, and on the whole the quality of most products had improved. Also important was the fact that until 1973 Venezuela suffered far less inflation than most other countries; thus because of price, comparisons between Venezuelan and U.S. products, for instance, were becoming less unfavorable to the local manufactures. An extreme example of this type of comparison was the case of the international soft drinks such as Pepsi-Cola, which in Venezuela was sold at slightly less than 6 U.S. cents per bottle until the mid 1970s, a price which, to say the least, compared favorably with the United States and Europe, a fact that was soon noticed by Venezuelans traveling abroad.

This normalization of the situation as far as consumer goods was concerned meant that towards the end of the 1960s packaging design assessment began to be taken into account in the research plans of manufacturers. This state of affairs lasted until 1973, and research organizations such as the research department of Carton de Venezuela* and the independent research organization Datos, C.A. were becoming increasingly involved in research projects covering package design.

At the end of 1973, petroleum prices quadrupled, and although most readers are only too aware of the effect of this price change on the economies of the importing countries, less is known about the effects on the economies of the petroleum exporters themselves. This chapter is not the place, nor this writer the authority to comment in detail on the phenomenon. Suffice it to note that the years 1974–1979 have been typified by rapidly increasing demand for goods and services of all kinds, inflation, shortages not only of goods but also of the factors of production with the exception of capital, increases in imports of all kinds of goods, including mass consumer products for which local industry was unable to meet demand, and finally a massive but unquantified immigration of labor from neighboring countries.†

Private consumption according to Central Bank figures rose 2.3 times in the years 1973–1977.* While this increase is calculated in Bolívares† even more striking are the following figures derived from the Datos National Retail Audit service. In terms of units or volume, consumption of packaged food products increased by 27% between 1976 and 1978, and consumption of household maintenance products increased by 35% during the same period.‡

These kinds of growth rates, for which government, manufacturers, and distributors alike were wholly unprepared, led to almost unmanageable shortages and production bottlenecks. On top of this, the government, in a not completely unsuccessful attempt to prevent runaway inflation, imposed drastic price controls on most mass consumption goods although it had at the same time decreed across the board wage increases, which put increasing pressures on manufacturers' profit margins.

‡Research studies carried out by the Datos organisation for H.J. Heinz

*"Carton de Venezuela" is a subsidiary of Container Corporation of America, and its research department "Centro de Diseño y Mercadotecnia" is well equipped to carry out sophisticated packaging evaluation studies.

†By 1979 capital was also becoming a scarce factor.

*Banco Central de Venezuela Informe Económico

†U.S. $1 = Bs.4.3.

‡Increases from 1973–1975 were even higher. Datos National Store Audit.

This has meant that the problem for most manufacturers during the past 5 years has not been the marketing of their goods but the production in sufficient quantities within the limits on profits imposed by price controls. In some ways the situation has been reminiscent of the problems faced by manufacturers during a war economy. Clearly, in circumstances such as these consumer and retailer acceptance of package design was not a high priority. Indications from mid 1979 are that this frenetic expansion is slowing down, and marketing as opposed to production problems are beginning to occupy a more important part in the decision making process of manufacturers.

So much then for the demand side constraints on the development of packaging research. On the supply side, while existing, the constraints are far less important. As has been mentioned above, whereas the demand for research services of any kind in small markets such as Venezuela is of necessity limited, on the whole the relative prosperity and stability of the country compared with the rest of Latin America has meant that research facilities do exist and have developed to a greater extent than elsewhere in the continent. Apart from economic reasons, one important factor has been that for the past 21 years the country has enjoyed democratic government. It seems almost certain that the general consuming public is more prepared to answer questions of any kind, even on their preferences for toilet soap brands, in a free society than in totalitarian or authoritarian societies. Consumer surveys had been conducted in Venezuela since the late 1940s. Significantly, the first was conducted during the short democratic interregnum of 1947–1948. In 1954 Datos, C.A. was established as a full service marketing research institute offering a wide range of consumer, trade, and communication research services.

Since that time other companies and organizations have been set up in the field, and some manufacturing companies have established their own (albeit small) research departments.*

Reverting to the fundamental problem of small markets, the research organizations that exist tend not to specialize in any one area of marketing. Thus at present the only organization that is dedicated principally to package design assessment will also carry out other kinds of marketing research.† This state of affairs leads to another constraint on the supply side; that is, it is unlikely that techniques have been developed in Venezuela in this area that contribute to an overall improvement in the state of the art. The kind of packaging assessment work that is carried out is derived from experiences acquired from the literature and from practical developments in more advanced markets. There is nothing particularly wrong with this kind of dependency, after all there is not a specifically Venezuelan way to carry out research, nor a U.S. way, only a correct way based on scientific method.

One final point concerning Venezuelan research practice—and one that is fairly common in most small and developing markets—is that for cost and practical reasons a large proportion of marketing research tends to be carried out in the capital cities. In Latin America in general the capital cities tend to account for a far higher proportion of consumption even of mass consumer products than their population would appear to

*The current edition of the Venezuelan Marketing Handbook "Guía de Publicidad y Mercadeo" lists 25 companies offering marketing research services, of which five specifically offer packaging evaluation services. Among leading manufacturers with research departments are Procter & Gamble, Colgate, British American Tobacco, Quaker and Unilever.

†Centro de Diseño y Mercadotecnia, Carton de Venezuela

warrant. Costs of research tend to be lower closer to the research facilities, which for obvious reasons are usually located in the capital cities. This comment is particularly relevant in the area of packaging assessment, and the author can only recall one case of a specific packaging research study in the past 20 years that was conducted outside Caracas.*

PACKAGE DESIGN ASSESSMENT STUDIES IN VENEZUELA

As has been mentioned above, the technology of package design assessment in Venezuela has largely been adopted from more sophisticated marketing economies. There are, however, certain broad generalizations that can be made about the procedures before entering into a discussion of particular cases.

In the first place the studies that have been carried out have almost in their entirety been concerned with consumer reactions and no serious research has been carried out to investigate the distributive trades, even with regard to the functional considerations of packaging. In the second place, with very few exceptions the bulk of the consumer research has been directed to problems of imagery rather than to problems of function. In the third place, outside of the work carried out by the "Centro de Diseño y Mercadotecnia" of Carton de Venezuela, very little use has been made of technical devices such as tachistoscopes or eye movement tracking.

To some extent these tendencies are a result of the lack of specialization in the field of packaging of the research companies involved, but they are also a result of the priorities of manufacturers discussed previously. By and large the principal problem of the package designers has been that of conveying an image of quality to the consumer.

It is clear that because of this emphasis much has been lost. Not nearly enough research has been carried out on the functional uses of the package. In Venezuela for example, as in many poorer countries, packages are often used for storing other products long after the original contents have been used. It is probably more common in the case of many food products for the actual package to be used on the table than would be the case in Europe or the United States. Climatic conditions and the prevalence of insects and rodents, particularly in poorer homes, would suggest that packaging methods different from those of Europe or the United States might well be beneficial. A high rate of illiteracy* would also suggest that greater importance should be attached to symbols, not only in terms of product identification but also in the area of usage instructions.

Even more surprising is the absence of research on the package as presented to the public by means of television commercials. Although all manufacturers have grasped the importance of this medium for mass consumer goods in a market with high illiteracy rates,† few if any have attempted to assess the communications effectiveness of their package when displayed on the small screen.

In all of the areas discussed in the preceding, it would be fair to say that virtually no serious research has been conducted. Manufacturers and designers rely heavily in this area on tradition,

---

*For example, based on the Datos organisation studies of purchasing power, the metropolitan area of Caracas accounts for approximately 36% of consumption. Census figures estimate the population of the same area as 26% of total population.

*1976 Census estimated that 23% of the population was illiterate.

†Approximately 40% of advertising expenditure is in television, and this proportion is considerably higher for mass consumption packaged goods.

intuition, and occasionally on a report from their sales force.

Given that imagery has been the main preoccupation of packaging research in Venezuela, what techniques are used? Probably the most common is the concept association test, in which various projective techniques are used to link desired and undesired concepts to the package. In a September 1978 study a manufacturer of a leading infant cereal product commissioned research to assess the risks involved in marketing his product in a less expensive and more readily available container that he was using already for another, lower priced cereal product. (This is an interesting case of a package assessment study stimulated by the constraints of costs and shortages discussed at length above). The problem was whether or not the proposed container, despite the label design of a well known and well accepted quality product, would be considered:

1 To contain less product.
2 To contain inferior product.
3 To be less attractive on the shelves.
4 To give less value for money.

It must be noted that the proposed container had been used for many years for a cereal product that was well accepted but that had a lower retail price and, more important, was not directed at the baby cereal market.

The research was carried out in the central location testing facility of the Datos organization among a sample of 150 mothers of babies under 2 years of age, half of whom were current consumers of the product in its existing packaging. Respondents were subjected to three different stimuli. In the first place they were exposed to simulated supermarket shelf displays of various products, including the test product in its current and proposed form. Second, they were shown, and allowed to handle, the alternative containers in rotated sequences and they were asked to rate each container on a series of concepts. Finally they were shown both containers together and asked to make direct comparisons between them.

The concept associations were scaled on a 1–7 scale and Table 2 indicates the mean rating for each container on each concept.

As frequently occurs in research studies, the structured questions and summarized tabulations often obscure some important findings, as in the apocryphal story of the product test in which 99% of respondents preferred product A but 1% died of food poisoning. In this less dramatic case it was found that the major reason for preference for Can A was found in its method of closing rather than in the concepts under study that were of concern to the manufacturer.*

*For permission to quote some of the findings, we are grateful to Productos Alimenticios Venezolanos Pralven C.A.

Table 2   Concept Associations

	Container A	Container B
Inadequate/adequate for product	6.2	4.6
Poor quality/good quality	6.4	6.2
Unattractive pack/attractive pack	6.1	4.4
Cheaper product/expensive product	4.4	4.0
Inadequate quantity of product/ adequate quantity	5.6	5.3
Not interested in buying/very interested in buying	6.0	4.7

A frequently used technique is that of the simulated or pseudo product test where, as in the concept association test described in the preceding, the manufacturer is able to outline his sampling requirements in regard to the target audience to which the product is directed and also in regard to the desirable and undesirable characteristics that should be measured. The following example is from a study carried out in May 1969 for a leading multinational cigarette manufacturer.

The study was designed to obtain reactions of target group consumers to two alternative package designs for a filter cigarette. The same brand name was well known as a nonfilter cigarette, in which form it had been marketed for many years.

The study was based on a sequential monadic in-home test procedure involving three visits to some 150 respondents in the target group. The first visit established the conditions of respondents to participate in the test. At the conclusion of this interview, qualified respondents were given two packs of the test design, half of the sample receiving one design, half the other. Second and third visits coupled with interviews and delivery of the untested package design were made at intervals of 4 days.

The analysis of the test results covered three principal areas of the findings:

1 *The Cigarette (Smoking) Characteristics* Strength, flavor, and so on that, by the logic of the method used, could be taken to be indicators of believed product differences suggested by the competing package designs.

2 *The Cigarette Images* Again, functions of the package designs, which influence to some extent the appeal and perception of the brand.

3 *Preference* Between the (identical) blends in the two package designs.

In an examination of the findings relating to cigarette characteristics four sets of data were compared. These referred to the ratings on smoking qualities described for the smokers "ideal" brand, to his usual brand, and to the cigarettes in each of the two test package designs. A similar procedure was adopted for the image characteristics. Table 3 summarizes the principal results as far as smoking characteristics are concerned.

As can be seen, Test Pack A was perceived as closer to the ideal cigarette on strength and smoothness characteristics than Test Pack B.

On image characteristics Table 4 shows the mean ratings on a 7 point scale for each test pack.

Finally respondents were asked to state which of the two packages they preferred. While there was no significant

Table 3 Smoking Characteristics

	Ideal	Regular Brand	Test A	Test B
Very strong/strong	4%	16%	6%	11%
Slightly strong/not at all strong	96	84	94	84
Smooth/fairly smooth	92	71	85	79
Slightly sharp/very sharp	8	29	15	21
Pronounced flavor	86	69	79	78
Fairly pronounced flavor/ not at all pronounced flavor	14	31	21	22
High satisfaction	99	69	79	75
Little or no satisfaction	1	31	21	25

Table 4   Image Characteristics

	Test A	Test B
Modern pack/old fashioned	1.8	2.8
For everyday smoking/for special occasions	4.2	3.3
For young people/for older people	3.2	3.7
Mainly for men/mainly for women	3.6	3.4
Smoked by few people/smoked by many	4.1	5.5
Smoked by city people/smoked by country people	1.8	2.4
Smoked by upper class/smoked by lower class	2.4	2.8
Smart, elegant/dull, ordinary	1.6	2.5
Smoked by sophisticated people/ not smoked by sophisticated people	2.1	2.6
Smoked by manual workers/smoked by white collar workers	5.7	4.9
My kind or cigarette/not my kind of cigarette	2.1	2.3

preference on this measurement between the two test packs, there was a significant *position* preference, that is, a preference for the package design placed first. Table 5.

Frequently design assessment studies are carried out using very small samples. In most cases the design or designs are used as stimuli in three or four focus group sessions, each consisting of 8–10 respondents. The objective of this exercise is presumably to enable the discussion leader or analyst to discover "negatives" in the package design and to enable the manufacturer to feel that he has gone beyond asking the opinions of his board chairman's wife and the sales manager.

Package design assessment research that is fully integrated with the on-going design process is not very common in Venezuela. The research department of Carton de Venezuela, whose principal business is the design and manufacture of containers, does attempt to persuade its clients to carry out integrated research at least as far as imagery is concerned. A typical "Carton de Venezuela" study will cover various stages as can be seen from the following investigation into packaging for a Sardine Cannery.

The first stage of the investigation consisted of four group sessions in which respondents discussed attitudes towards canned products in general and canned

Table 5   Preference*

	150	A First 75	B First 75
Preferred A	41%	49%	33%
Preferred B	39	28	51
No preference	20	23	16

*We are grateful to C.A. Cigarrera Bigott Sucs.

Table 6   Analysis of a Tachistoscope Study (20 respondents)

	Brand Identification			Product Identification			Basic Color Identification		
	$1/100$	$1/50$	Total	$1/100$	$1/50$	Total	$1/100$	$1/50$	Total
Test Design A									
Satisfactory	5	16	21	10	18	28	9	20	29
Unsatisfactory	15	4	19	10	2	2	11	0	11
Test Design B									
Satisfactory	16	20	36	15	15	30	20	20	40
Unsatisfactory	4	0	4	5	5	10	0	0	0

sardines in particular. Results of the groups indicated that the container should be flat and wide rather than narrow and thick, which had heretofore been the traditional "Portuguese" type can. Furthermore the group sessions indicated that the quality image would be enhanced by providing an outer wrapping for the can rather than by supplying merely the lithographed can. Also from these groups came the indication that the product should be described as fresh and mild in flavor and that the basic design should be traditional rather than modern. It was also found that serving suggestions should be included as part of the design on the back of the can. The preliminary indications from these groups was that color combinations could be left to the judgment of the designer as there were apparently no specific attitude stereotypes in this area. In the sessions respondents were asked to suggest and comment on possible names. With the findings of these sessions the design group was set to work to produce preliminary designs and to come up with suitable names for testing.

The second stage of the study consisted in evaluating six alternative names. This was done by interviewing consumers using free association, preference, and memory tests. With the results of these tests the design group produced various designs in combination with the most adequate names. Following this exercise two designs were selected for evaluation at consumer level. The third stage of the study then consisted of an evaluation of these two designs in terms not only of communication of the graphic elements but also of the attitudinal values evoked by the designs.

The first part of this stage consisted of exposure of the designs using a tachistoscope. Respondents were asked to write down what they could remember having seen after each exposure. The analysis of the tachistoscope study (Table 6) produced the above results. (The first two columns indicate exposure speeds in fractions of a second.)

The table presented in the preceding is merely a sample of many different tables that were obtained from the tachistoscope research which were later subjected to comparisons and significance tests.

The researchers also conducted attitudinal evaluations of the designs under study in which three basic areas were covered:

1 Product attributes.
2 Affective connotations.
3 Graphic design evaluation.

For this part of the study respondents were asked to rate, on a seven point scale, characteristics such as:

High quality product
Would serve to guests
Distinctive design
Low quality product
Would not serve to guests
Ordinary design

These characteristics were ranked in order of magnitude of the standard deviation in order to bring out the outstanding attitudinal features of each design.*

One final example of a packaging evaluation study is the following carried out by the company Mercanalisis S.R.L. for a leading manufacturer of dehydrated soups. The study was designed to guide the manufacturer in the choice of two alternative designs for the updating of his product, which had occupied a leading position in the market for a long time. Earlier research based largely on group session techniques had led to the development of two strikingly different designs which the company wished to evaluate. The first design was basically traditional, giving emphasis to the ingredients of the product; the second was based on the face of a child who was eating the soup.

The study consisted of two parts, both of which were conducted simultaneously. The first part of the study was based on 200 personal interviews with housewives, using photographs of the alternative designs (100 respondents were shown Design A, 100 Design B, as well as photographs of the two leading competitive brands). The second part of the study was based on group sessions with housewives, two groups discussing Design A and competition and two groups Design B and competition. At the end of each group the new design that had not been discussed was shown as a

*We are grateful for this example to Carton de Venezuela C.A.

final stimulus to obtain reactions to a comparison of both packages.

Among other measurements the quantitative study covered the following concept associations:

Table 7   Concept Associations

	(Child) Design A	(Ingredients) Design B
Initial Reactions		
Tenderness, beauty of child	57	—
Attractive presentation	37	54
Agreeable colors	22	4
Product is seen to be good	—	
Ingredients		23
Appetizing		22
The packet seems larger	—	10
Other mentions	6	11
Description of Product		
Noodle soup	44	49
Appetising, tasty	26	40
Chicken flavor	21	12
Nutritious	11	—
Well spiced	—	53
Other mentions	29	20
Consumer Prototypes		
All the family	55	76
Children	38	13
Adults	6	8
Other mentions	1	3

Respondents were then asked to rate the package on various characteristics using a four point verbal scale ranging from excellent to deficient. Finally, using a three point verbal scale, they were asked to rate the design in comparison with the two competitive brands and then to state a preference for one of the three designs shown.

Table 8   Comparing the Design with Competition

	Design A	Design B
Competitor X	20%	14%
Competitor Y	21	44
Design (A or B)	54	39
Indifferent	5	3

Finally they were shown both designs under study and asked to state their preferences.

Table 9 Preference

	Basic Test	
	Design A	Design B
Preferred Design A	72	68
Preferred Design B	25	29
No preference	3	3

The group session research covered in a different way the dimensions measured in the quantitative stage, and produced similar results.*

CONCLUSIONS

The presence of various constraints has limited the development of packaging

*We are grateful to Lever S.A. for permission to use this example

evaluation research in Venezuela. In the main the constraints have been associated with market size and the special economic circumstances associated with the petroleum boom, although there are also attitudinal constraints that are of interest in the context of developing economies. Research facilities are more adequate than in most markets of similar size, although techniques used will usually be adapted from more advanced economies.

REFERENCES

1 Schlackman & Dillon, "Packaging & Symbolic Communication," in *Consumer Market Research Handbook,* McGraw-Hill, London, 1972.

2 Schlackman & Dillon, p.448.

PART SEVEN

Advanced Package Evaluation Techniques

CHAPTER FORTY-SEVEN

Brain Wave Analysis: The Beginning and the Future of Package Design Research

Sidney Weinstein

Brain wave analysis is an outgrowth of the early form of electroencephalographic (EEG) recording, first discovered by Hans Berger in the early part of the twentieth century. Berger found that the brain has certain rhythms; these he labeled in order of his discovery, "Alpha," "Beta," and so on.

The electroencephalogram is a recording of the composite electrical activity of a person's (or animal's) brain, which is detected from electrodes placed on the scalp. The method requires highly technical knowledge to know where and

how to place the electrodes, elaborate and complex equipment for amplification and recording, computers for analysis, and much training to interpret the data.

It is not the function of this chapter to teach basic electroencephalography but to demonstrate its utility in package design research. For the reader who wishes to delve into the basic procedures of the method there are many textbooks available on neurophysiology and electroencephalography. A simple manual[1] may familiarize those who wish to become acquainted with the basics of the method.

OTHER PHYSIOLOGICAL METHODS

Several physiological methods are currently employed to various extents in assessing the potential of communication. A brief word on several of them may enable us to compare them with brain wave analysis.

Galvanic Skin Response

Perhaps the oldest of the physiological methods for assessing efficacy of advertising, packaging, and so on is the electrodermal skin response (EDR), sometimes referred to as the galvanic skin response (GSR) or, from an earlier time "psychogalvanic reflex" (PGR). The method was brought to this country more than a half century ago, but has not changed in all these years. The procedure employs placing electrodes on a palm (or finger), in order to measure sweat gland activity. Those who employ it maintain that the autonomic activation signalled by an EDR reflects some "involvement" by the subject. There are, of course, many types of irrelevant stimuli that can produce EDRs—a loud sound, sexually relevant stimuli (e.g., a scantily clad girl), increases in room temperature, and so on. Furthermore, there are many attendant problems: for example there is a long and very variable latency between stimulus and response; there is a very long and quite variable recovery time which fluctuates spontaneously; and there are indiscriminate responses to all sorts of stimuli such as intrusive sounds and private thoughts. The EDR does not differentiate between positive or negative valences of the stimuli so that something hated and something loved may show the same level of EDR. A recent study[2] has also intro-

duced another complexity: the hands from which the recordings are made respond differently to various types of stimuli, which, interestingly enough, are processed differently by the two cerebral hemispheres.

Pupillometry

Some years ago the use of pupillometry as a research tool in advertising burst on the scene, purporting to distinguish between stimuli that induce positive and negative effects in the viewer. Pupillometry, the procedure that measures the diameter of the pupil, employs various methods for this purpose. The latest ones utilize an infrared beam that is directed at the eye, absorbed by the pupil, and reflected by the iris. Various electronic devices then convert the electric signal of the reflected infrared beam into a diameter. The simplest criticism of the method is that the electronic arts are still not capable of measuring pupillary responses with the precision required by those who first reported it as a useful method. To achieve the necessary precision, cameras must be employed to photograph the pupil continuously, and subsequent hand measurements of the pupillary diameters must be made from images projected on screens. The electronic devices that automatically record pupillary diameter respond with 10 times less precision than the published requirements specify. This instrumentation problem is a minor one, however, when one considers other limitations such as the fact that the pupil responds primarily to variations in the degrees of illumination. Thus, packages printed with varying amounts of darker and lighter colors of ink will primarily be subject to this artifact. Finally, it must be pointed out that the method is limited in application, time-consuming, expensive, and perhaps most important, is subject to the criticism of most leading experts in pupillometry that there has never been a convincing demonstration of pupillary constriction to negative stimuli.

Voice Spectrography

The method of using spectra of frequencies of the voice to determine the truth of one's utterances has no applicability to package design. However, should some ingenious designer find some potential manner of applying the system it should be emphasized that the validity, reliability, and precision of the results are in question. The manufacturers of one such device claim that their machines "detect inaudible voice-modulation changes caused by psychological factors such as guilt, fear, or anxiety." They also refer to the responses as "voice stress" and "microtremors." Aside from the conscious factors that enable singers and actors to change pitch, vibrato, and so on at will, there is very little likelihood that "lying" about a package one prefers will produce sufficient "guilt, fear, or anxiety" to be measured by any device. To check on these claims we telephoned an executive of one of the most active companies (following their failure to respond to a written request for data on validity, reliability, etc.). We learned that they have no statistical data on reliability or validity, that they don't know how the device works, but that they "leave it to the customers." Much of the printed brochures also contain testimonials from "satisfied customers."

Electromyography

The electromyogram (EMG) is a recording of the electrical activity that appears in all muscles when they are activated. Several studies have demonstrated that "laughing" and "frowning" activities can be detected by placing appropriate electrodes on the muscles of the face. If these recordings are made while the individual is watching a comedy or dramatic show on television, the authors maintain that mirthful or angry feelings are detectable. The application of this procedure to package design is quite remote, even if the procedure is a valid one.

Visual Fixation

Recent developments enable the identification of stimuli on which individuals fixate. By directing weak infrared beams on the sclera, it is possible to determine the direction of a person's gaze by the reflected beams. Thus one can secure a record of the regions of a package at which the individual looks. By placing a package in front of a person, the device enables one to determine where, on the package, the gaze is being directed. Such a technique may enable the package designer to determine whether or not some graphic detail has indeed attracted attention. One of the problems attendant on such a procedure, however, is to determine whether the stimuli that attract the attention produce positive or negative affect or whether the person is indeed focusing on the stimulus in question or is resting his eyes and merely "looking through" the object while concentrating on irrelevant, private thoughts.

BRAIN WAVE ANALYSIS

Having briefly discussed the questionable applicability of other physiological techniques to package design, let us now consider the utility of brain wave analysis to package design. Since the application of brain waves has been greatest in analyzing television and radio programs, commercials, and magazine and newspaper advertising, it may be valuable to describe briefly the application of this latest method in these areas related to package design.[3-6]

Beta Waves

As indicated previously, Berger was the first one to record the EEG in a human (it had been recorded in animals 50 years prior to Berger by Caton). The first wave that Berger found had a frequency range of 8–12Hz (i.e., it beat from 8 to 12 times per second), and was named the "Alpha" wave. In general, when Alpha is present, the individual is not responding to stimulation in his environment, although he may respond to a sudden change of stimulation. When this stimulation occurs, there is a sudden awareness of the environment and the Alpha is "blocked." Alpha blocking refers to the replacement of the 8 Hz wave by a faster wave (13–40 Hz) which has been termed "Beta." When the Beta wave is present, the individual is responsive to the stimulation.

Thus the degree of "involvement" of an individual with a given stimulus can be quantified by determining the percentage of Beta brain waves during a given period of time. By visually exposing various samples of packages, the degree of involvement that the subject demonstrates to each is determined by analyzing the percentage of time that the brain waves are showing Beta activity rather than Alpha. The greater the percentage of Beta, the more likely it is that the subject is interacting with, is responsive to, or is showing involvement with the stimuli (e.g., the package).

Right Versus Left Cerebral Hemisphere Activity

It has long been known by means of clinical observations, but only recently clarified experimentally, that the right and left cerebral hemispheres have rather distinctive functions. Some 25 years ago I studied with my colleagues the effects of injuries to the two cerebral hemispheres. Our studies and those of subsequent investigators determined that the left hemisphere is concerned with the processing of verbal, intellectual, cognitive, and sequential material, while the right is concerned with the processing of nonverbal, spatial, holistic, pictorial, emotional, and musical material.[7-15]

Such discoveries have provided invaluable insights into how various forms of advertising are communicated to the observer.[16-20] Indeed, the insights acquired were so dramatic that Dr. H. Krugman entitled his talk dealing with the use of brain wave analysis of television commercials and print advertising "Toward an Ideal TV Pre-Test."[17]

Let us first briefly discuss how brain wave analyses for television, radio, print, or package design are accomplished. The target audience is first selected according to such demographic or psychographic variables as: sex, age, socioeconomic status, educational level, and use of a given product or service. The subjects are then instructed that there will be *no* interviewing, questionnaire completion, knob or dial turning, and so on required of them. They are merely asked to observe the materials presented to them and to look at them as they normally would in the stores at which they shop, in their homes, and so on. These instructions usually have the effect of permitting the subjects to relax, since no complex performances are required of them.

Small electrodes (referred to as "sensors") are then attached to the subject's scalp (Fig. 1). These electrodes are quite small—about the size of a shirt button (Fig. 2). To prepare the placement, two small areas on the scalp (above the parietal region of the brain) are first gently swabbed with alcohol saturated cotton balls in order to reduce the amount of

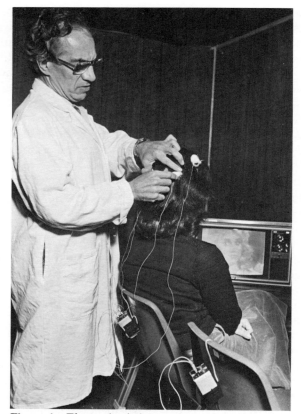

Figure 1   Electrodes being applied to respondent.

skin oil and thus reduce the skin imped-
ance recorded. A small amount of water
soluble electrode paste is then placed on
the cleaned site and the electrode placed
on the paste. The thin wire that connects
the electrode is usually held in place by
a bobby pin. (Following the session the
electrode is removed and the paste easily
dissolved with water applied by means
of a cotton ball, with no effect on the
coiffure.)

The jacks at the end of the electrode

Figure 2   Size of electrode shown in comparison to 10 cent piece.

wire are plugged into small unobtrusive boxes attached to the back of the subject's seat. These boxes contain amplifiers that amplify the brain wave and then conduct it to the computer where the percentage of Beta activity is computed and printed out separately for each hemisphere.

These print-outs are either made for the entire commercial, package, and so on or for segments thereof (e.g., for each 5 second epoch of the commercial, for each side of a package).

Let us consider some recently reported results of brain wave analysis of some corporate television commercials analyzed for General Electric. In this family of commercials, the actor Pat Hingle portrayed the role of Thomas Edison.[18] All the Edison commercials open with print superimposed (i.e., the name Thomas Edison and the year). In the audience, each individual reacted differently; some responded to the character of Edison (with their right hemisphere) or the print "super" with their left hemisphere. That is, different consumers react to different aspects of a commercial (or package). In this case the same part of the commercial was *seen* by some, *read* by others; heard by some as *words,* heard by others as *auditory images.*

The same commercial ended with video alone, but others added "voice-over." In these comparisons, it was found that the presence of the "voice-over" *diminished* the intensity of the brain response. This result most certainly was not anticipated by the creator of the commercials; one might have expected that the addition of a "voice-over" would have enhanced the brain response. Perhaps the lack of synchrony between the visual and the auditory caused conflict and the resultant diminution of the brain response.

Some other highly unexpected results appeared in studying this family of commercials. For example, the superimposed print on video and the final logo itself resulted in right rather than left brain response. Apparently, the words were not read; they were just looked at—as a word picture.

Similarly, the spoken names of unknown people resulted in right, not left, brain responses, whereas familiar names, with their attendant associations, were likely to cause left brain responses.

Krugman put it well, "The right brain looks at things, dwells on them, inspects them, and it all registers in a rather leisurely pattern until an idea creates a thought, which is relatively lightning fast. For example, there was a left brain "startle" reaction to Edison's mention of his deafness—which set off a host of respondent speculations."[18]

Several stories in these General Electric commercials evoked vivid *pictures* in the mind—they were imagined *scenes* and resulted in right brain activity. In others the stories evoked *ideas* and thus left brain activity. These latter examples are representative of insight, learning, and so on. One might quote Luria and say that intentional learning takes place in the left brain and incidental learning in the right.

It is thus apparent that analyzing packages by means of this procedure can provide valuable insight into how the consumer accepts and processes the information provided to him. Is he being entertained by the pretty designs on the package or is he avidly reading the verbal material? Is he reacting to the brand name with his right brain as though the name were only a "word picture," or is he reading it for the attendant associations it provides? Does the name cause left brain thought and appraisal or right brain "unfamiliarity processing"? Is the Beta activity of both hemispheres low, moderate, or high? If it is low, the package has been poorly designed since it won't attract attention or cause "ideas" (left) or "aesthetic association" (right).

In a recent[19] study conducted for the Simmons Market Research Bureau, and sponsored by *Family Circle, Newsweek, Reader's Digest, Time,* and *TV Guide,* my colleagues and I studied television commercials for 10 products. Each product had one television commercial that scored relatively high in "day-after recall" and another that scored relatively low in this measure. The brain wave activity was found to be significantly higher for the commercials that yielded the higher recall. Thus it seems reasonable that packages that result in higher total Beta activity will have a higher recall and higher recognition than ones with lower Beta activity.

We also recently[6] reported the results of a brain wave study that dealt with the Beta waves (and Cortical Evoked Potentials, discussed in the following) in television commercials that were viewed before or after exposure to paired radio commercials. The results clearly indicated that prior exposure to a paired radio commercial enhanced the percentage Beta of all subsequent paired television commercials. The implications of these results for package design are clear: prior advertising (print, radio, or television), which may be paired to a package's design, may have implications for its attention-provoking properties. Thus appropriately keyed prior advertising may make packages in a supermarket more outstanding. The analysis of the Beta waves produced by a package prior to the respondent's exposure to print, television, radio, and so on could serve as the base line to enable the determination of the degree of enhancement of Beta waves when viewing of the package is preceded by the paired advertising.

Summary: In general it has become clear that the efficacy of a package design can be reliably and validly determined by the percentage of Beta activity

recorded from the right and left cerebral hemispheres. The effects of size, copy, color, shape, prior advertising, and so on can all be assessed by what has now become a routine method.

Cortical Evoked Potentials

Although the written record of the EEG seems impossible to decipher by visual examination, highly sophisticated computer techniques have enabled us to look at miniscule aspects of the record and to determine their meaning. Such a small wave within the EEG is the cortical evoked potential (CEP), a highly specific wave that results from the stimulation of a respondent (e.g., with a sudden light). The CEP is a series of waves, alternately negative and positive in electrical sign, producing a very small voltage (in the microvolt range).

Many years of basic scientific study have revealed that there is a part of this series of waves that reflects interest in the stimulus being observed. Early research demonstrated that if an electrode is placed in a particular region of an animal's brain, repeated visual stimulation results in very consistent brain waves that are precisely "time-locked" to the stimulus; that is, they occur with such time precision that repeated recordings of brain activity follow almost exactly in time.[3-6]

This paradigm is a very useful one in determining degrees of interest. In one such study a cat was repeatedly stimulated by a bright light while its CEPs were recorded. The amplitude levels of the CEP were consistent and high. However, when a small mouse in a glass jar was introduced, the competition for attention between the repeated probe light stimulation and the mouse resulted in a total attenuation of the brain wave to the probe stimuli. Thus when there was nothing of intrinsic interest to the cat, the probing stimuli produced strong and consistent brain waves. When the mouse was brought into competition with the mere flashing lights, the cat's attention to the mouse abolished the CEP to the uninteresting flashing lights.

This paradigm was employed by the author in a study of the comparison of the brain waves resulting from observation of a sonar oscilloscope in contrast to reading an interesting magazine.[3] The use of the probe demonstrated that interesting material diminishes the brain wave to probe stimuli, while dull material has no such effect.

Subsequent studies have employed this paradigm of the probe with excellent predictability of the levels of interest. In one recent study,[21] the brightness level of the television screen was electronically controlled to permit sudden, brief increases to be made every few seconds during any television program. The author made several interesting comparisons of the CEP to the probe of the brightness change. Thus comparisons were made of CEPs to irrelevant probe stimuli during the following television or motion picture presentations: (a) erotic movies versus the television show "Face the Nation," (b) an actor and actress conversing about music versus the actor and actress engaging in erotic activities, (c) high versus low interest television programs personally selected by the viewers (e.g., the popular show "MASH" versus the much less popular "Meet the Press").

In all these comparisons the hypothesized reduction of the irrelevant flicker probe occurred consistently when the programs presented were rated to be of high personal interest to the viewers.

In several (proprietary) studies, pilot and television programs were studied using this probing approach to the CEP. Two of the television networks provided us with pilot films for study. The procedure enabled us to provide scene by scene analyses of the percentage of in-

terest generated, as well as an overall index of the interest level. The validity of the procedure was enhanced when one network failed to act on our report that three of the shows demonstrated low interest levels by this procedure and aired the shows. The method was vindicated when the subsequent Nielsen ratings were low enough to cause them to be withdrawn. Another television network contracted with us for a "casting" study to select the potentially best guest stars for a prime time television show. The CEP data were quite high for interest for a controversial entertainer. When the networks own verbally based studies produced conflicting results, the entertainer was not used. However, another network used the same entertainer in a "Special" production with good success. The results again indicated that verbal responses are not as good a predictor of a respondent's true interest levels as is his CEP.

These "casting" studies have also been conducted for advertising agencies that are seeking spokespersons for their clients. Recent (proprietary) studies have been very successful in appropriately selecting the person to represent large corporations.

APPLICATION OF BRAIN WAVE ANALYSIS TO PACKAGE DESIGN

There are numerous psychological attributes that the package with the ideal design posesses. First, it must attract *attention*. A given package immersed in a sea of competing packages fights for attention. Sometimes the very nature of what is considered competing makes the package less noticed. Thus if Package A, which contains a breakfast cereal, is to compete with others with highly colorful designs, it would of course be wise to invoke the von Restorff Effect, and print the designs in monochromatic inks. A black and white design would "stand out" in a sea of reds and greens. But how is one to know what the competition plans? The best approach would be to design a package based on the existing designs. Comparing the brain waves elicited by Package A with those of the current competitors would indicate the comparative level of attention produced by it.

A second psychological aspect a package should provide is *interest*. It would provide no useful purpose to attract attention only to find that the attention was short-lived, and that there was no intrinsic interest generated by the shape, color, graphics, or verbal content. A large splash of red color might momentarily cause a head to turn in a supermarket (assuming that the competition did not also have large red packages). But after the momentary glance what then? If there is no intrinsically interesting aspect to the package it will be ignored very rapidly for another one that is interesting.

A third psychological characteristic is *memory;* this general term can be subdivided into verbal recall (left hemisphere) and visual recognition (right hemisphere). Remembering what a given package resembles may consist of verbal recall if the name is on the box, some other verbal content, recognition of the logo, recognition of the shape or size, or even recognition of the design of the name (e.g., block, italic, or script letters printed in a triangular or circular shape, lying over or under a picture of a cow or a tower).

A fourth characteristic of package design consists of the *somesthetic characteristics*. First, the size must be considered relative to the human body. Is it small enough to be held between the thumb and fingers (e.g., a ring case), does it require holding in the palm of the hand (e.g., a cigarette case or a box of margarine), does it have to be held by both hands (a typewriter case)? Is the external surface smooth or rough? Is it polished to a gloss—is the surface smooth but deglazed—is it slippery? Is the surface smooth and soft (as in velvet) or is it rough and soft (as a woolen blanket)? Is the surface smooth and hard (a polished jewel case) or smooth and soft (a satin pillow)? Is the package lightweight or heavy? Does it appear cool to the touch (polished metal or foil) or warm (unfinished wood)? Are there alternations of smooth and dull (or rough) or cool and warm? Does it emit sounds when palpated (as in crinkly paper) or is it silent when touched?

These somesthetic and attendant auditory characteristics indicate that psychological considerations of packages are greater than those contained in print advertising in which the only characteristics are: verbal and pictorial, color versus monochromatic, size, margins or bleed, position (left/right or front/back of magazine). The considerations concerning packages are more complex than print advertising and even rival or exceed the complexity of television commercials with their attendant movement and use of the additional modality of audition.

Method of Analysis of Package Designs

Before a package design can be analyzed, the decision must be made concerning which psychological attributes are of interest to the client, (e.g., attention, interest, memorability, or somesthetic quality). Once this decision has been made, the procedures are determined for the analysis to proceed.

For most purposes it is best to videotape various aspects of the package, either singly or in a cluster. The single package analysis is most appropriate to items sold singly rather than being stacked on shelves, as is the case of supermarket items. The analysis of single packages is most appropriate for products such as perfume and jewelry, while the clustering concept is best for items like soaps and detergents, breakfast cereals, cat and dog foods, and beverages.

Once this decision is made, the single package or the group of packages are photographed with a color video camera for subsequent playback. The camera may be directed at the dominant side (e.g., front) of the package or the package may be rotated so that several sides (the top, etc.) are photographed.

The various alternative designs or competitor's packages are also photographed in a similar manner, for the same period of time (e.g., 30–60 seconds). The video cassettes are then organized so that the order of presentation of the packages is balanced in various subgroups of the target group.

The subjects are first interviewed to determine conformity to the target group (e.g., age, sex, socioeconomic level, and use of products). The procedure of recording is then explained to them, emphasizing that there are no questionnaires to complete, no dials or knobs to turn, and so on. They are merely to observe the television screen and look at the packages, while "the computer does the work of analysis."

In cases in which somesthetic analyses of how size, texture, weight, temperature, and so on affect the brain waves are desired, the respondents are instructed that they will be asked to examine (visually and tactually) various packages and consider them as they do when considering purchases.

The electrodes are then placed over the midparietal areas of the scalp, and the television set turned on. Initially, all respondents are asked to observe the blank television screen so we can "calibrate our equipment." This procedure enables us to acquire norms of right and left Beta based on stimulation with a formless, motionless, colorless, low contrast stimulus. While the respondents observe the screen, we initiate a series of sudden brightness changes in order to determine the CEP and Beta to a screen devoid of any content of interest. Beta and CEP thus recorded are considered those for "zero" 'interest and involvement, and all subsequent Beta and CEP levels for that subject are calculated relative to these initial base line levels.

Following the acquisition of the control right and left Beta and CEP levels, the appropriate TV cassette is employed and the various package designs are exhibited while the brain waves are recorded. The CEPs and the levels of Beta activity are recorded and analyzed separately for each package design.

For those package designs for which the somesthetic qualities are important, base lines are recorded while the respondent palpates and visually observes a blank, white cardboard box, similar in size and weight to the test packages. The brain waves are first recorded during this examination in order to achieve base line levels.

The actual test then requires the respondent to palpate and visually to examine each package mockup for approximately one minute, while his brain waves are recorded. The comparisons are made with the baseline levels determined when the control packages were examined.

REPORT TO THE CLIENT

Each client receives a report containing the percentage interest levels and the percentage of right and left Beta activity evoked by each package (activation and involvement).

These data are usually in the form of comparisons of all analyzed package designs. Table 1 provides the results in a typical summary comparison of six po-

Table 1 Comparison of Six Designs for a Package of _____

Package	% Interest (CEP)	% Increased Activation	
		(Beta) 100	$\frac{\text{(Package-Control)}}{\text{Control}}$
		R	L
A	11	4	7
B	13	4	2
C	9	2	13
D	28	8	0
E	47	18	12
F	36	7	16

tential packages being considered for adoption by a client. The report has been modified to eliminate any identification with the client, and the analyses according to the demographic and psychographic variables (e.g., sex, age, socioeconomic level, education, and usage of product) are not included because of space considerations.

Brief Summary of Report to Client

Figure 3 presents the results of analysis of five package designs. The means are ordered from highest percentage interest (Design E) to lowest (Design C). The percentage of increased activation in right and left hemispheres is shown in bar graphs below the percentage interest for each design.

The statistical comparisons of the designs according to interest show three distinct clusters: E, (F,D), and (B,A,C). That is, E yielded significantly greater interest than F and D, which did not differ significantly. F and D, in turn, produced significantly more interest than B,A, and C, all of which did not differ significantly one from the other.

Concerning percentage increased activation (increased Beta waves in both hemispheres) it can be seen that there was a general correspondance between Interest and Activation-Involvement. Design E showed the highest mean Beta

level with the right 6% greater than left. This finding indicates that there is a relatively high left Beta level due to the copy on the package face. The high right Beta indicates that there is a pleasant emotional aspect that is even greater than the degree of verbal involvement. Design F is somewhat lower in Beta than E, with left greater than right. D shows less than moderate right Beta and no left Beta. This result stems from the minimal amount of verbal stimuli on the package and a pleasant use of color. B,A, and C are all low in mean Beta; however, the rise in left Beta for C reflects the extensive copy printed on several sides of the package.

In general, the results point to E as the overall best package from the viewpoint of high interest (CEP) and moderately high activation (right and left Beta). The other designs in order (F and D) would have been considered adequate if E were not in contention. B,A, and C should not be considered as contenders because of their low interest and activation levels.

SUMMARY

The advent of brain wave research in the marketing, advertising, entertainment areas has had a dramatic effect. In contrast to highly subjective and fre-

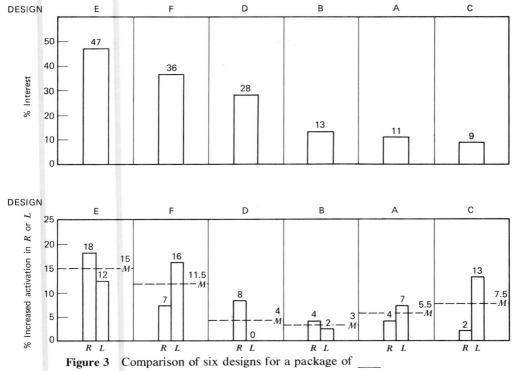

Figure 3 Comparison of six designs for a package of _____

Interest is expressed as a percentage. Activation in right *(R)* and left *(L)* hemispheres is expressed as an increase over the respondent's base. *M* equals the means of *R* and *L*. The six designs are given in decreasing order of interest.

quently invalid verbal approaches and physiological methods of dubious validity, brain wave analysis has become the standard method of choice in detecting the most effective commercials, print advertising, packages, and so on. Its applicability seems to increase with the demands of the industry.

The procedures are so adaptable that research directors have been able to design studies to answer questions of importance in many areas of marketing, advertising, and so on. The fact that there have been increasingly more brain wave studies conducted each successive year of the 1970s indicates that the trend has become an established procedure, and will apparently continue to grow more frequent in the coming years.

REFERENCES

1   A.R. Craib and M. Perry, *EEG Handbook,* 2nd ed., Beckman Instruments Inc., 1975.

2   J.M. Lacroix and P. Comper, *Psychophysiology,* **16** (1979), 116.

3   S. Weinstein, Paper presented at 16th Annual American Marketing Association Advertising Research Conference, Physiological and Attitudinal Measures of Advertising Effectiveness, at the Hotel Biltmore, New York, May 16, 1978.

4   S. Weinstein, Paper presented at the First Advertising Research Seminar of the Canadian Advertising Research Foundation, Toronto, January 16, 1979.

5   S. Weinstein, *Marketing Review,* **34** (1979), 17.

6   S. Weinstein, Paper presented at the Association of National Advertisers/Radio Advertis-

ing Bureau Radio Workshop, at the Waldorf-Astoria Hotel, New York, June 19, 1979.

7   S. Weinstein, "Differences in Effects of Brain Wounds Implicating Right or Left Hemispheres: Differential Effects on Certain Intellectual and Complex Perceptual Functions," Chapter 8 in V.B. Mountcastle, Ed. *Interhemispheric Relations and Cerebral Dominance,* Johns Hopkins University, Baltimore, 1962.

8   S. Weinstein, "Functional Cerebral Hemispheric Asymmetry," Chapter 2 in M. Kinsbourne, Ed., Cambridge University, Cambridge, England, 1978.

9   S. Weinstein, Functional asymmetries of right and left cerebral hemispheres, Presidential Address to Division of Physiological and Comparative Psychology, American Psychological Association Meetings, Los Angeles, 1964.

10 R.J. Davidson, G.E. Schwartz, E. Pugash, and E. Bromfield, *Biological Psychology,* 4 (1976), 119.

11 J.C. Doyle, R. Ornstein, and D. Galen, Psychophysiology, 11 (1974), 567.

12 R. Dumas and A. Morgan, *Neuropsychologia,* 13 (1975), 219.

13 D. Galin and R.R. Ellis, *Neuropsychologia,* 13 (1975), 45.

14 D. Galin and R.E. Ornstein, *Psychophysiology,* 9 (1972), 412.

15 A.H. Morgan, H. MacDonald, and E. Hilgard, *Psychophysiology,* 11 (1974), 275.

16 H.E. Krugman, *Journal of Advertising Research,* 11 (1971), 3.

17 H.E. Krugman, Paper presented at the 16th Annual American Marketing Association Advertising Research Conference, at the Hotel Biltmore, New York, May 16, 1978.

18 H.E. Krugman, Paper presented at the Annual Conference of the Advertising Research Foundation, Waldorf Astoria Hotel, New York, October 16, 1978.

19 V. Appel, S. Weinstein, and C.D. Weinstein, *Journal of Advertising Research, 19,* (1979), 7.

20 S. Weinstein, V. Appel and C.D. Weinstein, *Journal of Advertising Research, 20,* (1980), 57.

21 E.W.P. Schafer, *International Journal of Neuroscience,* 8 (1978), 71.

CHAPTER FORTY-EIGHT

Psychophysical Approaches to Package Design and Evaluation

Howard R. Moskowitz

INTRODUCTION

Psychophysics is the science that relates perceptions to stimuli. Psychophysics allows the designer to interrelate consumer's subjective perceptions with design characteristics of a package. Consequently, the psychophysical approach to design achieves several objectives:

- Quantitative measurement of the impact of design factors on perception.
- Sensitive scales of differences among

alternative design options, which facilitate marketing decisions.
- Optimization, or fine tuning a final design according to specific criteria, whether these be acceptance ratings or economic factors (e.g., cost of goods).

This chapter concerns several aspects of package design from the viewpoint of psychophysical measurement:

- Scaling consumer reactions.
- Developing quantitative rules of law or laws interrelating physical factors to perceptions.

The author wishes to express his thanks to Mrs. Maria Sabatel, who was indispensable in the preparation of this manuscript.

- Actionable assessment of designs, with the aim of directional modification using quantitative consumer feedback.
- Experimental designs approach for design studies.
- Implementation of a screening/optimization study for fine tuning a selected design.

PACKAGE DESIGN AS A MEASURABLE PERCEPTION

Designs comprise various perceptual elements. Often (but not always) a package or a product can be analyzed into constituents including the container size and shape, color and graphics, and specific embellishments. The design components interact to produce an overall impression to which consumers react. On the basis of these impressions, consumers may accept/reject the product or by-pass it in the store. As with perfumes and flavors, design constituents interact to generate an overall perceptual gestalt or 'whole.' By means of appropriate techniques, a designer can engineer the package components. The psychophysical or analytic approach coupled with the 'gestalt' or underlying artistic approach to the initial design provide two complementary points of view on which to base a marketing decision.

Simple Sensory Constituents of Designs

The sensory elements of design comprise color, shape, size, and so on. In many studies of packages for marketing decisions, the designer or market researcher simply may solicit a free descriptive analysis of consumer perceptions. With a more disciplined approach, the consumer may quantitatively scale perceptions, such as the depth of color and the size of the package. The con-

sumer may also rate the degree of overall liking/disliking of the specific elements, as well as rate the overall package/design/product acceptability.

During the analysis of sensory constituents, the panelist becomes a *measuring instrument*. The panelist accepts or rejects, notices or fails to notice what the designer has already incorporated into the package design and reports specific aspects of satisfaction and dissatisfaction. Perhaps the area occupied by graphics needs modification. Panelists might feel that the package is too large or too small for the product. Perhaps the colors are too bright or too subdued. Straightforward package profiling on a variety of characteristics, coupled with a self-designed "ideal" profile, enables the designer to uncover aspects of dissatisfaction and discern design characteristics that can be modified.

Sensory Laws

Psychophysicists have discovered that sensory reactions do not relate linearly to physical magnitudes of stimuli. These nonlinearities show up reliably when the panelist evaluates stimuli using an unbiased scale. Magnitude estimation scaling (discussed in the following as the most sensitive and preferred procedure for psychophysical testing of packages), allows the panelist to use any range of numbers that he or she feels is most comfortable.

With the magnitude estimation scaling, psychophysicists have discovered the following rules governing perception:

1   The relation between perceived sensory intensity and physical magnitude estimated by instruments is nonlinear, as shown in Fig. 1. When presented with shapes of different areas, with colors of different reflectances, or shades of green of varying

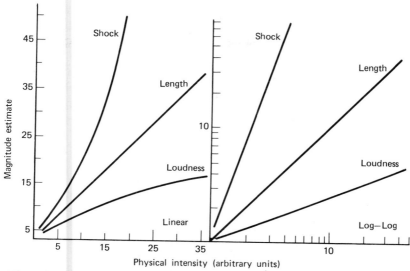

Figure 1   The relation between physical measurements and magnitude estimates of perceived intensity. The left panel shows three curves, representing different relations between what is measured and the numbers that panelists assign to reflect 'intensity.' These curves can be straightened out in logarithmic coordinates, as shown on the right. Such curves allow the designer to learn how specific changes in physical characteristics will be perceived. In most cases the change is not linear (viz., doubling the physical level may more than double the perceived intensity or less than double the perceived intensity).

saturation, the magnitude ratings usually "track" the stimulus variations to trace out a psychophysical curve.

2   The curves in Figure 1 follow a power law, as suggested by Stevens.[1] The power law describes magnitude estimates of the loudness of tones (versus sound energy level of the tone, instrumentally measured), the brightness of lights (versus level of luminance), and so on. The power law of sensory perception is expressed by a simple equation:

sensory intensity = k (physical intensity) ^{n}

(Sensory intensity is the average magnitude estimate rating assigned by panelists, whereas the physical intensity is the instrumental measure.) Furthermore, the power law (usually a curve) can be made more tractable by a logarithmic conversion that transforms the curve shown in the left panel of Fig. 1 into straight lines, shown on the right in Fig. 1.

3   The exponent of the power law, n, informs how changes in *sensory intensity* (the subjective, perceptual reaction) vary with changes in the physical stimulus. If n is 1, then a tenfold change in the physical stimulus level produces a tenfold change in perceived sensory intensity. Perceived length versus physical length is governed by a power law with exponent 1.0. If n is less than 1.0 (as it is for apparent area, with a 0.7 value of n), then we perceive the larger physical range as a smaller sensory range. For perceived area or volume, a tenfold change in the area of a shape will produce only $10^{0.7}$ (or approximately fivefold) change in perceived area or volume. Conse-

quently, if the designer wishes to double the perceived design area, then the actual area must be increased by more than a factor of 2. The solution to this problem of modification or "sensory engineering" is to increase the actual area by the factor X, so that X is the ratio (New/Old)$^{0.7}$ = 2. Effectively, $X^{0.7}$ = 2, or X, the actual ratio, is 2.69. This ratio will produce a perceived ratio of areas of 2.0. (Coincidentally, cartographers have known for years that to double perceived area on a map requires an increase of 2.7 in physical area, not 2.0). Other sensory continua, such as perceived depth of color (versus physical color saturation) are governed by exponents greater than 1.0. Doubling the physical stimulus level *more* than doubles the perceived sensory level. (Perceived roughness versus grit size is an instance where our sensory system magnifies small physical changes to larger perceptual changes.)

4   Table 1 lists representative values for different sensory continua.

Cognitive Factors

Sensory laws governing perception are not entirely fixed and immutable. Panelist instructions regarding the criteria for evaluation will change ratings, be they ratings of package, product, or both. Ask a panelist to scale the *apparent subjective area* occupied by a variety of shapes, and these ratings will follow the

Table 1 Exponents of the Psychophysical Power Law* of Perception and Their Utilization in Modifying Stimuli

		Physical Change Needed	
Continuum	Exponent(N)	To Double Perception	To Increase by a Factor of 10
Roughness (touch)	1.50	1.6	4.6
Heaviness	1.45	1.6	4.9
Fingerspan	1.30	1.7	5.9
Sweetness (sugar)	1.30	1.7	5.9
Duration of time	1.10	1.9	7.9
Area (panelist told to be "exact")	1.00	2	10
Hardness	0.80	2.4	18
Coffee (flavor)	0.70	2.7	27
Volume	0.67	2.8	32
Area (perceived)	0.67	2.8	32
Coffee (aroma)	0.55	3.5	66
Saturation (versus dye level)	0.55	3.5	66
Brightness (point target)	0.50	4	100
Viscosity	0.50	4	100
Brightness (big target)	0.33	8	1000

*Sensory Intensity = K (Physical Intensity)^n

power law with a 0.7 exponent. Change the instructions so that the panelist now rates "actual area" (as he/she perceives it) and the data become linear with exponent 1.0 Teghtsoonian.[2] Our perceptions of simple stimuli can be modified by judgmental criteria. Similar, and more severe, distortions can and do occur when panelists rate more complex, ecologically valid stimuli.

THE TYPICAL PSYCHOPHYSICAL EXPERIMENT

Psychophysical experiments, whether designed to assess visual perception of packages or measure attitude towards stimuli (e.g., attractiveness of graphics, liking of overall appearance) are executed in a standard, straightforward manner. The approach outlined here provides quantitative answers pertaining to package perceptions. There are six steps:

STEP 1   Stimulus development/selection, attribute selection.

STEP 2   Panelist selection.

STEP 3   Orientation in scaling and explanation of attributes.

STEP 4   Product/package/concept ratings.

STEP 5   Calibration of ratings.

STEP 6   Analysis of results.

Developing the Stimuli

The stimuli should be well defined in terms of physical characteristics. Sometimes psychophysical scaling can be used effectively to screen acceptable designs from a set of alternatives. If so, then the stimuli will vary along many different dimensions. On other occasions, psychophysical scaling may be a tool by which to determine how specific variations in the stimulus influence design perceptions. If so, then the design should be systematically modified along a small set of dimensions. (Examples include modifications of the shape, package size, the position, type of graphics, etc.)

One should bear in mind that the best data are generated when the stimuli have been correctly selected. If the study aims at screening, then the data yield answers as to which design or product is 'best' and what characteristics (color, design type) correlate with the winning product. If the design characteristics are varied in a prespecified systematic manner it is possible to interpolate and predict reactions to design combinations not originally tested.

Proper development of test stimuli plays a key role. Many studies yield poor data because of poor stimuli. Similarly, proper stimulus selection allows, but does not guarantee, a successful study. In this first stage, one should assess many different alternative stimuli should the study goal be to screen package alternatives in preparation for further work. Should one wish to develop the interrelation in this package between design elements and perception to guide reformulation of package and product (in anticipation of subsequent optimization), then again the recommendation is to study a wide range of colors, shapes, sizes, and so on.

Panel—Types and Sizes

Many designers/product developers/researchers prefer to use in-house experts, whereas others use external, paid panelists. There is no consensus as to which panelist selection strategy is better. For technical and scientific reasons, psychophysicists prefer to use consumer panelists both for multiproduct screening and for optimization. The consumer ultimately accepts or rejects the product.

The conservative, more risk-free strategy suggests testing with the consumer target group, rather than testing with expert in-house panelists as surrogates for the ultimate consumer. If only an expert detects the proposed design changes, then the expert does not reflect the consumer. On the other hand, if contemplated changes in package design could dramatically affect acceptability, then the expert panelists can act only as rough indicators of the consumer response.

In terms of sample size, the psychophysical study should be viewed in terms of overall project criticality. A final go/no-go decision regarding package and product acceptability requires a broad sample of the target consumer population. Broadscale testing becomes economical with a few final package alternatives. Multiproduct screening or fine-tuning optimization of design alternatives become economical and feasible only with a smaller base of 25–50 panelists per stimulus sample. Screening and optimization studies illustrate how product variations impact on perceptions. They generate feedback about product or design modifications that enhance acceptability.

Attribute Selection and Explanation of Attributes

Attributes capture nuances of perceptions. Attributes comprise three major groups.

Sensory/Perceptual: (e.g., darkness of shade, perceived area of a design, perceived volume of the package, perceived heaviness of the package with the product included.) These sensory/perceptual attributes are often 'physical' in nature. They usually involve no explicit evaluation in terms of 'good/poor.' Most of the data on which the power law is based have been obtained for simple stimuli and ratings of the magnitude of a sensory/perceptual attribute (see Table 1).

Evaluative/Performance: (e.g., overall liking/disliking, liking/disliking of specific sensory characteristics, purchase interest, purchase interest at specific price levels.) Evaluative attributes integrate many impressions of a product into a single number. Liking/disliking ratings apply to the entire product or package. Panelists also can scale their liking/disliking of specific design or product characteristics (e.g., the graphics, package/design shape, specific parts of the package, color, etc.). When probing for purchase interest, one should recognize that purchase interest ratings often reflect an unknown combination of factors, such as overall product liking, liking/disliking of specific product or package components, the current need for the item, perceived product utility, and, finally, product cost.

Image Attributes: (e.g., elegant, unique, mass appeal product, appropriate for various age groups, that is, for adult or for child, etc.) Image attributes are based on complex judgments, which integrate many perceptual elements of the product along with cognitive factors, such as expectations about the appropriate product usage occasion. (For instance, a yellow pattern on bathroom tissue might be perceived as appropriate for the home. The same design might be perceived as 'too expensive,' and 'inappropriate for commercial bathrooms,' and so on.)

Orientation in Scaling

Orientation in magnitude estimation scaling occurs during a short instruction period, with an interviewer, or by oneself with a self-administered questionnaire. Orientation shows panelists how

Table 2   Orientation Questionnaire

NAME_____   I.D. #_____

Training Section

Please pick up the card deck which has been given to you. *Do not look* through the deck, but pay attention only to the top card, which has a shape labeled. Now, you are going to assign numbers to show how large the shapes you will see in this card deck seem to you. Give the first area any number you wish. Write this number on the line labeled "Size" under the first shape at the bottom of the page. *Remember:* You will be using this first number to compare the size of the first shape to the size of other shapes, which could be larger, smaller or the same size as the first shape. Therefore, there are no upper limits on the size of the number you use, but the number should not be so small that you cannot easily divide it into smaller portions (not smaller than "10" for instance). Now, flip to the second shape on the second card. Give it a number that represents the area or size of the second shape as compared to the first shape. For instance: If you give a number of 28 to the first shape and the second shape seemed to have approximately the same size or area, you would also give the second shape a rating of 28. If the second shape seemed only one-half as large as the first shape, then you would give it a rating of 14. If the second shape appeared to you to be four times as large as the first shape, then you would give it a rating of 112. Now, flip to the third shape on the third card and evaluate the size of this shape in the same manner as above. Then do the same for all the other cards in the deck. Please *do not* look ahead, but evaluate each shape as you come to it.

Rating of Shapes

Shape A	Shape D	Shape G	Shape J	Shape M
Size____	Size_____	Size_____	Size_____	Size_____
Shape B	Shape E	Shape H	Shape K	Shape N
Size____	Size____	Size_____	Size_____	Size_____
Shape C	Shape F	Shape I	Shape L	Shape O
Size____	Size_____	Size_____	Size_____	Size_____

NAME_____   I.D.#_____

Training Section

As another exercise, we would like you to express your liking or disliking of different words. A list of words is shown below. Using another scale of your own design, show how you feel about each word. You will use numbers to show how you feel about each word. If you like a word, write an (L) next to it or a (D) next to it if you dislike it. Then indicate just how much you either like or dislike the word by also writing in a number. A large (L) number means you like it a lot, while a large (D) number means you dislike it a lot. A small (L) number means you like it a little, while a small (D) means you dislike it a little. If you feel indifferent or neutral about a word, give it a zero (0). Let's begin. Give the first word an L or a D to show if you like it (L) or dislike it (D). If you like it a lot, give it a large L number. If you dislike it a lot, give it a large D number. *Example:* Say you give the first word an "L10" to show how you feel about it, but like the second word twice as much. You would then give the second word an "L20." If you dislike the third word, you should give it a D and a number to show

Table 2   Orientation Questionnaire (Continued)

how much you dislike it. If you dislike it just a little, you might give it a "D3" but if you dislike it a lot, you might give it a "D150." Match numbers to show how you feel about the words. Please remember that the scale that you use is entirely your own. There are no limits on the size of the scale that you use, and no one's scale is more "right" than anyone else's. You may use any number you wish to show how much you like or dislike a particular word. Simply assign each word the number that you feel represents how much you like (L) or dislike (D) the word. Please keep in mind that large "L" numbers mean you like a word a lot and large "D" numbers mean you dislike a word a lot.

WORD	YOUR RATINGS	
	L/D	How Much?
Flowers	____	_____
Mud	____	_____
Perfume	____	_____
Murder	____	_____
Sex	____	_____
Cigar	____	_____
Spaghetti	____	_____
Rattlesnake	____	_____
Love	____	_____
Sun	____	_____
Hate	____	_____
Worm	____	_____
Kiss	____	_____
Puppy	____	_____
Pollution	____	_____

to use a wide range of numbers to reflect perceptions. Orientation breaks respondents out of fixed 1–5 or 1–9 scale limits. Table 2 shows the orientation used for central location evaluations. Orientation occurs either during a one-to-one interview (one interviewer questioning one panelist) or in a classroom type situation. In the latter, a group of 15–35 panelists are prerecruited to evaluate products and/or packages during a single period. A moderator instructs panelists in magnitude estimation scaling.

Orientation stresses specific points:

- *Free Number Matching*   The panelist may use a continuous, unbounded scale. A continuous, unbounded scale enhances sensitivity and allows the panelist to rate stimuli according to his/her own frame of reference.

Table 3 Magnitude Estimates*

A. ASSIGNED TO SHAPES OF VARYING AREAS

Circles		Triangles		Squares	
Area (cm²)	Rating	Area (cm²)	Rating	Area (cm²)	Rating
7.1	7.8	2.0	4.7	10.1	13.2
19.7	10.0	7.4	10.7	17.9	18.8
43.0	31.1	24.8	22.5	72.4	47.6
91.6	52.2	64.9	49.7	123.0	68.5
145.3	69.5	104.0	65.4	123.0**	69.5
216.4	106.9	322.0	92.4	203.0	97.4
Power Law Exponent	.81		.62		.67

B. ASSIGNED TO WORDS

Sex	138
Love	126
Kiss	84
Money	82
Sun	73
Flavors	58
Puppy	47
Spaghetti	41
Perfume	37
New York City	0
Worm	− 5
Mud	− 25
Cigar	− 31
Rattlesnake	− 40
Pollution	− 72
Hate	− 81
Murder	−140

*22 Panelists.
**Reliability test.

The instructions make no specific statements about wrong/right numbers to use.

- *Hedonic Categorization For Liking/ Disliking* Orientation emphasizes that hedonic ratings comprise a two phase process. During the first phase, the panelist categorizes the overall feeling as liking or disliking. During the second phase, the panelist scales degree of feeling. Again, there is no limit on the size of numbers.

These instructions, along with a set of shapes and words, illustrate how to assign numbers both to perceptions and to liking/disliking of those perceptions.

Table 3 shows some areas and word stimuli used in orientation, as well as average ratings of liking/disliking from one group of panelists. Note the wide range of numbers and how the ratings track actual areas. (Data from numerous consumer panels comprising individuals who have never done magnitude estimation before, but who are being oriented in the technique for product evaluation, conform to the same power law of perception that characterizes laboratory test situations reported by Teghtsoonian. We can feel comfortable about the ability of panelists to scale their perceptions when we correlate their ratings against physical area and reproduce scientifically reported results.)

The Assessment of the Product/Stimulus

The stimuli (design, total product, etc.) are usually scaled singly (monadically) on the different characteristics. When more than one stimulus is being evaluated, the first stimulus may or may not be removed after being rated before the second stimulus is rated. (From time to time researchers have compared the utility of testing the second stimulus while the first stimulus remains present for comparative purposes. By and large

there is no added utility of side-by-side presentation in visual and tactual product assessment. There may be some unwanted effects of side-by-side presentation when one must taste or smell the product. Adaptation diminishes sensitivity. Adaptation does not occur in package design evaluation.[3]

Calibration

Each panelist uses his or her own scale. Ratings for the same test stimulus vary from one person to another due both to differences in perception and to differences in scale or number usage. To remove the scale differences, we index or calibrate each person's ratings. The panelist rates a hypothetical product which is extremely liked, liked very much, and so on. Table 4 shows the calibration questionnaire and indexing procedures.

Benefits of calibration include:

- Reduction in the variability of rating.
- Assignments of verbal descriptors, (extreme, very much, moderate, slight, none) to numerical magnitude estimates, which enhances communication.
- Determination of the percent of panelists who rate a specific stimulus as being 'extreme', 'very much', etc., on any attribute.

Types of Stimuli Scaled

In product and package studies, panelists often scale a variety of packages and products during a single session. After all packages/products have been rated, the panelist may be asked to scale an 'ideal product' or 'ideal package' (e.g., for a specific end use) with the same attributes and the same scale used to rate the actual packages. Table 5 compares calibrated magnitude estimation ratings for three beer can designs and

Table 4   Calibration Questionnaire and Indexing Procedures

A.   CALIBRATION QUESTIONNAIRE

What number on your scale reflects . . .

Liking/Disliking		Amount		Purchase Intent	
Like Extremely	L	Extreme	____	Definitely Buy	____
Like Very Much	L	Very Much	____	Probably Buy	____
Like Moderately	L	Moderate	____	Might/Might Not Buy	____
Like Slightly	L	Slight	____	Probably Not Buy	____
Neutral	0	None	0	Definitely Not Buy	0
Dislike Slightly	D				
Dislike Moderately	D				
Dislike Very Much	D				
Dislike Extremely	D				

B.   EXAMPLE OF "DATA" CALIBRATION FOR A PANELIST

Typical Magnitude Estimation Ratings For a Panelist For
Uniqueness, Attractiveness and Eye Catching

BEFORE CALIBRATION

Package "017"

Calibration		Product Rating	
Extremely	160	Unique	47
Very	128	Attractive	83
Moderately	90	Eye Catching	94
Slightly	35		
None	0		

AFTER CALIBRATION

New M.E. = (Old M.E.)/(Pivot) × 100

Pivot = (Extremely + Very + Moderately + Slightly)/4

Calibration		Product Rating	
Extremely	155	Unique	46
Very	124	Attractive	80
Moderately	87	Eye Catching	91
Slightly	34		
None	0		
Pivot	103.25	Pivot	43.25

C.   EXAMPLE OF "DATA" CALIBRATION FOR A PANELIST

"LIKING/DISLIKING"

BEFORE CALIBRATION

Calibration

Like Extremely	+200	Dislike Extremely	−160
Like Very Much	+140	Dislike Very Much	−100
Like Moderately	+60	Dislike Moderately	−70

Table 4   Continued

Like Slightly	+ 20	Dislike Slightly	− 30
	Neutral 0		

Package Rating

Graphics Liking/ Disliking	+ 20	(− = Dislike)
Overall Liking/ Disliking	+ 10	(+ = Like)

Example Rating For A Disliked Package Rated as −40

AFTER CALIBRATION

$$\text{new M.E.} = (\text{old M.E.})/(\text{pivot}) \times 100$$

Pivot = absolute value of $\left[\dfrac{\text{extremely} + \text{very} + \text{moderately} + \text{slightly}}{8}\right]$ (liking)

absolute value of $\left[\dfrac{\text{extremely} + \text{very} + \text{moderately} + \text{slightly}}{8}\right]$ (disliking)

Calibration

Like Extremely	+287.5	Dislike Extremely	−213.0
Like Very Much	+187.0	Dislike Very Much	−133.0
Like Moderately	+ 80.0	Dislike Moderately	− 93.0
Like Slightly	+ 27.0	Dislike Slightly	− 40.0
	Neutral 0		

Package Rating

Graphics Liking/ Disliking	27.0
Overall Liking/ Disliking	13.0
Dislike package rating =	− 53.0

the ideal. By profiling both actual products and the ideal product, a panelist communicates to the designer:

- Comparative product/package performance on a perceptual basis.
- Where the consumer would like the product/package to be on the key attributes.
- The gap between current designs on products and desired ideal.

Furthermore, this scaling procedure enables the panelist to point out those test variations that reflect the ideal level of key attributes. This capability enhances communications and shortens the development cycle.

PSYCHOPHYSICAL SCREENING EXPERIMENTS

Let us consider a typical screening study. The aim is to determine which design variations are most promising, which are least acceptable, and which correlate with high and low acceptability ratings.

In the screening study there were six different packages for a processed sausage product. The study determined (1) which package was most attractive for sausage and (2) what sensory impression about sausage each package conveyed.

Panelists evaluated each of the six packages according to the following procedures:

Table 5 Comparison of Three Beer Can Designs and the
Consumer Rated "Ideal"*

Attribute	Design Prototype			Ideal Desired
	A	B	C	
Liking/Disliking	72	118	68	Not Asked
Purchase Intent	62	84	50	Not Asked
Unique	49	78	60	128
Colorful	90	60	48	94
Memorable	40	66	45	115
Size of Lettering	80	42	78	52
Richness of Colors	95	59	45	66
"A Great Tasting Beer"	54	80	71	145
Rich Taste	50	89	65	138

Calibration

159 = Extreme
122 = Very much
 75 = Moderate
 36 = Slight
  0 = None at all

*Data from 120 male respondents, beer drinkers (ages 18–49)
in New York, Philadelphia, and Baltimore.

- Panelists were 64 housewives (18–49) who regularly buy sausages.
- Panelists participated in a 3 hour test session.
- Panelists first rated four different sausage products, representing part of the wide spectrum of product variation. Panelists profiled each of the four sausage products on key sensory, image, and hedonic characteristics.
- Panelists then scaled all six sausage packages in random order. They rated each package separately (monadically). They rated the packages in terms of sensory characteristics that panelists *thought* the sausage would possess, based on visual inspection of the package. In a sense, this design is a *projective test* of package communication, using psychophysical scaling procedures. Since the only test stimuli available to the panelists in package profiling are the packages, we obtain a sensitive measure of what the package communicates about the product and can anchor those communications in terms of actual sausage products.
- Panelists then calibrated their magnitude estimation ratings.

Table 6 shows the study results. The key results are:

- Package U conveys the impression of the most acceptable sausage. There is a smaller cluster of acceptable packages, specifically packages V, T, and R.
- Consumers agree most as to how well the package conveys "good flavor." (There is low variability around the average ratings for good flavor.) Consumers disagree as to which package best communicates "healthful/nutritious" (none of the packages addressed the concept of nutrition directly). There is substantial interpanelist variability on the attribute of healthful/nutritious. Even though we calculate average ratings,

Table 6   Product and Package Profiles—Sausage

	Product				Package					
	A	B	C	D	Q	R	S	T	U	V
Liking	38.5*	22.6	40.3	67.4	50.7	59.2	48.1	60.4	78.5	65.5
	(3.9)**	(5.2)	(6.6)	(4.1)	(4.8)	(5.1)	(4.2)	(5.3)	(5.1)	(4.9)
Good flavor	42.1	40.3	54.1	79.2	50.2	49.3	60.2	84.3	95.2	49.8
	(2.8)	(2.6)	(2.9)	(3.1)	(3.1)	(4.2)	(3.8)	(4.3)	(4.5)	(3.7)
Healthful/ nutritious	30.7	26.8	37.5	40.5	40.3	29.2	30.8	50.5	45.3	65.1
	(4.9)	(5.4)	(6.4)	(5.2)	(8.3)	(7.1)	(8.0)	(6.9)	(7.3)	(6.2)
Appropriate for children	42.4	48.3	60.2	70.3	48.3	50.9	35.2	60.4	61.6	52.5
	(3.9)	(4.1)	(4.2)	(6.0)	(4.3)	(5.0)	(3.9)	(5.8)	(5.5)	(4.9)
Appropriate for adults	50.5	59.3	75.4	68.2	57.2	40.8	49.5	40.1	60.0	48.3
	(5.0)	(4.2)	(4.5)	(5.2)	(4.9)	(4.6)	(4.7)	(4.0)	(5.9)	(4.5)
Spiciness	48.3	59.8	60.5	39.6	35.6	50.5	49.2	60.3	48.4	50.8
	(3.1)	(3.5)	(3.1)	(2.4)	(3.2)	(4.0)	(3.5)	(5.1)	(4.6)	(4.8)
Greasiness	60.2	48.3	38.2	50.5	60.3	29.2	50.1	30.8	52.5	71.4
	(2.3)	(1.9)	(2.0)	(1.8)	(4.1)	(2.8)	(5.7)	(3.4)	(2.9)	(4.3)
Purchase intent	48.4	60.3	81.8	92.3	75.8	80.2	64.8	78.3	88.9	70.1
	(5.0)	(4.1)	(4.8)	(5.1)	(6.2)	(6.0)	(6.4)	(5.3)	(7.1)	(7.3)

*First number is the average.
**Second number (in parenthesis) is the standard error.

Calibrated Scale

Extremely	174
Very Much	138
Moderate	92
Slight	39
None	0

the variation around the average is so high that we conclude that these packages did not convey unambiguous visual signals about nutrition.

- We can correlate the attributes of the package perception with each other and discover which characteristics most highly correlate with liking. Here, the attributes are appropriate for children (correlation = .80), good flavor (correlation = .63), and healthful/nutritious (correlation = .47).

Advantages and Disadvantages of Screening Studies

Screening studies provide a valuable opportunity to assess consumer reactions to many design options in an efficient, scientific manner. After evaluating many alternatives and obtaining a range of ratings, the designer or manufacturer learns which package elements improve overall acceptability and purchase interest and which elements diminish them. One ever-present problem with screening studies, however, is the limited range of product variation. One cannot adequately determine the contribution of different design factors to product/package perceptions, since in a screening study interest is not focused on systematic variation stimulus with a wide range of each design factor. We can obtain diagnostics as to which factors have an impact on acceptance but it is hard to use fully the screening information to design subsequent packages adequately. The information provides hints, but not solid actionable direction. That direction is left to systematic product variation and experimental design.

THE EFFECTS OF PACKAGE DESIGN ON PERCEPTION

Packaging often modifies product perception. Although we usually think of our sensory systems as objective registrators of stimuli, cognitive effects exerted by packages modify perceptions. They can change perceptions of foods, cosmetics, and fragrances. Furthermore, the changes can be measured in a simple experiment during which the panelist evaluates products on two occasions. On the first occasion, the products are evaluated 'blind,' that is, unbranded. The only cue is the product itself. On the second occasion, the products are evaluated in their packages. All other study factors are held constant. The dif-ference between the ratings blind versus branded (or packaged) reflects the net package effect.

Among the important results from studies such as these are:

- Product performance (indexed by liking or purchase interest ratings) either increases or decreases. Packaging exerts negative as well as positive effects on product perceptions.

- Specific characteristics may fluctuate as well. These characteristics are often image attributes, but they may also be such "physical" characteristics as sweetness and flavor intensity, which we often consider not subject to disturbing cognitive influences.

Table 7 shows the results of studies on chocolate bars and perfumes, evaluated by various panelists. All products were purchased in markets and evaluated on

Table 7   Effect of Branding (Packaging) on Product Perceptions

A.   PERFUMES

Brand	Liking	Intensity	Uniqueness	Sensuousness
Pavlova (blind)	96	90	85	91
Pavlova (branded)	60	98	76	55
Courreges (blind)	31	95	89	49
Courreges (branded)	57	81	88	68
Vivre (blind)	69	78	69	47
Vivre (branded)	68	67	72	68
YSL (blind)	18	86	81	49
YSL (branded)	15	91	70	51
De Lubin (blind)	25	90	78	47
De Lubin (branded)	44	88	80	47

B.   CHOCOLATE BARS

Branding	Liking	Sweetness	Crispness	True Milk Taste
Nestlé (blind)	32	65	35	55
Nestlé (branded)	56	77	45	69
Poulain (blind)	26	62	30	54
Poulain (branded)	27	55	36	49
Beaumont (blind)	70	80	31	66
Beaumont (branded)	50	66	40	69

two occasions by the same groups of panelists. In all cases panelists profiled the unbranded products first.

"DESIGNED" EXPERIMENTS

Designed experiments illustrate how variations in design factors impact on perceptions and overall acceptability. There are two different types of designed experiments. We first consider the type I (continuous variable) experiment, which uses continuously variable design factors. For instance, when designing a package for soap, the design factors could be package area, the size of the graphics, and so on. The color, the type of the graphics, and what the graphics say are constants. The design factors (area dimensions, relative size of graphics) are varied in a continuous manner, within feasible limits. Type I experiments determine:

• How the design factors and their interactions generate specific sensory impressions and contribute to overall acceptability.

• What combinations of design factors, at what levels optimize a criterion consumer reaction (e.g., acceptability ratings), subject to specific design constraints.

• What combination of design factors, at what levels, maximize the criterion reaction, subject to externally imposed cost constraints and so on.

We considered a new package in depth. Package area, width of stripes on the package, and relative size of graphics were to be varied. Package color and the specific graphics were previously selected and were constant.

PACKAGE OPTIMIZATION/"FINE TUNING"

The three design factors were varied according to a 'central composite experimental design.' This specification design investigates a wide range of design variations using a small but efficient set of design elements. Table 8 shows the central composite design, as well as the actual design variations. Panelists eval-

Table 8   Experimental Design of Soap Package (Central Composite)

	Package Area	Schematic Graphic Size	Stripe Size	Actual Design Levels Package Area	Actual Design Levels Graphic Size	Actual Design Levels Stripe Size
1	H	H	H	10.63	3.0	1.00
2	H	H	L	10.63	3.0	.25
3	H	L	H	10.63	2.0	1.00
4	H	L	L	10.63	2.0	.25
5	L	H	H	5.25	3.0	1.00
6	L	H	L	5.25	3.0	.25
7	L	L	H	5.25	2.0	1.00
8	L	L	L	5.25	2.0	.25
9	H	M	M	10.63	2.5	.75
10	L	M	M	5.25	2.5	.75
11	M	H	M	8.44	3.0	.75
12	M	L	M	8.44	2.0	.75
13	M	M	H	8.44	2.5	1.00
14	M	M	L	8.44	2.5	.25
15	M	M	M	8.44	2.5	.75

uated the 15 soap packages, rating each package on the three types of attributes: sensory, performance, and image.

Table 9 (p. 522) shows the average ratings using magnitude estimation scaling. Table 9 becomes the data base.

We now compare alternative design options on different characteristics. We can determine which designs are acceptable and which are not, as well as quantitatively ascertain what each design communicates. Furthermore, because the designer systematically varied the three design factors, one can interrelate the design factors with reactions to develop a model of soap package perception. The model answers several design issues:

- The likely consumer reaction to specific design modifications. If we develop equations interrelating these design factors to consumer perceptions, and if we "plug in" to the equation the new design values for area, stripe width, and graphics size, then we can *predict* the most likely perceptual profile of the new design.
- The likely hedonic/purchase interest rating for a modified design, comprising different levels of the design factors.
- Within the limits of physical variation that we tested, the specific combination of design factors (area, stripe size, and graphic size) generating the highest acceptance rating.
- If area, stripe size, and graphics size each entail specific costs, then the specific combination of design factors when maximizes acceptance (measured by liking or by purchase interest ratings), while at the same time maintaining product total cost within bounds.
- If one can predict consumer perceptions of changed design factors, then if one specifies a desired sensory profile of perceptions, what combi-

nation of design factors generates that desired perceptual profile?

To understand optimization, first consider how design factors relate to sensory, image, and performance ratings. The easiest way is to relate the three design factors to perceptions by means of simple *linear equations*. Linear equations are expressed as:

$$\text{Perceptual Attribute Rating} = k_0 \\ + k_1 \text{ (area)} + k_2 \text{ (graphics size)} \\ + k_3 \text{ (stripe size)}$$

The linear equation summarizes the data in a simple quantitative way. The linear equation allows no interaction among the three package factors. If k_1 is greater than 1, then doubling area, for instance, will *more* than double the attribute rating. Conversely, if k_2 is less than 1.0, then doubling area will less than double the attribute rating. A linear equation summarizes the interrelation between the independently varied design factors and the consumer reactions. Keep in mind that:

- The equation is an abstraction that summarizes the data and provides a way to eliminate sensory reactions to design modifications.
- The consumer rating data are sometimes very well fit by the equation but sometimes not so well fit (even though the equation is the *best* we can do). [Goodness of fit of an equation to empirical data is indexed by the multiple correlation (R). The R^2 value $(R \times R)$ times 100% measures the proportion of variability in the rating data accounted for by variation in design factors. Ideally, $R^2 \times 100$ should be 100, meaning 100% of the variation is predicted by the equation.]

Table 10 shows the equations relating each attribute to the three design factors. As we might expect, the sensory ratings

Table 9 Average Panel Rating for 15 Soap Package Variations: Data Base

Soap Package			Rating							
Package Area	Graphic Size	Stripe Size	Percvd Length	Percvd Width	L/D Package	Percvd Area	Graphic Size	App. As Beauty	App. As Cleaning	App. As Bathroom
10.63	3.0	1.00	97.0	87.0	−40.0	108.0	77.0	17.0	73.0	24.0
10.63	3.0	.25	73.0	61.0	−21.0	83.0	76.0	19.0	50.0	35.0
10.63	2.0	1.00	83.0	66.0	−37.0	96.0	28.0	3.0	57.0	30.0
10.63	2.0	.25	98.0	80.0	−81.0	118.0	37.0	25.0	55.0	34.0
5.25	3.0	1.00	35.0	31.0	42.0	49.0	98.0	42.0	55.0	51.0
5.25	3.0	.25	33.0	25.0	48.0	55.0	71.0	87.0	24.0	85.0
5.25	2.0	1.00	28.0	27.0	30.0	53.0	44.0	33.0	22.0	47.0
5.25	2.0	.25	25.0	22.0	10.0	36.0	24.0	42.0	17.0	45.0
10.63	2.5	.75	102.0	77.0	−11.0	110.0	56.0	31.0	78.0	54.0
5.25	2.5	.75	39.0	28.0	46.0	46.0	55.0	61.0	23.0	68.0
8.44	3.0	.75	59.0	51.0	52.0	71.0	66.0	55.0	55.0	65.0
8.44	2.0	.75	44.0	37.0	29.0	68.0	58.0	44.0	54.0	46.0
8.44	2.5	1.00	51.0	42.0	46.0	66.0	53.0	44.0	40.0	53.0
8.44	2.5	.25	99.0	91.0	37.0	106.0	51.0	27.0	54.0	94.0
8.44	2.5	.75	54.0	45.0	35.0	65.0	50.0	47.0	78.0	74.0

Soap Package			Rating						
Package Area	Graphic Size	Stripe Size	App. for Dishwashing	App. for Hotel	L/D Length	L/D Width	L/D Area	L/D Size Graphics	Purchase Intent
10.63	3.0	1.00	49.0	11.0	17.0	23.0	− 6.0	−13.0	64.0
10.63	3.0	.25	43.0	21.0	−27.0	−27.0	−25.0	3.0	21.0
10.63	2.0	1.00	39.0	12.0	−18.0	−22.0	−39.0	−63.0	15.0
10.63	2.0	.25	51.0	8.0	− 3.0	− 4.0	−36.0	−80.0	15.0
5.25	3.0	1.00	59.0	52.0	32.0	31.0	37.0	50.0	60.0
5.25	3.0	.25	43.0	84.0	42.0	33.0	42.0	66.0	93.0
5.25	2.0	1.00	41.0	55.0	30.0	32.0	30.0	29.0	37.0
5.25	2.0	.25	28.0	57.0	4.0	14.0	− 3.0	5.0	36.0
10.63	2.5	.75	44.0	15.0	17.0	22.0	− 4.0	.0	54.0
5.25	2.5	.75	26.0	74.0	40.0	36.0	42.0	78.0	86.0
8.44	3.0	.75	70.0	32.0	81.0	78.0	84.0	71.0	90.0
8.44	2.0	.75	62.0	40.0	74.0	71.0	43.0	−13.0	67.0
8.44	2.5	1.00	55.0	31.0	65.0	63.0	69.0	59.0	66.0
8.44	2.5	.25	95.0	22.0	44.0	44.0	28.0	2.0	49.0
8.44	2.5	.75	65.0	37.0	66.0	59.0	54.0	36.0	64.0

Package Area	Graphic Size	Stripe Size	Value If Cost (Cents/Bar) =			
			19¢/Bar	25¢/Bar	35¢/Bar	45¢/Bar
10.63	3.0	1.00	107.0	71.0	51.0	31.0
10.63	3.0	.25	66.0	56.0	41.0	29.0
10.63	2.0	1.00	73.0	55.0	38.0	24.0
10.63	2.0	.25	69.0	45.0	35.0	18.0
5.25	3.0	1.00	49.0	44.0	22.0	13.0
5.25	3.0	.25	50.0	24.0	18.0	13.0
5.25	2.0	1.00	52.0	38.0	27.0	11.0
5.25	2.0	.25	49.0	28.0	22.0	9.0
10.63	2.5	.75	104.0	70.0	57.0	26.0
5.25	2.5	.75	61.0	29.0	19.0	12.0

Table 9 Continued

Package Area	Graphic Size	Stripe Size	Value If Cost (Cents/Bar) =			
			19¢/Bar	25¢/Bar	35¢/Bar	45¢/Bar
8.44	3.0	.75	89.0	76.0	47.0	25.0
8.44	2.0	.75	90.0	42.0	27.0	20.0
8.44	2.5	1.00	87.0	70.0	54.0	21.0
8.44	2.5	.25	68.0	53.0	35.0	20.0
8.44	2.5	.75	62.0	56.0	30.0	26.0

(which are based on physical variations) are described by equations that show relatively high R^2 values. Liking/disliking and purchase intent attributes are less well fit by the linear equation and show lower R^2 values. This lack of fit of the linear equation to overall liking and purchase intent ratings results from either poor data (always a possibility) or the more plausible possibility that liking/disliking and purchase intent are related to design factors in a *nonlinear* fashion. Too large and too small areas, size of graphics, and stripes could be equally unacceptable. Acceptability could be high only at the mid-range of size where the design factors are "just right." [A parallel situation comes from the case of

Table 10 Linear (or Straight Line) Equations that Relate Each Sensory Attribute to a Weighted Combination of Area, Graphic Size, Stripe Size

Attribute	Intercept	Package Area	Graphic Size	Stripe Size	Multiple R^2
Perceived length	−28.84	10.78	3.80	−10.10	0.77
Perceived width	−25.81	8.82	4.60	− 8.79	0.73
Liking of package	35.15	−12.54	26.00	20.72	0.59
Perceived area	1.36	10.15	− 1.00	− 8.69	0.77
Perceived graphic size	−43.85	− 0.70	39.40	10.90	0.77
Appropriate for beauty soap	58.76	− 6.07	14.60	−11.42	0.59
Appropriate for cleaning soap	−40.36	6.55	10.40	15.40	0.67
Appropriate for bathroom soap	69.15	− 3.92	11.60	−19.09	0.35
Appropriate for dishwashing soap	18.55	1.69	8.60	− 3.58	0.10
Appropriate for hotels	103.98	− 9.51	5.60	− 6.20	0.88
Liking of length	22.20	− 4.79	11.60	27.82	0.20
Liking of width	25.52	− 4.57	9.40	27.35	0.20
Liking of area	− 1.22	− 8.34	27.40	32.08	0.43
Liking of graphics size	−42.03	−13.50	59.80	25.91	0.76
Purchase intent	4.63	− 4.84	31.60	15.08	0.49
Value @ 0.19/bar	− 5.01	6.04	5.60	20.68	0.64
Value @ 0.25/bar	−36.57	5.21	12.60	19.99	0.77
Value @ 0.35/bar	−22.22	4.30	6.00	10.78	0.71
Value @ 0.45/bar	−18.49	2.66	5.80	3.51	0.90

coffee and sugar. If an individual likes coffee with sugar, then with increasing amounts of sugar, perceived sweetness will increase, generally according to a straight line. Overall *liking* of the coffee will probably not conform to the straight line. Rather, with increasing sugar, liking first increases, reaches its highest level (or the bliss point), and then diminishes. A century ago, the German psychologist Wilhelm Wundt suggested that an inverted U-shaped curve relates sensory (or physical) intensity to liking for most, if not all, sensory perceptions. Some attributes are more curvilinearly related to sensory level than others.[4] Figure 2 shows this general rule.]

USING OPTIMIZATION IN PRACTICAL DESIGN WORK—SOME QUERIES AND ANSWERS

Query 1: What combination of design factors maximizes acceptability? The design factors must lie within the limits that were tested. One can, however, interpolate to find intermediate values of the three design factors not directly tested but which should be optimally acceptable according to the model.

Answer to Query 1: We first relate overall liking (or purchase interest) to the three design factors. We must take into account nonlinearities and interactions of design factors versus purchase intent or liking (e.g., liking may increase with graphics size, maximize, and then diminish; liking may be related to the interaction of graphics size and package area, etc.). Table 11 compares the linear and the nonlinear models (i.e., quadratic model) that relate design factors to liking. Liking ratings are described better by a nonlinear equation than by a linear equation. [Not only is the multiple R value higher for the nonlinear equation, but also other relevant statistics are higher (e.g., the F ratio). This difference in goodness-of-fit values means that the curvilinear model of liking versus design factors containing additional nonlinear and interaction terms better summarizes the interrelation between design factors and liking.]

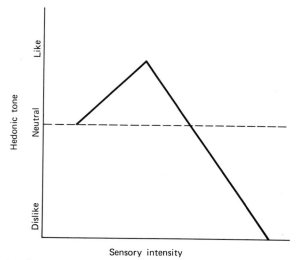

Figure 2 The relation between sensory intensity (abscissa) and liking (ordinate). According to Wundt, pleasantness or liking first increases with sensory intensity, reaches a bliss point, and then decreases with further increases in sensory intensity.

Table 11 How Liking Ratings of Soap Package Vary with Design Factors

A. Linear Model (Straight Line or Plane)

Liking $= 35.15 - 12.54$ (Area) $+ 26.00$ (Graphics size) $+ 20.7$
(Stripe Size)

$$R^2 = (.77)^2 = 0.60$$

B. Curvilinear Model (With Interactions) ($R^2 = 0.99$)

Liking $= -626.7 + 74.5$ (Area) $+ 266.3$ (Graphics size) $+ 221.2$ (Stripe size)
$- 5.7$ (Area)$^2 + 0.7$ (Area) (Graphics size) $+ 1.91$ (Area) (Stripe Size)
$- 41.5$ (Graphics size)$^2 - 58.8$ (Graphics size) (Area) $- 61.5$ (Area)^2

	Sensory Attributes		
Attribute	Unconstrained Optimum	Cost Less Than 0.68	Cost Less Than 0.60
Perceived area length	50.6	43.2	35.7
Perceived width	43.4	37.3	31.2
Perceived area	62.8	55.8	48.7
Perceived graphics size	75.7	75.1	75.5
Appropriate for beauty soap	56.3	60.6	65.0
Appropriate for cleaning soap	42.3	37.5	32.7
Appropriate for bathroom soap	68.2	71.5	74.5
Appropriate for soap dish	54.1	52.9	51.8
Appropriate for hotels	53.6	60.4	67.2
Liking	67.7	64.9	56.4
Cost	0.76	0.68	0.60
	Design Levels		
Package area	6.75	6.05	5.34
Graphics size	3.00	3.00	3.00
Stripes size	0.470	0.459	0.045

Thus we should use the better fitting nonlinear equation shown in Table 11. Optimization procedures ascertain in a straightforward way the design factor values at which liking reaches its highest level. The surface described in Table 11 is complicated. Nonetheless, straightforward mathematical procedures exist that tell us where, on this surface, the highest liking or purchase interest level can be reached, as well as what combination of area, stripe size, and graphics size generates that highest liking. The answer is 6.75 for area, 3.00 for graphics size, and 0.470 for stripe size. Liking is estimated to be 67.7.

Query 2: What other sensory perceptions are generated by this optimal set of design factors?

Answer to Query 2: Since Table 10 relates sensory/image/hedonic impressions to design factors, we can use these equations to estimate (in a simple linear manner) the likely profile of design perceptions. Table 11 illustrates the *estimated* sensory, image, and performance perceptions of the optimal design, based on the linear equations which relate design factors to perceptions.

Query 3: What is the optimal configuration of physical variables, within specific sensory or cost constraints?

Answer to Query 3: Query 3 deals with the concept of constrained optimization. Total cost can be represented as a linear combination of area, stripe size, and graphics size. Table 10 shows the linear equation for cost. Virtually 99% of the cost is the result of package area. We can determine the optimal combination of design factors for two different costs, as shown in Table 11. Table 11 shows the combination of design factors that provide the highest liking rating and yet deliver a package whose cost lies within specific limits.

Query 4: If perceptions are relatable to design factors, can we then prespecify a profile of package perception and ascertain the combination of design factors which produce that profile? Suppose we wish to match a specific consumer perception of area, graphic size, and so on. What combination of design factors should we use? We are constrained to use our three design factors and to make sure that each design factor lies within the limits that were tested.

Answer to Query 4: To match a desired profile by modifying design factors requires a different type of optimization. Previously we maximized the nonlinear liking curve. Here we are attempting to match a profile. We try to maximize the closeness of fit (or minimize the difference) between a prespecified perceptual profile and a profile generated by three design factors. To ascertain the appropriate levels of our three design factors, we use the procedure of "goal programming." We 'reverse' the linear equations that relate design factors to perceptions (these equations appear in Table 10). Were we given levels of the three design factors, we could easily predict the sensory profile using Table 10. The inverse problem (which requires additional mathematical treatment to insure feasible answers) allows us to set the perceptual profile levels and ascertain the values of the three design factors that will generate the desired perceptual profile. (See Table 12.)

The 'desired' perceptual profile might emerge from several sources, including:

- The consumer perceptual profile of a competitive package, using one's own materials and graphics.
- A profile, based on one's own 'ideal' or based on the profile of a product/concept. The consumer panelist first profiles actual products and packages before profiling the ideal or a concept. This procedure anchors the ideal or concept profile in reality.

The utilitarian benefit of this optimization procedure is that the designer or manufacturer narrows down quickly to the optimal package design in those situations wherein the specific design elements have been chosen. Fine tuning can be then accomplished efficiently, holding in mind business constraints and key consumer feedback. Optimization does not replace the creative designer, however. Optimization is only a tool, as good as the user and as creative as the problem posed to it.

Table 12 Development of a Set of Design Factors to Match a Desired Profile

Design Factors	Design Factor Limits		Design Factor Levels Should Be
	Lower Bound	Upper Bound	
Package area	5.25	10.6	6.4
Graphics size	2.00	3.0	2.9
Size of stripes	0.25	1.0	1.0

Sensory Attributes	Desired Profile	What the Best Matching Profile Will Be
Perceived length	66.0	41.4
Perceived width	35.0	35.5
Perceived area	55.0	55.0
Perceived graphics size	64.0	64.0
Appropriate as cleaning soap	70.0	47.4
Appropriate for hotels	60.0	52.9

OPTIMIZING THE SELECTION OF DISCRETE OPTIONS IN PACKAGES, USING CONSUMERS

There are many design problems where the design options include different colors, different shapes, different sizes, and so on. The design factors do not vary in a continuous manner but represent many alternatives (Type II). How can one select the optimal combination of design options to maximize consumer acceptance, knowing that with as few as three design factors (e.g., color, shape, dispenser, each having three options), there are already 27 alternatives? Is there a procedure for efficient screening of alternatives?

Let us consider the case of a woman's cologne bottle. There were four color options for graphics (pink, red, orange, magenta), two types of dispensers (A and B), and three proposed bottle shapes (circular, narrow oval, wide oval). The stimuli were actual cologne bottles, containing a fragrance about to be launched. The requirement was to

select the best combination. Twenty-nine consumers tested 11 of the 24 possible combinations. (To test all combinations would have been too costly and time consuming.)

The panelists tested each bottle and cologne combination, profiling their perceptions of the entire package/fragrance combination in terms of specific characteristics:

- Overall liking/disliking of the bottle.
- Appropriate for a young woman.
- Appropriate for a mature woman.
- Appropriate for a night fragrance.
- Appropriate for a day fragrance.
- Elegant product.
- Functional appearing product.
- Perceived expensiveness of the product.

These characteristics tap a wide variety of image characteristics for a fragrance bottle (although they do not address specific sensory characteristics of color, shape, and so on; those sensory characteristics are integrated into the image perceptions).

Analytical Strategy for Selecting the Best Combination

The strategy to solve this problem differs from the previous soap-package problem. Previously, the design factors were varied continuously. Here we must select the optimally acceptable combination of the options as presented without testing all of them.

The first analytical step is to relate each attribute as rated by consumers to the *presence/absence* of each design option. We develop these interrelations by an equation that relates each design option (graphics, color, shape of bottle, dispenser, all binary coded) to the consumer attribute. The design package options are transformed into a binary code: '1 = present' or '0 = absent.' (See Table 13B.)

Table 13A shows the ratings for the eleven design options as originally coded. The left side of Table 13A shows a code of the design options. The right shows average magnitude estimates from the 29 consumer respondents. Just below

Table 13 Transformation of Design Package Options into a Binary Code

A. COLOGNE BOTTLE DESIGN OPTIONS AND PROFILE OF PERCEPTIONS

Design Options			Bottle Perceptual Profiles							
Graphics Color	Dispenser Type	Bottle Shape	Overall L/D	Approp Yng Wm	Approp Mat Wm	Approp Nights	Approp Day	Ele- gant	Functl Applng	Expen- sive
2	0	2	− 8.6	53.0	57.5	70.2	64.6	65.0	84.7	64.0
2	1	2	66.8	92.5	97.0	94.2	85.7	73.1	85.2	88.0
3	0	1	50.7	93.6	102.6	78.7	55.0	107.3	95.8	84.1
1	1	0	− 30.1	66.3	67.9	64.7	29.7	90.2	91.8	83.0
0	1	2	− 50.0	28.1	30.5	43.5	58.3	27.4	53.0	51.7
1	1	2	57.4	87.9	80.2	91.5	99.5	70.9	89.4	71.5
0	1	0	− 81.1	43.1	38.8	35.0	23.9	62.3	64.9	55.4
3	0	1	− 21.9	86.3	86.0	53.4	25.9	107.0	94.7	108.0
1	0	0	− 41.1	56.0	45.0	29.0	13.8	79.0	95.9	84.7
2	0	1	− 12.9	78.5	77.5	64.5	21.5	105.8	101.9	74.7
2	0	2	72.2	94.9	78.7	90.9	75.7	65.7	76.7	62.1

0 = Pink   "A" Rectangular
1 = Red   "B" Small oval
2 = Orange   Large oval
3 = Magenta

B. BINARY CODING OF THE BOTTLES
(Continued)

Pink Graphics	Red Graphics	Orange Graphics	Magenta Graphics	Dispenser Type A	Dispenser Type B	Rectangular	Small Oval	Large Oval
0	0	1	0	1	0	0	0	1
0	0	1	0	0	1	0	0	1
0	0	0	1	1	0	0	1	0
0	1	0	0	0	1	1	0	0
1	0	0	0	0	1	0	0	1
0	1	0	0	0	1	0	0	1
1	0	0	0	0	1	1	0	0
0	0	0	1	1	0	0	1	0
0	1	0	0	1	0	1	0	0
0	0	1	0	1	0	0	1	0
0	0	1	0	1	0	0	0	1

Table 13 are the same data, but this time recoded to 'binary.' (Table 13B.) Each design option is either a '1' to show its presence, or '0' to show its absence. There were four colors, two dispenser types and three shapes. Thus originally there were three different design factors (color, dispenser, shape). Converting these to binary now produces 4 + 2 + 3 = 9 different binary design options. Each design option becomes an independent option that can be either present or absent in a product. (When we recode these options in binary form, then if the bottle has pink graphics, it cannot have orange, red, or magenta graphics.) For example, consider the first product in Table 13A. It is coded 202. This means graphics type 2 (orange) dispenser type 0 (dispenser "A") and shape 2 (large oval).

Making An Equation

We can always develop an equation to relate the presence/absence of each of the nine design options to a consumer rated characteristic. (This is straightforward statistics, which can be done with currently available "canned programs" well known to every practicing statistician.) The form of this equation is interesting. (See Tables 14A and B.) It differs from the equation that we used previously to relate design factors of area, stripes, and graphics size to soap package perceptions. For our current cologne bottle problem we use an equation in which the predictors are '1' (to denote that the design option is present) or '0' to denote that the option is absent. The equation for overall liking looks like this:

overall liking = 0.0 (pink) + 80.7 (red) + 127.7 (magenta) + 0.0 ("A" dispenser) + 30.4 ("B" dispenser) + 0.0 (rectangle shape) + 11.5 (small oval shape) + 57.5 (large oval shape) − 124.9

Table 14 Cologne Bottle Data—Summary of Equations Fit to Consumer Reaction Versus Presence/Absence of Each Design Option

A. OVERALL LIKING/DISLIKING

Multiple R squared	=	.75
Multiple R	=	.87
Intercept constant	=	−124.8
F for analysis of variance	=	2.18
N.D.F. numerator	=	6.
N.D.F. denominator	=	4.

Design Option	Raw Weights
Red	80.7
Orange	100.4
Magenta	127.7
B Dispenser	38.4
Small oval shape	11.5
Large oval shape	57.7

B. APPROPRIATE FOR YOUNG WOMAN

Multiple R squared	=	.767
Multiple R	=	.87
Intercept constant	=	14.9
F for analysis of variance	=	2.189
N.D.F. numerator	=	6.
N.D.F. denominator	=	4.

Design Option	Raw Weights
Red	41.3
Orange	55.4
Magenta	66.9
B dispenser	18.9
Small oval shape	8.1
Large oval shape	3.4

C. APPROPRIATE FOR MATURE WOMAN

Multiple R square	=	.92
Multiple R	=	.95
Intercept constant	=	5.20
F for analysis of variance	=	7.29
N.D.F. numerator	=	6.
N.D.F. denominator	=	4.

Design Option	Raw Weights
Red	39.5
Orange	61.1
Magenta	77.9
B dispenser	28.5
Small oval shape	11.1
Large oval shape	1.8

Table 14   Continued

D.   APPROPRIATE FOR NIGHT

Multiple R squared	=	.84
Multiple R	=	.91
Intercept constant	=	2.43
F for analysis of variance	=	3.50
N.D.F. numerator	=	6.
N.D.F. denominator	=	4.

Design Option	Raw Weights
Red	34.7
Orange	52.2
Magenta	53.8
B dispenser	25.9
Small oval shape	9.7
Large oval shape	21.7

G.   FUNCTIONAL APPEARANCE

Multiple R squared	=	.96
Multiple R	=	.98
Intercept constant	=	61.37
F for analysis of variance	=	20.94
N.D.F. numerator	=	6.
N.D.F. denominator	=	4.

Design Option	Raw Weights
Red	32.6
Orange	28.3
Magenta	21.7
B dispenser	1.6
Small oval shape	12.1
Large oval shape	− 8.1

E.   APPROPRIATE FOR DAY

Multiple R squared	=	.90
Multiple R	=	.94
Intercept constant	=	−5.45
F for analysis of variance	=	6.01
N.D.F. numerator	=	6.
N.D.F. denominator	=	4.

Design Option	Raw Weights
Red	22.0
Orange	20.6
Magenta	40.0
B dispenser	19.7
Small oval shape	6.2
Large oval shape	53.5

H.   EXPENSIVE

Multiple R squared	=	.79
Multiple R	=	.89
Intercept constant	=	47.79
F for analysis of variance	=	2.61
N.D.F. numerator	=	6.
N.D.F. denominator	=	4.

Design Option	Raw Weights
Red	28.1
Orange	31.6
Magenta	53.0
B dispenser	11.7
Small oval shape	−4.7
Large oval shape	− 12.0

F.   ELEGANT

Multiple R squared	=	.98
Multiple R	=	.99
Intercept constant	=	46.69
F for analysis of variance	=	49.24
N.D.F. numerator	=	6.
N.D.F. denominator	=	4.

Design Option	Raw Weights
Red	34.5
Orange	43.4
Magenta	44.8
B dispenser	11.1
Small oval shape	15.6
Large oval shape	−25.9

In this equation, and by definition, one design option of each design factor was always given a coefficient or multiplier of 0. Relative to that base line 0, we can relate liking to the presence/absence of the other options. For instance, for pink graphics, rectangular bottle with type B dispenser (each of whose coefficients are 0), liking equals the value of the intercept, or −124.9. By substituting 1's for specific design options, we can estimate the likely consumer reaction. For instance, if we want to estimate consumer liking of a magenta, B dispenser, small oval shape bottle, the equation for liking is:

Liking = 127.7 (Magenta Value) + 30.4
("B" Dispenser Value) +
+ 11.5 (Small Oval Shape Value)
− 124.9 (Intercept Value)
= 44.7

An equation is useful because it determines:

- The specific quantitative contribution of each design option to overall liking (or to other perceptual characteristics, albeit with a different equation).
- The estimated liking for tested as well as untested design combinations. The equation summarizes the data *and* allows the designer to estimate the consumer reaction to untested designs. The equation is an integrated *model* describing consumer reactions to package designs. Tables 14A–H show the actual equations fit to these cologne bottle data.

Measuring the Validity of the Package Equation

Equations are statistical abstractions which may or may not describe the data. Previously, in the case of the soap package, we developed linear equations to relate sensory reactions to levels of design factors. We tested how good the equations were, using the multiple correlation (R). Here we can also determine how good each equation is by looking at the multiple correlation.

Partwise "Utilities" of Package Options

In this cologne bottle study we developed an *additive linear* equation to predict liking ratings. (See Table 13A.) The equation estimates the liking rating provided we know which design options are present or absent. We can conceive of the coefficients or 'raw weights' of the equations (in Table 14A) as partwise contributions of each design to liking. Specifically:

Pink color (coefficient = 0) contributes '0' utility or '0' partial liking to a consumer's overall reaction to the bottle. Pink neither adds nor subtracts. The '0' for pink is the base line value against which other options will be compared. (In these studies one option of each package variable must always obtain base line value of '0'. The other utilities or partwise contributions of specific options are compared to that '0' base line.)

Red color (coefficient = 80.7), generates 80.7 units of liking and is a more acceptable package option than pink. Orange color (coefficient = 100.4) generates 100.4 units of liking. Magenta color (coefficient = 127.7), generates 127.7 units of liking.

Furthermore, not all coefficients, or partwise utilities are positive or '0'. Some are negative. Design options with negative partwise utilities diminish the specific characteristics. Specifically, in terms of appearing "functional," (Table 14G) the rectangular shape is by definition given a utility of '0.' The small oval shape contributes 12.1 units to the perceived functionality of appearance, which means that the small oval shape gives the bottle an appearance of being more functional. A large oval shape generates a −8.1 or a negative contribution to the appearance of functionality. (This information on functional appearance may be useful in choosing the correct shape of bottle to insure the perception of a specific image.)

Optimizing Combinations of Design Options

By means of the additive model in which design options separately contribute to "liking," (Table 14A) one can optimize the selection of color, bottle shape, and dispenser. One selects that combination of color, shape, and dispenser to maximize a prespecified criterion (e.g., overall liking or perceived expensiveness).

For example, consider the problem of maximizing liking. To maximize liking of the bottle one chooses those design options that have the greatest coefficients (or partwise contributions of utilities). (See Table 14A.)

Color: Magenta (utility or partwise contribution = 127.7)
Shape: Large Oval bottle (utility = 57.7)
Dispenser: "B" (utility = 30.4)

To estimate liking we add together these partwise contributions (sum = 215.9), and add in again the intercept of the equation (−124.0), to estimate the total liking as +91.

As another example, let us maximize the perception of "expensiveness" of the bottle (Table 14H).

Color: Magenta (utility = 53.1)
Shape: Rectangular (utility = 0)
Dispenser: "B" (utility = 11.8)
Intercept constant = 47.8
Total = 112.7

Game Playing and Simulation of Alternative Designs Combinations

With a model to interrelate perceptions (or liking) to design options, one can estimate what will happen to ratings of specific characteristics (e.g., liking or expensiveness) for less than optimal combinations of design options. Specifically, the net effect of changing from dispenser B to dispenser A, with a rectangular shaped bottle of magenta graphics, is to diminish liking from 33.2 to 2.8 and to diminish perceived expensiveness from 112.6 to 100.8.

Problems in Optimizing Discrete Combinations of Design Options

One deficit of discrete option optimization is that one cannot take into account interactions among design factors. No additive model can do so. For instance, one cannot predict the synergistic interaction between magenta color and each dispenser type, over and above the partwise contributions of the two factors alone. This failure to account for interactions is a by-product of the mathematics. To overcome it requires very expensive analyses of many more design combinations. Still, with all that additional testing one could account for only a *general* interaction between two design factors, not specific interactions between *one* color and *one* dispenser.

Optimizing the Choice of Package Options With Constraints

As discussed previously, the simple optimization problem was to select that combination of dispenser, color, and bottle shape that maximized a specific perception (e.g., overall liking/disliking). We did this by using linear liking equations and selecting one design option from the colors, the shapes, and the dispensers. Not all optimization problems are so straightforward. Sometimes the most liked combination of design options may be too expensive or may generate specific undesirable perceptions.

Query: Maximize consumer liking of the package but make sure that the perception of 'elegance' of the optimized combination of design options exceeds a specific minimum criterion. Competition has a fragrance package that scores 90.0 on elegance with the same scale.

We want our cologne bottle to be perceived as having as much or more elegance. For "insurance" purposes one can optimize liking, with a constraint that its elegance scale value must exceed 95.0.

Answer: The most liked design combination (magenta, large oval, B dispenser) generates an estimated liking of 91.1 (according to the liking equation) and a perceived elegance of 76.7. Although liking was maximized, perceived elegance of this bottle is lower than desired. In order to fulfill the constraint, one must generate a higher level of perceived elegance, albeit at the risk of diminishing overall liking. A search through the many combinations of design options is required to find the highest liking possible with elegance exceeding 95.0. The most liked combination of design option with elegance higher than 95 is magenta, dispenser B, and small oval shape. The liking rating diminishes to 44.8, but the bottle achieves an acceptable level of perceived elegance, which is 118.3.

An Overview of Discrete Options Optimization

Options optimization provides a different approach to psychophysical package design. What is sacrificed is the capability to vary in a continuous manner those design factors which impact on perception. What is gained is the ability to work with many alternative design options, in terms of color, shape, size, graphics, pricing, and the like, without a continuum to connect alternative colors, shapes, sizes, and so on. We cannot interpolate between an orange and a red bottle to get a reddish-orange, as we could previously, when the design factors were systematically varied. *One can only select alternative combinations of design options from the set that is tested.*

In terms of the utilization of the two types of optimization, each answers a different set of questions. Discrete options optimization is most useful when dealing with 'what will we put on the package to generate a specific perception.' Options optimization can be used initially in order to screen dozens of alternatives. Continuous variable optimization fine tunes the design to accord with cost, production, and other physical/business factors.

AN OVERVIEW—PSYCHOPHYSICS IN ACTION

Psychophysics is a dynamic branch of psychology, whose measurement procedures, experimental design approach, and optimization methods have provided, and will continue to provide, powerful aids both to product/package developers and to marketers. The approaches discussed here show the gamut of psychophysics, from the simplest measurement procedure to business-oriented optimization procedures currently in use. These approaches are grounded in academia, where the procedures have been tested and validated by scientists, and in business, where the decisions based on the results of these procedures have seen action in the marketplace. The next decade of product and package design will see increasing use of psychophysical procedures, with these ensuing benefits:

- Better measurement of consumer reactions.
- More efficient evaluation of package/product alternatives.
- Better utilization of consumer feedback for business decisions.

REFERENCES

1 S.S. Stevens, *Psychophysics: An Introduction to Its Perceptual Neural and Social Prospects,* Wiley, New York, 1975.

2 M.A. Teghtsoonian, "The Judgement of Size," *American Journal of Psychology,* **78** (1966), 392–402.

3 S.S. Stevens, "The Direct Estimation of Magnitude—Loudness," *American Journal of Psychology,* **69** (1956), 1–25.

4 J.G. Beebe-Center, *The Psychology of Pleasantness and Unpleasantness,* Van Nostrand Reinhold, New York, 1932.

CHAPTER FORTY-NINE

Determining Conspicuity and Shelf Impact through Eye Movement Tracking

Elliot C. Young

The utilization of eye tracking in packaging research is comparatively new. For a number of years, eye tracking had served primarily as a diagnostic tool utilized in a laboratory setting, on university campuses, and in hospital research centers. In 1973, Perception Research Services began implementing eye tracking on an everyday basis within the consumer research area. Though eye tracking had long been available, the basic technology required participant set-up time of almost 45 minutes prior to implementing experimentation. This was due to the utilization of a bite bar. A bite impression of the subject's teeth was made and allowed to harden. This mold was attached to an eye tracking unit and the participant was instructed to bite into his/her impression to eliminate head movement. Eye tracking behavior was recorded on 16mm film that was then played back at slow speeds with detailed tracking recorded for later processing. This procedure was cumbersome, time-consuming, and uncomfortable for participants. Accordingly, this "bite" procedure was unacceptable for the conducting of day-to-day consumer research.

As the sophistication of eye tracking technology improved, and the eye tracking "hardware" became less confining, its utilization in the communications research area began to grow rapidly. Eye tracking offers a unique capability—the ability to observe consumer behavior as it occurs. This documentation of behavior, with computer sophistication and accuracy, now provides the means for in-depth exploration of primary marketing and packaging research questions. Over the past few years, eye tracking studies have been implemented to determine:

- Do shoppers search for familiar brands, and does this discovery of familiar brands discourage examination of competitive products within the category?
- Do regular users of a product have difficulty finding their brand on the shelf?
- What is the effect of shelf positioning on a package's shelf impact?
- Are there preferred shelf positions, and do these positions vary by product category?
- Do shoppers scan left to right, top to bottom, or randomly?

The ability to track consumer behavior through eye movements allows marketers to generate an understanding and a definition of the word "impact." The initial step in the communications process is the ability of materials to gain and hold attention and to generate consideration. What better way to document attention, or the absence of stopping power, than to record and observe the visual experience through eye tracking?

Package communication is basically a two-step procedure. Initially, a package on the shelf must generate involvement and consideration for the brand. Secondarily, but of equal importance, this package must communicate the desired product imagery, price/value relationship, aesthetic appeal, product efficacy, and purchase interest. A wide variety of question and answer procedures has been developed to uncover consumers' attitudes toward packaging *if* consumers were to take the time to consider products for purchase. However, prior to the availability of eye tracking technology, little scientific information was available to document the ability of packaging to gain and hold attention and to generate consideration. Market researchers have long recognized the possibility that an aesthetically pleasing package might "test well" in a question and answer interviewing procedure and yet in the "real world" suffer from a deficiency in generating awareness and consideration in the in-store competitive environment.

For many years, a number of researchers have attempted and continue to attempt to measure shelf visibility through the use of a device called the tachistoscope. The tachistoscope allows the researcher to administer quick flash exposures to an individual—exposures designed to uncover the speed with which consumers claim to find a package on the shelf. Resulting tachistoscopic information allows researchers to rank

Table 1   T-Scope Brand Summary

	Package Design	
Correct Brand Identification	A	B
At $1/5$ second	83%	66%
By $1/2$ second	94	78
By 1 second	100	89

alternative packages. This ranking is based on the package's ability to communicate fastest in its competitive environment. Speed of communication, however, is only one shelf visibility attribute. Effective packaging must generate both attention and involvement within the time a consumer chooses to consider a product category; specifically, within the time the consumer, not the researcher, takes to make her purchase decision.

Eye tracking research indicates categories are examined differently. The shampoo section might be examined in "quick hitter" fashion. In this category, the shopper might tend to find her brand and quickly move on. A food category may work in a different manner. When "shopping this category," the consumer may spend a good deal of time examining all products and judging the price/value relationships as conveyed by both the packaging and the pricing information. The implementation of eye tracking enables researchers to document scientifically the steps consumers take as they make their purchase decisions. Eye tracking can be used to document:

- *Shelf Visibility* The ability of packaging to gain consumer attention and involvement in the store environment, in the time consumers choose to give a category.
- *Speed of Communication* The ability of packaging to draw consumer attention quickly. Eye tracking records the speed with which an individual is drawn to an item.

Table 2 Average Viewing Time Summary

	Seconds
Shampoo	4.3
Deodorant	4.8
Instant potatoes	5.4
Salad dressing	5.9
Tea bags	6.1
Frankfurters	7.6

- *Involvement* The ability of packaging to generate dwelling time and reexamination—the ability of the packaging to encourage the consumer to ponder the product's value.

Eye tracking also documents those packages within the competitive environment that are totally overlooked or quickly by-passed.

Table 3 Shelf Impact Summary

Brand	Consumers Noting	Consumers Not Noting
A	92%	8%
B	86	14
C	80	20
D	79	21
E	62	38

Its information becomes two-fold, identifying those packages that work effectively and, conversely, those that appear to be deficient.

Market researchers have long recognized the limitations of asking consumers "What did you see?" The pitfalls are:

1 Consumers may only play back those brands that they recall, though many others may have been examined.

2 Consumers tend to guess. When a shopper is briefly shown a shelf display and later questioned on brands she recalls seeing, familiar or high awareness products tend to generate higher levels of memorability. This recall bias is logical, since recognition of a category leads the consumer to name as many familiar products as possible.

On numerous occasions products are identified that never appeared within the cluttered store environment.

Market researchers are forced to categorize these phantom brands as "misidentified." The eye tracking of the visual experience, linked with brand name recall, documents a good number of brands that consumers claimed to have seen but actually overlooked.

Table 4 Identification of Products Not Included in Display

Not Included in Display	Unaided Recall
Minute Rice	6%
Mueller's	5
Kraft	3

Table 5 Brands Consumers Overlooked But Claimed to Have Seen

Brand	Consumers Noting	Unaided Recall
A	56%	66%
B	49	58
C	46	52

The preceding indicates consumers oftentimes guess correctly. Guessing is widespread, yet guessing can lead to incorrect marketing decisions. For example, a packaging design for a low awareness product may perform poorly on a shelf visibility test utilizing recall as its primary measurement. A marketer relying on a recall score may eliminate a packaging design from consideration, yet by utilizing eye tracking this marketer may have discovered that the packaging prototype did generate higher levels of attention and involvement than primary competitors. (See table 6)

Table 6 Level of Visual Involvement Versus Recall Score

Brand	Consumers Noting	Unaided Recall
A	86%	35%
B	84	4
C	80	29
D	79	39

Eye tracking research has oftentimes uncovered new product packaging that generated strong levels of visual involvement, yet poor recall scores, since the product was unfamiliar to target audience consumers. One wonders how many strong packaging designs have been discarded due to reliance solely on recall measurements.

How does eye tracking work? Computerized eye tracking technology became available in the mid 1970s. Computer based systems enable researchers to document visual behavior to an accuracy of within ½ degree. By linking computer and video technology, the viewer's point of gaze is superimposed onto a video screen, enabling his focal point to be displayed directly on top of the stimulus being examined. (Fig 1).

Computer/video technology enables researchers to observe the visual experience as it occurs. Hardware improvements also eliminate the need for "bite bar" confinement. Little, if any, confinement is required. As the visual experience proceeds, the point at which the individual is looking is recorded instantaneously onto magnetic tape at a rate of 30 readings per second. Recording onto magnetic tape enables eye tracking information to be processed by computer within 24 hours from the conclusion of the project (Fig. 2).

The eye tracking "logic system" perceives the viewing screen as a grid (256 × 256) and records each participant's viewing point on that grid at the rate of 30 times per second. Primary elements within the viewing field are defined by grid coordinate location; consumer viewing behavior as recorded by coordinate location can thus be translated through computer processing into visual elements within the test stimulus.

The implementation of shelf visibility testing utilizing eye tracking is accomplished as follows: Qualified participants, specifically target audience con-

Figure 1 The respondent's focal point is shown on a large video monitor—it is superimposed on top of the shelf display the respondent is viewing on the screen. The focal point can be identified as the location where the horizontal and the vertical lines intersect.

sumers for the test category, are seated at an eye tracking unit and informed they will be viewing a series of scenes they might ordinarily encounter when shopping in a local store. Each participant is allowed to move through these scenes at his or her own speed, spending as much or as little time as desired on each. The participant is not alerted as to the number of scenes he will be viewing, nor is he aware of the test items being studied. He need only press a button to move on to a new scene in the sequence.

The eye tracker automatically documents onto magnetic tape the amount of time spent on the scene, but most importantly, the tracker records those items within the display that were noted, that were dwelled on, and that were reexamined, and those items that were quickly by-passed or totally overlooked. (See table 7)

Eye tracking not only documents the shelf visibility of the test packaging, but it also provides comparable information for competitive packaging. (See table 8) By exposing the test categories in 35 mm slide format, packaging materials can be displayed in a number of shelf locations within the competitive environment. Thus comparisons of the shelf visibility of test packaging versus competition negate the influence of shelf positioning.

Eye tracking research has been im-

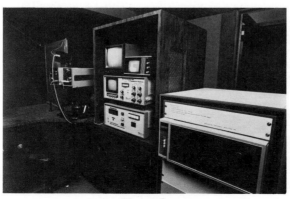

Figure 2

Table 7 Shelf Visibility of Test Product

% Noting	Test Product
At all	56
Within 1 second	14
Within 2 seconds	28
Within 3 seconds	39
More than once	25
Brand first noted	2
% of time spent	7
Average viewing time of total shelf scene	11.7 seconds

Table 8 Shelf Visibility: Test Product Versus Competition

% Noting	Test Product	Competition
At all	56	79
Within 1 second	14	64
Within 2 seconds	28	69
Within 3 seconds	39	76
More than once	25	61
Brand first noted	2	11
% of time spent	7	12

plemented to uncover optimum shelf position and, conversely, those areas within categories that should be avoided at all cost. Eye tracking has also been utilized to document consumers' viewing behavior within categories to uncover those categories that engender logical scanning patterns. The tracking has clarified the benefits of product line horizontal versus vertical displays.

The impact of color on shelf visibility is still another area where eye tracking has been used. Contrary to what many people believe, the color red is not always appropriate, especially when a number of competitors are using a red format. Eye tracking information reminds both packaging designers and marketing directors that the packaging that may appear to generate shelf visibility when viewed alone in an office environment may function quite differ-

ently when positioned in the cluttered store environment. Behavior often speaks more eloquently and descriptively than might have been imagined.

The ability of a package's label to communicate effectively is an essential ingredient in insuring sales success. Eye tracking provides the tool to document whether consumers will take the time to attend to labeling material. The tracking documents visibility of brand name and manufacturer identification. It indicates the ability of primary elements to generate readership. Label readability is often the determining factor in brand selection. The labeling must communicate quickly and easily.

The tracking of visual behavior indicates that consumers often examine a single element on a label and leave the product. The tracking has also uncovered the unique ability of labeling to stimulate extensive involvement and attention to primary product attributes. The tracking of eye movements for labeling material uncovers those elements that are read, that are reexamined, that are dwelled on and, most important, those items within the design that are quickly by-passed or totally overlooked. It is fair to assume all elements within a packaging design are essential, though the manufacturer may place greater emphasis on a selective few. Eye tracking documents whether or not marketer priority is compatible with consumer emphasis. That is, do consumers logically attend to the packaging, and are key elements primary or subordinate within the viewing experience? (See table 9)

The ability to track consumer behavior in response to packaging design elements enables packaging designers to experiment with layouts and copy treatments. The eye tracking provides instant output of consumer behavior based on modifications in design. Elements that may be dominating the visual experience

Table 9 Detailed Eye Movement:
Visibility of Package Elements

	% Noting	% Noting Within 3 sec	% Noting More Than Once	% of Total Time Spent
Product illustration	93	85	80	39
Background design	89	62	49	15
Brand name	82	76	72	23
Flavor	44	20	18	10
Manufacturer name	30	14	15	3
Ingredients	22	9	5	5
Weight	15	6	4	5

Total time spent on package: 5.3 seconds

Examination Sequence: Product Illustration → Brand Name →
Background Design

may also be perceived by consumers to be of little importance in the ultimate purchase decision. Designers can eliminate or modify emphasis on these elements. The following eye tracking information indicates speed of visual perception of key product information for an existing and a modified packaging design. (See table 10)

These data indicate that the Modified Design maintains an effectiveness in drawing attention to brand name, but is somewhat slower in guiding consumers to the illustration. The objective for the

modification in design, however, was to place greater emphasis on the list of ingredients displayed on the packaging. These data indicate that the Modified Design provides a vast improvement in directing consumer attention to the list of ingredients. Within 1 second, 48% of the participants noted the ingredients on the Modified Design, while only 12% reached this element on the Original Design.

The availability of behavioral data enables packaging designers to dispense with hypotheses that "this must have

Table 10 Speed of Visual Perception

Respondents Who Viewed	Original Design	Modified Design
Brand name		
Within $1/5$ second	100	100
Key illustration		
Within $1/5$ second	38	22
Within $1/2$ second	48	46
Within 1 second	54	62
List of ingredients		
Within $1/5$ second	—	—
Within $1/2$ second	6	18
Within 1 second	12	48

Table 11   Shelf Impact and Label Readability Versus Package Appeal

	Test Product	Competitor A	Competitor B
SHELF IMPACT	%	%	%
% noting . . .			
At all	93	79	65
Within 1 second	44	12	8
Within 2 seconds	62	49	35
Brand noticed first	35	18	3
% of total time	14	7	4
LABEL READABILITY			
% noting			
Brand name	100	99	100
Product illustration	94	97	100
List of ingredients	90	85	89
CONSUMER ATTITUDES			
A *package* that is . . .			
Attractive	47	79	92
Easy to read	43	77	70
Makes product seem appetizing	21	86	84
Colorful	10	67	91
A *product* that is . . .			
High quality	27	64	68
A good value for the money	22	59	61
For someone like me	20	66	79

occurred'' or ''that undoubtedly occurred'' or perhaps ''the data were mistabulated.''

The ultimate objective of packaging research is to provide a thorough understanding of the communications effectiveness of packaging materials. The eye tracking capability documents shelf visibility, thus insuring that a package will not be lost in the cluttered store environment. It also documents the ability of executions to generate attention and consideration from prospective buyers — attention to labeling, attention to primary product attributes. Tracking cannot and should not be utilized as a substitute for question and answer procedures to uncover consumers' attitudes toward packaging *if* consumers

were to consider the product for purchase. Table 11 demonstrates a package that generates unique shelf visibility and label readability, yet is considered by consumers unattractive and inappropriate for the category.

As a consumer research tool eye tracking provides an unbiased documentation of the ability of packaging materials to ''deliver'' on the first step in the communications process — to gain and hold attention and to generate consideration. A package may contain a unique and viable product. If this product is ''lost'' on the shelf or fails through its package design to communicate product attributes, the product may ultimately be a failure in the marketplace.

CHAPTER FIFTY

VOPAN: Voice Pitch Analysis in Testing Packaging Alternatives

Glen A. Brickman

In testing the effectiveness of package alternatives, there are two variables that must be evaluated in order to assess the impact of a package. First, did the package attract the consumer's attention? Would it be seen on the shelf? Second, the persuasiveness factor—if it was seen, or if the consumer was looking for it—did the package increase the consumer's desire for the product? This desire is measured by asking the consumer an attitudinal question about the package: Based on this package, are you interested in purchasing (product)? or: Is this an appropriate package for (product)? Diagnostic questions can also be asked to determine which elements of the package contributed to positive attitudes

about the product and which elements were ineffective.

The measurement of consumer attitudes towards packaging alternatives is often a very difficult task. Because the differences between the alternative packages are usually subtle, it takes a sensitive measure to scale what will be subtle differences in the consumer's attitudes about the packages.

In the marketing research industry, practitioners have long relied on verbal and semantic differential measures to rate consumer attitudes although the sensitivity and accuracy limitations of these measures have long been known. However, it has always been difficult to develop and implement new techniques which can be proven to the satisfaction of research practitioners. Once proven, it is also difficult to motivate the use of a new technique.

The field of measuring consumer attitudes through studying human physiology has taken great strides forward since 1965. Based on the premise that expressions of strongly committed attitudes will cause physiological changes, researchers have studied these changes and used them to gain insights into consumer behavior.

Because of the difficulties of measuring these changes, however, the application of the physiological techniques have been severely limited. Most of the problems stem from the need to bring consumers into a laboratory setting and wire them to a measuring device.

Since 1975, a new technique for measuring physiological response has been introduced to the marketing research community. While most people can accept the premise that voice inflections indicate a speaker's true feelings, this new technique—VOPAN—measures much more subtle changes, which are more reliable indicators of true feelings.

The advantages of VOPAN over other physiological measurement devices are numerous. Because the medium for all measurement is a tape recorder, there is no wiring of the respondent. This allows for a natural interviewing environment that can be used in almost all types of research and in any location. This flexibility of location becomes crucial in conducting research on nationally distributed products.

BACKGROUND OF VOPAN

It has long been recognized by several prestigious governmental agencies such as the CIA and the Defense Department

that the change in pitch of a person's voice (the difference in the relative vibration frequency of the human voice) reveals the degree to which emotion is felt and expressed. Extensive work has also been done in this area by the psycholinguistics departments at several universities, most notably Brown University[1] and MIT.

Most of the early work that took place in the 1950s and 1960s, however, used the technique for lie detection. It was thought that measuring emotions as they manifest themselves in the voice would be superior to the standard polygraph technique—the advantage being that the respondent would not be aware of the measurement process.

In the early 1970s, a research team at Brown University explored the use of voice pitch analysis as a tool for measuring consumer attitudes. Dr. Philip Liebermen—one of the early pioneers in voice analysis—and myself as well as several Brown faculty members from interdisciplinary fields, worked towards the development of research procedures and computer programs to apply voice analysis to the measurement of consumer attitudes. Several years of data gathering and validation work formed the basis for the VOPAN technique.

In 1973, a commercial enterprise— VOPAN Marketing Research—was started to apply the advances of the Brown research team to the marketing research field. Immediate interest in the VOPAN technique caused a vast number of research applications to be explored. VOPAN Marketing Research now employs its technique in all areas of marketing research, not as a methodology in and of itself, but as a more objective measure of consumer attitudes to be included in most types of marketing research procedures. Package testing has become an especially useful application of the technique.

THE NEED FOR A PHYSIOLOGICAL MEASURE IN MARKETING RESEARCH

In market research it is always difficult to determine the degree of commitment behind attitudinal responses. Many scaling devices such as numeric and semantic differential scales are not as objective as we would like because each respondent interprets points on a scale differently.

The problem is to determine how strongly the consumer feels about his attitude. Is he giving lip service or is he committed to what he is saying? When we discuss degrees of feeling, we are really addressing the emotions behind attitudes.

Psychologists have long known that most attitudes are emotionally based and that there are two aspects to emotion:

- *The Affective Aspect* Defined as the individual's verbalizations of how he feels. These verbalizations are subject to the respondent's moods on any given day. Traditional research measures only this aspect.

- *The Physiological Aspect* The involuntary changes in the body in response to the emotions—heartbeat, pupillary dilation, and changes in voice pitch. All emotions felt in the mind are expressed by the body. These reactions cannot be consciously altered by the individual.

In order to understand an attitude, it is necessary to study both aspects of emotion. An objective physiological measure is required to determine the degree of emotional commitment behind an attitude. The individual's verbalizations of his attitudes are required to understand whether the emotion felt is positive or negative. Validation data have shown that attitudinal research can become far more predictive of consumer behavior by measuring both aspects of emotion.

THE PHYSIOLOGY OF VOPAN

The premise of all physiological measurements is that what we feel in our minds, we express with our bodies. Thoughts, and outside stimulations toward which we have emotions, cause activity in a part of the brain called the hypothalamus.

The hypothalamus is the emotional center of the brain; it is here that positive or negative emotional reactions to external stimuli or internal thoughts take place. In the majority of cases, outside stimulations or internal thoughts have no relevance to the individual and do not cause emotions. If no emotion is registered and the respondent is forced to make an attitudinal expression, a lip service reply usually ensues. This type of reply can confuse research findings.

When a consumer has a positive or negative feeling about a product, a package, a TV commercial, or any stimulus, the emotional center stimulates the nervous system and various physiological reactions occur. Bodily changes familiar to most researchers are increased heartbeat, perspiration, and pupillary dilation or contraction.

One additional physiological response probably not well known is variation in voice pitch. Although our ears are sufficiently sensitive to hear dramatic changes in voice pitch, such as the notes of the musical scale, subtle variations are not perceptible to the human ear.

HOW EMOTIONS AFFECT THE VOICE

In the vocal system, sound is created at four different places. Each location makes a unique sound with its own pitch or tone. When we hear the voice, however, we hear a blending of the four sounds much like a stereo system, with each speaker making a different sound. In the voice, the "speakers" are the oral cavity, the nasal cavity, the pharynx, and the vocal cord.

The center for emotional influence of voice pitch is in the vocal cords, which are analogous to guitar strings. The tighter the string, the higher the pitch (tone) of the sound. The more relaxed the string, the lower the pitch.

When a respondent has emotional conviction behind an attitude, the "string" is made abnormally taut. Therefore, the pitch of the vocal cord sound is higher than normal. These slight variations are camouflaged, however, by other vocal sounds produced by the oral and nasal cavities. Consequently, voice pitch reactions are inaudible to the human ear. A specially designed and programmed computer is required to screen out the extraneous sounds, calculate a base line range for the vocal cord pitch, and then determine which attitudinal expressions have emotions behind them.

PHYSIOLOGICAL RESEARCH: A PROGRESS REPORT

The measurement of physiological response to advertising or packaging is not a new concept. Other techniques that have been used are: galvanic skin response, pupillary dilation, changes in perspiration, heartbeat and respiration. Although each of these techniques provides a measure of the autonomic nervous system's response to stimuli, the techniques all share some rather serious problems.

First, the need for a respondent to be wired to cumbersome apparatus may cause the consumer to be nervous and have reactions not to the stimulus but to the interviewing environment. Also, the equipment is usually available at only one location, thus it is impossible to get geographic representation in the sample.

Other disturbing aspects of physiological measures used in the past were response and measurement errors due to the environment, apart from the apparatus itself. For example, pupillary response could be biased if the lighting in the room was not perfectly controlled, or if the lighting in the film was not carefully controlled. The problem is, of course, that by treating the film to control lighting you destroy the creative values that make for a successful commercial. Even the measurement of galvanic skin response could be biased if the temperature of the room was not perfectly set at a constant level.

In addition to these shortcomings, a serious pragmatic problem with available physiological techniques has been the inability to determine whether the emotional response was favorable or unfavorable. It was not known whether a dilated pupil meant that the individual liked or didn't like what he saw.

Most physiological measures are useful only as scaling devices, and in no way do these measures indicate the positive or negative nature of the attitude. The only way to determine the positive/negative direction is to ask the respondent how he feels. Because of the lag between emotional response and verbal probes, however, a great margin for error existed.

VOPAN OVERCOMES THESE PROBLEMS

VOPAN solves many of the disturbing problems of physiological measurement because the basic measuring tool is the tape recorder, which is quite portable.

VOPAN does not require wiring an uncomfortable apparatus. This means the consumer doesn't have to be interviewed in a laboratory—instead interviewing can be done at any established field location as well as in door to door situations.

With VOPAN it is possible to detect whether the emotional response is a favorable or an unfavorable one because the verbal and emotional responses are one and the same and therefore take place simultaneously.

THE VOPAN PROCEDURE

All VOPAN interviews are conducted on a one-to-one basis. The first step in the VOPAN testing procedure is to have a short conversation with the respondent about nonanxiety topics such as the weather, hobbies, vacations, and shopping. This conversation is tape recorded and analyzed later by computer to establish the respondent's normal, unstimulated range of voice pitch. The computer is specially programmed to filter out all extraneous sounds coming from the oral and nasal cavities. Each respondent's range, average, and standard deviation of voice pitch from the vocal cord is computed.

The respondent is then given a normal series of attitudinal questions. These questions are very directed—asking for a "yes" or a "no" response. Later, the computer analyzes the voice pitch of the verbal response to see if there is a significant increase in voice pitch over the base line level. Since the positive or negative verbal reply occurs simultaneously with the voice pitch, it is used to pinpoint whether the attitude expressed is positive or negative. The voice pitch data are used to determine the degree of commitment behind the attitude.

Because many responses have no emotion behind them, it is possible to determine whether the consumer actually means what he says or whether he is merely giving us polite "lip service." Only favorable buying intentions that are accompanied by a statistically significant emotional response are accepted by VOPAN as evidence of really wanting to buy the brand. In other words, if a positive verbal response to purchase interest occurs simultaneously with significant heightening of voice pitch, then the respondent is expressing an objective positive emotional response.

The veracity of the response is established if the change in voice pitch falls within an empirically defined range. Studies conducted at Brown University and several government agencies document that changes in voice pitch greater than that empirically defined range indicate a lie (either conscious or unconscious) or a confused response. Usually not more than 5% of any sample fits into this pattern. Since it is not known what causes these atypical responses, they are dropped from the analysis. Sometimes when a question borders on some social mores where it is not in vogue to admit to a certain attitude, the consumer will consciously give a response that is the opposite of the way he feels. In these cases, it is important to study these "lie" reactions.

THE VOPAN SCALE

In summary, VOPAN evaluates two types of responses:

1 *Lip Service Responses* These are not counted as *committed* responses.
2 *Emotionally Committed Responses* These are accompanied by a significant change in voice pitch over base levels and are counted as committed responses.

The result is a verbal response weighted by voice pitch intensity to yield a complete picture of consumer attitudes. The intensity of a voice pitch reaction is expressed as the amount of change over the base line range.

Range of Response

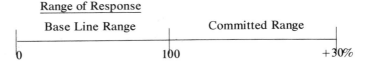

These changes usually are from +1 to +30% of the base line range. The exact level is recorded as the rating of commitment. For any given attitudinal question, VOPAN will look at the percentage of respondents who said "yes" to the question and showed a significant increase in voice pitch. This will be weighted by the average percent increase in their pitch indicating the intensity of commitment.

VALIDATION

Voice pitch analysis has been proven to be more accurate than standard verbal measures in predicting consumer behavior. A previous research study conducted by General Motors corporate research department and reported in June 1976[2] showed voice pitch analysis to be more accurate in determining which product benefits cause a consumer to purchase a brand. This led to greater accuracy in predicting buying behavior.

TEST MARKET VALIDATION

Recently, a large advertiser validated the VOPAN technique in predicting whether consumers would buy a new product. Using a new product that was about to be introduced in test market, VOPAN was to predict which consumers in a controlled group would actually buy the

product. The category was one wherein the package was a highly influential variable in trial purchase.

400 carefully screened target audience respondents were interviewed in the test market prior to the introduction of the product. Each respondent was shown a picture of the package with copy about the product. A trial interest question was then asked as follows: Based on the message that you just read, are you interested in trying new (product name)? Respondents gave a "yes" or "no" response which was used for VOPAN analysis. A seven-point trial interest scale was then administered.

The objective of this study was to evaluate each respondent's trial interest by VOPAN and a standard verbal scale. Each technique would make a determination as to whether the respondent would try the product.

VOPAN predicted that those respondents who said "yes" to the trial interest question and showed emotional commitment (increase in voice pitch) would buy. Respondents were called back at the conclusion of the test market to see if they actually bought. (Table 1)

The results show that of the 400 respondents, 124 gave the "definitely will try" rating. 99 of these people were reached for call back, and 50 actually bought the product, indicating that 51% of the verbally committed respondents would try the product.

Using voice pitch analysis, 88 of the 400 respondents showed committed trial interest. Of this group, 70 were reached for call back and 61 actually bought. This indicated that 87% of the positive voice pitch reactors would by the product. Among those respondents who VO-

Table 1 Purchase Prediction Accuracy

	Number of Predictions	Reached for Call Back	Actually Purchased	Percent Correct
Verbal	124	99	50	51
VOPAN	88	70	61	87

PAN predicted would not purchase, only 2% tried the product.

The results of this validation study showed that verbal techniques will accept more lip service than VOPAN and are therefore less accurate in predicting consumer buying behavior. VOPAN seems to be a more accurate scaling device for consumer attitudinal research. It should also be noted that in this test a series of diagnostic questions were asked to determine what attracted each consumer to the new product. VOPAN found that the package was a crucial factor in the purchase decision.

TEST OF PACKAGE ALTERNATIVES

A good example of the application of VOPAN to package testing occurred when a major national advertiser tested five package alternatives for an existing product. One of the packages was currently being used and, because of the high market share, was widely recognized.

The new alternative packages were subtle design variations on the current package. Previous research had shown no clear differences. The packages were evaluated by a user group and a nonuser group. Standard verbal measures as well as VOPAN data were collected for comparative purposes.

PROCEDURE

Each package was shown to 50 users and 50 nonusers in locations across the country. After looking the package over,

the respondent was asked a purchase interest question: Would you be interested in purchasing this package of (product)? This question called for a "yes/no" response for VOPAN analysis. A five-point purchase interest scale was also admininstered as follows: Would you be interested in purchasing this package of (product)? Would you say you are:

- Very interested?
- Somewhat interested?
- Neither interested nor uninterested?
- Not very interested?
- Not at all interested?

A series of product benefit questions were then asked to determine which package was most effective in influencing attitudes towards the product.

After all the benefit questions were asked, the respondent was shown the other four package alternatives. A question of which package was most appropriate was asked followed by probes of the next most appropriate. In research, this is typically called a forced choice measure. Given all the alternatives, which would be chosen? Usually this measure has difficulty when the alternatives are similar because respondents get confused.

RESULTS

In Table 2, we see a comparison of the verbal five-point scale results (expressed as mean ratings for the sample) and VOPAN. For the user group, respond-

Table 2   Purchase Interest—Users

Package	VOPAN	Mean of 5-Point Rating	Forced Choice
A (Current)	252*	4.5	48%
B	73	4.0	26
C	179**	4.1	11
D	50	3.6	9
E	94	4.0	10

*Significantly highest at 99% level of confidence.
**Significantly greater than the three lowest scores.

ents gave the highest verbal ratings for the current package. After that, Packages B, C, and E were all equal—with Package D scoring lowest.

VOPAN showed that the current package was significantly most effective. However, Package C was also very effective in stimulating purchase interest. VOPAN gave a clear ranking for the alternatives scoring them E, then B, followed by Package D.

The forced choice question showed the current package chosen most often as the most appropriate. Package B was second with the other alternatives equal.

Among the nonuser group, both VOPAN and the verbal scale agreed that Package E was most effective. However, VOPAN showed that the difference was significant and one could have confidence in choosing Package E for this group. VOPAN also showed that the current package was second followed by Package B. Verbal ratings alone showed all others equal. (Table 3)

The forced choice measure disagreed with VOPAN and the verbal scale, and again showed the current package to be most appropriate. This is probably because when respondents saw all of the packages together, they tended to pick the one that was familiar. When shown individually, the alternatives could be measured more fairly.

CONCLUSIONS

In this test, VOPAN proved to be more sensitive than standard verbal measures in evaluating package alternatives with subtle differences. The added sensitivity was extremely useful to the advertiser and made possible a decision about a package change.

Table 3   Purchase Interest—Nonusers

Package	VOPAN	Verbal 5-Point Mean	Forced Choice Appropriateness
A (Current)	180	3.9	48%
B	158	3.8	29
C	120	3.6	5
D	109	3.7	7
E	206	4.2	11

In general, VOPAN is a very useful tool in package design testing because of its ability to scale consumer attitudes accurately. The technique has gone a long way in solving the problem of determining the persuasiveness factor of packaging alternatives.

REFERENCES

1   Philip Lieberman, ''Perturbations in Vocal Pitch,'' *The Acoustical Society of America,* **33,** No. 5 (May 1961), p. 597.

2   Glen Brickman, 'Voice Analysis,'' *Journal of Marketing Research,* **16,** No. 3 (June 1976), p. 43.

Board of Contributors

Anthony Armer is a partner in Armer Research Counsel, a Boston-area marketing research and counselling firm specializing in custom-designed qualitative research. Much of his company's work involves unusual samples such as children, executives, physicians, expectant mothers.

Armer started in market research at Procter & Gamble after being graduated from Oberlin College. Sixteen years later he left P&G as Associate Manager of Advertising Research to develop the research function at Clairol, Inc., where he became Manager of Advertising and Sales Research. He then moved to Boston as Research Director of the Kenyon & Eckhardt advertising agency. In 1975 he established Armer Research Counsel, fulfilling a long-felt need to be involved in all the steps of research projects.

Cheryl N. Berkey is currently the Senior Research Associate of Analysis/Research Limited, a west coast firm of research consultants with fully bilingual capabilities on all levels, which she joined in 1976 after being graduated with a B.S. Econ (Hons.) from the University of South Wales and Monmouthshire (UK), where she studied psychology, sociology, and social administration in addition to research methodologies.

Berkey has extensive experience in data collection and attribute sampling as well as the designing, recruitment, moderation and analysis of focus group research, in addition to other research methodologies such as attitude and image studies, consumer and company profiling, and usage and investigative research projects.

Milton I. Brand is President and Principal of Brand, Gruber and Company and its subsidiary, General Interviewing Surveys of Southfield, Michigan. Before organizing these firms, Mr. Brand was Director of Product Planning Services of the Burroughs Corporation; earlier he served successively as Project Director and Associate Director of Client Relations of the consulting and research firm of Nowland and Company. His background also includes positions on the research staff of the Digital Computer Laboratories and on Project Lincoln of the Massachusetts Institute of Technology.

Brand received his B.S. degree from Trinity College and conducted his graduate studies at the University of Connecticut, Northeastern University, and Boston University. He is active in the American Management Associations, where he has been a seminar leader and guest speaker on over 300 occasions; he is also a member of the American Marketing Association and several other professional groups within the marketing profession.

Glen A. Brickman is President of VO-PAN Marketing Research in Boston. He received a B.A. in motivational economics from Brown University and in 1975 pioneered the application of voice pitch analysis to consumer research.

As a leader in this field of consumer attitude measurement, he and his company are involved in the pretesting of packaging and product concepts and of advertising and television commercials.

Donald Cesario is Vice President of the Center for Family Research, a research group specializing in the surveying of adults, teens, and children, both individually and as interacting family units.

Cesario is a member of the American Marketing Association and the American Association of Public Opinion Research and has over 15 years experience in the marketing research field. Among the companies he has worked for are Appel, Haley, Fourezios; Oxtoby-Smith, Inc., and Young & Rubicam International, both in New York and in Europe. He has conducted seminars on various aspects of market research at Brooklyn

Community College and has acted as consultant for J. Walter Thompson, Don Wise Advertising, Holt, Rinehart and Winston, and others.

Louis Cheskin is credited with being the first market research professional to make marketers aware of the importance of packages and symbols, design, and color for promoting the sale of consumer products because of the effect forms, colors, and design have on the unconscious mind. He has also been largely responsible for developing the concept that consumers are often motivated by irrational factors such as the transfer of sensations from a package to its contents.

As Chairman of Cheskin Associates and the Color Research Institute, specializing in predictive research of marketing communications in packaging and other media, he has authored hundreds of articles and more than a dozen books. His contributions to the field of design research have been given world-wide recognition; he is listed in *Who's Who in America,* the *Royal Blue Book of Great Britain, Dictionary of International Biography, The Two Thousand Men of Achievement, Intercontinental Biographical Association,* the *Directory of British and American Writers,* and *Community Leaders of America.*

Priscilla Douglas is President of Embryonics New Product Workshop, Inc., a company she organized in 1970 for market research of new products commencing with needs/wants/gaps surveys through ideation, development, and screening of product concepts, rough package design concepts, and introductory advertising, with special emphasis on the development and testing of brand names (with which her chapter deals).

After being graduated from the American Institute of Banking, she was a copywriter/supervisor at Young & Rubicam, NCK, and Cunningham &

Walsh, where she worked on major package goods accounts such as P&G, General Foods, American Home Products, and others.

Dr. Ernest Dichter is widely recognized as the conceptualist, and the leading exponent and practitioner, of motivational research both here and abroad. He is Chairman of Ernest Dichter Associates International, Ltd. and President of the Institute for Motivational Research with offices in the United States, Zurich, Paris, London, Milan, Barcelona, Johannesburg, and Caracas.

He is a member of the American Psychological Association, American Marketing Association, and American Sociological Association, and he is the publisher of *Findings,* a monthly psychological report for business and advertising.

Dr. Dichter has authored more than six books in English, Italian and German, the most widely distributed of which *(Strategy of Desire)* has been published in nine different countries. His international reputation is based on the humanistic bias that every product designed, advertised, packaged, and sold should be presented with the idea that a person is involved in its purchase and use.

Richard S. Ernsberger, Jr., has been closely associated with issues of package research ever since graduating with B.S. and M.S. degrees in packaging from Michigan State University.

As packaging consultant for NASA, he conducted investigations in food packaging technology and associated container design research generic to both the "Skylab" series and ASTP flight programs. In 1975 he joined the International Paper Company, where he is Manager of Packaging Products Research at the Science and Technology Corporate Research Center.

Ingrid Flory is a consultant in marketing and public affairs and President of the Swedish Packaging Research Institute in Stockholm. She also heads the Public Affairs Department of the PLM Group and is their coordinator of environmental and consumer affairs. As such, she is the Group's industry representative on the board of the Swedish National Agency for Consumer Affairs and works on establishing a better rapport between business and consumerism through public speaking, seminars, and the publication of articles and presentations.

Flory is a graduate of the Stockholm School of Economics and a former exchange student at the University of Washington at Seattle, United States, with experience in market research, marketing, and strategic market planning.

George M. Gaither is President of Gaither International, Inc., a broad-spectrum marketing and opinion research group that conducts research only outside of the United States.

Gaither has a background of over 20 years in international research beginning with his presidency of International Research Associates after being graduated from the University of Missouri in journalism. His professional background includes the direction of more than 1000 studies in Europe, Asia, the Middle East, and Latin America.

He is a member of the Market Research Council, the European Society of Market Research, and the World Association for Public Opinion Research.

Burleigh B. Gardner is President of Social Research, Inc., which he formed in 1946, and has often been credited with having been nationally influential in gaining business acceptance of the value of the social sciences' application in the field of consumer research.

Dr. Gardner has published extensively, is listed in *Who's Who in America* and *American Men and Women of Science,* is a member of the American Marketing Association and the Society for Applied Anthropology, and is a director of the Duncan Medical Center YMCA.

Professor Gardner was graduated from, and subsequently studied anthropology at, the University of Texas. He earned a Ph.D. in social anthropology at Harvard and after conducting research in personnel counseling at Western Electric Company helped organize (and became executive director of) the Committee on Human Relations in Industry at the University of Chicago.

Irving Gilman, the founder and President of the Analytical Research Institute, has been responsible for much innovative research methodology that will provide more exacting qualitative and quantitative measurements of consumer response; his group's findings have been applied in markets throughout the United States, Canada, Mexico, and Europe. His work on "Motivation Research and the Canadian Consumer" and his company's research projects in Canada provide unique qualifications for the author of our chapter on design research in Canada.

Gilman was graduated from Columbia and New York Universities and is, in addition to managing his group's professional research work, frequently called on to address business, management, labor, and market research groups, exploring the complex manufacturer/personnel/consumer/government relationships and discussing the methods that can be applied to elicit responses providing positive direction for product, packaging, marketing, and communications development.

Larry S. Krucoff is President of Contemporary Studies, Inc., specializing in communications research. His previous

research experience was with Comlab, Inc. and with Post-Keyes-Gardner, Inc.

He is a B.A. graduate of Wesleyan University and holds an M.A. degree from the University of Chicago. For several years his firm has been heavily involved in package design research to study package communications effectiveness.

Norman B. Leferman is President of Leferman Associates, Inc., which he founded in 1977. Holding a B.S. from Rensselaer Polytechnic Institute, he began his career in junior level positions at BBDO. With an M.B.A. from the University of Connecticut, Mr. Leferman switched to the retail side of marketing research as the manager of consumer research of Supermarkets General Corporation. Returning to BBDO as Associate Research Director, he was ultimately named Manager of Marketing Information Services. His most recent position before he formed his own consulting firm was as Executive Vice President of Consumer Response Corporation.

A firm believer in knowing the "real" objectives of a study before designing a research project, Leferman has published papers in many professional journals and has lectured frequently at the American Psychological Association and several universities.

Clifford V. Levy, whose chapter on "Depth Interviews" is considered definitive in that area, is the President of Far West Research, Inc., which he founded in 1956.

He is the author of *A Primer for Community Research* and the editor of *Viewpoints,* the journal for data collection for the Marketing Research Association. Mr. Levy took an M.A. in communication and did extensive work in depth interviews prior to founding his own organization. At present his group is concerned with researching product and packaging concepts and feasibility, advertising research, and research of the black and hispanic markets.

Arline M. Lowenthal has over 25 years in public opinion research and brings to Analysis/Research Limited, the research group of which she is President, an impressive variety of experience in designing research for the public and private sectors.

In addition to her extensive background in design research, she is uniquely qualified to discuss the pitfalls of bilingual research because her company's operational units include Spanish/French/English capabilities used in special research projects for the North and South American continents.

Lowenthal has held office, headed committees, served on committees, and has been active in the American Marketing Association, the Marketing Research Association, and the Association of Public Opinion Researchers. She is listed in *Who's Who in American Women*.

Sanford G. Lunt is Senior Vice President of the S.R.Leon Company, an advertising/marketing group specializing in packaged goods.

In addition to his intimate working knowledge of market research methods pertaining to the product/package/consumer relationship, he has an impressive depth of experience in product and marketing management as former Director, Vice President, and General Manager for Feudor Limited (a division of Swedish Match Company) and as Brand Manager at Whitehall Laboratories, Norcliff Laboratories, and Bristol-Myers.

He holds a B.A. degree from Syracuse University and an M.B.A. from the New York University Graduate School of Business Administration and has lectured extensively at seminars on pack-

aging and marketing topics at the American Management Associations, New York University, the Society for the Advancement of Management, and at Pace College.

Bonnie Lynn specializes in test design and computerized test result tabulation in the area of color perception relating to packaging and graphic communication, and she is Vice President of the Color Research Institute.

Her principal area of interest is in the implementation of Color Ratings, Prescribed Colors, and Color File Search Reports; in addition she monitors consumer attitudes toward colors in association with specific packaging or products.

Lynn majored in psychology, sociology, and anthropology at Johnston College, University of Redlands, with a thesis on "The Effect of Color on Human Behavior."

Davis L. Masten is President of Cheskin Associates and has degrees in marketing and psychology from Johnston College and the University of Redlands.

He has an unusually extensive background in all phases of research as it pertains to corporate communications in the areas of identification, packaging, advertising, and marketing aids. His broad view of the synergistic way in which these corporate statements influence consumer behavior has enabled him to give frequent seminars on design research at the American Management Associations, the New York Advertising Club, and the Packaging Institute.

Richard McCullough is President of Winona, Inc. in Minneapolis, a family organization in which he grew up. He graduated from the University of Minnesota and spent 2 years with Hunt-Wesson Foods before rejoining Winona, whose specially developed product and package research techniques he discusses in his chapter.

Herbert M. Meyers, a managing partner of the firm of Gerstman & Meyers, has been a specialist in marketing and package design for over 25 years and has been responsible for innumerable corporate identification and package design projects, and the research evaluating them, for major corporations both in the United States and in Europe.

Meyers is a frequent lecturer and writer on packaging assessment and package design planning; he is a member of AIGA and of the Board of Directors of the Package Designers Council.

Donald Morich is President of Donald Morich, Inc., a firm specializing in custom designed consumer and industrial research studies. Prior to establishing his own firm in 1978 he was Director of Market Research for Britt & Frerichs, Inc. and has also held key marketing research positions with Ketchum, MacLeod and Grove, Young & Rubicam, Hills Bros. Coffee, and the Procter & Gamble Company. Morich is a graduate of the Wharton School, University of Pennsylvania, and Xavier University; he holds a B.S. degree in economics and an M.B.A. degree in marketing.

Howard R. Moskowitz received his Ph.D. in experimental psychology from Harvard University, where he specialized in applying the methods of psychophysical scaling to the chemical senses.

During a 7 year tenure as Government Scientist at the U.S. Army Natick R&D Command, he worked on the development and supervision of research programs in taste, smell, and texture perceptions—creating measurement techniques that he later applied to the marketing research and marketing communities after he founded MPI Sensory

Testing, Inc. These psychophysical assessment techniques have led to important reductions in the time necessary to evaluate and optimize products and packaging.

Dr. Moskowitz is Executive Vice President and Technical Director of the Weston Group New Products, Inc., specializing in marketing and package/product optimization of new and current products.

Lawrence E. Newman is President of Newman-Stein, Inc. and has over 20 years of practical experience in the marketing research profession, both on the client and the consultant side. As a cofounder of Newman-Stein, his focus is on studies conducted in the areas of product testing, concept testing, package design research, test market tracking, attitude and image studies, and segmentation research.

Newman is a frequent speaker at professional meetings and has given a number of seminars for the American Management Associations on topics pertaining to package design research.

Nicholas T. Nicholas is a marketing and communications consultant specializing in retail consumer programs and strategies for business, industry, and media. He is also President of Retail Marketgroup, Inc., a marketing-communications company.

Nicholas's background has involved him intimately with all phases of retail planning and marketing, advertising, and product and package development and all aspects of product and package evaluation. He is also Editor/Publisher of *The Retail Marketletter* and has recently authored a book on Retail Marketing.

Lorna Opatow is President of Opatow Associates, an organization handling consulting and survey research assignments for major manufacturing, service, and retail organizations, emphasizing investigation and problem solving in the areas of new product and package development, design programs for products, packages, and exhibits, and in all aspects of corporate identification.

Opatow has appeared before business, professional, and academic groups throughout the country to discuss various facets of marketing, research, packaging, new products, and consumer affairs. Her by-line articles have appeared in a number of professional and trade publications.

David K. Osler is President of David Osler Design, a firm specializing in marketing communications design consultation. Prior to establishing his own consultancy, he was a senior vice president for the Raymond Loewy design firm. Because his principal area of interest for over 16 years was the administration and implementation of multinational and international package design programs for such firms as Nabisco, Shell, TWA, Pitney Bowes, and many others, he has outstanding qualifications to discuss all facets of this unusual and complex area. His activities required him to deal directly with package evaluation and optimization problems not only in the United States but also in England, France, Italy, Spain, Germany, and Scandinavia as well as in Australia, New Zealand, Japan, and Central and South America.

Osler has lectured extensively on all aspects of package design development both here and abroad and has been the recipient of many awards for design excellence.

J. Roy Parcels is a founding member of the New York-based marketing/design firm of Dixon & Parcels Associates, Inc. A recipient of numerous design awards and honors, he has been a frequent contributor of articles to the advertising, marketing, and packaging publications

and has lectured on the research and marketing aspects of packaging and trademark design to classes at New York University and The New School.

A member of the board and past President of the Package Designers Council, he is also a former board member of the United States Trademark Association and is on the Packaging Council of the American Management Association. He is a member of the Marketing Executives Club and of the Packaging Institute USA.

John E. Pearl, who contributed a chapter about testing product imagery in Latin America, studied economics at the Universidad de los Andes in Bogotá and became a professor of market research and statistics.

After having served as Director of Market Research at Esso Colombiana for 6 years, he established Interamerican Research in 1966, of which he is now President. He is much in demand as a guest lecturer at international conferences and meetings.

Lester Rand was graduated from Brown University with a B.A. in economics. He began his career as a public relations account executive in New York City, a position that made him increasingly aware of the developing youth market. After studying marketing research and statistics at Columbia University, he formed the Rand Youth Poll in 1953. Because that organization conducts consumer and opinion research studies among teen-agers, college students, young marrieds and young singles, Rand is superbly equipped to discuss the unique problems and opportunities for packaging in the teen market in his chapter "Testing the Teens."

Over the years, Rand has been a leading figure in focusing attention on the important role young people play in stimulating business activity.

Jack Richardson has held leadership positions in market research and marketing services at Bristol-Myers and American Cyanamid, and for a number of years managed his own firm of consultants in these areas. He is a graduate of Upsala College and has done graduate work at New York and Washington Universities.

Roy Roberts is President of the Home Arts Guild Research Center, which he joined in 1957 after being associated with market research and advertising planning at Ruthrauff & Ryan and Bozell & Jacobs.

Roberts took a B.A. at Stanford University and has been closely involved in market research ever since, starting as an interviewer in his teens. He was instrumental in developing the miniature store concept that his chapter discusses.

Robert A. Roden is Market Research Manager for the Household and Hardware Products Division of the 3M Company, (their consumer products marketing organization.) He took an M.B.A. from the University of Minnesota in 1973 and is currently a Ph.D. candidate in marketing at that university. He is President of the Minnesota chapter of the American Marketing Association and an instructor in marketing at St. Thomas College in St. Paul.

Daniel D. Rosener is a Principal Associate of Technomic Consultants, an international, marketing-oriented management research organization.

After taking a B.A. in Economics at Marietta College and an M.B.A. at the University of Pittsburgh, Rosener joined Technomics where he has been directing their packaging group for the past 5 years, conducting proprietary and multiclient studies in virtually every major packaging area but concentrating heavily on packaging and product concept evaluation. He is a certified management

consultant and a member of their institute.

Michael Sarasin who developed the chapter on "The Importance of Package Design in the EEC Marketing Community" is European Marketing Director for Ralston Purina's Consumer Products Division based in Brussels, Belgium.

Because Ralston Purina operates on a world-wide basis as the world's largest producer of pet foods and commercial feeds, Sarasin's activities give him exceptional opportunities to observe the packaging/marketing interrelation in various European countries.

Prior to working at Ralston Purina, he was Marketing Director for CPC International in the United Kingdom and Belgium. He holds degrees from Trinity College Cambridge (M.A.) and Harvard Business School (M.B.A.).

Paul B. Schipke is Associate Director of Advertising/Publicity at Ralston Purina USA, a company with which he has been associated for 20 years, with the exception of a 4 year term as Art Director and Producer at Gardner Advertising.

Because of the unique way in which his company administers all communications areas including advertising, publicity, and the design of all corporate statements, Schipke has been involved with all phases of Ralston Purina's marketing strategies including packaging and the administration of its corporate identity program.

He is the recipient of numerous design awards; his accomplishments include the logo design and advertising graphics for the Busch Gardens family entertainment centers in Tampa, Houston, and Los Angeles.

Thomas T. Semon is Research Director of BSI/Business Science International and a former chairman of the American Marketing Association's Research Techniques Committee. He has served for many years on the Editorial Advisors Board of the Journal of Marketing and has himself authored more than 30 articles on a variety of research-related subjects.

Semon is trilingual, has translated two books from German to English, and has traveled extensively in Europe and North and South America. He is a graduate of Columbia College.

Robert G. Smith is Manager, Product Development and Design, and Chairman of the Design System Group of the J.C. Penney Company, Inc., with which he has been affiliated for almost 20 years. J.C. Penney was one of the first companies to communicate the image objectives for a fundamentally new market strategy through a highly comprehensive design system that embraced all corporate visual communications such as exterior and interior store architecture, displays and merchandising aids, packaging and labeling, advertising, and product design. This coordination is achieved consistently and continually through a group of departmental representatives that Smith chairs.

Prior to his association with J.C. Penney, he was a Vice President for Product Development and Design for such outstanding U.S. design organizations as Lippincott & Margulies, Raymond Loewy, Harley Earl, and Carl Otto, for whom he directed design programs in all facets of graphic and product design for major international clients.

His presentations and articles have been frequently published in the professional press; he has served as Chairman of the New York chapter of the Industrial Designers of America as well as having been Regional Vice President and Executive Vice President of that organization's national office. He served as National Representative of IDSA at international congresses in Japan and Ireland.

Glenn Sontag, as National Art Director for Milprint, Inc., has, for the past 24 years, been devoted exclusively to packaging design and to the merchandising and marketing of packaging.

In addition to his position at Milprint, Sontag has been active in the local art directors' club and the advisory board of Milwaukee Area Technical College in the graphic arts division. He has won many prestigious awards in the packaging field, and his experience of over 30 years in that field was an important factor in his recent election to the advisory board of the National Confectioner's Association. He is frequently a speaker at seminars given by the American Management Associations and at professional conferences in the packaging and graphic arts fields.

Walter Stern is President of the Walter Stern Consultancy Ltd., a consultant group of packaging generalists in the areas of package design research, engineering, and design planning. He has a widely diversified background in packaging and in visual corporate communications and has solved problems and developed comprehensive programs in these fields for such clients as Coca-Cola, Warner-Lambert, Exxon Corporation, the U.S. Postal Service, Mobil, Shell, R.J. Reynolds, Canada Dry, Swift, Hallmark, Nabisco, and many others. He has lectured and written extensively on packaging topics both here and abroad and has authored several books on the subject.

Stern is past President and a Fellow of the Package Designers Council, an invitational member of the Industrial Design Society of America, a member of the Packaging Institute USA and its Professional Seminars Advisory Council, and of the Chicago Artists Guild. He has served as contributing editor to Food and Drug Packaging magazine and recently served as Consultant Editor for Huethig Verlag, Heidelberg in the preparation of a six-language dictionary of packaging terms. He is listed in *Who's Who in the East.*

George I. Suter is New Products Director of EMS European Marketing Surveys Ltd. and of its parent company, NCK (United Kingdom).

He has a broad and varied background in marketing management and new product development for such companies as Bird's Eye Foods, Watney's Brewery, Brooke Bond, Tate & Lyle, Standard Brands, Weetabix, Crown, Shell, and others.

Lee Swope is Manager of Design for the Package Products Company, a major packaging converter that designs, researches, and produces flexible, and folding box packaging for national accounts. His daily contact with a vast variety of marketing problems has given an exceptional opportunity to utilize package assessment techniques. He has received numerous awards and is a frequent speaker at professional meetings and seminars.

Andrew J. Templeton is Managing Director of Datos C.A., Latin America's largest research organization, established in 1954. In addition he was, from 1974–1978, a professor of marketing at the Universidad Metropolitana in Caracas, Venezuela. His main interest is in the area of public opinion research and electoral surveys, a topic on which he delivered a paper at the 1975 WAPOR conference in Lausanne.

Francis P. Tobolski, Sr. is Director of Market Research for the Design and Market Research Laboratory of Container Corporation of America. The laboratory is comprised of a staff of marketing, design, and design research professionals in all parts of the country as well as in Latin America and Europe.

For over 20 years, Tobolski has been involved in all aspects of communications, marketing, and behavioral re-

search. He took his B.A. and M.A. in psychology at Illinois Institute of Technology.

Tobolski has worked and published in all of the media and has lectured at colleges and universities throughout the United States. He is presently an Instructor of Marketing Communications, and of Marketing and Distribution at the Institute for Management of the Illinois Benedictine College. He is also an Associate of the Consumer Psychology Division and the Society of Engineering Psychologists of the American Psychological Association, and he was a member of the Packaging Committee of the American Marketing Association for the entire tenure of that committee.

Lawrence J. Wackerman, a packaging consultant, served previously as Marketing Communications Manager for the St. Regis Company, Bag Packaging Division and has over 25 years of practical experience in the use of packaging to achieve marketing objectives. He was a member of the Board of Trustees and the Task Advisory Council for the Center for Marketing Communications, and he later served on the steering committee of the Business Advertising Research Council of the Advertising Research Foundation. He is a member of the Packaging Institute USA and serves as a representative at PMMI.

Doris G. Warmouth is Vice President of Associates for Research in Behavior, Inc. of Philadelphia. Her responsibilities include all field work and tabulation within the firm as well as major project responsibilities. Her areas of particular interest include package testing and new product research. Much of the material contained in her chapter is the result of the extensive amount of package design research done by her firm at the Springfield Mall interviewing facilities in suburban Philadelphia. Warmouth, who has a degree in economics, has been with "Associates" for more than 10 years.

Sidney Weinstein Ph.D. is President of NeuroCommunication Research Laboratories, the company that conducts the brain wave research described in his chapter. His academic and professional credentials are impressive. He now holds full professorships in Neurology and Psychology at four universities, is the editor-in-chief of the *International Journal of Neuroscience,* a consultant at a Veterans Administration hospital and at the Institute for Crippled and Disabled in New York, has privileges at the Danbury CT hospital, and is the director of the Neuropsychological Research Foundation.

Author of over 100 professional publications, he has found time to be President of the Division of Physiological and Comparative Psychology of the American Psychological Association and President of the Division of Academic and Research Psychology of the New York State Psychological Association. He has held various elected and appointed positions in numerous professional societies.

Julie H. Williams is co-founder and Manager of the Medical Research Bureau, a research group specializing in marketing research for the health care field. The company's constant intimate contact with the health care professions and hospital professionals enables Williams to evaluate packaging effectiveness in hospitals by reviewing actual case histories in the packaging of medical devices and medications.

In addition to the management of MRB, Williams moderates medical group discussions and is active in the American Medical Surgical Market Research Group.

Elliot C. Young is President of Perception Research Services, Inc. in Englewood Cliffs, N.J. He attended New York University Graduate School of Business Administration and holds a B.S. degree in business administration from Boston University.

Prior to establishing his present firm in 1972, Mr. Young spent 15 years within the Interpublic Group of Companies as Administrative Vice President of Marplan Research, Inc., and within McCann-Erickson's Dataplan, and the Institute of Communications Research.

He is a member of the American Marketing Association, the Pharmaceutical Advertising Club, and the Advertising Research Foundation, and is frequently a speaker at seminars and presentations for the Advertising Research Foundation, American Marketing Association, and the American Management Associations.

Index